AF820845

THE SCIENTIFIC SATELLITE PROGRAMME
DURING THE INTERNATIONAL MAGNETOSPHERIC STUDY

ASTROPHYSICS AND SPACE SCIENCE LIBRARY

A SERIES OF BOOKS ON THE RECENT DEVELOPMENTS OF SPACE SCIENCE AND OF GENERAL GEOPHYSICS AND ASTROPHYSICS PUBLISHED IN CONNECTION WITH THE JOURNAL SPACE SCIENCE REVIEWS

Editorial Board

J. E. BLAMONT, *Laboratoire d'Aéronomie, Verrières, France*

R. L. F. BOYD, *University College, London, England*

L. GOLDBERG, *Kitt Peak National Observatory, Tucson, Ariz., U.S.A.*

C. DE JAGER, *University of Utrecht, Holland*

Z. KOPAL, *University of Manchester, Manchester, England*

G. H. LUDWIG, *NOAA, National Environmental Satellite Service, Suitland, Md., U.S.A.*

R. LÜST, *Max-Planck-Institut für Physik und Astrophysik, Garching-München, Germany*

B. M. MCCORMAC, *Lockheed Palo Alto Research Laboratory, Palo Alto, Calif., U.S.A.*

H. E. NEWELL, *NASA, Washington, D.C., U.S.A.*

L. I. SEDOV, *Academy of Sciences of the U.S.S.R., Moscow, U.S.S.R.*

Z. ŠVESTKA, *American Science and Engineering, Cambridge, Mass., U.S.A.*

VOLUME 57

PROCEEDINGS

THE SCIENTIFIC SATELLITE PROGRAMME DURING THE INTERNATIONAL MAGNETOSPHERIC STUDY

PROCEEDINGS OF THE 10TH ESLAB SYMPOSIUM,
HELD AT VIENNA, AUSTRIA, 10–13 JUNE 1975

Edited by

K. KNOTT

Space Science Department (ESLAB),
European Space Research and Technology Centre,
Noordwijk, The Netherlands

and

B. BATTRICK

ESA Scientific and Technical Information Branch,
c/o ESTEC, Noordwijk, The Netherlands

D. REIDEL PUBLISHING COMPANY

DORDRECHT-HOLLAND / BOSTON-U.S.A.

Prepared for publication by
ESA Scientific and Technical Information Branch,
ESTEC, Noordwijk, The Netherlands

ISBN-13: 978-94-010-1894-4 e-ISBN-13: 978-94-010-1892-0
DOI: 10.1007/978-94-010-1892-0

Published by D. Reidel Publishing Company,
P.O. Box 17, Dordrecht, Holland

Sold and distributed in the U.S.A., Canada, and Mexico
by D. Reidel Publishing Company, Inc.
Lincoln Building, 160 Old Derby Street, Hingham,
Mass. 02043, U.S.A.

All Rights Reserved
Copyright © 1976 by D. Reidel Publishing Company, Dordrecht, Holland
Softcover reprint of the hardcover 1st edition 1976
No part of the material protected by this copyright notice may be reproduced or
utilized in any form or by any means, electronic or mechanical
including photocopying, recording or by any informational storage and
retrieval system, without written permission from the copyright owner

TABLE OF CONTENTS*

---

* Contributions presented by first author, unless specified otherwise.

3. QUASISTATIC MAGNETOSPHERIC PHENOMENA AND PARTICLES

4. PARTICLES AND MAGNETOSPHERIC INSTABILITIES

5. SUBSTORMS AND CONJUGATED PHENOMENA

6. NEAR-EARTH EXPERIMENTS RELATED TO THE IMS SATELLITE PROGRAMMES

PREFACE

The 10th ESLAB Symposium was held at Grossenzersdorf near Vienna on 10-13 June 1975 under the title 'The Scientific Satellite Programme During the International Magnetospheric Study'. The Symposium was attended by an invited audience of 60 scientists from the ESA Member States, the United States, Japan, Canada and Austria.

Following a report by the joint COSPAR-IUCSTP Special Working Group, the International Magnetospheric Study (IMS) is proposed as an international co-operative enterprise of limited duration, having as its principal objective the achievement of a comprehensive, quantitative understanding of the dynamical processes operating in the Earth's plasma and field environment. In order to accomplish this objective, it is thought to be necessary to carry out simultaneous measurements with nearly identical instrumentation at various points in space. These measurements will need to be made in combination with appropriate observations at or near the Earth's surface.

Besides near-Earth observations by ground-based, rocket- and balloon-borne instrumentation, satellite investigations are expected to make an important contribution to the IMS. A number of satellites assigned to magnetospheric research have recently been launched, or will be launched shortly, to be operational during the IMS. The European Space Agency has devoted two of its forthcoming scientific satellites – GEOS and ISEE-B – to magnetospheric and interplanetary research. It was therefore quite natural for the Agency, and its Space Science Department (ESLAB) in particular, to organise a Symposium for the purpose of discussing and identifying the opportunities offered by the magnetospheric and interplanetary satellite coverage in the period 1976–78, and to highlight outstanding questions which may be solved with the help of GEOS, ISEE and other spacecraft.

In retrospect, it can be said that the 10th ESLAB Symposium has achieved its objectives and can be considered as the first step towards active co-ordination of IMS satellite programmes. It is hoped that the present book will fulfil a double purpose. It should serve as a reference to all those already familiar with, and engaged in, IMS-related activities, and it should also provide information on the aims of the IMS to those geophysicists who are not directly involved with, but are generally interested in, the objectives of the IMS.

The Editors acknowledge with pleasure the support of all those who helped to make this Symposium a success, and thank the Austrian Space Agency for its hospitality, many ESLAB colleagues for general support, and the ESA Scientific and Technical Information Branch for assistance in producing these proceedings.

OPENING ADDRESS

Ladies and Gentlemen,

The Director General of the European Space Agency has asked me to welcome you to this, the Tenth ESLAB Symposium. It has become a tradition that the Space Science Department of ESTEC – formerly ESLAB – should hold one symposium per year. The fact that this is the tenth indicates that we have recently celebrated the tenth anniversary of ESLAB.

After almost exactly 10 years of running this laboratory, I have recently left it. But I look back at these ESLAB years with a certain feeling of nostalgia, remembering the period when we, particularly Edgar Page and some of the more senior staff, literally built it up with our own hands.

We have used our symposia in the past for different purposes: sometimes for generating new ideas, as we did successfully a few years ago in the field of infrared astronomy. We have also organised symposia to highlight particular satellite missions or to discuss and advance new experiments and space instrumentation. Also, I think the idea of developing multisatellite programmes was first considered during our second symposium, in Noordwijk, at a time when we were fairly new in the field and as such were not much involved and had a slightly different outlook from most. I distinctly remember that it came to my attention during that symposium that more than one satellite is needed if anything really interesting is to be found out about the magnetosphere, which does not necessarily mean that nothing sensible has been found out using data from single satellites.

I hope that this symposium, which is the third devoted to magnetospheric satellite projects, will be equally successful, and I thank you all for coming.

E.A. Trendelenburg

Director of Scientific and Meteorological Programmes, ESA

LIST OF PARTICIPANTS

Abdalla, M.	CNET, 3 ave. de la République, 92 Issy-les-Moulineaux, France.
Amata, E.	Space Science Department, ESTEC, Noordwijk, Netherlands.
Balogh, A.	Physics Dept., Imperial College, London SW7 2LB, England
Balsiger, H.	Physikalisches Inst., Universität Bern, Sidlerstrasse 5, 3000 Bern, Switzerland.
Battrick, B.	ESA Sci. Tech. Info. Branch, c/o ESTEC, Noordwijk, Netherlands
Borg, H.	Space Science Dept., University of Umea
Boswell, R.	Space Science Department, ESTEC, Noordwijk, Netherlands
Bryant, D.A.	Appleton Laboratory, Ditton Park, Slough SL3 9JX, England
Candidi, M.	Lab. Plasma Spazio, CNR, CP 27, 00044 Frascati, Italy
Cole K.D.	NASA, Goddard Space Flight Center, Greenbelt, Maryland 20771, USA
Coleman, P.J., Jr.	University of California, Los Angeles, Calif. 90024, USA
DeForest, S.E.	University of Alabama, PO Box 1248, Huntsville, Alabama 35807, USA
Dobrowolny, M.	CNR, CP 27, 00044 Frascati, Italy
Domingo, V.	Space Science Department, ESTEC, Noordwijk, Netherlands
Dryer, M.	Space Environment Laboratory, Boulder, Colorado, USA
Durney, A.C.	Space Science Department, ESTEC, Noordwijk, Netherlands
Ejiri, M.	Institute of Space & Aeronautical Sciences, University of Tokyo, Tokyo, Japan
Etcheto, J.	CNET, 3 ave. de la République, 92 Issy-les-Moulineaux, France
Fälthammar, C.	Royal Institute of Technology, S-100 44 Stockholm 70, Sweden

Fan, C.Y.	Dept. of Physics and Astronomy, University of Maryland, College Park, Maryland 20742, USA
Fiszleiber, F.	21c Rue Faidherbe, 94130 Nogent's Marne, France
Formisano, V.	CNR, CP 27, 00044 Frascati, Italy
Gonfalone, A.	Space Science Department, ESTEC, Noordwijk, Netherlands
Haskell, G.P.	ESA Headquarters, 114 ave. Charles-de-Gaulle, Neuilly-sur-Seine, France
Hills, H.K.	Rice University, Houston, Texas 77036, USA
Hynds, R.J.	Physics Dept., Imperial College, London SW7 2LB, England
Jones, D.	Space Science Department, ESTEC, Noordwijk, Netherlands
Kennel, C.F.	Centre de Phys. Théorique, Ecole Polytechnique, 17 rue Descartes, Paris 5, France
King, J.	Appleton Laboratory, Ditton Park, Slough SL3 9JX, England
Knott, K.	Space Science Department, ESTEC, Noordwijk, Netherlands
Korth, A.,	Max Planck Institute, 3411 Lindau/Harz, Germany
Leubner, M.	Inst. für Theor. Physik, Universität Innsbruck, Austria
Lyons, L.R.	Environmental Research Laboratory, Boulder, Colorado 80302, USA
Manka, R.H.	National Science Foundation, Washington DC 20531, USA
Mozer, F.	Space Sciences Laboratory, University of California, Berkeley, USA
Obayashi, T.	Institute of Space & Aeronautical Sciences, University of Tokyo, Tokyo, Japan.
Olson, W.P.	McDonnell Douglas Astronautics Co., Huntington Beach, California 92647, USA
Ortner, J.	Austrian Space Agency, Garnisongasse 7, 1090 Vienna, Austria
Page, D.E.	Space Science Department, ESTEC, Noordwijk, Netherlands
Pedersen, A.	Space Science Department, ESTEC, Noordwijk, Netherlands

Perron, C.	Space Science Department, ESTEC, Noordwijk, Netherlands
Reeve, C.	Space Science Department, ESTEC, Noordwijk, Netherlands
Roederer, J.G.	University of Denver, PO 10127, Denver, Colorado 80301, USA
Rostocker, G.	Dept. of Physics, University of Alberta, Edmonton, Canada T6G 2E1
Russell, C.T.	University of California, Los Angeles, Calif. 90024, USA
Rycroft, M.J.	Dept. of Physics, University of Houston, Houston, Texas 77004, USA
Sanderson, T.R.	Space Science Department, ESTEC, Noordwijk, Netherlands
Signorini, C.	CNR, CP 27, 00044 Frascati,Italy
Shea, M.A.	Air Force Cambridge Research Laboratory, Hanscom Air Force Base, Bedford Mass. 01730, USA
Stoffregen, W.	Alvagen 29a, 752-45 Uppsala, Sweden
Stössel, F.	Inst. für Theor. Physik, Universität Innsbruck, Austria
Stüdemann, W.	Max Planck Institute, 3411 Lindau/Harz, Germany
Sugiura, M.	NASA, Goddard Space Flight Center, Greenbelt, Maryland 20771, USA
Theile, B.	Inst. für Geophys., Univ. Braunschweig, Braunschweig, Germany
Trendelenburg, E.A.	ESA Headquarters, 114 ave. Charles-de-Gaulle, Neuilly-sur-Seine, France
Vermeer, S.	ESA Sci. Tech. Info. Branch, c/o ESTEC, Noordwijk, Netherlands
Vette, J.I.	NASA, Goddard Space Flight Center, Greenbelt, Maryland 20771, USA
Voigt, G.H.	Inst. für Theor. Physik, Universität Braunschweig, Germany
Wenzel, K.P.	Space Science Department, ESTEC, Noordwijk, Netherlands
Wilcox, J.M.	Radioscience Lab., Stanford University, Stanford, California 94305, USA
Wilhelm, K.	Max Planck Institute, 3411 Lindau/Harz, Germany
Wrenn, G.L.	Mullard Space Science Laboratory, Holmbury St. Mary, Surrey, England
Young, D.T.	Physikalisches Inst., Universität Bern, Sidlerstrasse 5, 3000 Bern, Switzerland

PART 1

OPENING SESSION

A SUMMARY STATUS REPORT ON THE INTERNATIONAL MAGNETOSPHERIC STUDY

Juan G. Roederer

Chairman of the IMS Steering Committee

It is encouraging to see international scientific meetings on the IMS being organised on a regular basis by the ESA. Each one of them brings the scientific community interested in this programme closer together, helps identify more precisely the problems of interest, and generates new ideas for their solution, thereby increasing the chances of a successful programme.

The International Magnetospheric Study (1976–78) was conceived with the prime objective of studying the magnetosphere as a whole, single system of interacting plasmas, fields, waves and energetic particles. This requires tightly co-ordinated observations from spacecraft, ground-based facilities, aircraft, balloons and research rockets, conducted simultaneously at many different spatial locations under carefully pre-planned circumstances.

The ensuing requirements of international co-ordination were quickly recognised to be quite different from those prevailing during the IGY and IQSY era. Indeed, international geophysical co-operations in the past mainly involved a somewhat 'passive', long-term co-ordination, based mainly on an *a priori* information exchange on major observational plans, followed by an *a posteriori* exchange of data obtained from individual projects run more or less independently. The success of such a mode of co-ordination was fundamentally determined by the occurrence of fortuitous coincidences of measurements made individually during interesting periods of time. A global interdisciplinary programme such as the IMS, however, demands 'active', real-time co-ordination with mid-course correction capabilities based on an effective, quick-response feedback information system. Indeed, the IMS is the first programme in the history of international scientific co-operation [1] the backbone of which is a set of complex spacecraft missions flown by several agencies, with participation of many countries, which will dictate to a large extent the time-frame and actual scientific possibilities of the IMS as a whole. It is this spacecraft programme, with long lead-times on the one hand and unpredictable 'emergencies' on the other, with which a vast programme of planned

Knott & Battrick (eds.), The Scientific Satellite Programme During the International Magnetospheric Study, 3–8.
All Rights Reserved. Copyright © 1976 by D. Reidel Publishing Company, Dordrecht-Holland.

ground-based, balloon, rocket and aircraft observations has to be tightly interlocked in order to make the IMS a success.

There will be six spacecraft specially dedicated to the IMS; 14 spacecraft launched during the IMS that will carry one or several instruments of interest to the IMS; and approximately 25 spacecraft, with instrumentation of interest, launched before the IMS period, expected to be at least partially operational during the IMS. More than 20 countries have so far submitted national programmes of ground-based, balloon, rocket and aircraft studies, and there are over 1000 individual research programmes registered in detail on the computerised file of the IMS. The order-of-magnitude value of total financial investment in published IMS programmes is estimated at 100 million dollars (not including launch vehicles) – way over ten times that for solar-terrestrial programmes in the IGY!

The IMS organisation has proceeded in several stages [2]. At this time, individual countries, research groups, and individual scientists are defining their own programmes in support of the IMS. The official IMS Organisation (presently the Steering Committee and the Satellite Situation Center) is defining the actual needs in co-ordination and information exchange.

In particular, at the Steering Committee meeting that took place prior to this Symposium, several concrete tasks were identified that typify the systems requirements of the IMS programme. The following are examples:

1. One of the key IMS projects is the Mother-Daughter ISEE-A/B mission. When this pair of satellites is in orbit, their (common) line of apsides will drift through the magnetosphere with a one-year period. As different local-time intervals are covered, different regions or boundaries of the magnetosphere will become accessible to special in-situ scrutiny (e.g. dayside boundary, dawn-side mantle, plasma sheet, dusk-side mantle, etc.). As they do, it will be higly desirable to adjust, or intensify accordingly, ground-based, balloon, and rocket programmes, as well as adjust data retrieval modes of other IMS-related spacecraft. On the basis of orbital information to be provided by the Satellite Situation Center, a detailed IMS phase plan should be worked out in close consultation with ISEE project scientists, other satellite experimenters, and ground-based, balloon and rocket (GBR) scientists, and recommended to the whole IMS scientific community.
2. Many of the spacecraft to be operated during the IMS have extremely flexible data-gathering modes. The IMS Organisation ought to help investigators from different spacecraft projects, run by agencies in different countries, to use this flexibility responsibly, in a way that ensures optimisation of all possibilities of co-ordination, satisfying as many participants as possible.
3. It is probable that after some months of orbit, data from IMS satellites may be discontinuous for technical or financial/operational reasons. Frequently, the 'off' periods can be controlled (e.g. during a power-sharing exercise when a number of selected experiments can be operated). It may well be that for maximum efficiency the period of data collection and the ensemble of experiments being operated will have to be varied within an orbit, and as that orbit in the course of a year varies its situation in the magnetosphere (see 1 above). The

IMS Organisation should help with the required detailed planning, based on knowledge of what is going on in all the projects and on a fast and efficient exchange of information.

4. The IMS Organisation, with the co-operation of relevant principal investigators, should work out and recommend to the appropriate agencies a scheme of solar-wind data acquisition and distribution that will maximise the time-coverage of information on solar-wind perturbations. This is particularly important until the heliocentric spacecraft ISEE-C becomes operational in the late part of the IMS.
5. Based on past experience, it is expected that a small but potentially important part of magnetospheric research will be conducted under bi- or multi-lateral agreements between institutions or governmental agencies from different countries, or in campaigns or projects organised 'privately' by informal *ad hoc* groups of participants. The IMS can help exert 'friendly persuasion' on the institutions, agencies, and individuals thus involved to channel information on their activities, or even to adjust the latter, to the IMS programme.

All this would require a systems approach to international co-ordination. Indeed, it is easy to identify in the IMS programme each of the basic stages of systems analysis: (i) Identification of all interacting subparts which form a single, comprehensive entity (the magnetospheric system); (ii) Formulation of objectives and subgoals on the basis of system-modelling and prediction-making (IMS objectives as determined by the current state of theoretical knowledge and observational systems); (iii) Specification of alternatives and treatment of uncertainties (mainly related to the aleatory circumstances in the spacecraft programme, and to the magnetospheric perturbation events expected to occur); (iv) Presentation to decision makers and assistance in cost-effective implementation (recommendations to IMS participants); (v) Performance assessment and evaluation of decisions (real-time scientific evaluation and mid-course correction capability of short-term projects and satellite data acquisition modes).

It was recognised that neither members of the IMS Steering Committee, nor the Satellite Situation Center, nor the SCOSTEP Secretariat have at their disposal the time or the machinery necessary to follow through the details of a co-ordinated programme along the above lines. The IMS Steering Committee has thus decided to discuss the possibility of recommending the establishment of an 'IMS Operations Co-ordination Center' (OCC) that would act as a 'clearinghouse for information' provided by national IMS co-ordinators, space agencies, the Satellite Situation Center, CCOG, and individual IMS scientists. It *would not,* however, be an executive organisation, and *should not* dictate a policy to the scientific community. The OCC should employ three or four full-time scientists with adequate secretarial assistance and communications facilities, bear a truly international character, and have a suitable neutral location. Its actions should be accountable to the IMS Steering Committee and to the agencies that provide financial support, in a way somewhat similar to that of the Satellite Situation Center.

There were many other important items discussed at the recent Steering Committee meeting. Reports were presented on the status of some of the major IMS spacecraft programmes, on major co-operative GBR programmes, and on the Satellite Situation Center. Questions regarding short-term co-ordination procedures and information exchange, spacecraft data acquisition and distribution, publications, etc., were thoroughly discussed. The detailed minutes of this meeting will be published by the SCOSTEP Secretariat.

Some of the formal IMS Resolutions adopted at this Steering Committee meeting of particular relevance to the participants of this Symposium, are summarised below:

1. A recommendation that all possible effort be made by the relevant space agencies and national scientific foundations to ensure that the goals of the GEOS mission are accomplished in all respects, including the following: (a) That the sensitivity of the instruments in the whole frequency range of interest not be degraded; (b) That enough resources be allocated for GEOS data-handling capabilities to maximise the benefits of this mission for the whole scientific community; (c) That a financial solution be found in order to fill the gaps in the associated ground programmes as defined by the CCOG, and to set up a good warning system.
2. Urging NASA to take steps to carry out the Electrodynamic Explorer mission at the earliest possible date in order to allow simultaneous co-ordinated observations from the EE spacecraft and other IMS spacecraft, and extensive ground-based observations to be made, at least during the early part of the EE mission.
3. Encouraging the agencies who control deep-space telemetry antennae and the scientists who have operable experiments on currently operable heliocentric spacecraft to develop a data-acquisition plan consisting of discrete intervals throughout the three years 1976–78 in order to optimise the understanding of the conditions of the interplanetary medium during these discrete intervals, and asking that these discrete intervals be communicated to the IMS Steering Committee.
4. Urging NASA to continue the operation of IMP-H and -J until ISEE-C is in orbit and operating properly, and inviting the IMP-H and -J experimenters to participate in IMS and make data available to IMS in a manner similar to ISEE data.
5. Urging all agencies launching applications satellites to include suitable particle detectors and, wherever possible, related experiments on their spacecraft. In particular, the Steering Committee believed that it would be valuable if the ESA and Japanese meteorological satellites each carried a particle monitoring experiment.
6. Recommending that the GEOS experimenters consider the following points when reconsidering the GEOS longitude shift plan in the light of the revised launch date:
   (i) It would be particularly valuable for GEOS to be located on the longitude of Scandinavia, where extensive co-ordinated GEOS and GBR measurements can be made, at the time when the ISEE-A/B line of apsides passes through the plasma sheet.
   (ii) In view of the fact than an extensive programme of GBR experiments is planned for Syowa during the IMS, GEOS should be located near the Iceland/Syowa longitude for some appropriate period, particularly during the first year after launch.
   (iii) The revised shift plan should be announced as soon as possible so that GBR experimenters can adjust their programmes in good time.

7. Recommending that the possibility be explored that the longitude shift plans of the ATS-6 and GEOS spacecraft be co-ordinated in such a way as to maximise the scientific return.
8. Recommending to SCOSTEP that urgent steps be taken at the ICSU level to ensure that enough time be added to the IMS period to allow its objectives to be fully realised.

Finally, a few *very personal* remarks.

The IMS *has not* enjoyed the widespread attention and until now *has not* relied on the sophisticated propaganda machine of previous international scientific ventures. There are still many scientists – even magnetospheric physicists – who are not fully aware of the circumstances under which the idea of the IMS has emerged and under which its planning is proceeding. We must join forces and urgently engage in effective national and international public-relations campaigns. As scientists, we may have some disdain for this, but our future as scientists may strongly depend on it!

The IMS comes at a time when many science administrators, and even many scientists in collateral fields, believe that the days of magnetospheric research are over, that all important discoveries belong to the past, and that any future magnetospheric programme would only contribute to an increase in the already indigestible volume of accumulated data. True, the IMS does come at a time when more 'glamorous' space physics programmes such as the planetary fly-bys are capturing public and official interest. But would it make sense to give up the quest for a basic, quantitative understanding of our handy, well-behaved Earth magnetosphere, just when that basic understanding is dearly needed to digest so many fascinating and puzzling facts discovered in the distant magnetospheres of the planets? We must join forces in an effective campaign to bring forth all the many convincing arguments which show that true scientific interest in magnetospheric research is still very much alive!

The IMS comes at a time when, because of the constraints of research budgets, any programme that requires new or additional funding is likely to bite into some ongoing programmes, or is likely to decrease the probability of renewal for some existing research. We must join forces in working out programmes that are realistic, if possible interdisciplinary, and convincingly cost-effective.

Finally, the IMS comes at a time when a new generation of scientists is taking over the actual work and the administration of research, a generation of hard-working people with a no-nonsense approach toward solving problems. A generation not interested in formalities, a generation not interested in international co-operation unless it promises to yield results on a short time scale with an absolute minimum of effort on their own part. But let us make a little effort and *not forget* that as scientists we do have a great responsibility not just toward ourselves, our sponsors, and our countries, but also to the world as a whole. *The IMS is an international programme: therefore, not only ask what the IMS can do for you – also ask what you can do for the IMS!*

References

1. International Magnetospheric Study; First Report on Spacecraft Missions, Special Publication, IUCSTP Secretariat, 1971.
2. Roederer, J.G., The International Magnetospheric Study 1976/78, ESRO Special Publication SP-107, February 1975.

THE IMS SATELLITE PROGRAMME: SCIENTIFIC OBJECTIVES

Christopher T. Russell

Institute of Geophysics and Planetary Physics,
University of California,
Los Angeles, USA

Abstract

The International Magnetospheric Study is a proposed international co-operative enterprise during the years 1976–1978 whose aim is a quantitative understanding of the dynamic plasma and field environment of the Earth. Satellite programmes play a key role in this study, together with co-ordinated ground-based, balloon and rocket observations. Herein we review the presently announced IMS satellite programmes, and illustrate many of the objectives of the IMS satellite investigations with examples drawn from current research.

1. Introduction

The key to success during the International Magnetospheric Study lies in the ability to perform synoptic studies, i.e. co-ordinated observation programmes, including ground-based, balloon, rocket and satellite measurements. Satellites form a vital element of this programme for without them we could do little but repeat the IGY. Their role is both to provide the *in-situ* data required to understand magnetospheric behaviour and, just as importantly, to act as monitors of the state of the solar wind and the magnetosphere, in order that the ground-based, balloon and rocket data may be placed in their proper context.

Before discussing the specific objectives of the IMS satellite programme, we will first examine what satellites are scheduled for launch during the IMS and what instruments will be carried on board these satellites. Next, we will present a brief overview of the magnetosphere and how it works. Then we will present a sample of problems to be attacked during the IMS period with these satellites. Some of these problems are illustrated with examples of data obtained during fortuitous alignments of two or more satellites. The fact that these few multisatellite studies

Knott & Battrick (eds.), The Scientific Satellite Programme During the International Magnetospheric Study, 9–42.
All Rights Reserved. Copyright © 1976 by D. Reidel Publishing Company, Dordrecht-Holland.

have each provided a new level of understanding of magnetospheric processes bodes well for the co-ordinated programme of IMS satellites.

2. The IMS Satellites

During the International Magnetospheric Study, there will be a series of IMS-related satellites launched by the ESA, Japan, the USA and the USSR. In addition to these spacecraft, there will be a number of satellites still operating carrying one or more instruments capable of contributing to IMS investigations, and satellites launched primarily for nonmagnetospheric applications, such as communications, or weather monitoring. We will not discuss these latter spacecraft here, but defer this discussion to the paper by Vette [1975]. We have no guarantees that the pre-IMS magnetospheric spacecraft will still be operational during the IMS. Further, the applications spacecraft do not carry sufficient instrumentation to perform the detailed investigations required during the IMS. Certainly, their measurements will be of some use, but to list them as IMS satellites suggests an opulence that the IMS does not possess.

TABLE 1
IMS Satellite Programme

	Launch Date	Orbit
GEOS	February 1977	Geosynchronous
ISEE-A	October 1977	Apogee 23 R_e
ISEE-B	October 1977	Apogee 23 R_e
ISEE-C	July 1978	Solar, 1 AU
EXOS-A	January 1978	350-4500 km
EXOS-B	August 1978	Apogee 5.7 R_e
ISS	February 1976	1000 km circular

Table 1 lists the satellites whose launch during the IMS seems assured. The first satellite listed is GEOS, scheduled for launch by the ESA in February 1977. Although there have been several slips in the scheduled launch date of GEOS, there is good reason to be confident in this launch date. This will be the first IMS satellite launched into the outer magnetosphere and the first strictly scientific spacecraft launched into geostationary orbit, at a geocentric distance of 6.6 Earth radii (R_e).

The next three satellites listed form a joint NASA-ESA programme called 'ISEE', the International Sun Earth Explorers. The first two spacecraft ISEE A and B will be launched in October 1977 by the same vehicle into the same highly elliptical orbit, having an apogee of 23 R_e. The orbit is very similar to that of the OGO-1, 3 and 5 spacecraft. ISEE A and B will have a variable separation in this orbit ranging from 100 km to 5000 km near apogee, with a much greater separation near perigee where the satellite velocities are greater. The separation distance is increased or decreased by firing gas jets on the smaller of the two spacecraft, ISEE-B, and allowing the spacecraft to slowly drift apart or together from orbit to orbit. Enough gas will be carried to cycle between 100 km and 5000 km, two or three times per year. The launch window for this mission repeats every six months. The initial line of apsides is planned to be on the dawn side of local noon and apogee will be in the northern hemisphere. For the April launch window, the initial line of apsides remains fixed, but apogee switches to the southern hemisphere.

The next satellite, ISEE-C, will serve as a monitor of the behaviour of the Sun and the solar wind as well as making measurements fundamental to understanding the Sun and the solar wind themselves. Unfortunately, this satellite will not be launched until mid-1978, whereas it is needed throughout the IMS period.

The last three satellites form the Japanese contribution to the IMS. The first, EXOS-A, will be launched into a low-altitude, high-inclination orbit. The second, EXOS-B, will be launched into a moderate-eccentricity, low-inclination orbit. The final satellite listed, ISS, is principally an ionospheric sounder, launched into a low-altitude moderate-inclination orbit. If it is launched on schedule, it will be the first 'IMS' satellite, preceding the second 'IMS' satellite by over a year.

Although the plans have yet to be announced, there will also be a series of Russian spacecraft. Attractive possibilities to complement the existing programme would include a solar-wind monitor much earlier in the IMS than ISEE-C, a mission similar to ISEE A/B, but at high latitudes to study the polar cusp, and more low- and mid-altitude polar orbiting satellites to study the coupling of the magnetosphere to the ionosphere and the upper atmosphere.

3. The IMS Satellite Instrumentation

Having satellites in the proper orbits is a necessary condition for success during the IMS, but it is not sufficient. These satellites must also carry the proper instrumentation. Fortunately, as we will see, the planners of the IMS satellites have shown much wisdom in this regard. In this section we examine the instrumentation of each of the IMS satellites in turn. This will assist us in assessing the capabilities of the satellite programme.

3.1 GEOS

The first ESA contribution to the IMS will be GEOS, located in a geostationary orbit on the Earth's geographic equator at a geocentric distance of 6.6 R_e. Most of the time it will be stationed so that its northern conjugate point will be in Northern Scandinavia, but manoeuvres are planned to co-ordinate with ground-based and balloon campaigns from Iceland.

Table 2 shows the payload of GEOS. There is a plasma wave experiment covering both electric and magnetic components. Of the 100 kbit/s data rate, 90 kbit/s are assigned to the plasma wave experiment. There are electrostatic analysers covering

TABLE 2
GEOS Instrumentation

Instrument	Organisation
Search coil Electric dipoles	GRI/CNET, ESTEC, DSRI
Electrostatic analysis (Thermal)	Mullard
Crossed fields analyser	Univ. of Bern / MPE
Electrostatic analysers (0.2-20 keV)	Kiruna
Energetic particles	MPA
Electron gun	MPE
Fluxgate magnetometer	Univ. of Rome

Orbit: Geostationary

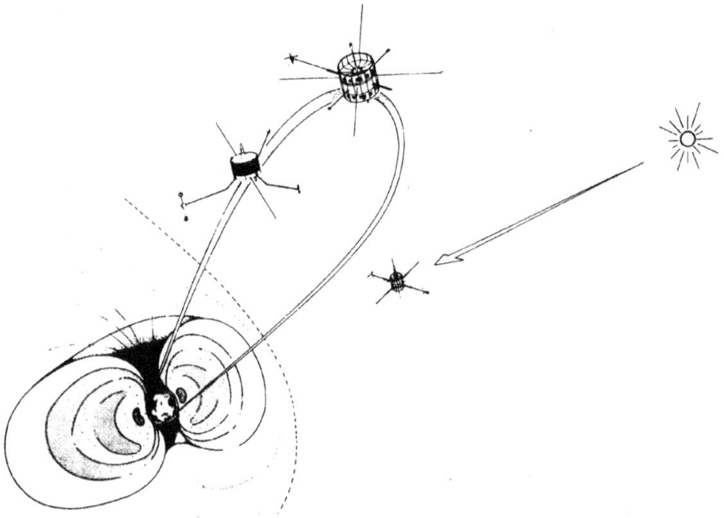

Figure 1. Artist's impression of the ISEE mission. The ISEE-C, or Heliocentric spacecraft, monitors the solar wind while the ISEE-A and -B spacecraft, or Mother and Daughter, orbit the Earth, passing through the Van Allen belts, the magnetopause and shock front.

the energy range from thermal to 20 keV. There is an energetic-particle experiment, an electron gun which will be used to measure the electric field, and finally a flux-gate magnetometer. This is an excellent and very complete set of wave particle and field experiments. For a more complete description of the GEOS payload, operation and shift plan, the interested reader is referred to the paper by Knott [1975].

3.2. ISEE A-B AND C

Figure 1 shows an artist's impression of the ISEE mission. While GEOS is circling the Earth well inside the magnetosphere every 24 h, the ISEE-A and -B spacecraft, also called 'Mother' and 'Daughter', will be making measurements of the radial gradients in tandem every 2½ days in a near-equatorial, highly elliptical orbit with an apogee of 23 R_e. These spacecraft will leave the magnetosphere whenever apogee is on the dayside of the magnetosphere and make measurements of the magnetopause, magnetosheath and solar wind. The use of two satellites close together in the same orbit permits the separation of temporal from spatial changes, a determination of the velocity of propagation or convection of wavefronts and boundaries, and thus permits the conversion of the time profile of these boundaries into thicknesses, current densities, etc. While Mother and Daughter are probing the magnetosphere, ISEE-C, or the 'Heliocentric' spacecraft, will be in orbit about the Sun in synchronism with the Earth's motion about the Sun, thus providing knowledge of the boundary conditions imposed on the magnetosphere by the solar wind.

TABLE 3
ISEE-A Instrumentation (Mother)

Instrument	Organisation
Quadrispherical analysers	LASL
LEPEDEA	Univ. of Iowa
Triaxial electrostatic analysers	GSFC
Energetic particles 1-200 keV	UCB
Energetic particles 20-2000 keV	NOAA, MPI Lindau
Particle telescopes 5-6000 keV	MPI, Garching
Fluxgate magnetometer	UCLA
Electric dipole	GSFC
Electric dipole	UCB
Electric dipole and search coils	Univ. of Iowa
Plasma resonance and propagation	Observatoire de Paris
Ion mass spectrometer	Lockheed, Univ. of Bern, MPE

Orbit: Eccentric, Apogee 23 R_e

Table 3 shows the payload of the ISEE-A, or Mother, spacecraft, which is the larger of the two vehicles. First, there is a series of low-energy plasma experiments.

The first listed experiment will be capable of measuring the solar wind. Next, there are a set of energetic particle experiments covering up to low-energy cosmic-ray energies. There is a fluxgate magnetometer, two long dipole antennas for measuring the DC electric field, and an AC electric and magnetic plasma wave experiment. The cold plasma density will be measured in two ways: first with a plasma resonance and radio propagation experiment supplied by Meudon, and second with a modification of the Lockheed heavy-ion experiment to allow it to also measure the thermal protons.

TABLE 4
ISEE-B Instrumentation (Daughter)

Instrument	Organisation
Energetic particles 1-200 keV	UCB
Energetic particles 20-2000 keV	NOAA
Quadrispherical electrostatic analyser	MPI, Garching
LEPEDEA	Univ. of Iowa
Solar-wind experiment	Univ. of Rome
Electric dipole and search coils	Univ. of Iowa
Plasma propagation	Observatoire de Paris
Fluxgate magnetometer	UCLA

Orbit: Eccentric, Apogee 23 R_e

TABLE 5
ISEE-C Instrumentation (Heliocentric)

Instrument	Organisation
Solar-wind	LASL
Vector helium magnetometer	JPL
Solar X-ray/solar electrons	UCB
Energetic protons 30-1500 keV	Utrecht, IC, ESTEC
Solar-wind composition	GSFC
Particle telescopes 5-6000 keV	MPE
Cosmic-ray isotopes	UCB
Cosmic-ray isotopes	CalTech
Solar and galactic cosmic rays	GSFC
Radio emissions	Observatoire de Paris
Plasma waves: E and B	TRW

Orbit: Solar, 1 AU

Table 4 shows the ISEE-B, or Daughter, instrumentation. There are two energetic-particle experiments, two low-energy plasma experiments, a solar-wind experiment, a plasma-wave experiment, the receiver for the radio-propagation experiment, and a fluxgate magnetometer.

Table 5 shows the ISEE-C, or Heliocentric, instrumentation. There is a solar-wind experiment, a vector helium magnetometer, a solar X-ray experiment, energetic-particle experiments, solar-wind composition experiments, cosmic-ray experiments, a radio-emission experiment and a plasma-wave experiment. For further details, see the paper by Wenzel [1975].

TABLE 6
EXOS-A Instrumentation

Instrument	Principal Investigator
Plasma probe	Oyama
Energetic particles	Mukai
UV auroral TV	Kaneda
Plasma wave	Yoshino
UV spectrometer	Nakamura

Orbit: Apogee, 4500 km; Perigee, 350 km

EXOS-B Instrumentation

Instrument	Organisation
Plasma resonance 3 kHz-10 MHz	Tohoku Univ.
VLF, LF and MF electric field	Tohoku Univ.
VLF propagation	Kyoto Univ.
DC electric field	Tokyo Univ.
Fluxgate magnetometer	Tokai Univ.
Low-energy particles (0.05-20 keV)	Tokyo Univ.
Electron gun	Tokyo Univ.

Orbit: Eccentric, Apogee, $5.7R_e$

ISS Instrumentation

Instrument	Organisation
Swept-frequency sounder	RRL, Tokyo
Radio noise	RRL, Tokyo
Retarding potential trap	RRL, Tokyo
Bennett, ion mass spectrometer	RRL, Tokyo

Orbit: 1000 km circular

3.3. EXOS-A, EXOS-B AND ISS

The Japanese contributions to the IMS satellite programme are shown in Table 6. The first satellite, EXOS-A, will be launched into a low-altitude high-inclination orbit. It will carry a plasma probe, an energetic-particle experiment, a UV auroral TV, a plasma-wave instrument, and a UV spectrometer, but no DC magnetometer or electric-field probe. There is still much to be learned about the low-altitude convection pattern and so a DC electric-field experiment is a high-priority item for low-altitude high-inclination spacecraft. A DC magnetometer is essential for probing the field-aligned currents in the magnetosphere.

The EXOS-B satellite will probe the inner magnetosphere - plasmasphere region. It does include a magnetometer and a DC electric-field experiment, but it does not include an energetic-particle experiment or a cold-plasma measurement. On the other hand, the plasma-resonance experiment may provide a measure of the density, but as far as I am aware this technique has yet to be proven in the outer magnetosphere. For more information on this technique, which is also being implèmented on GEOS and ISEE-A, the interested reader is referred to the paper by Bitoun *et al.* [1975].

The last satellite listed, ISS, is an ionospheric sounder and is not directed towards the same objectives as the other two missions. I should like to point out that because of the limited size of these spacecraft, many compromises have had to be made and a full complement of experiments could not be flown. The mission planners should not be faulted for that, but it is important to point out at this stage what those compromises were.

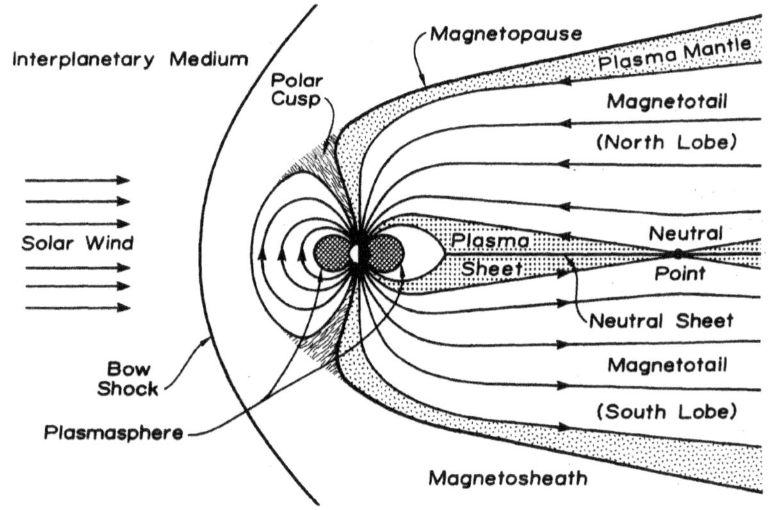

Figure 2. Noon-midnight meridional cross-section of the magnetosphere [*after Russell, 1972*].

4. The Magnetosphere

Having seen what we have to work with, let us quickly review the morphology of the magnetosphere and examine a simple model for how it works, to prepare for our later examination of a sampling of what we can do during the IMS. Figure 2 shows a noon-midnight meridian cross-section of the magnetosphere. There is a bow shock in front which slows, deflects and heats the solar wind and allows it to flow around the magnetosphere. There are phenomena upstream from the shock: plasma waves and counter-streaming particles, but they little affect the macroscopic properties of the solar wind. Behind the shock is the magnetosheath, and between the magnetosheath and the magnetosphere is a layer of current called the magnetopause. The magnetopause is the most important boundary in the magnetosphere and is the most poorly understood, both observationally and theoretically. Understanding the nature and behaviour of the magnetopause is one of the primary objectives of the IMS.

The magnetosheath plasma gains direct access to the ionosphere in the polar-cusp region. This plasma then expands and flows down the sides of the lobes of the tail to form the plasma mantle. It appears that the mantle plasma and the magnetosheath plasma, entering on the flanks of the magnetosphere, both supply the plasma sheet. There is a neutral point sitting on a reconnection line somewhere deep in the tail. There is a current sheet, which is seldom neutral, residing in the plasma sheet which reverses the field from the north to south lobes. Deep in the magnetosphere is a torus of cold dense plasma maintained by the ionosphere and called the plasmasphere. Figure 3 shows a representation of the three-dimensional magnetosphere [Heikkila, 1972].

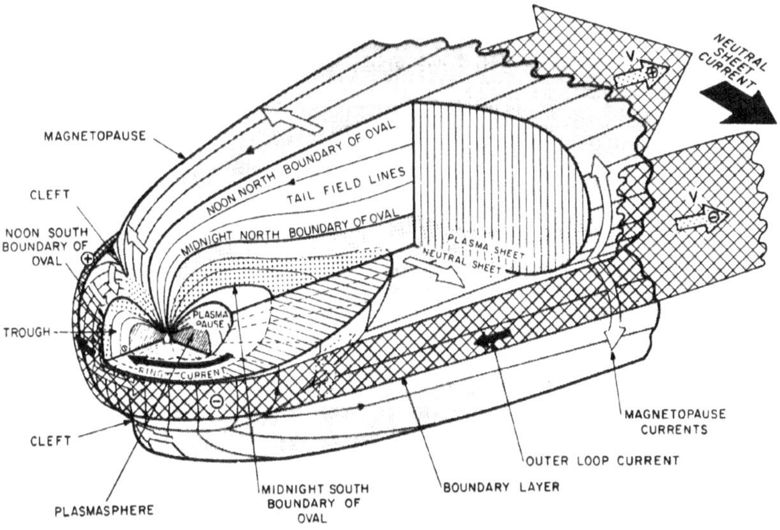

Figure 3. Cutaway sketch showing the various magnetospheric regions [Heikkila, 1972].

The most important process occurring in the magnetosphere is reconnection or merging, in which two opposing magnetic field lines join together to form two new field lines. Figure 4 shows Dungey's sketches of the topology of merging [Dungey, 1963]. I should stress that these sketches show only the topology of merging and are not to scale. Dungey [1965] fully realised the tail would be of the order of 500 R_e long. The top panel illustrates merging between a south pointing interplanetary field and the magnetosphere. Interplanetary field lines join with magnetospheric fields and are carried back behind the Earth. This process removes magnetic flux from the dayside magnetosphere and adds it to the tail. To maintain a steady state, field lines must then reconnect in the tail and add closed field lines to the night-time magnetosphere. These field lines then flow around both sides of the auroral oval to the dayside.

For an exactly northward field, the situation is quite the reverse. Field lines merge with tail field lines and add flux to the dayside and then there is flow antisunward around the auroral oval. If this merging takes place with open field lines, it reduces the number of open field lines. If all the field lines in the tail were closed, it essentially moves field lines from the nightside to the dayside.

In general, the interplanetary field is not exactly north-south and the field line that connects to the north tail lobe does not connect to the south tail lobe. This process is sketched in Figure 5 for merging with the north lobe only [Russell, 1972]. Other field lines must connect with the south lobe. The connected field lines exert a stress on the magnetosphere which causes a circulation in the tail and in the polar cap, but no exchange of open and closed fields. The sense of flow is down along the flanks towards the plasma sheet and away from the plasma sheet in the centre of the tail. In the ionosphere, the flow is sunward in the centre of the polar cap, and antisunward at the edges. We note that only in this process which replaces an open field line with another, can the merging rate in the two tail lobes be unequal. In merging or reconnection processes which form open lines from closed lines, or vice versa, both tail lobes are affected equally.

Figure 6 illustrates schematically the possible polar-cap flows. If there is dayside merging as illustrated in the top left panel, closed field lines open up and are connected into the polar cap near noon. The polar cap increases in area, i.e. the area enclosed by the 'last closed field line' (LCF) is increasing. If flux is reconnected at the neutral sheet, the open polar-cap flux closes and is convected out of the polar cap near midnight. Then the polar cap defined by the last closed field line decreases. Neither process can proceed alone indefinitely, for the polar cap cannot expand or shrink indefinitely. However, they can proceed simultaneously in the steady state. If the interplanetary field is close to being perpendicular to the magnetospheric equator and directed southward, the expected flow is essentially directly over the polar cap. This is simply because the north polar cap is connected to the solar wind north of the magnetosphere and the south polar cap is connected to the solar wind south of the magnetosphere. In this situation, newly reconnected field lines are piled on top of old field lines, resulting in a sinking of field lines into the lobes of the tail.

If the interplanetary field is northward and there is no dayside merging, then the situation sketched in the middle two panels exists, in which the size of the polar cap remains fixed, but an asymmetric convective flow exists. The sign and strength of this asymmetry depends on the magnitude and direction of the east-west component of the interplanetary magnetic field.

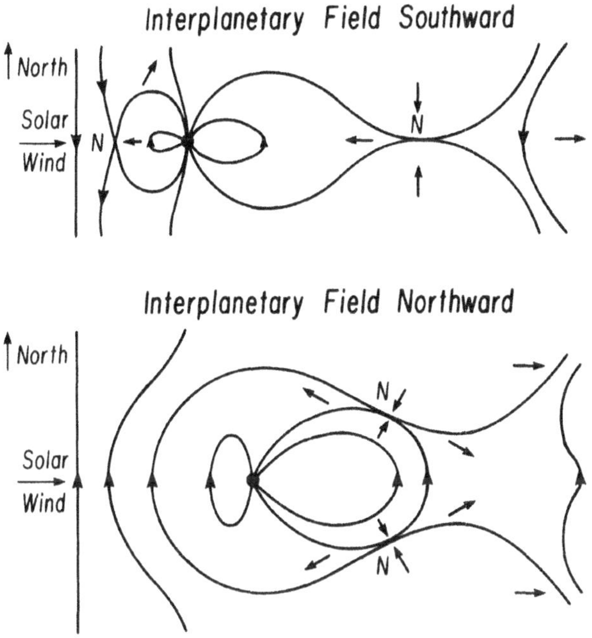

Figure 4. Topology of merging for southward fields (top panel) and for northward fields (bottom panel) [Dungey, 1963].

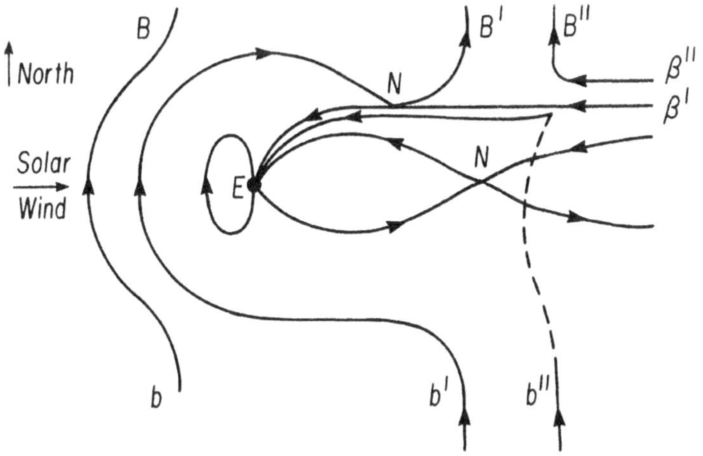

Figure 5. Temporal sequence of events occurring when field line B-b merges with the north-tail lobe, but does not connect to the southern tail lobe [Russell, 1972].

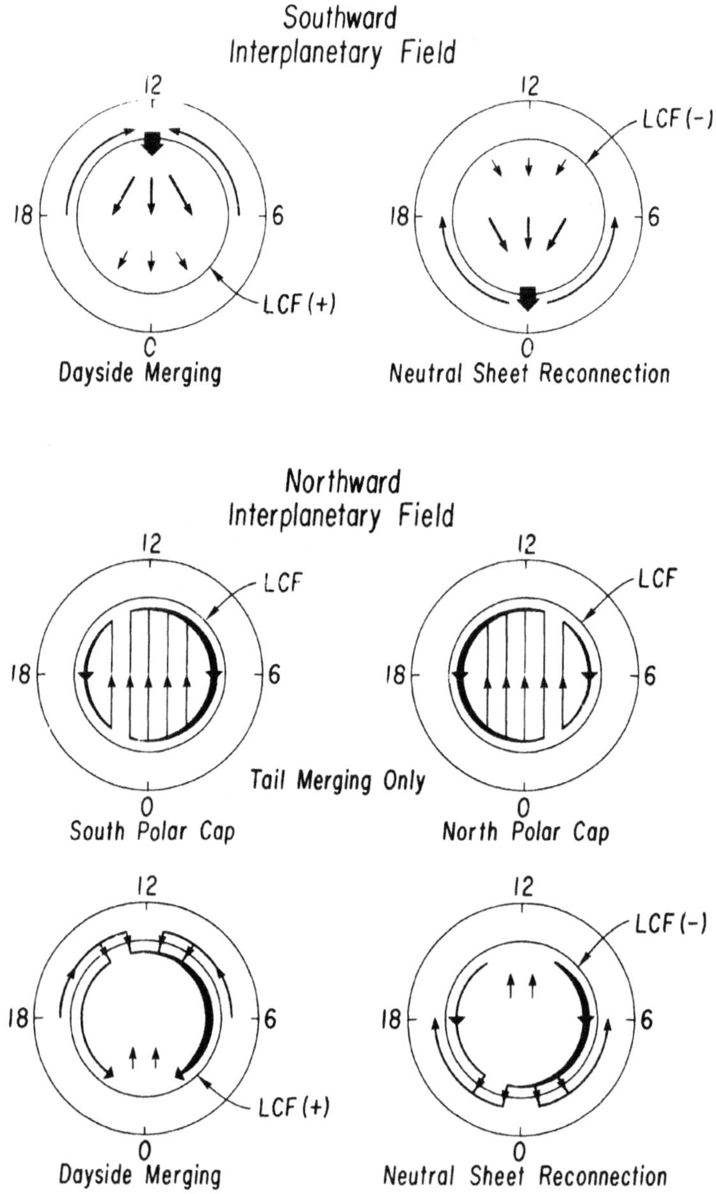

Figure 6. Expected flow patterns in the polar cap associated with merging on the dayside, closed lines reconnection between the open tail lobes and merging between the interplanetary lines and the open tail lobe field lines [Russell, 1972].

Finally, the bottom left panel shows the case for dayside merging with interplanetary fields having sizeable east-west components. As in the previous case, there will be a dawn-dusk asymmetry in the flow. However, here the polar cap increases in size and thus night-time reconnection must eventually occur. Thus, the average convection pattern would resemble the sum of the left-hand and right-hand bottom panels, and no sunward polar-cap flows would be present. The only situation in which we expect sunward directed flows on open field lines is when the interplanetary magnetic field merges with the tail-lobe field.

Many of these expectations based on the geometry of the merging process have been confirmed. The asymmetry of the polar-cap convection pattern and its dependence on the east-west component of the interplanetary magnetic field has been observed directly by Heppner [1972] and indirectly using magnetic records by Friis-Christensen & Wilhjelm [1975] and Maezawa & Obayashi [1975]. The dependence of the strength of the polar-cap electric field on the interplanetary magnetic field has been studied by directly observing the polar-cap electric field with balloon measurements [Mozer *et al.*, 1974; Mozer, 1975] and indirectly from magnetic records by Friis-Christensen & Wilheljm [1975] and Maezawa & Obayashi [1975]. Finally, the existence of sunward flow for northward interplanetary fields has been deduced from ground-based magnetic records by Maezawa & Obayashi [1975].

Our next task is to examine how merging and reconnection leads to substorms. Figures 7 and 8 illustrate this schematically. In Figure 7, the magnetospheric flux has been arbitrarily divided into three regions: the dayside closed flux Φ_{Day}, the tail lobe open flux, I_{Lobe}, and the nightside closed flux, Φ_{PS} [McPherron, 1974].

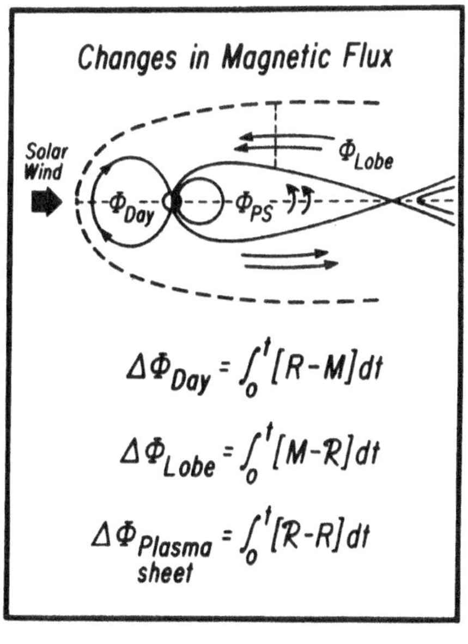

Figure 7. Flux bookkeeping in the open magnetosphere [*McPherron, 1974*].

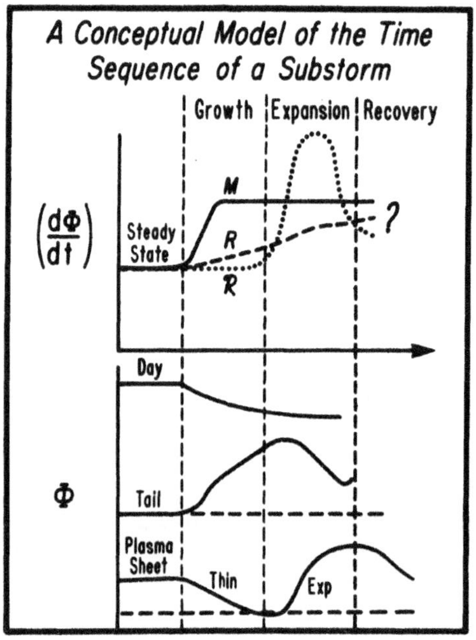

Figure 8. Conceptual model of a substorm based on the temporal behaviour of the amount of magnetic flux on the dayside, the nightside and in the tail lobes [McPherron, 1974].

The change in flux on the dayside is the integral of the difference between the return of flux from the nightside and the dayside merging rate. The change in the lobe flux is the integral of the difference between the merging rate and the reconnection rate in the tail. The flux in the nightside magnetosphere is the integral of the difference between the reconnection rate and the return rate to the dayside.

Figure 8 shows how these integrals may be associated with a substorm. The top panel shows the rate of change of the flux in the three regions. Initially, there is a steady state, and in this simple model we suddenly increase the merging rate, the return rate R increases slightly to replace the lost flux, but this rate is controlled by the ionosphere and magnetospheric pressure gradients which build up in proportion to the amount of missing flux. In the bottom panel we see that the amount of flux in the dayside decreases, that in the tail lobe increases and that on the nightside decreases. Eventually, the stresses on the tail become too great and reconnection begins. Perhaps this is due to the fact that the plasma sheet becomes too thin. Then the tail lobe loses flux and the nightside magnetosphere gains flux.

These ideas are somewhat controversial, although I believe there is strong observational evidence in favour of every one of them. In fact, there are even some who still maintain that the magnetosphere is closed [Michel & Dessler, 1975]. However, many of the differences between the open and closed 'camps' are more apparent than real.

5. IMS Objectives

5.1. MACROSCOPIC BEHAVIOUR OF THE MAGNETOSPHERE

The above model of the magnetospheric substorm process leads to our first set of objectives for the IMS. How does the merging rate depend on interplanetary parameters? Why is reconnection in the tail delayed with respect to dayside merging? What is the trigger that starts the substorm? Does the magnetospheric substorm sketched here correspond one-to-one to the auroral substorm seen from the ground?

The heliocentric spacecraft can aid in attacking these problems by monitoring the solar-wind conditions while we observe the magnetospheric response. Figure 9 shows an example of what can be done using Explorer 33 and 35 magnetic-field and solar-wind data [Burton *et al.*, 1975]. Here we have plotted the rate of change of the ring current as a function of the dawn-to-dusk component of the interplanetary electric field. This is simply the product of the north-south component of the interplanetary magnetic field times the radial component of the solar-wind velocity. When the electric field is from dusk to dawn, the D_{st} index does not change, but when it is from dawn to dusk it does change and in fact appears to be linear in E. Thus, the magnetosphere seems to be an efficient rectifier of the solar-wind electric field.

We have used this relationship together with the known dependence of D_{st} on the solar-wind dynamic pressure and with a constant 8 h decay time for the ring current to predict D_{st} from solely the solar-wind dynamic pressure and dawn-dusk electric field. Figure 10 shows the results for a double main phase storm. The top panel shows the square root of dynamic pressure, the middle panel the electric field, and the bottom panel the observed and predicted D_{st} values. The dashed line is the prediction.

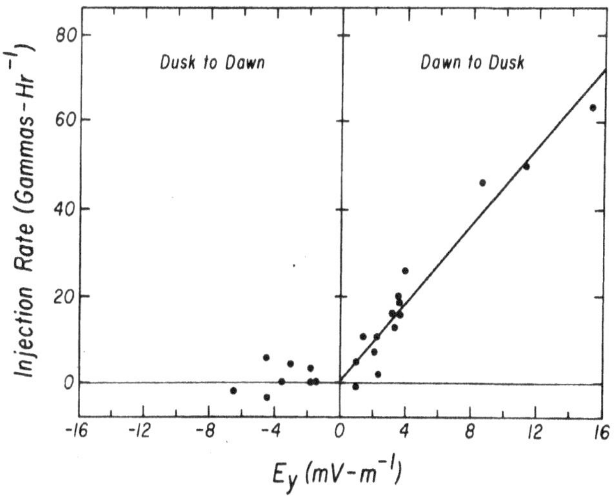

Figure 9. Ring-current injection rate as a function of the dawn-dusk component of the interplanetary electric field [*Burton et al., 1975*].

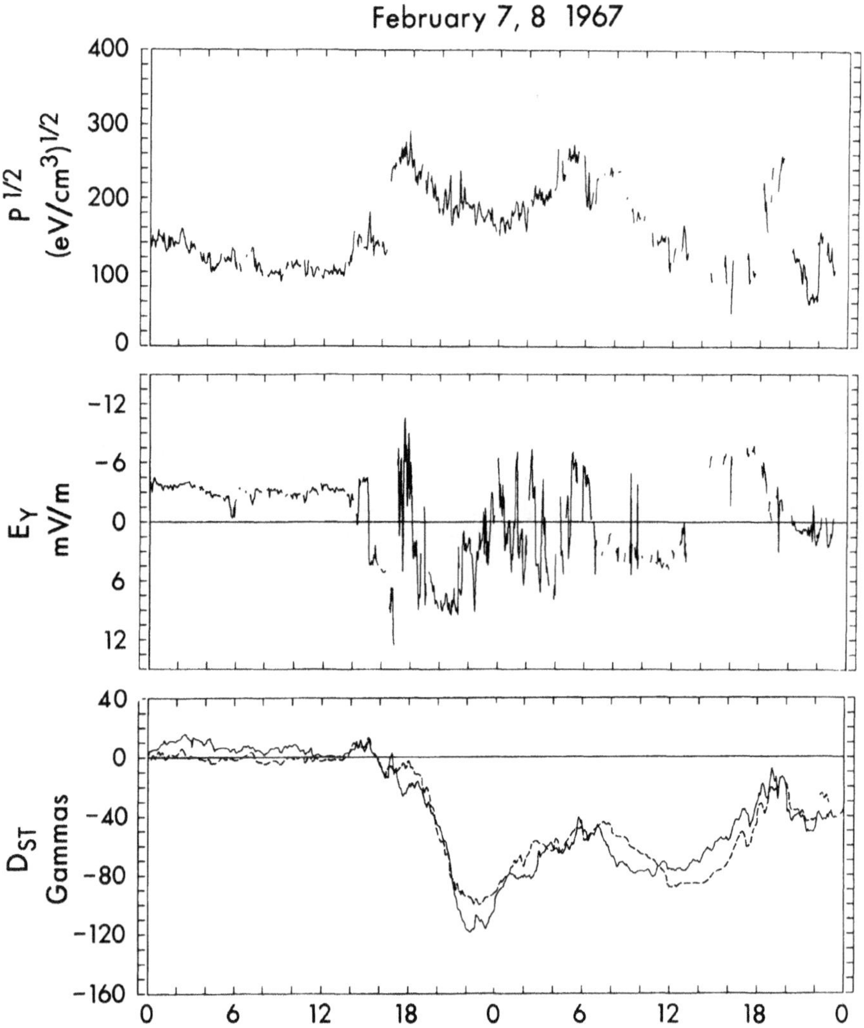

Figure 10. *Square root of the solar-wind dynamic pressure (top panel), the dawn-dusk component of the interplanetary electric field (middle panel) and the predicted and measured D_{st} index (bottom panel), during a complex geomagnetic storm on February 7 and 8, 1967. The predicted curve is the dashed line [Burton et al., 1975].*

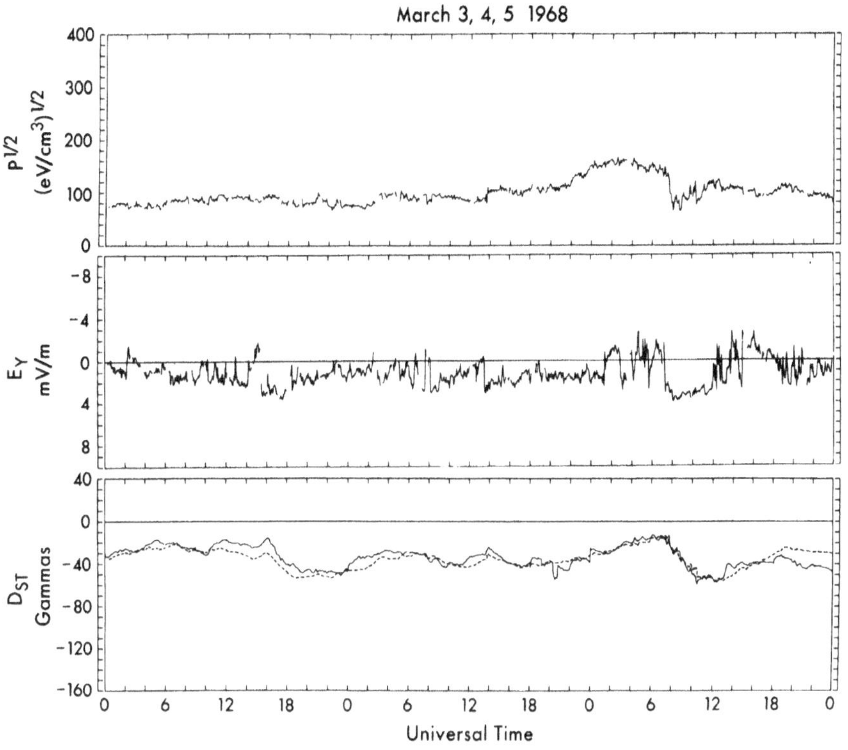

Figure 11. Solar-wind dynamic pressure, interplanetary electric field and predicted and measured D_{st} index during a moderately disturbed period from March 3-5, 1968.

Figure 11 shows the results for an interval in which there was no storm but rather a continuous period of disturbance. Again good agreement. These results show that the ISEE-C measurements can be used as a predictor of the state of the magnetosphere if one requires such knowledge one hour in advance and replaces the need to acquire ground-based data from around the world, if only a rough measure of activity is required.

This is only one of the possible monitor functions of ISEE-C. It will also provide the interplanetary cosmic-ray flux measurements essential to polar particle entry studies, for example. It is also essential for monitoring solar-wind conditions while we are probing the structure of the shock front and the magnetopause.

5.2. MICROSCOPIC PROCESSES

In addition to examining the macroscopic response of the magnetosphere to external conditions, we are also interested during the IMS in the microscopic processes and how they vary when solar-wind conditions change. The ISEE A-B, or Mother-Daughter, spacecraft are ideally suited for this purpose. In addition to having sufficiently high data rates so that the structure of the boundaries can readily be resolved, they permit an assignment of dimensions to this structure. In the past, there have been a limited number of boundary studies when two or more spacecraft by chance assumed a favourable configuration for such studies. However, with ISEE A and B, these studies will be undertaken on a regular basis.

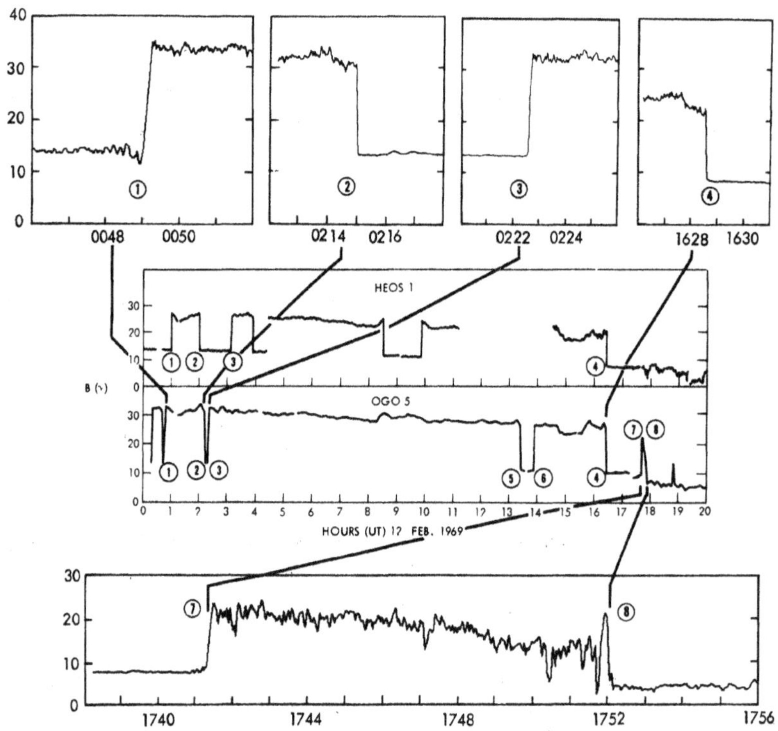

Figure 12. Magnetic field strength observed at HEOS-1 and OGO-5 (middle two panels) while nearly aligned along a radius vector from the Earth as the bow shock repeatedly moved back and forth, first crossing one and then the other. The inserts show high-resolution snapshots of the field as measured by the OGO-5 satellite [Greenstadt et al., 1975].

The Bow Shock

Figure 12 shows magnetic field measurements taken during a chance alignment of OGO-5 and HEOS-1 along a radius vector in the vicinity of the Earth's bow shock [Greenstadt *et al.*, 1975]. The middle two panels show the magnetic field strength at HEOS and OGO-5. The panels surrounding these data show high-resolution snap shots of the shock crossings using the OGO-5 data. The inner two panels illustrate the motion of the shock: it moves back and forth crossing HEOS then OGO; OGO then HEOS, etc. We can then use the relative timing to get shock velocities and thence thicknesses. At this time, the velocities ranged from 10 to over 100 km/s and the thicknesses roughly twice the ion inertial length, c/ω_{pi}, or about 300 km. The durations of the crossings were about 10 s.

The shock studied was a 'quasi-perpendicular laminar' shock. By 'quasi-perpendicular', we mean that the local angle between the interplanetary magnetic field and the shock normal was between 50° and 88°, and by 'laminar' we mean that the solar wind Mach number was below a critical value close to 3 and the magnetic energy density was much greater than the thermal energy density, i.e.$\beta << 1$. The results for the thickness obtained for this one class of shocks should not be applied to other conditions. For example, Formisano *et al.* [1975a] have studied, with a single satellite, a shock crossed in less than 0.3 s. If this had a thickness of 300 km, then the velocity must have been greater than 1000 km/s. Such high velocities are not expected theoretically [Völk & Auer, 1974] and thus the thickness must be much less than 300 km. The determination of dependence of thickness on solar-wind conditions must await the ISEE A-B investigations.

If one is interested in only the qualitative dependence of the shock structure on the upstream solar-wind conditions, and not in the scale length of the structure, then two satellite studies with one satellite at the shock and another somewhere in the solar wind in the near vicinity of the Earth will suffice. A number of such studies have been performed. They show that the bow shock does not possess a well-defined structure when either the thermal pressure becomes much greater than the magnetic pressure [Formisano *et al.*, 1975b] or when the interplanetary field is parallel to the shock normal [Greenstadt *et al.*, 1975]. In the latter case, a pulsation region at least $2R_e$ thick has been identified using simultaneous HEOS-1 and OGO-5 measurements. Defining the dimensions of this pulsation region and its dependence on the field direction is ideally suited to the Mother-Daughter mission. The existence of this broad pulsation region suggests that a wide range of interspacecraft separations should be used in shock studies.

The Magnetopause

Although the physics of the magnetopause is quite different from the physics of the bow shock, the same types of studies must be performed at the magnetopause also. Figure 13 shows the three components of the magnetic field and the total field for a 6 h period during which there were many magnetopause crossings, as indicated by the reversals in some of the components at the crossings [Aubry *et al.*, 1970]. The multitude of crossings at this time is due both to an average inward displacement of the magnetopause roughly matching, averaged over the 6 h period, the inward motion of the satellite, and to oscillations of the boundary about this inward moving equilibrium position at a velocity much exceeding the spacecraft velocity. There are only two unambiguous means of deducing the relative velocity between the spacecraft and the boundary, the use of gradients in the energetic particle populations [Kaufman *et al.*, 1973], and the use of two satellites. The former method which uses only one satellite is limited to times when large fluxes of energetic magnetospheric or solar protons are available as tracers. We note that this

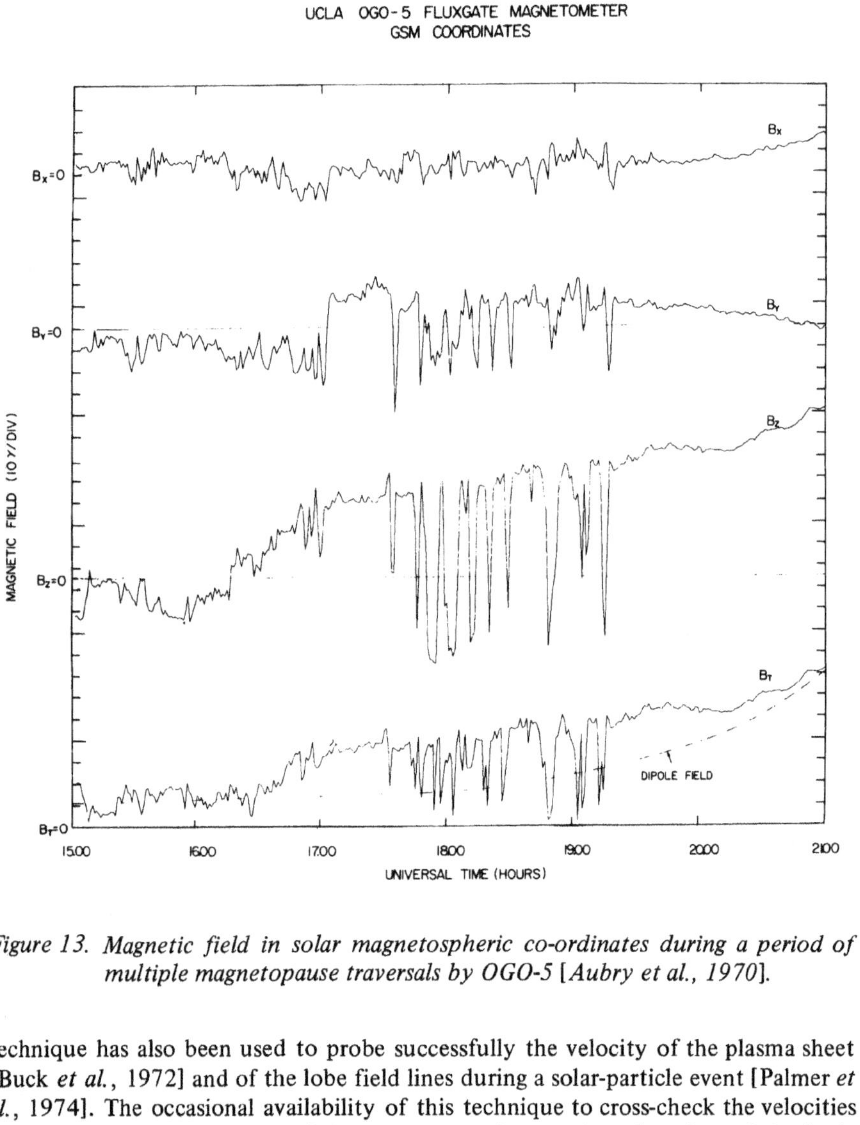

Figure 13. Magnetic field in solar magnetospheric co-ordinates during a period of multiple magnetopause traversals by OGO-5 [Aubry et al., 1970].

technique has also been used to probe successfully the velocity of the plasma sheet [Buck *et al.*, 1972] and of the lobe field lines during a solar-particle event [Palmer *et al.*, 1974]. The occasional availability of this technique to cross-check the velocities derived from two-satellite studies permits a testing or the relaxation of the basic assumptions used in reducing the two-satellite data, i.e. that the boundaries are planes with known normals.

It is becoming increasingly evident that the magnetopause is a complex boundary whose structure is very sensitive to solar-wind conditions. As observed in the plasma data, the magnetopause is not thin, but can be up to an Earth radius thick [Bezrukikh *et al.*, 1975; Haerendel & Paschmann, 1975; Formisano, 1975]. On the other hand, the magnetopause at times can assume a simple structure. Figure 14 shows the

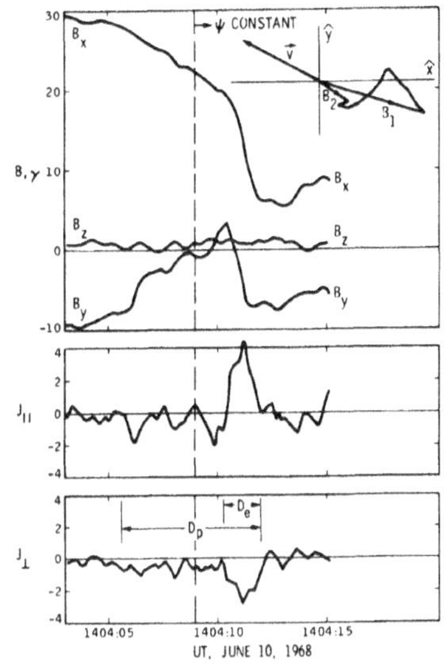

Figure 14. Magnetic field in boundary system during magnetopause crossing (top panel) and current parallel and perpendicular to the field [*Neugebauer et al., 1974*].

field structure through such a simple magnetopause [Neugebauer *et al.*, 1974]. In the uppermost panel, the top and bottom traces are the field variation in the boundary. The middle trace is the variation along the normal to the boundary. The bottom two traces show the derived current densities along the field and perpendicular to it. The parallel component shows a region of negative current and a region of positive current. This is most easily interpreted in terms of the proton turning radius and the electron turning radius. Since they are not equal, the charge separation electric field must have been largely shorted out. The current densities are given in arbitrary units because we have no measure of the velocity as would be possible with ISEE A-B.

Substorm-Associated Boundaries

Figure 15 shows a conceptual sketch of a popular model of the expansion and recovery phase of an isolated substorm [Russell, 1972; McPherron *et al.*, 1973; Nishida & Nagayama, 1973; Hones *et al.*, 1974]. In this model, the neutral point forms in the tail close to the Earth and then moves rapidly away from the Earth. Two satellites closely spaced can test whether the predicted temporal sequence of events is in fact seen as a function of distance from the Earth.

A fortunate alignment of the OGO-5 and ATS-1 spacecraft in the night-time magnetosphere permitted detection of another feature of this model, the inward-moving compressional wave associated with the return of magnetospheric field lines to a more dipolar shape. The three components of the magnetic field in solar magnetospheric co-ordinates together with the total field as measured at both OGO-5 and ATS-1 are given in Figure 16 [Russell & McPherron, 1973]. At this time both OGO-5

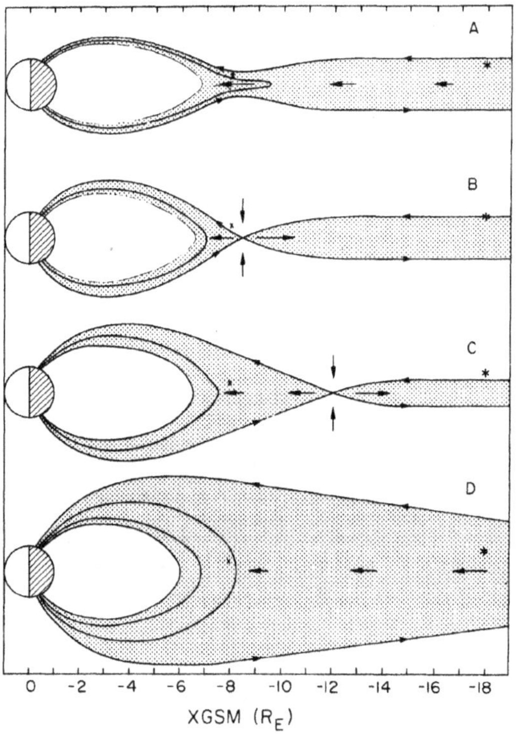

Figure 15. Conceptual model of the initiation of the expansion phase and recovery phase of substorms in which a neutral point is formed near the Earth and then recedes [*Russell, 1972*].

and ATS-1 were near midnight: OGO-5 at 8 R_e and ATS-1 at 6.6 R_e. Shortly after the initiation of a substorm expansion phase, a compression of the field occurs first at OGO-5 and then at ATS-1, 94 s later. Without the two satellites, we would not have known the propagation direction of the disturbance, its velocity, or the thickness of the boundary.

Wave-Particle Interactions

One of the highest priority objectives of the IMS is to determine what wave-particle interactions are operative, what is the relative importance of each one, and how this relative importance changes with time during a storm with spatial location in the magnetosphere, and with changing plasma conditions. One of the most fruitful means of studying wave-particle interactions is the examination of the change in plasma-wave properties in response to the rapid change in a single parameter of the magnetospheric plasma. There are two ways in which this occurs at moderately frequent intervals in the magnetosphere.

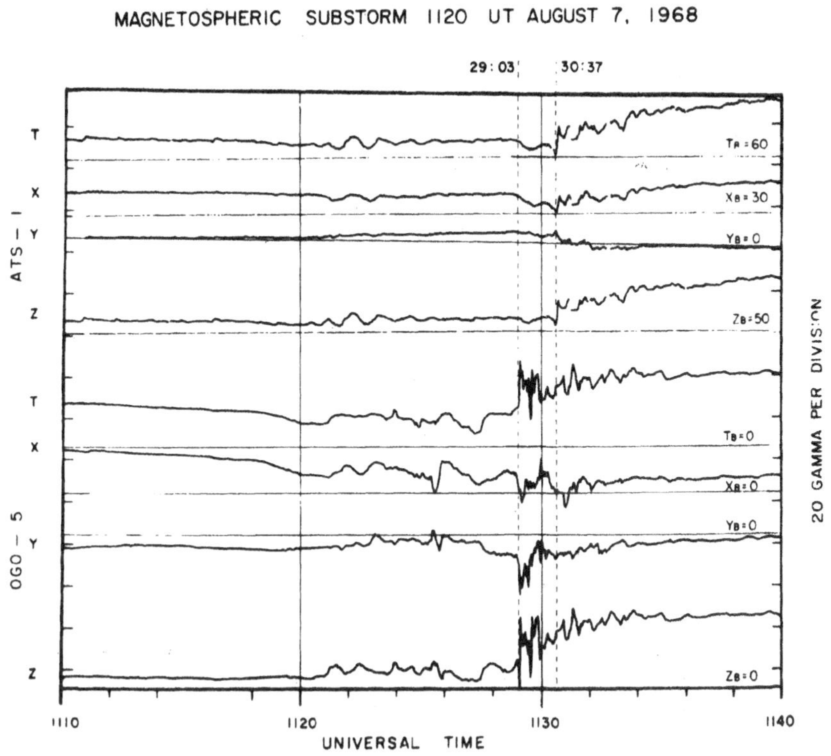

Figure 16. Inward-moving compressional wave detected first at OGO-5 at 8 R_e near midnight local time and then at ATS-1 at 6.6 R_e. Magnetic field data are expressed in solar magnetospheric co-ordinates [Russell & McPherron, 1973].

First, sudden compressions of the field such as that shown in Figure 16 alter the pitch-angle anisotropies of energetic particles. Kivelson *et al.* [1973] have studied the pitch-angle anisotropy changes in >50 keV electrons during an event similar to the event discussed in Figure 16. When the pitch-angle anisotropy was in such a direction that $T_{\parallel}$ was greater than $T_{\perp}$, i.e. there were larger fluxes at small pitch angles than at large (near 90°) pitch angles, then no electromagnetic waves were observed. However, after the passage of the compression, the pitch-angle anisotropy reversed. There were then larger fluxes of 90° pitch-angle particles, and ELF chorus emissions resonant with the energetic electrons occurred, as would be expected for Doppler-shifted cyclotron resonance.

Another means of stimulating emissions is to alter the wave-particle resonance by increasing the density of the plasma. Although it is difficult to raise the density of the magnetospheric plasma rapidly at any one location, isodensity contours are seldom parallel to the drift shells of energetic particles. Thus, the energetic particles

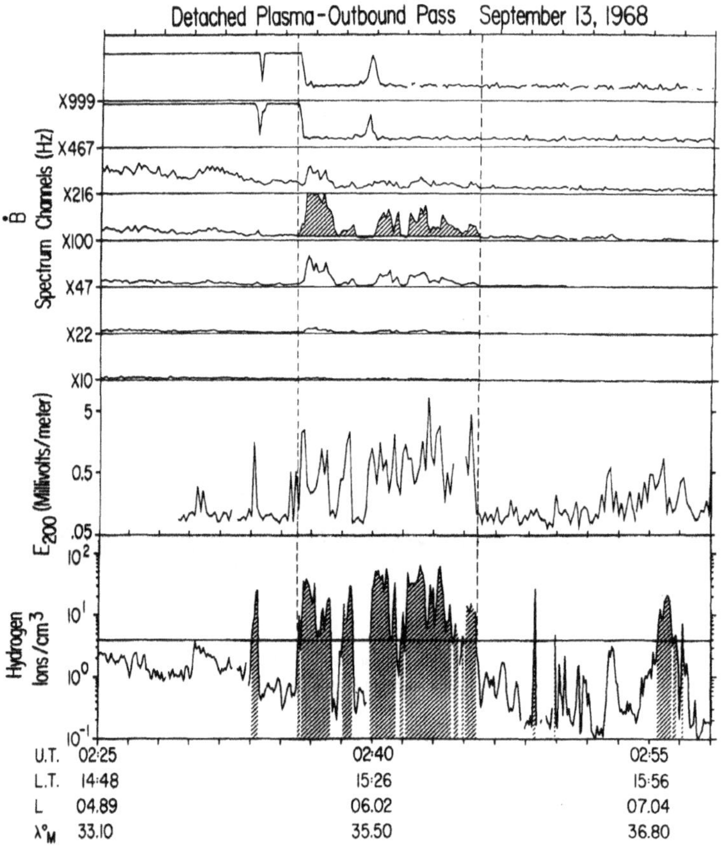

Figure 17. OGO-5 search-coil and electric-dipole measurements at ELF frequencies during a traversal of detached plasma regions as indicated by the ion density measurements in the lower panel [Kivelson & Russell, 1973].

drift into regions where the plasma wave speed and hence the resonant energies are different. Figure 17 illustrates such an occurrence in the outer magnetosphere [Kivelson & Russell, 1973]. The lower trace shows the ion density measured by the OGO-5 ion mass spectrometer. The upper traces show the wave amplitudes at frequencies from 10-1000 Hz. Initially the satellite is in a low-density region of 2-4 ion/cm^3 with a high-frequency hiss present that is characteristic of the high-latitude dayside outer magnetosphere. Then, when enhanced plasma densities are encountered, new frequencies become unstable. Not all density enhancements cause plasma wave enhancements. Presumably this is due to the narrowness of these enhancements, resulting in an inability to duct the waves and lead to amplification or to the energetic particles fluxes being below the stably trapped limit. With the complement of wave, particle and plasma instrumentation on Mother-Daughter and GEOS we should be able to further investigate these regions in much finer detail.

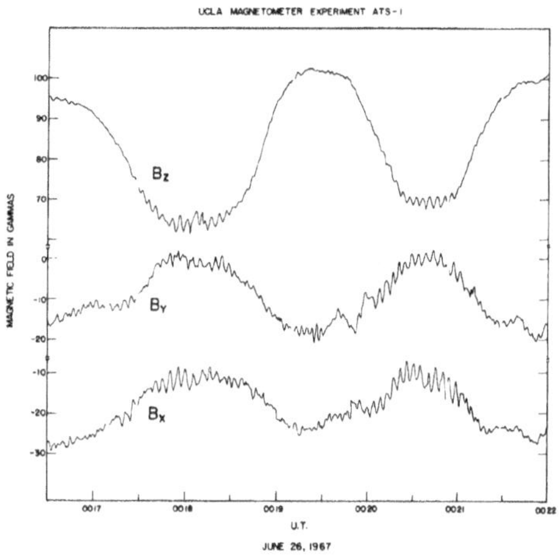

Figure 18. *Simultaneous occurrence of Pc1 and Pc5 micropulsations at synchronous orbit [Barfield & Coleman, 1972].*

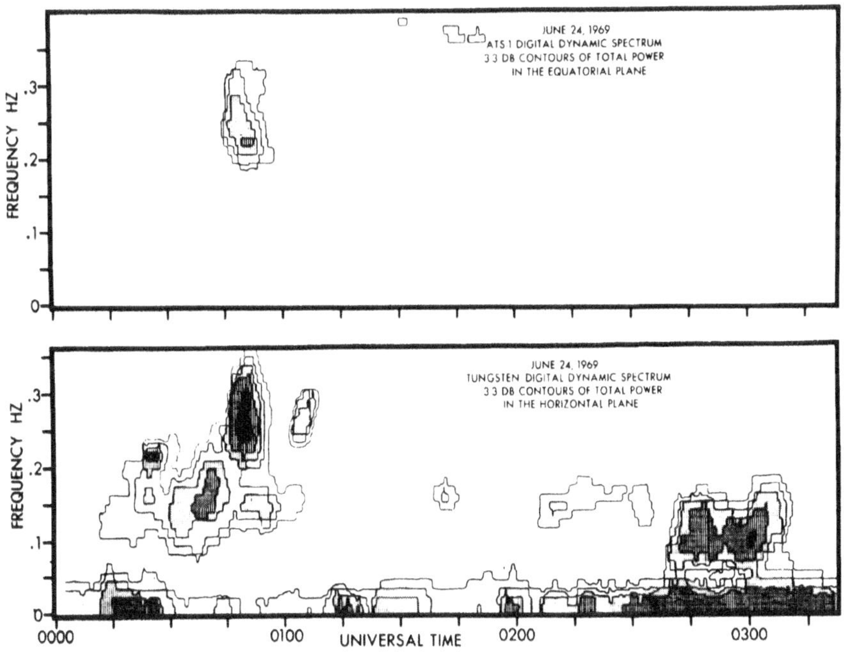

Figure 19. *Simultaneous observations at synchronous orbit (ATS-1) and at the nominal conjugate point (Tungsten, NWT) during an IPDP or sweeper event [Bossen et al., 1975].*

There is much to be learned about ULF waves too. Figure 18 shows Pc5 and Pc1 waves observed simultaneously with the ATS-1 magnetometer [Barfield & Coleman, 1970]. While ATS-1 can measure the magnetic component of the waves very well, there was no low-energy or thermal-plasma instrumentation with which to measure the plasma conditions during these events. Thus, the source of these waves must to some extent be left to speculation. We note that ATS-6 has a good complement of plasma diagnostics although it has no electric-field experiments. Thus some but certainly not all of these problems may have been attacked before the start of IMS.

A very strong aspect of the GEOS programme will be ground-satellite correlations. Figure 19 shows a simultaneous dynamic spectrum of an IPDP or sweeper seen at ATS-1 and its conjugate point [Bossen *et al.*, 1975]. In the ATS-1 data at the top, no upward sweeping in frequency is observed. Furthermore, much less wave activity is seen over the entire period. This raises the question of whether the waves which are not seen at ATS-1 are generated at a different longitude or radial distance. Perhaps some of the fluctuations are generated in the ionosphere. The possibility of having several satellites in synchronous orbit, such as ATS-6 and GEOS, plus having radial cuts with Mother-Daughter when combined with a network of ground magnetometers should provide answers to these questions.

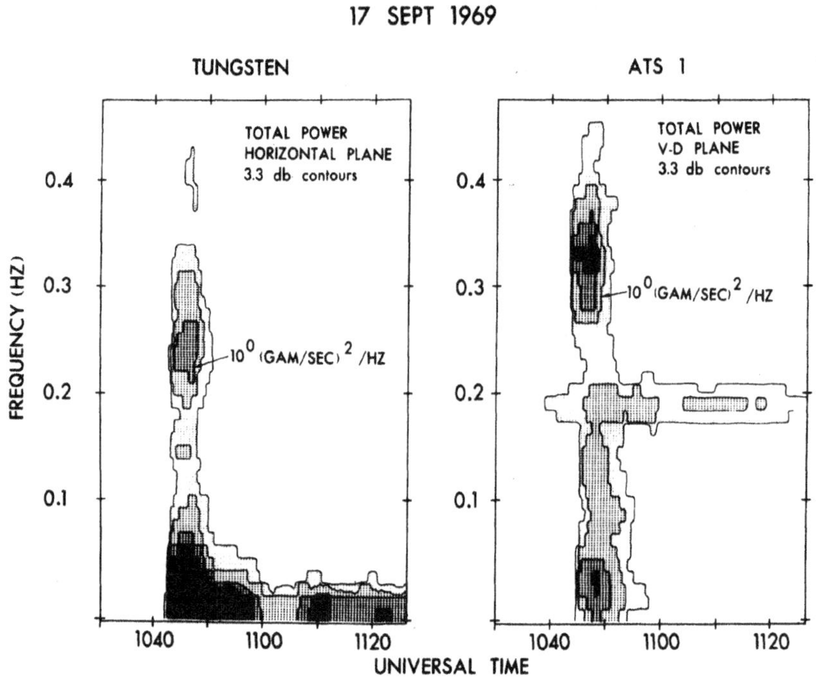

Figure 20. Simultaneous observations at synchronous orbit and the nominal conjugate point during a Pi2 micropulsation.

Figure 20 shows a simultaneous ground-satellite Pi2 burst studied by Arthur & McPherron [1975]. Both show a high-frequency enhancement in the burst, but not at the same frequency. The tone just below 0.2 Hz in the ATS-1 record is spacecraft-generated. Again, at low frequencies there are emissions not seen in space. Here we have the problem of why the space signal was not seen on the ground. Where is the correct conjugate point of ATS-1? Apparently, not a single station, but a network of ground stations surrounding the expected conjugate point is required. The extensive ground network around the GEOS conjugate point should be very helpful in this regard.

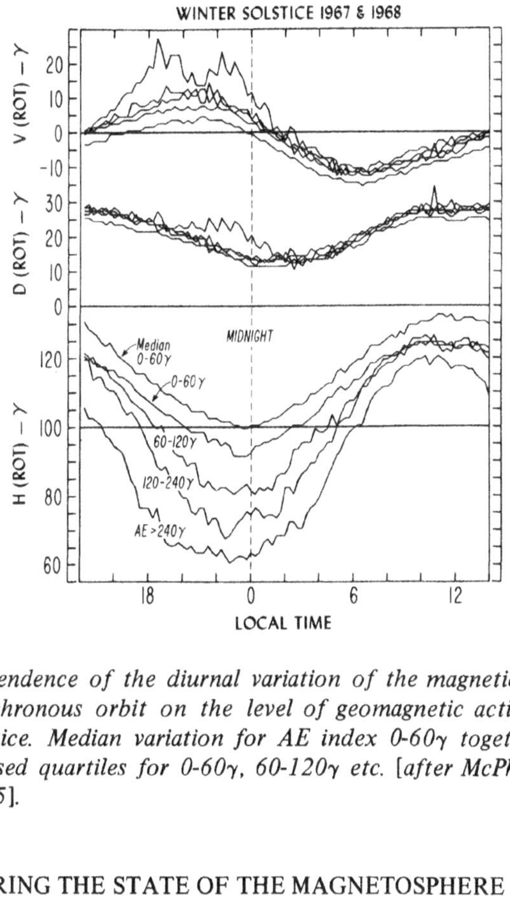

Figure 21. Dependence of the diurnal variation of the magnetic field at ATS-1 in synchronous orbit on the level of geomagnetic activity during winter solstice. Median variation for AE index 0-60γ together with most depressed quartiles for 0-60γ, 60-120γ etc. [after McPherron & Coleman, 1975].

5.3. MONITORING THE STATE OF THE MAGNETOSPHERE

In addition to providing detailed *in-situ* data on magnetospheric processes, the magnetospheric satellites can serve as monitors of the state of the magnetosphere. This role could be quite useful in the timing of rocket and balloon launches. Figure 21 shows the field as measured by ATS-1 at synchronous orbit during summer solstice as a function of geomagnetic activity [McPherron & Coleman, 1975]. V is vertical away from the Earth, D is azimuthal in the direction of the Earth's rotation, and H is horizontal northward. Examining the H-component, we see that the field becomes more and more depressed during the night as geomagnetic activity increases, as measured here by the AE index. Measurements by GEOS in the midnight sector will thus be a sensitive indicator of the state of the magnetosphere.

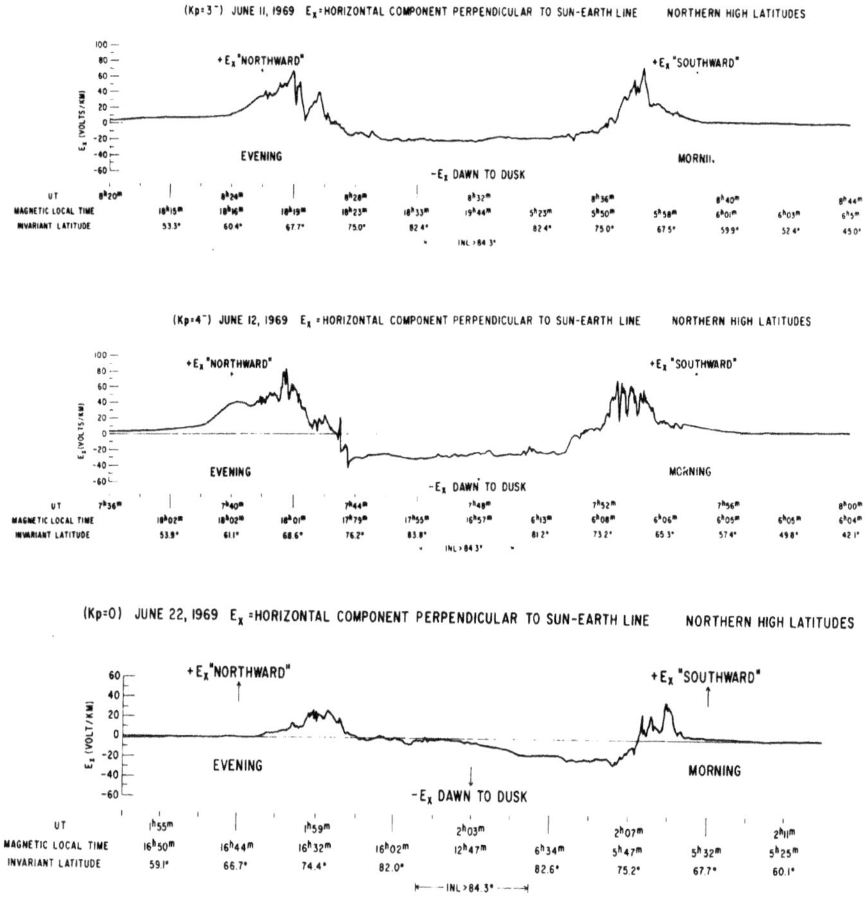

Figure 22. OGO-6 observations of dawn-dusk component of ionospheric electric field during transits of the polar cap in the dawn-dusk meridian [Heppner, 1972].

5.4. MAGNETOSPHERE - IONOSPHERE COUPLING

One of the most interesting, yet least understood, regions of the magnetosphere is its lower boundary, the ionosphere. In particular, we need to determine the mechanisms by which energy and momentum are transferred from the outer magnetosphere into the ionosphere and then to the neutral atmosphere. Many of these mechanisms have been identified but their role is only qualitatively understood. Precipitating particles, wave damping and Joule heating in current systems all can heat the ionosphere. We must determine which is more important in a given situation. Similarly, since the ionosphere acts as a drag on the magnetospheric plasma in most situations, stresses must be transmitted from the magnetosphere to the ionosphere if the ionosphere is moving with the magnetosphere as it appears to do. Field-aligned currents can transmit this stress. If, on the other hand, the ionospheric plasma is not flowing with the magnetospheric plasma, then the field lines cannot be equipotentials and there must be potential drops, i.e. parallel electric fields along these field lines. Since these processes are occurring in and just above the ionosphere, we must fly low-altitude satellites to observe them.

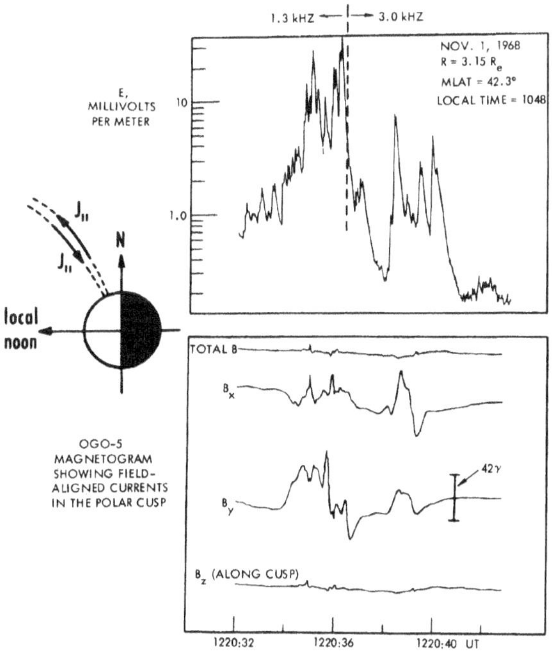

Figure 23. OGO-5 observations of field-aligned currents and VLF electrostatic noise in the polar cusp [Fredricks et al., 1973].

Figure 22 shows the ionospheric electric field measured on three passes of the OGO-6 satellite across the polar cap. The broad depression in the electric field in the centre of each pass corresponds to antisunward flow. The asymmetries in these flows have been shown by Heppner [1972] to be controlled by the direction of the dawn-dusk component of the interplanetary magnetic field. However, we do not yet have a quantitative correlation between the potential drop across the polar cap and interplanetary conditions, nor do we know how to relate the size of the asymmetry to the magnitude of the dawn-to-dusk interplanetary magnetic fields.

Figure 23 shows an example of field-aligned currents observed in the polar cusp [Fredricks *et al.*, 1973]. Similar results apply to the night-time plasma sheet [Scarf *et al.*, 1973]. These currents are often accompanied by VLF electrostatic waves which in turn can cause anomalous resistivity leading to potential drops which are able to accelerate particles to auroral energies. As mentioned above these field-aligned currents also provide the coupling of magnetospheric stresses to the ionosphere. For example, the role of the field-aligned currents in the auroral oval studied by Zmuda & Armstrong [1974 and refs. therein] seems to be to drag the ionosphere along with the magnetospheric plasma as it convects from the plasma sheet to the dayside magnetosphere. Unfortunately, to date no satellite has been launched with the comprehensive set of instrumentation required to effectively investigate other than the morphology of these currents. Perhaps if the EXOS-B orbit carries it over the auroral oval it will be able to investigate this problem. Perhaps one of the Russian spacecraft will cover this. If not, then we may have to wait until the launch of the Electrodynamic Explorers which have the study of these currents as one of their primary goals.

6. Concluding Remarks

In closing we note that this review of objectives of the IMS satellite programme is far from being a comprehensive or exhaustive list. Rather, we have seen only a sampling of the many possible problems to be attacked and that sample is strongly biased to the problems on which the author has worked prior to IMS. Each of the investigators associated with the programmes discussed at the beginning of this article certainly has his own list of objectives just as large as mine, and many of these objectives we have scarcely even mentioned. In the area of the magnetospheric plasma, the most noticeable absence from our discussion was the problem of the acceleration of magnetospheric plasma. Which is more important: acceleration by parallel electric fields or cross-equipotential curvature and gradient drift? Also, we skipped over the use of energetic particles such as solar protons as probes of magnetospheric structure. Much has been done in this regard [Morfill & Scholer, 1973], but much remains to be done. Finally, our brief discussion of wave-particle interactions does little justice to the myriad of plasma-wave phenomena observed in the magnetosphere, many of whose sources remain merely objects of speculation.

In short, there is much to be done during the International Magnetospheric Study. This fact should not be construed to mean that magnetospheric physicists have been lax in their efforts over the last decade. On the contrary, their efforts have been quite successful, and we have benefitted tremendously from them. Our large list of objectives consists of problems which are well defined and which can be solved with the IMS programme. If it were not for the pioneering efforts of the Explorer, OGO, ATS, and similar programmes, we would not be in this position today.

7. Acknowledgments

Much of the research discussed within this article was performed in collaboration with colleagues to whom I am much indebted. At one time or another these included: R.K. Burton, C.R. Chappell, P.J. Coleman, Jr., F.V. Coroniti, V. Formisano, E.W. Greenstadt, P.C. Hedgecock, R.E. Holzer, C.F. Kennel, M.G. Kivelson, R.L. McPherron, M. Neugebauer, F.L. Scarf, E.J. Smith, R.J. Walker and H.I. West, Jr. The National Aeronautics and Space Administration provided support for the preparation of this review under Research Grant NGR 05-007-004.

References

Arthur, C.W. & R.L. McPherron, A preliminary study of simultaneous ground-satellite observations of substorm-associated Pi-2 micropulsations and their high frequency enhancements, in preparation, 1975.

Aubry, M.P., C.T. Russell & M.G. Kivelson, Inward motion of the magnetopause before a substorm, *J. Geophys. Res.* **75**, 7018 (1970).

Barfield, J.N. & P.J. Coleman, Jr., Storm-related wave phenomena at the synchronous, equatorial orbit, *J. Geophys. Res.* **75**(10), 1943 (1970).

Bezrukikh, V.V., T.K. Brens, M.I. Verigin, P.A. Maysuradze, A.P. Remizov & E.K. Solomatina, Dependence of the earth's magnetopause and bow shock positions on the solar wind parameters and the magnetopause plasma structure observed by charged particle traps aboard the Prognoz and Prognoz-2 satellites, Institute for Space Research Report D-192, Academy of Sciences of the USSR, Moscow, 1975.

Bitoun, J., J. Etcheto & B. Higel, Diagnostics of spatial plasmas with a relaxation sounder, Proc. 10th ESLAB Symposium, this volume, 1975.

Bossen, M., R.L. McPherron & C.T. Russell, Substorm associated Pc1 micropulsations at ATS-1 and its conjugate point, in preparation, 1975.

Burton, R.K., R.L. McPherron & C.T. Russell, An empirical relationship between interplanetary conditions and Dst, *J. Geophys. Res.* **80**, in press, 1975.

Dungey, J.W., The structure of the exosphere or adventures in velocity space, in Geophysics, The Earth's Environment, (Eds. C. DeWitt, J. Hieblot & A. Lebeau), Gordon and Breach, New York, 1963.

Dungey, J.W., The length of the magnetospheric tail, *J. Geophys. Res.* **70**(7), 1753 (1965).

Durney, A.C., Present status of the ISEE A and B Mission, Proc. 10th ESLAB Symposium, this volume, 1975.

Formisano, V., Personal communication, 1975.

Formisano, V., C.T. Russell, J.D. Means, F. Scarf & M. Neugebauer, The perpendicular bow shock, in preparation, 1975a.

Formisano, V., C.T. Russell, J.D. Means, E.W. Greenstadt, F.L. Scarf & M. Neugebauer, Collisionless shock waves in space: A very high β-structure, *J. Geophys. Res.* **80**(16), 2013 (1975b).

Fredricks, R.W., F.L. Scarf & C.T. Russell, Field-aligned currents, plasma waves, and anomalous resistivity in the disturbed polar cusp, *J. Geophys. Res.* **78**(13), 2133 (1973).

Friis-Christensen, E. & J. Wilheljm, Polar cap currents for different directions of the interplanetary magnetic field in the Y-Z plane, *J. Geophys. Res.* **80**(10), 1248 (1975).

Greenstadt, E.W., C.T. Russell, F.L. Scarf, V. Formisano, & M. Neugebauer, Structure of the quasi-perpendicular laminar bow shock, *J. Geophys. Res.* **80**(4), 502, (1975a).

Greenstadt, E.W., C.T. Russell, V. Formisano, P.C. Hedgecock, F.L. Scarf, M. Neugebauer & R.E. Holzer, Structure of the quasi-parallel, quasi-laminar bow shock, in preparation, 1975b.

Haerendel, G. & G. Paschmann, Entry of solar wind plasma into the magnetosphere, Proc. Nobel Symposium, Kiruna, April 1975, Plenum Publ. Corp., New York, 1975.

Heikkila, W.J., Penetration of particles into the polar cap regions of the magnetosphere, in Critical Problems of Magnetospheric Physics (Ed. E.R. Dyer), p. 67, National Academy of Sciences, Washington DC, 1972.

Heppner, J.P., Polar cap electric field distributions related to the interplanetary magnetic field direction, *J. Geophys. Res.* **77**, 4877 (1972).

Hones, E.W., Jr., A.T.Y. Lin, S.J. Bame & S. Singer, Prolonged tailward flow of plasma in the thinned plasma sheet observed at $\gamma \approx 18$ R_e during substorms, *J. Geophys. Res.* **79**(10), 1385 (1974).

Kaufmann, R.L. & A. Konradi, Speed and thickness of the magnetopause, *J. Geophys. Res.* **78**(28), 6549-6568 (1973).

Kivelson, M.G. & C.T. Russell, Active experiments, magnetospheric modification and a naturally occurring analogue, *Radio Science* **8**(11), 1035 (1973).

Kivelson, M.G., T.A. Farley & M.P. Aubry, Satellite studies of magnetospheric substorms on August 15, 1968. 5. Energetic electrons, spatial boundaries and wave-particle interactions at OGO-5, *J. Geophys. Res.* **78**(16), 3079-3092 (1973).

Knott, K., Agreed plans for the payload operations and orbital manoeuvres of GEOS, Proc. 10th ESLAB Symposium, this volume, 1975.

Maezawa, K. & T. Obayashi, Magnetospheric convection induced by the interplanetary magnetic field variations, Proc. 10th ESLAB Symposium, this volume, 1975.

McPherron, R.L., Current status of the growth phase controversy, *EOS Trans. AGU* **55**(11), 994 (1974).

McPherron, R.L. & P.J. Coleman, Jr., Substorm observations of magnetic perturbations and ULF waves at synchronous orbit by ATS-1 and ATS-6, this volume, 1975.

McPherron, R.L., C.T. Russell & M.P. Aubry, Satellite studies of magnetospheric substorms on August 15, 1968. 9. Phenomenological model for substorms, *J. Geophys. Res.*, **78**(16), 3131-3149 (1973).

Michel, F.C. & A.J. Dessler, On the interpretation of low-energy particle access to the polar caps, *J. Geophys. Res.* **80**(16), 2309 (1975).

Morfill, G. & M. Scholer, Study of the magnetosphere using energetic particles, *Space Sci. Rev.* **15**(213), 267-354 (1973).

Mozer, F., Magnetospheric DC electric fields and outstanding problems to be solved, Proc. 10th ESLAB Symposium, this volume, 1975.

Mozer, F.S., W.D. Gonzalez, F. Bogott, M.C. Kelley & S. Schutz, High-latitude electric fields and the three-dimensional interaction between the interplanetary and terrestrial magnetic fields, *J. Geophys. Res.* **79**(1), 56 (1974).

Neugebauer, M., C.T. Russell & E.J. Smith, Observations of the internal structure of the magnetopause, *J. Geophys. Res.* **79**(4), 499 (1974).

Nishida, A. & N. Nagayama, Synoptic survey for the neutral line in the magnetotail during the substorm expansion phase, *J. Geophys. Res.* **78**(9), 3782-3798 (1973).

Palmer, I.D., Gradients of solar protons in the high latitude magnetotail and the convection electric field, (Abstract) *EOS Trans. AGU* **55**(12), 1168 (1974).

Russell, C.T., The configuration of the magnetosphere, in Critical Problems of Magnetospheric Physics (Ed. E.R. Dyer), National Academy of Sciences, Washington DC, 1972.

Russell, C.T. & R.L. McPherron, The magnetotail and substorms, *Space Sci. Rev.* **15**, 205 (1973).

Scarf, F.L., R.W. Fredricks, C.T. Russell, M. Kivelson, M. Neugebauer & C.R. Chappell, Observation of a current-driven plasma instability at the outer-zone plasma sheet boundary, *J. Geophys. Res.* **78**(13), 2150 (1973).

Vette, J.I., Identification of satellites possibly active during the IMS and their orbital configurations, Proc. 10th ESLAB Symposium, this volume, 1975.

Volk, H. & R.D. Auer, Motions of the bow shock induced by interplanetary disturbances, *J. Geophys. Res.* **79**(1), 40 (1974).

Wenzel, K.-P., IMS and the ISEE-C mission, Proc. 10th ESLAB Symposium, this volume, 1975.

Zmuda, A.J. & J.C. Armstrong, The diurnal flow pattern of field-aligned currents, *J. Geophys. Res.* **79**(31), 4611 (1974).

DISCUSSION

M. Dryer: Have there been any studies regarding the magnetopause thickness as a function of distance from the subsolar point?

C.T. Russell: At the present time there are two studies in progress, one by Formisano and co-workers using the HEOS-1 data, and one by Gringauz and co-workers using Prognoz data, but the results are not in yet.

G. Rostoker: You have spoken of most of the scientific community accepting an open magnetospheric model. However, there is strong evidence in the work of McDiarmid and Burrows that the plasma sheet (the site of auroral phenomena) lies on closed but distended field lines. Can we therefore talk in terms of a closed magnetosphere when we deal with auroral oval activity?

C.T. Russell: With the exception of the occasional observation of loops of field lines in the plasma sheet (X and O neutral point pairs), I think most advocates of the open magnetosphere propose that the plasma sheet is on closed field lines. Auroral arcs in this picture are associated with processes occurring on closed field lines, such as field-aligned currents and associated parallel electric fields.

G. Rostoker: I think that there is little evidence for neutral points in the tail. The most compelling evidence is southward B_z. However that may well be caused by the end effects of Birkeland currents in the tail. How strong do you feel the evidence for neutral points is?

C.T. Russell: If the tail lobes are open, that is the field lines eventually connect with the interplanetary magnetic field, then there must somewhere be a neutral point between the tail lobes on simply topological grounds. The evidence for open tail lobes is very strong: solar-particle entry, and the many aspects of the response of the magnetosphere are most readily explained in terms of this model. The question of whether the observed southward fields in the plasma sheet are caused by the appearance of the neutral point close to the Earth is a separate question, which cannot be answered by examining the magnetic data alone. However, the simultaneous behaviour of the tail lobes, in which the flux decreases after the appearance of the southward fields, the behaviour of the plasma sheet, which thins and then thickens, and the observed flows away and then later towards the Earth, require that the neutral points approach the Earth at substorm onset and then move away.

PART 2

IMS SATELLITES: ORBITS AND SCIENTIFIC POTENTIAL

IDENTIFICATION OF SATELLITES POSSIBLY ACTIVE DURING THE IMS AND THEIR ORBITAL CONFIGURATIONS

James I. Vette, Robert H. Hilberg & Michael J. Teague

IMS Satellite Situation Center,
Goddard Space Flight Center,
Greenbelt, Maryland, USA

Abstract

A survey has been made to identify presently operating as well as planned spacecraft which may contribute measurements of the magnetosphere or interplanetary medium during the IMS. Fifty-one have been identified. Twenty-three are presently operating. The others are planned for launch prior to 1979. The total set consists of 25 scientific Earth-orbiting, 9 heliocentric and 17 communication, navigation or meteorological spacecraft, which carry environmental monitoring experiments. The orbital elements of each are displayed. In addition, techniques for the display of positions in various orbits are presented to illustrate how interesting configurations can be determined during the IMS. The positions of presently orbiting heliocentric spacecraft are shown at various intervals with respect to the nominal interplanetary magnetic field spiral and the Earth. Plots of the latest ISEE-A orbit in solar magnetospheric co-ordinates are presented to determine neutral-sheet crossing. The angular separation if IMP-H and -J approaches zero around the end of 1975. Periods where the separation is less than $10\,R_E$ are given, other characteristics of the spacecraft orbits are discussed, and plans for the distribution of this information through the Satellite Situation Center are presented.

1. Introduction

Orbiting spacecraft will play an important role in the IMS programme by providing *in-situ* measurements over the whole magnetospheric domain and much of interplanetary space. Once in orbit and checked out, spacecraft usually continue to operate for many years. With the exception of satellites with a portion of their orbit at altitudes below 300 km, it is possible to predict their position several years in the future. Consequently, the identification of satellites that may be active during the

Knott & Battrick (eds.), The Scientific Satellite Programme During the International Magnetospheric Study, 45–59.
All Rights Reserved. Copyright © 1976 by D. Reidel Publishing Company, Dordrecht-Holland.

IMS period is important to the planning of the integrated rocket, balloon, aircraft, and ground-based programme, since one knows the kind of measurements that are possible as a function of time and space. In this paper we will identify those spacecraft we consider likely to be making measurements of a solar-terrestrial physics nature during the IMS, give their orbit parameters, display some techniques for determining especially interesting simultaneous positions of two or more satellites, and finally present some plans by the IMS Satellite Situation Center to produce and distribute graphical information on the positions of these spacecraft.

We restrict ourselves to the present official IMS period of 1 January 1976 - 31 December 1978, although we recognise that this interval may be changed in the future because of launch dates of important spacecraft. For those spacecraft presently operating, we have attempted to determine how long data will continue to be taken by contacting various offices responsible for such decisions. We have used the 'Report of Active and Planned Spacecraft and Experiments', produced semi-annually by the National Space Science Data Center/World Data Center A-Rockets and Satellites, to determine those spacecraft carrying appropriate experiments to measure IMS-related phenomena. For planned spacecraft we have indicated launch dates that are current as of 1 May 1975. It should be recognised that many of these dates will change by several months as various difficulties or launching conflicts arise.

2. Possibly Active Spacecraft

The result of this identification effort is displayed in Tables 1, 2, and 3. Table 1 identifies those spacecraft whose prime purpose is to conduct scientific investigations. These are placed in the three scientific categories shown in the last column of the table and based on the major emphasis of their experiment complement. For details of the specific experiments one should refer to the 'Report of Active and Planned Spacecraft and Experiments'. The orbit parameters are shown to one decimal in the indicated units, and most are epoch January 1975. For presently orbiting spacecraft, the elements are generally known to a higher accuracy; however, the purpose of these parameters is to indicate the approximate region of space that each traverses. Some parameters, particularly those of planned spacecraft, are shown only to the accuracy of units. We have not shown spacecraft such as ISIS 1 and AE-C, which may be capable of operation in the IMS time period, because of priorities for ISIS 2 and AE-D and AE-E data acquisition. For synchronous satellites we have shown the longitude in the remarks column. Some recently launched spacecraft have been added although orbit parameters were not available.

Applications spacecraft which carry some experiments of interest to the IMS are presented in Table 2. We have indicated the nature of these experiments in the remarks column.

The heliocentric spacecraft are given in Table 3. The amount of telemetry acquisition for Pioneers 6-9 is infinitesimal at the present time because of priority for Pioneer 10 (not listed since it is moving out of the solar system) and Pioneer 11. If there are particularly important positional configurations of these spacecraft during the IMS, it may be possible through timely requests to the Pioneer Project Office at NASA/Ames Research Center to obtain some data reception.

TABLE 1
Possible Scientific Spacecraft

Spacecraft	Launch date	Inc. (deg)	Perigee alt. (km)	Apogee alt. (km)	Period (min)	Remarks	Sci. type
OSO-I	06/ /75	33.	550.	550.	96.	Solar pointed & wheel expts.	SOLAR
D2-B	07/ /75	–	–	–	–	Some geocoronal expts.	
SOLRAD 11A	11/ /75	0.	127622.	127622.	8140.	Several energetic particle expts. $R = 20R_E$	
SOLRAD 11B	11/ /75	0.	127622.	127622.	8140.	Same payload as SOLRAD 11A	
P78-1	1Q/ /78	–	–	–	–	Launched with Solwind expt.	
VELA 6A	04/08/70	33.5	111139.0	117959.0	6701.1	VELAs may be operated during some IMS periods	MAGNETOSPHERIC FIELDS & PARTICLES
VELA 6B	04/08/70	33.3	111073.0	118169.0	6698.2	6A & B; $R = 18.4 - 19.5\ R_E$	
IMP-H	09/23/72	12.7	204593.0	237410.0	17984.0	$T = 12.5$d; $R = 33.1 - 38.2\ R_E$	
IMP-J	10/26/73	18.1	169037.0	263116.0	17403.0	$T = 12.1$d; $R = 27.5 - 42.3\ R_E$	
HAWKEYE 1	06/03/74	89.8	2145.1	125274.0	3077.7	Only highly eccentric polar spacecraft for IMS	
SESP 74-2	1Q/ /76	90.	200.	8000.	164.5		
GEOS	02/ /77	1.	35700.	35700.	1436.2	Only synchronous S/C dedicated to IMS. 25°W-35°E-0°E-35°E long.	
ISEE-A & -B	10/ /77	28.	280.	133852.	3270.	The two S/C have controlled spacing for spatial & temporal studies	
EXOS-B	08/ /78	30.	500.	30000.	480.		
ISEE-C	07/ /78	Earth-Sun Libration Point				To measure undisturbed interplanetary medium	
ISIS 2	04/01/71	88.2	1355.8	1422.9	113.5	Many TM stations around world	ATMOSPHERIC & IONOSPHERIC
INTASAT	11/15/74	101.7	1440.0	1458.0	114.9	Beacon; will turn off prior to Jan. 1, 1977	
SRATS	02/24/75	31.6	2255.0	3136.0	120.3	Some solar experiments	
ARIABAT	04/19/75	50.7	568.0	611.0	96.5	India's spacecraft	
D5-B	05/25/75	–	–	–	–	Micro accelerometer	
SESP 73-6	3Q/ /75	90.	230.	900.	131.4		
AE-D	09/ /75	98.	150.	4000.	129.	Orbit manoeuvres to change perigee; final state-circular	
DADE-A &-B	11/ /75	90.	400.	1500.	122.	Dual air density experiment	
AE-E	12/ /75	22.	150.	4000.	129.	Similar manoeuvres to AE-D	
ISS	02/ /76	70.	1000.	1000.	135.1	First dedicated IMS S/C	
EXOS-A	02/ /78	60.	250.	3000.	140.5		

TABLE 2
Applications Spacecraft with Environment Experiments useful to IMS

Spacecraft	Launch date	Inc. (deg)	Perigee alt. (km)	Apogee alt. (km)	Period (min)	Remarks	Appl. type
ATS 5	08/12/69	2.2	35718.0	35858.0	1436.2	Fields & particles expts.; turned on for special operations only. 105°W long.	COMMUNICATIONS
ATS 6	05/30/74	1.3	35760.0	35820.0	1436.3	Fields & particles expts. 94°W to 35°E 5/17-6/26/75. 35°E long. until 7/76	
SIRIO-A	2H/ /76	0.	35700.	35700.	1436.2	Fields & particles expts. 15°W long.	
TRIAD 1	09/02/72	90.1	749.5	832.1	100.6	Magnetometer; data through 1976.	NAVI-GA-TION
NTS 2	09/ /76	63.	26561.	26561.	541.7	Magnetometer	
SCATHA	1Q/ /78	0.	35700.	35700.	1436.2	Fields & particles expts.	S/C TECH
DMSP	When nec.	90.	833.	833.	101.6	Auroral photos & precip. part. made available. Always 2 in orb. Dawn-dusk & noon-midnight.	USAF OPER S/Cs
DSP	/ /76	0.	35700.	35700.	1436.2	Energetic particles expts.	
NOAA 3	11/06/73	101.7	1448.0	1454.0	114.9	Solar proton monitor	METEOROLOGICAL
NOAA 4	11/15/74	101.7	1444.0	1458.0	114.9	Solar proton monitor	
SMS 1	05/17/74	1.9	35783.0	35792.0	1436.1	Particles, B field & X-ray monitor. 75.3°W long.	
SMS 2	02/06/75	1.1	35680.0	36685.0	1456.4	Same monitor as SMS 1	
ITOS-E2	07/15/75	103.0	1448.0	1453.0	114.9	Solar proton monitor. ITOS becomes NOAA after launch.	
SMS-C	10/ /75	0.	35700.	35700.	1436.2	Same monitor as SMS 1	
ITOS-H	10/ /76	103.	1448.	1453.	114.9	Solar proton monitor	
GOES-B	12/ /76	0.	35700.	35700.	1436.2	Same monitor as SMS 1	
GOES-C	05/ /77	0.	35700.	35700.	1436.2	Same monitor as SMS 1	
ITOS-I	1Q/ /78	103.	1448.	1453.	114.9	Solar proton monitor	

TABLE 3
Possible Interplanetary IMS Spacecraft

Spacecraft	Launch date	Ecliptic inc. (deg)	Perihelion (AU)	Apohelion (AU)	Period (d)	Remarks
PIONEER 6	12/16/65	0.160	0.8143	0.9360	311.3	Very little telemetry reception; Pioneers 10 & 11 have all the time
PIONEER 7	08/17/66	0.097	1.0100	1.1250	409.2	Same as Pioneer 6
PIONEER 8	12/13/67	0.058	0.9892	1.0880	386.6	Same as Pioneer 6
PIONEER 9	11/08/68	0.087	0.7542	0.9905	297.6	Same as Pioneer 6
PIONEER 11	04/06/73	15.353	3.7330	30.8480	26280.0	On trajectory to go inside inner ring of Saturn; mid-orbit corr. req.
VENUS 9	06/08/75	–	–	–	–	Fly by Venus
MJS 77A	08/ /77					Fly by Jupiter and Saturn
MJS 77B	08/ /77					Fly by Jupiter and Saturn
HELIOS-A	12/10/74	0.000	0.3080	0.9840	190.0	
HELIOS-B	01/ /76	0.000	0.3000	0.9800	192.0	

3. General Remarks on Spacecraft Coverage

The low-altitude regime seems to have adequate coverage by the atmospheric and ionospheric satellites. Although many of these have polar orbits, some mid- and low-latitude orbits are occupied. Since the Atmospheric Explorers (AE-D and -E) will carry out manoeuvres to reach perigees of 130 km, there is good coverage up to 4000 km altitude.

Until the launch of ISEE A and B (Mother and Daughter) the only highly eccentric orbit spacecraft is Hawkeye 1, and this is in a polar orbit. Consequently, the low-inclination sweeps through the magnetosphere provided by spacecraft such as OGO 1, 3, and 5 as well as IMP-A, -B, -C, and -F, which provided much of our understanding of the magnetosphere in earlier years, will be absent until the ISEE-A and -B launch in October 1977. EXOS-B will be placed in an orbit similar to S^3, which provided interesting measurements of the ring current and the plasmasphere. IMP-H and -J will provide the main information of the interplanetary medium in near-Earth space as well as tail field phenomena in the 30 - 40 R_E region. Of course ISEE-C (Heliocentric) will be the first spacecraft to measure this local interplanetary medium completely undisturbed by the Earth's bow shock. Unfortunately, this launch will not occur until the end of the presently designated IMS period. Hopefully Vela 6A and 6B data coverage can be increased during the IMS to provide information at approximately half the distance of IMP-H and -J. Because of the near-circular orbits of these four spacecraft, the tail region will be sampled several times per month.

Although GEOS is the only scientific spacecraft in synchronous orbit, and a cornerstone for the IMS, there are nine others in this orbit, and certainly ATS 5 and 6 have extensive experiment complements for studying magnetospheric phenomena. Clearly this special region of space will have excellent coverage. On the other hand, the inner- and outer-radiation belts will only receive coverage by SESP 74-2 in a polar orbit. Very little coverage will be provided at low latitudes for energetic particles; one experiment on ISEE-A and -B will provide traversals once every 2.2 d.

4. Techniques for Displaying Interesting Spacecraft Positional Configurations

Of primary importance to the conduct of the IMS is the determination of interesting positional configurations of spacecraft. This is a task of the IMS Satellite Situation Center. It is more difficult to accomplish for low-altitude satellites because of the numerous revolutions per day. One type of display of interest for polar orbits is illustrated in Figure 1, taken from the first Satellite Situation Center Report, where the projections of several revolutions of the ISIS 2 satellite on the surface of the Earth at two different periods in a polar co-ordinate system of magnetic latitude and magnetic local time are shown. Since the change in this configuration is not rapid and for Sun-synchronous orbits the configuration is practically stationary, this technique is useful to show the character of passes over the auroral regions and polar cap; noon-midnight and dawn-dusk traversals are of particular interest. Plots of this type spaced at weekly intervals should be adequate to determine at a glance the changing character of such satellite orbits. Since there are 15 low-altitude satellites

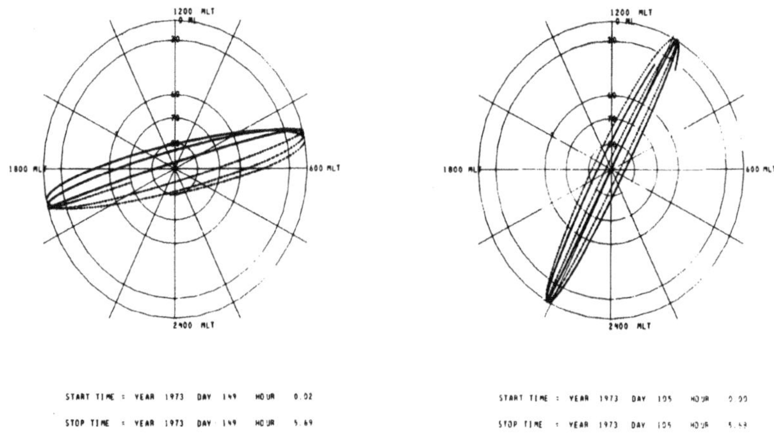

Figure 1. *ISIS 2 positions projected onto Earth's surface using magnetic latitude and magnetic local time as polar co-ordinates. Solid lines correspond to Northern Hemisphere, and dashed lines to Southern Hemisphere portions.*

with orbital inclinations greater than 60° listed in Tables 1-3, this would require up to 780 plots a year. This number could be reduced by a factor of 2 or 3 by choosing the proper satellites to show on a single graph.

Heliocentric orbits represent the other extreme; the interesting configurations can be determined adequately by one plot per year. This is illustrated in Figure 2 where all the presently orbiting spacecraft, except Venus 9, listed in Table 3 are shown in a heliocentric polar co-ordinate system rotating with the Earth. Superimposed on this grid are the interplanetary magnetic field lines for a uniform solar wind velocity of 400 km/s spaced at equal intervals of 2.5 d of corotation delay. The radial scale is broken at 1.25 AU to show Pioneer 11 on the graph without destroying the clarity of the positions of the numerous spacecraft orbiting within 1 AU.

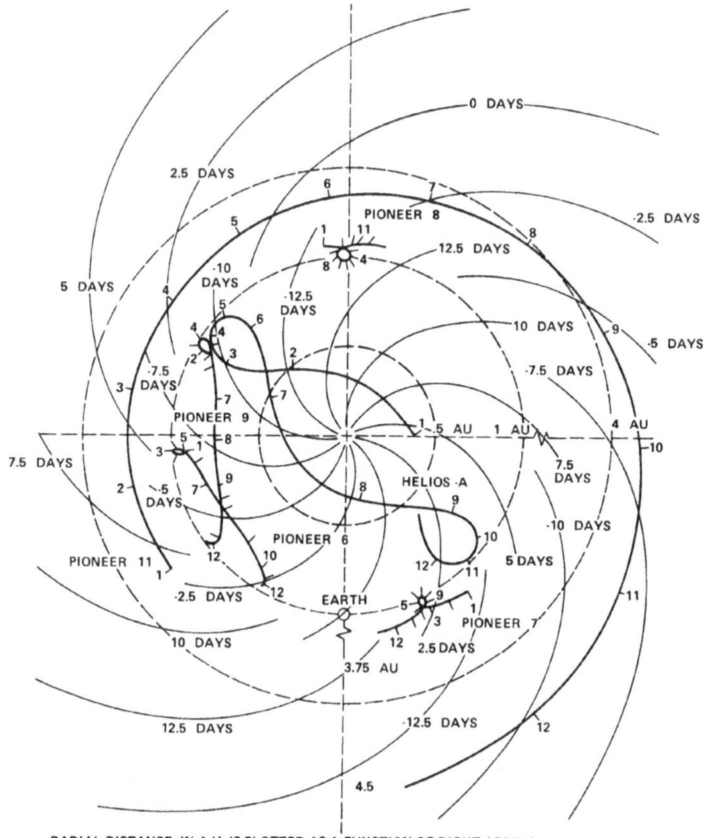

RADIAL DISTANCE, IN A.U. IS PLOTTED AS A FUNCTION OF RIGHT ASCENSION RELATIVE TO EARTH. THE CURVED GRID GIVES COROTATION DELAY TIMES IN DAYS. TIME TICKS ALONG ORBIT CURVES SHOW BEGINNING OF EACH MONTH OF 1976.

Figure 2. Heliocentric orbits projected onto ecliptic plane.

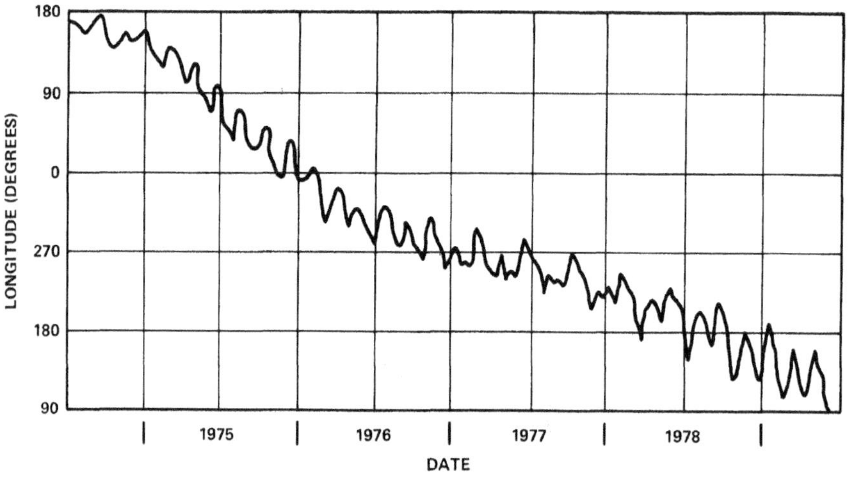

Figure 3. *Separation in ecliptic longitude between IMP-H and IMP-J during IMS period.*

The six high-altitude spacecraft listed in the tables can be displayed in a number of ways to show possibly interesting configurations. One such configuration occurs at the beginning of the IMS between IMP-H and -J. A plot of the longitude separation of these two spacecraft over the IMS period is shown in Figure 3. Around February 1976 it can be seen that this angle is near zero. During the time period October 1975 - February 1976 the separation of these spacecraft becomes less than $10 R_E$ at several time periods. This is shown in Figure 4. At these separation distances interesting measurements in the geomagnetic tail region could be possible. Since these close approaches only last for several hours, every effort should be made to ensure that data reception is achieved. The region of space where this interesting configuration occurs is illustrated in Figure 5 where the positions are shown in the geocentric solar magnetospheric (GSM) Y-Z plane. It can be seen that the time of close separation occurs in the tail region. This is true for all of these close approaches. The idealised position of the neutral sheet is shown by the Z_{GSM} = constant lines labelled with UT; this time change of the neutral sheet position is produced by the wobble of the geomagnetic dipole as the Earth rotates. The tick marks on the orbit track show year, day of year and hour; i.e. 18.41.12h.

For highly eccentric spacecraft such as ISEE-A, the local time of apogee, the passage through the magnetosheath, and the intervals in the interplanetary medium and in the magnetotail region are of interest. Figure 6 shows selected revolutions of ISEE-A in the geocentric solar ecliptic (GSE) X-Y plane where the position of the spacecraft is rotated into the X-Y plane about the X-axis for X negative and at constant ecliptic longitude for X positive. These rotations are performed so that the spacecraft position may be assessed in relation to the bow shock and magnetopause.

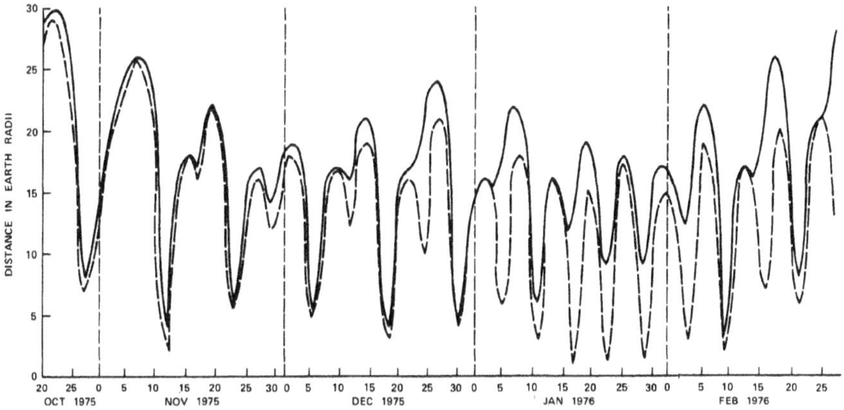

THE SOLID CURVE GIVES THE DISTANCE BETWEEN IMP–H AND –J. THE DASHED CURVE GIVES THIS DISTANCE PROJECTED ONTO THE GSM X–Z PLANE.

Figure 4. *Distance between IMP-H and IMP-J, October 1975 through February 1976.*

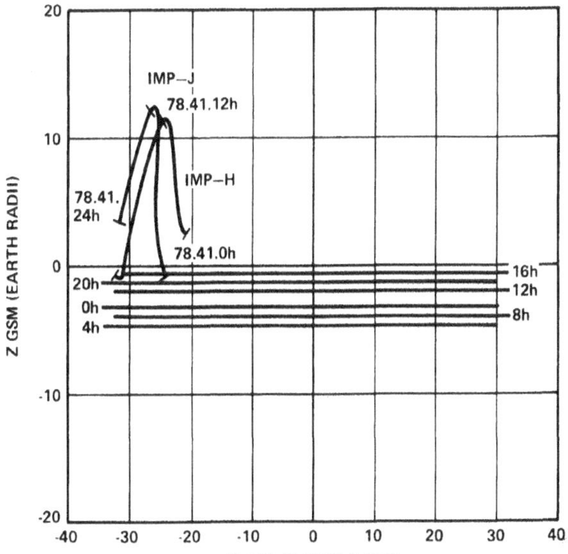

THE CURVES SHOWN ARE FOR 10 FEB 1978 WHEN THE TWO SATELLITES WILL BE ESPECIALLY CLOSE TO EACH OTHER. THE HORIZONTAL LINES SHOW APPROXIMATE POSITIONS OF THE NEUTRAL SHEET AS A FUNCTION OF TIME.

Figure 5. *IMP H and -J projected onto GSM Y-Z plane.*

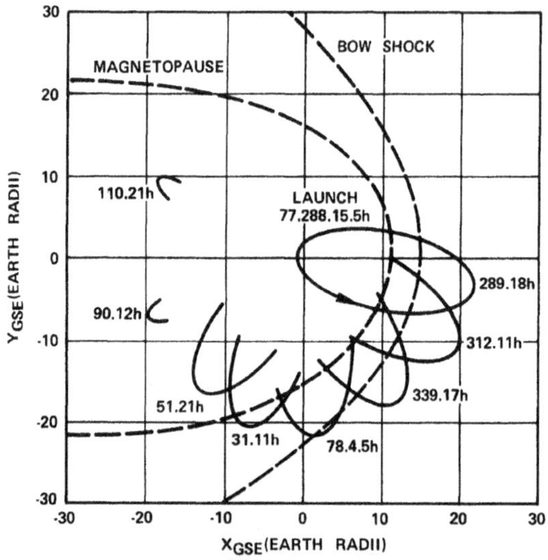

ORBIT PRECESSION IN THE GSE PLANE

EACH REVOLUTION IS ANNOTATED WITH THE UT OF APOGEE. IT MAY BE SEEN THAT ENCOUNTERS WITH THE BOW SHOCK OCCUR UNTIL EARLY JAN 1978 AND THAT ISEE HAS PROCESSED INTO THE REGION OF THE MAGNETOTAIL BY EARLY APRIL 1978.

Figure 6. ISEE-A orbit rotated into GSE X-Y plane.

The positions were determined from the latest planned orbit for ISEE-A. It can be seen that the spacecraft encounters the bow shock from launch until early January 1978 and then enters the magnetotail region around early April 1978. A determination of its possible traversal of the neutral sheet is illustrated in Figure 7, a projection of position in GSM *Y-Z* plane. To provide a rapid visual check of possible neutral-sheet intersections, the projection is shown as a dashed line if $X_{GSM} > -10R_E$. The time marks allow one to determine the approximate distance between the spacecraft and the neutral sheet.

Another presentation that may be useful to identify interesting configurations is given in Figures 8 and 9 where the magnetic latitude versus magnetic local time for Hawkeye 1, IMP-H, and IMP-J is shown. Nearly simultaneous positions at the same magnetic local time at different latitudes or the same latitude at different magnetic local times could provide some unique opportunities. In Figure 8 two consecutive passes of the Hawkeye 1 through the region of 5 h local time are shown. For the first pass IMP-J appears at the same local time, and for the second pass IMP-H appears at the same local time. Hawkeye 1 is at high northern latitudes while IMP-H and -J are at lower latitudes. All three satellites are in the magnetosheath for the time period shown. A configuration that is potentially interesting is shown in Figure 9. Hawkeye 1 is in the cusp region while IMP-H and -J are at local noon at a lower latitude. For the period shown, however, IMP-H and -J are in the interplane-

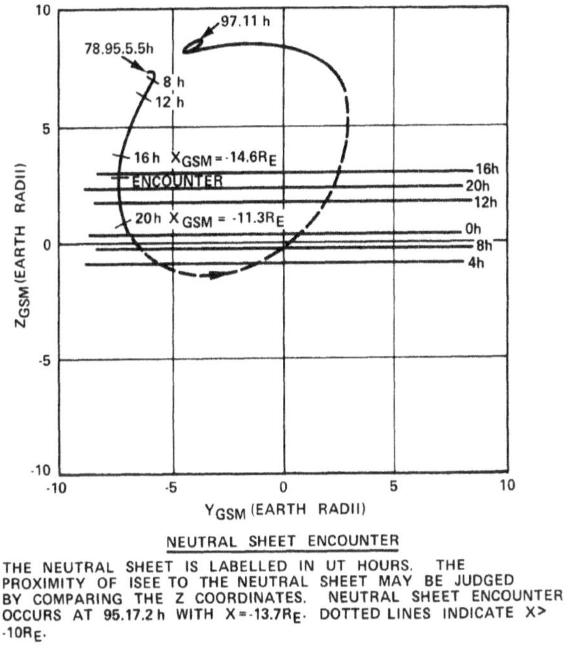

Figure 7. *ISEE-A orbit projected onto GSM Y-Z plane.*

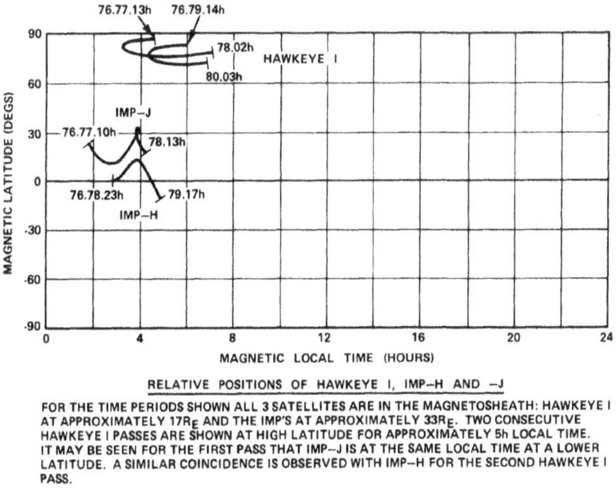

Figure 8. *Hawkeye 1, IMP-H and -J, magnetic latitude as a function of magnetic local time.*

tary medium, and Figure 9 is intended only to illustrate the useful configurations that may be demonstrated in this projection.

A coarser view of the position of several high-altitude satellites may be given by indicating in what region of the near-Earth space each is located as a function of time. This is illustrated in Figures 10 and 11 for the three satellites in Figures 8 and 9. The eight regions of space we have defined for this exercise are:

1. interplanetary medium
2. dayside magnetosheath
3. nightside magnetosheath
4. dayside magnetosphere
5. nightside magnetosphere
   $(-10R_E < X_{GSE} < 0)$
6. high-latitude magnetotail
   $(X_{GSM} < 10R_E, \mid Z_{GSM} - Z_{\substack{neutral \\ sheet}} \mid > 6R_E)$
7. mid-latitude magnetotail
   $(X_{GSM} < -10R_E, 2R_E < \mid Z_{GSM} - Z_{\substack{neutral \\ sheet}} \mid \leqslant 6R_E)$
8. neutral-sheet region
   $(X_{GSM} < -10R_E, \mid Z_{GSM} - Z_{\substack{neutral \\ sheet}} \mid \leqslant 2R_E)$

The first three of these regions are shown in Figure 10 for early 1976. Note that for purposes of clarity no distinction has been drawn between the dayside and nightside magnetosheath regions. All of the IMP magnetosheath passes are nightside passes. The vast majority of the time Hawkeye 1 is in the magnetosheath it is in the dayside region. Periods for which all three satellites are in the magnetosheath or the interplanetary medium may be readily seen. In addition, nearly simultaneous crossings of the bow shock and magnetopause are observed for two satellites. The five regions of the magnetosphere are illustrated in Figure 11 for a time period that is a subset of that shown in Figure 10. Neutral-sheet encounters are observed for IMP-H and -J. Displays of this type, coupled with tables that give entry and exit times for each region, should allow one to determine rapidly the possibility that certain spacecraft may provide interesting data because of their unique relative positions. Such possibilities can thus be examined in more detail by means of displays in various co-ordinate systems.

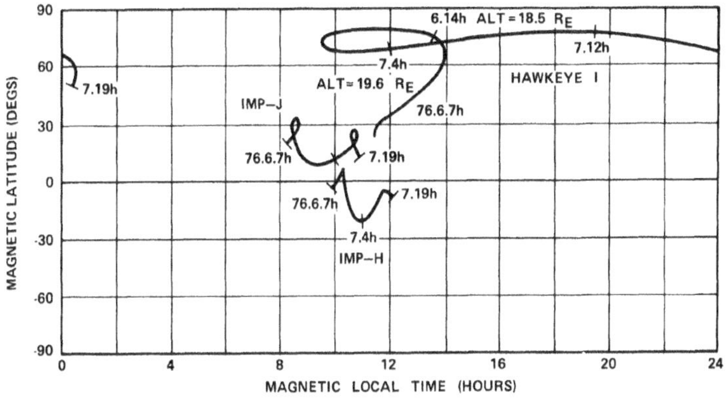

RELATIVE POSITIONS OF HAWKEYE I, IMP–H AND –J

HAWKEYE I MAY BE SEEN TO BE IN THE CUSP REGION AT THE END OF DAY 6 1976. AT THE SAME TIME IMP–H AND –J ARE AT APPROXIMATELY THE SAME MAGNETIC LOCAL TIME AT A LOWER LATITUDE.

Figure 9. *Hawkeye 1, IMP-H and -J, magnetic latitude as a function of magnetic local time.*

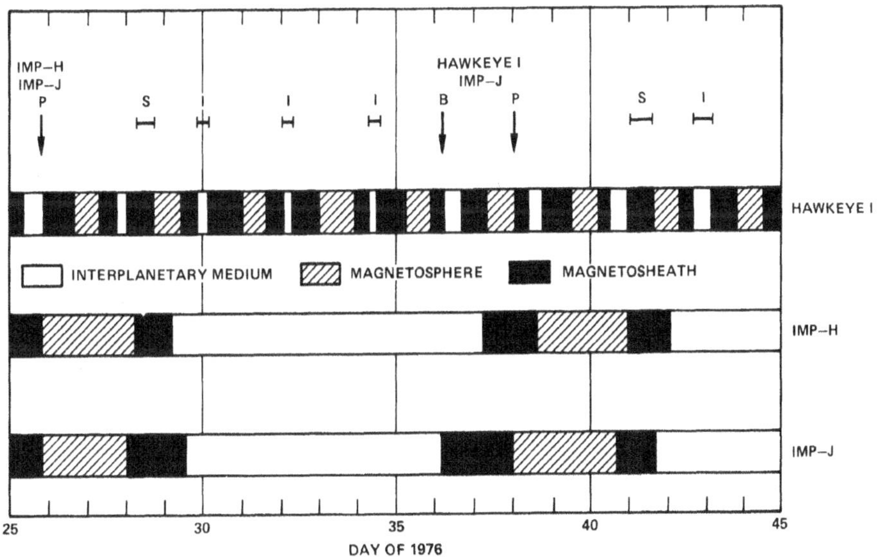

THE TIME PERIODS ARE SHOWN FOR WHICH HAWKEYE I, IMP–H AND IMP–J ARE IN THE INTERPLANETARY MEDIUM, THE MAGNETOSPHERE, AND THE MAGNETOSHEATH. PERIODS ARE IDENTIFIED BY BARS FOR WHICH ALL 3 SATELLITES ARE IN THE INTERPLANETARY MEDIUM (I) OR IN THE MAGNETOSHEATH (S). SIMULTANEOUS CROSSINGS OF THE MAGNETOPAUSE (P) AND THE BOW SHOCK (B) ARE OBSERVED FOR PAIRS OF SATELLITES.

Figure 10. *Regions of space for Hawkeye 1, IMP-H and -J.*

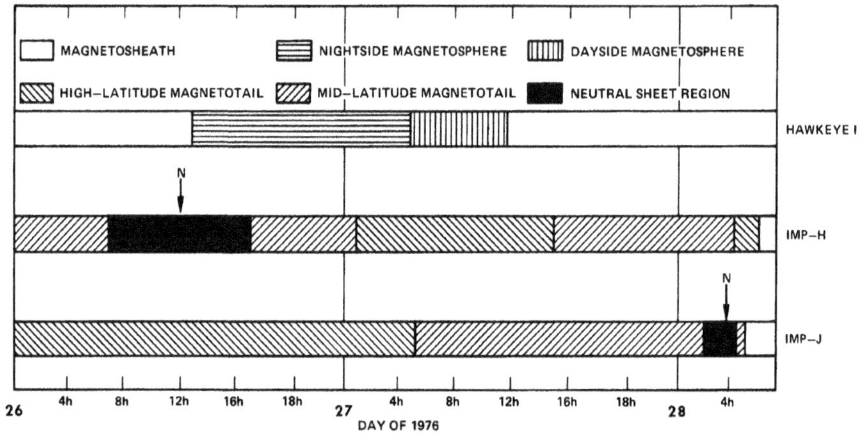

THE PASSAGE OF HAWKEYE I, IMP–H AND –J THROUGH THE MAGNETOSPHERE IS SHOWN. HAWKEYE ENTERS BOTH THE DAYSIDE AND NIGHTSIDE MAGNETOSPHERE. THE HIGHER ALTITUDE SATELLITES IMP–H AND –J ENTER THE MAGNETOTAIL REGION WHICH IS SUBDIVIDED INTO HIGH– AND MID–LATITUDE AND NEUTRAL SHEET REGIONS. THE TIMES ARE IDENTIFIED FOR WHICH IMP–H AND –J ENCOUNTER THE NEUTRAL SHEET (N).

Figure 11. *Regions of magnetosphere for Hawkeye 1, IMP-H and -J.*

5. IMS Satellite Situation Center Activities Prior to IMS

It seems appropriate here to indicate the plans of the Satellite Situation Center between June and December 1975 related to the previous discussion of satellite orbits and the display of spacecraft positions. Work is presently underway to produce the type of plots shown in this article for all the presently orbiting high-altitude spacecraft listed in Table 1 and heliocentric spacecraft in Table 3 for the 1976-1978 IMS time period. In addition, ISEE-A with its latest planned orbit will be included because of the importance of this spacecraft to the whole IMS programme. It is intended to have a limited number of copies of an extensive graphical report available in time for the IMS Workshop at the 1975 IUGG meeting in Grenoble. By October 1975, copies of this report and a questionnaire will be sent to all individuals listed in the IMS Bulletin No. 2. A complete publication of all the plots and for the six high-altitude and the six heliocentric spacecraft will be completed by December and will be sent to those who have ordered it through the questionnaire. The Satellite Situation Center in turn will request that scientists use the report to determine interesting time periods based on the position of these spacecraft. By providing those identified intervals to the Satellite Situation Center all of the IMS community can be made aware of such intervals, and the IMS Steering Committee can make recommendations for certain observing programmes.

DISCUSSION

C.T. Russell: The production of data such as the egress and ingress latitude of the field line through the ISEE A/B spacecraft can be performed by the project office, then the Satellite Situation Center can restrict itself to just the plotting and distribution of such data.

J.I. Vette: The calculation of the zero-altitude field-line co-ordinates for all ISEE-A/B spacecraft positions will depend upon obtaining an agreement between the experimenters on a suitable magnetospheric magnetic field model. If such an agreement can be reached and the project can afford the computational costs to provide this information, it should be possible for the Satellite Situation Center to produce the plots and handle their distribution.

A. Pedersen: What is the capability of the Satellite Situation Center to make use of existing and future magnetospheric models to give the best possible information on crossing of magnetospheric boundaries?

J.I. Vette: The positions of the bow shock and the magnetopause that are used in the present graphs were obtained from the IMP magnetometer data by Don Fairfield. It would be possible to use other models. However, since these positions are known to vary by several Earth radii depending upon the conditions of the interplanetary medium, I would not be in favour of using a complex model which would require a special calculation for each spacecraft. The best way to obtain realistic information on boundary crossings would be a rapid reporting of quick-look plasma and magnetic-field data on high-altitude IMS spacecraft to some co-ordination centre where interested persons could address their inquiries.

M. Dryer: I have a remark and a question. The *remark* is concerned with the STIP Intervals which have been approved by SCOSTEP in Varna last week. (STIP = Study of Travelling Interplanetary Phenomena). One Interval is the February-March 1976 period (which may slip one month because of Helios-B launch in Dec. 1975 or Jan. 1976) when Helios-A and -B will be in interesting configurations. This interval is coincident with very interesting configurations, IMS-speaking, of Hawkeye and IMP's H and J. If any of the attendees here are interested in further information, I will be glad to receive your names and addresses. The *question* is: Would it be possible to urge the launching agencies for Helios-B to launch the spacecraft into a heliocentric orbit which will maximise lifetime (in-keeping with the low-AU objective, of course!)? You noted that some thought is being given to a 0.28 AU perihelion which would have a three-month lifetime. I find this most disturbing because I doubt very much if the added science (such as plasma fluctuations within streams – however interesting!) will be worth it. In my opinion, it is far more important to sample low-AU (out to aphelion) data over much longer periods of time, especially since their solar-wind and cosmic-ray sampling will be especially crucial for IMS and STIP scientific objectives. Finally, I am especially grateful to Drs. Vette and Sugiura of the Satellite Situation Center for showing the excellent 'road maps' and interplanetary constellations. Such displays will be of tremendous use in orienting those of us who are planning interdisciplinary studies.

J.I. Vette: It is my understanding that the Helios-B project personnel and experimenters are aware of the various scientific trade-offs. A recommendation by the IMS or STIP Steering Committee favouring a long-lived mission would certainly be appropriate to register the feeling of that scientific community.

M.J. Rycroft: It is interesting to note that, for the remainder of 1975, the orbital planes of ISIS I and II are nearly coincident; thus the temporal or spatial resolution of variations observed in the topside ionosphere may be accomplished.

AGREED PLANS FOR GEOS PAYLOAD OPERATION AND ORBITAL MANOEUVRES

K. Knott

Space Science Department, ESTEC,
Noordwijk, Netherlands

Abstract

A high degree of operational flexibility and the possibility for experiment control by a real-time ground computer system make operation of the GEOS payload a delicate task which had to be defined well in advance of the spacecraft launch. It has been agreed to adopt three different modes of commanding:

(i) scheduled commands to all experiments
(ii) commands to an individual experiment based on the data received from this experiment, and
(iii) commands to an individual experiment, but based on the data received from other experiments.

Typical examples of all three modes are presented. The agreed shift plan for GEOS is presented and justified, and the orientation of the spacecraft in orbit is discussed with particular reference to directional observations of particles and waves.

1. Introduction

The GEOS payload consists of seven experiments, four of which measure particles and three fields. Particle energy coverage stretches from thermal energies up to a few 100 keV for electrons, and up to a few MeV for protons. Particles with higher mass numbers, up to 140 amu, are also covered up to energies of about 20 keV.

The viewing angles of the various particle experiments are shown Figure 1. A suprathermal plasma detector measuring electrons and protons up to 500 eV is mounted on a boom and views both parallel and roughly perpendicular to the spin axis. The medium-energy experiment (200 eV - 20 keV, electrons and protons) views

Knott & Battrick (eds.), The Scientific Satellite Programme During the International Magnetospheric Study, 61–67.
All Rights Reserved. Copyright © 1976 by D. Reidel Publishing Company, Dordrecht-Holland.

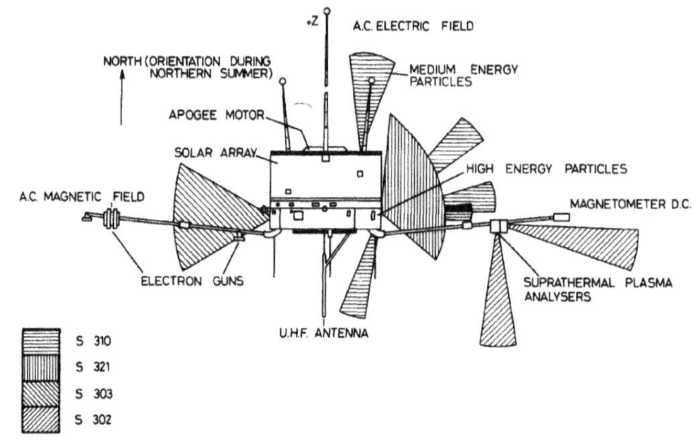

Figure 1. *Elevation aspect angles of the four particle experiments aboard GEOS.*

with high resolution into a direction about 15° off the spin axis – the anticipated direction of the loss cone – and has coverage with less spatial resolution in other directions. The high-energy experiment (20 keV - 200 keV electrons, 20 keV - 2 MeV protons) has a wide acceptance angle for electrons with a spatial resolution of about 12°, and a smaller acceptance angle and reduced resolution for protons. The acceptance angle of the mass analyser is also indicated.

Figure 2 shows the accommodation of the GEOS wave experiments. Their frequency coverage is 0 - 77 kHz for electric fields and 0 - 20 kHz for magnetic fields. Two experiments are aimed at detecting DC electric fields: a double-probe experiment with a baseline of more than 40 m, and an electron beam experiment. AC electric fields are measured by the long boom sensors and by meshed probes on axial booms. DC magnetic fields are measured by a fluxgate magnetometer. For AC measurements in the magnetic domain, a search-coil instrument is employed. The

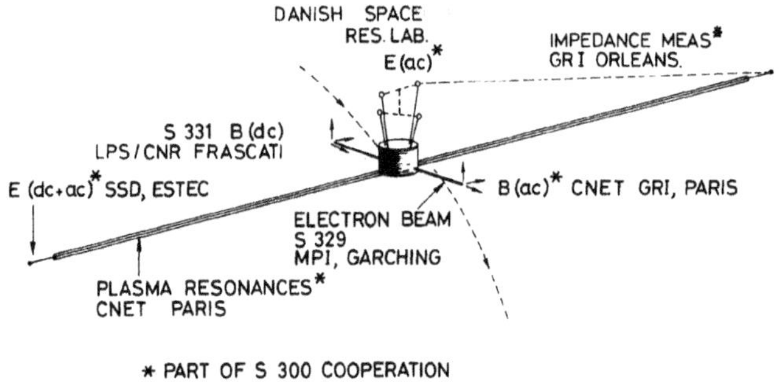

Figure 2. *Schematic of the accommodation of the various field and wave sensors carried by GEOS.*

wave experiment can be used in either of two active modes: as a plasma relaxation sounder and as a mutual or selfimpedance experiment.

2. Payload Operation

All experiments have a high degree of operational flexibility. With the particle experiments, for example, one can:

- emphasise proton or electron measurements
- trade off spatial coverage (number of detectors) against time resolution
- trade off time resolution against energy resolution by commanding the rate of energy stepping
- choose sampling periods to obtain statistically significant count rates from any one detector
- control count rates in various pitch-angle intervals
- select particular energy and/or mass ranges to be monitored
- adjust for different spacecraft potential.

All in all, one can adjust the operation of the experiment to the prevailing environmental conditions.

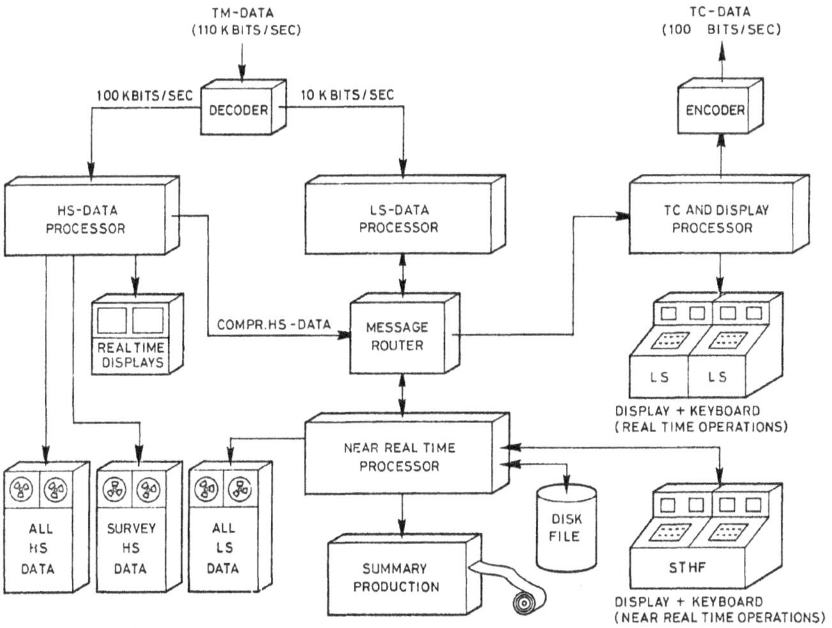

Figure 3. *Block diagram of the on-line computer facilities used in operating GEOS, and its payload in particular.*

Similarly, for the field experiments one can:

- select a particular sensor combination for wave observations
- select the frequency intervals to be explored with high resolution via the high-speed transmission system
- operate with fixed or automatically adjusted gain
- switch to various active modes (relaxation sounder, impedance)
- control the beam characteristics of the electron-beam electric-field experiment
- initiate various processing modes of the on-board correlator.

The operations and data-handling system being built for GEOS is shown in Figure 3. The satellite transmits continuously at 110 kbit/s. This data stream contains both low-speed (LS) and high-speed (HS) data, which are separated before any processing starts. The HS data (100 kbit/s) contain six digitised analogue wave-form channels and the output of a digital correlator. Each analogue channel has a bandwidth of 450 Hz. The correlator has a time resolution of 30 ms and a frequency resolution of 40 Hz. This data stream can be considered as a luxurious substitute for a wide-band analogue link for the wave experiment. The LS data stream (10 kbit/s) transmits the remaining data from the wave experiment and the data from all other experiments.

Telecommanding to GEOS is possible at a maximum rate of 100 bit/s. It should be noted that this uplink capability is comparable to the downlink capability of first-generation satellites.

The HS data processor has four outputs: one to a tape recorder where all HS data are archived, one to a tape recorder where 'survey data' are recorded, one to selected real-time displays, and one via which compressed HS data are passed on. Compressed HS data consist of averaged correlograms and information on wave amplitudes in different frequency intervals. The low-speed data do not undergo any compression in real time, and are available in this form for all subsequent computer processing, as outlined below.

2.1. REAL-TIME ACTIVITIES

Under this heading fall real-time displays and the generation of telecommands, the latter being performed manually, automatically, or via a scheduled command file.

Manual commanding is foreseen mainly for the early life of GEOS, when the experimenters can work in a conversational mode with their experiments using real-time displays and real-time manual commanding via a special keyboard, in order to check out their experiments and to determine their behaviour in different operating modes.

Automatic commanding will be used throughout the lifetime of GEOS. The LS data processor monitors a number of key experiment channels. Based on the geophysical conditions indicated by these channels, the experiments will be commanded into the most appropriate mode to monitor the ongoing phenomena. Typical examples of where an experiment will be automatically commanded on the basis of its own data are the following. The wave experiment uses the outputs of a fixed filter array to switch the swept frequency analyser into a frequency range where pronounced VLF signals are encountered. One particle experiment performs a routine systematic energy and mass scan. This scan is analysed and if any special features are observed in the data, an attempt is made to study them in more detail by narrowing down the energy and mass range to be scanned. Experiments are not only commanded on the basis of their own results, but also act upon results from other experiments. For example,

- the intensity of ULF waves registered by the wave experiment will switch the ion detector to high-resolution energy and pitch-angle scans
- the DC magnetic-field direction measured by the fluxgate magnetometer will be used to automatically initiate interrogation of those particle detectors that are looking at the most appropriate pitch-angle interval.

For a considerable part of its lifetime, the GEOS payload will be operated on a scheduled basis. All experimenters have agreed to switch their experiments simultaneously, for a few minutes every hour on the hour, to a predetermined survey mode, which will remain unchanged throughout the satellite's lifetime. The prime purpose of this exercise is to ensure maximum continuity in the GEOS data. The data recorded during these survey periods will be treated as separately as possible from all other data.

In the time between survey periods, some experimenters intend to operate on a strict scheduled-command basis, others will operate in a fully automatic mode, and still others are adopting a hybrid mode with scheduled programmes, but with automatic command generation routines built into the various programmes. The wave experiment, for example, will run on a scheduled basis to predefined programmes with various scientific aims, such as ELF studies, VLF studies, electromagnetic and electrostatic observations, etc. Each of these programmes has different built-in automatic command generation criteria. Each programme consists of a list of scheduled telecommands and a set of parameters and thresholds for the command-generating subroutines.

2.2. NEAR-REAL-TIME ACTIVITIES

All LS data and the compressed HS data are stored in a large disk file, which has a capacity of about 24 h of data. The data can be quickly assessed, further processed and displayed in various formats. From the real-time console, an experimenter can interact in a conversational mode with his experiment. From the near-real-time console the experimenter can interact in a conversational mode with his data, can try out different display formats, and can survey data on a longer time scale. It will also be possible to display data from different time periods next to each other. It is believed that this near-real-time facility will constitute an ideal complement to the real-time display facilities, when it comes to understanding the performance of an experiment or assessing the significance of any data on a short time scale.

2.3. SUMMARY PRODUCTION

The 24 h disk file also forms the basis for the production of the so-called 'daily summaries' from GEOS. Averaging and formatting is done in the near-real-time processor and the data are plotted in a Xerox-compatible display. The end products are two sheets of paper, one displaying data from the wave experiments, and one data from the particle experiments. The content and format of these summaries will be described in a subsequent presentation.

2.4. DATA ARCHIVING

The final point to be mentioned is the archiving of the GEOS data. The content of the disk file will be dumped onto tape once per day. The resulting tape is the LS archive tape from which the individual experimenter tapes are produced. The HS data are taped in real time (1 tape every 45 min). The resulting tapes enter a cyclic buffer and the wave experimenters have about three months in which to decide which HS tapes to keep and which HS tapes to recycle. Finally, only 10% of all HS data will be kept and archived. Included in this saved portion are the survey data, which are recorded separately and archived without selection.

3. Orbital Manoeuvres

Two aspects of the GEOS mission are to be covered here: (i) the longitudinal shift plan, and (ii) the orientation of the satellite in orbit. The longitudinal shift plan has to take into account:

- the excellent magnetic conjugacy between Northern Scandinavia and the geostationary orbit
- the dense network of observing stations (existing or planned) in Scandinavia
- the availability of rocket ranges at Kiruna and Andøya
- the possibility of balloon launches from Scandinavia and Iceland
- the three-point magnetic conjugacy: Iceland - GEOS - Syowa Base
- the observational facilities in and around Syowa Base
- the existence of a node between the magnetic and geographic equator somewhere between Scandinavia and Iceland
- the availability of the Søndre Strømfjord rocket range, at very high magnetic latitude, which at certain times is magnetically conjugated to the polar cap
- the distance of an observational network along the west coast of Greenland.

The launch of GEOS has been delayed until February 1977. From the day of launch, about one month is needed for initial orbit and attitude manoeuvres and for switching on and testing the payload. It is therefore uncertain whether there will be an opportunity for optical observations and auroral rocket campaigns in conjunction with a fully operational GEOS payload during the northern winter of 1976/77.

Given this situation, it will be more advantageous to chose an initial stationary position for GEOS over Iceland. GEOS should then be available for studies in conjunction with observatories in Syowa Base and Iceland during the early summer of 1977.

Around the summer soltice of 1977, GEOS will be placed in an orbit such that it will drift about 0.5° in longitude per day towards Scandinavia. This drift will provide an opportunity for balloon campaigns from Iceland and Scandinavia, and a slow crossing of the geomagnetic equator. The satellite will arrive around 1 September over the Scandinavian longitude, where it will remain stationed throughout the winter of 1977/78.

At this stage in the spacecraft's life, the shift plan for the second year will be discussed and decided upon. Further shift manoeuvres will depend on the fuel remaining. If there is sufficient propellant left at this stage, a second excursion toward the west, possibly as far as Greenland, will be carried out.

The orientation of the spacecraft in orbit will always be such that the spin axis is perpendicular to the orbital plane (to within 2°). The orbital plane will coincide with the equatorial plane to within 1.6°. With these restrictions, two orientations are possible: positive z-axis of the spacecraft pointing northward or pointing southward. Intermediate positions are not possible because of the narrow UHF antenna pattern.

Returning to Figure 1, it can be seen that the directional coverage for particles is not isotropic. In general, coverage is more pronounced along the $+z$-axis of the spacecraft. From the point of view of a ground-based observer in the Northern Hemisphere, one would want the satellite orientated with the $+z$-axis pointing to the south, because those particles that precipitate in the northern auroral zone are then seen by the satellite.

Unfortunately, this orientation is only possible during Northern Winter, because boom shadows on the spacecraft's solar array must be avoided. During Northern Summer, the satellite will be oriented with its $+ z$-axis pointing north. This situation is not very favourable for X-ray balloon campaigns, but it is hoped that during special campaign periods the desired spacecraft orientation can be arranged, if necessary at the expense of some loss of power and interference to the wave experiment.

DISCUSSION

L.R. Lyons: While the GEOS particle experiment is adjusting itself to measure specific energy and mass ranges, is the total energy range still being measured? I strongly recommend that the entire energy range always be sampled, or else interesting physical processes will not be able to be studied.

K. Knott: Where energy or mass resolution is increased, this will be at the expense of energy or mass coverage in other regions as the bit rate from the experiment stays constant.

D.A. Bryant: Could you say when we might expect a decision on the GEOS shift plan?

K. Knott: This plan will probably be established by September 1975.

PRESENT STATUS OF ISEE (A AND B) MISSION

A.C. Durney

Space Science Department, ESTEC, Noordwijk, Netherlands

Abstract

The Mother-Daughter part of the International Sun-Earth Explorer (ISEE) mission is described briefly, with summaries of the spacecraft, their capabilities and limitations, and the measuring ability of the payloads. The present state of the orbit and launch windows are discussed together with their impact on co-ordinated measurements for IMS. The possibilities of the separation strategy are reviewed with illustrations of how and where the ISEE mission can make effective contributions to the IMS.

1. Introduction

NASA's major contribution to the International Magnetospheric Study will be made by the International Sun-Earth Explorer Mission. It is clear that GEOS is the hinge-pin for the IMS, but investigators should be aware of the wide range of opportunities that will be provided by ISEE. This note will describe the spacecraft briefly and discuss some of the operational possibilities.

The ISEE mission consists of three spacecraft with the official names of ISEE-A, ISEE-B and ISEE-C. Formerly these spacecraft were known as the IME Mother, Daughter and Heliocentric, respectively. Figure 1 shows the operational lifetime of these spacecraft compared with those of GEOS and the IMS. The ISEE-A and ISEE-B spacecraft will be launched in tandem on a single 2914 Thor Delta vehicle from the Eastern Test Range, USA in the Autumn of 1977. It is intended that these spacecraft will operate and data will be taken from them for three years. ISEE-C will be built in parallel with the A spacecraft by essentially the same team, using slack time from the A programme until after the A launch, when the full attention of the team can be devoted to ISEE C. The launch of the C spacecraft will follow as soon

Knott & Battrick (eds.), The Scientific Satellite Programme During the International Magnetospheric Study, 69–80.
All Rights Reserved. Copyright © 1976 by D. Reidel Publishing Company, Dordrecht-Holland.

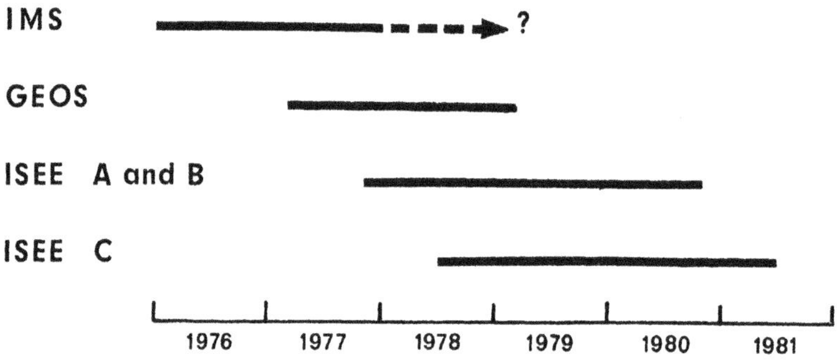

Figure 1. Comparison of ISEE mission schedules with those of IMS and GEOS.

as possible after the tandem launch. The original estimate of the gap between these two launches was nine months, but the C programme is ahead of schedule at present and it is hoped that this inter-launch delay can be reduced to 5-6 months. The planned lifetime for ISEE-C also is 3 years. The launch of GEOS has recently slipped to February 1977 and, with a nominal lifetime of 2 years, this gives an overlap of 15 months with ISEE-A and ISEE-B, and about 9 months with ISEE-C. The IMS starts officially at the beginning of 1976 and lasts nominally for 2 years. It has been planned originally to coincide with the operational time of GEOS, but there will now be no fully instrumented spacecraft available for the first year, and in addition the lifetime of these spacecraft now extends well beyond the IMS. There are clear advantages in lengthening the IMS to include 1978 and perhaps 1979. This possibility has been discussed by the IMS Steering Committee, with the result that the endpoint of the IMS is at present undefined.

2. The Spacecraft

The ISEE project is a combined operation. NASA will build ISEE-A and ISEE-C following the IMP-series design pattern; ESA will build the ISEE-B, which is an entirely new design. Figure 2 shows a sketch of ISEE-A. It is rather shorter and fatter than the original IMP design in order to accommodate ISEE-B, which sits on top of it during launch. The 16 flat solar-array panels can be seen; interestingly, the wire antenna for Heppners' DC electric-field experiments measure 215 m tip-to-tip. This presents a fair technical problem. A sketch of ISEE-B is shown in Figure 3. It has a round solar array with the object of cutting down array noise; it has three short rigid 2 m booms, and a long wire antenna measuring 30 m tip-to-tip. It is smaller than ISEE-A, with a mass of 164 kg, against about 340 kg. Having lower mass and inertia than ISEE-A, it is more nimble and so it carries the gas jet propulsion system by which the separation distance of the two spacecraft can be varied to fit in with the scale size of the features to be measured. ISEE-C will be placed well upstream in the solar wind in order to be away from the disturbance of the magnetosphere.

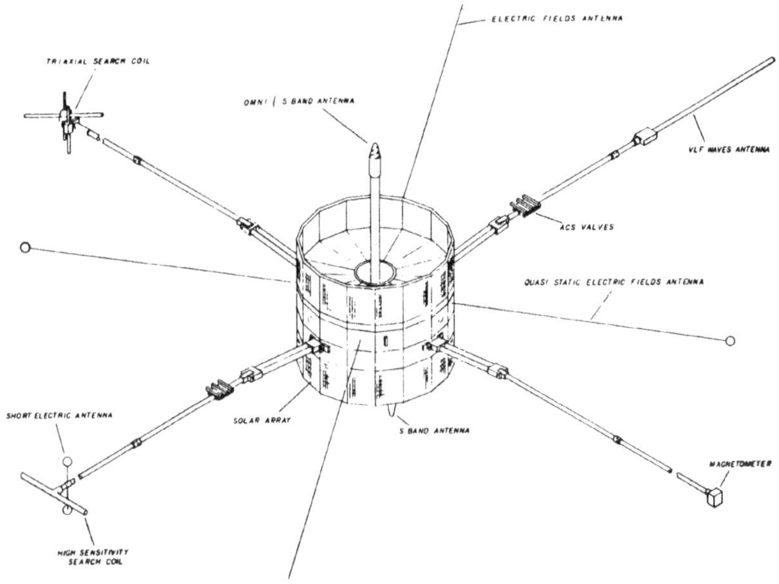

Figure 2. Sketch of ISEE-A spacecraft.

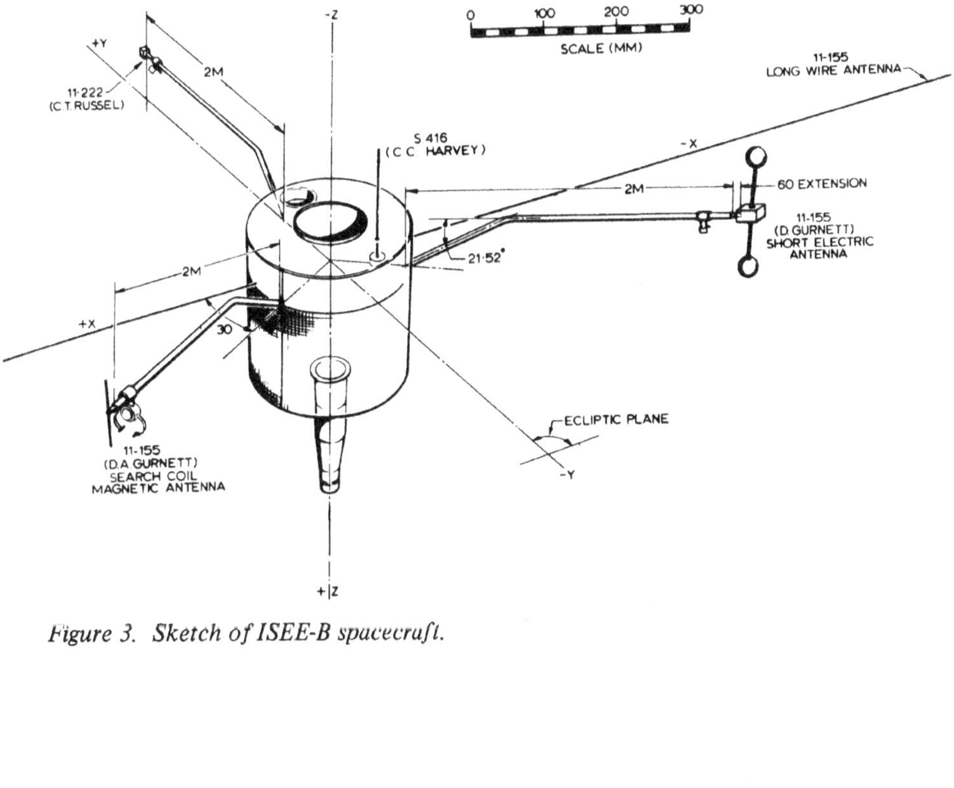

Figure 3. Sketch of ISEE-B spacecraft.

3. The Payloads

The chief purpose of this mission is to separate time variations in features from movements, and this will be done by flying two spacecraft (ISEE-A and B) a controllable, known distance apart so that the time-space ambiguities can be removed by a comparison of the times at which observations are made. To make good comparisons, both spacecraft carry similar payloads. In fact, some of the experiment packages are identical and can be interchanged between the two spacecraft. The nature of the payload is similar to that of GEOS. Again, some of the Investigators are involved with both missions, for instance Keppler and Pedersen. The payloads are shown in a 'global' fashion in Table 1 for quick reference. The top section shows the common payload measurement ranges, and it will be seen to be comprehensive. The ranges here have been pieced together from the capabilities of a number of different instruments and there are overlaps but no gaps. The cohesive nature of the

TABLE 1

'Global' presentation of ISEE-A & B measurement ranges. Top section shows measurements common to A & B; below are additional measurements made by ISEE-A alone.

PAYLOAD COMMON TO A & B:		
Electric waves		10 Hz - 2 MHz
Magnetic waves		DC - 10 kHz
Protons		1 eV - 2 MeV
Electrons		1 eV - 250 keV
Solar-wind ions		100 eV/n - 10 keV/n
Electron density		Integral A to B
ADDITIONAL EXPERIMENTS ON ISEE-A:		
Static & LF electric fields		
VLF propagation (passive)		
Electrons		6 eV - 10 keV
Electron density		Sounder
Plasma composition		< 40 keV/q
Particle composition		< 20 MeV/n, Z = 1 to 20
PARTICLE EXPERIMENT RESOLUTION:		
Time:		10 ms
Energy:		1%
Angular	azimuth	2.8°
	meridian	10°

payload was guaranteed by adjustments made during three three-day meetings of the mission investigators prior to official approval of the mission. The middle section contains the additional experiments that are carried by ISEE-A, which make it a powerful tool in its own right for measuring wave-particle interactions. DC electric-field measurements are made by two separate experiments using different techniques. This summary table does not illustrate the sophistication and complexity of the experiments and an attempt is made to do this at the bottom by presenting some resolution figures. These figures are taken from different instruments working in their highest-resolution mode.

Resolution depends upon the bit rate of the link, and in this respect ISEE does not compete with GEOS. However S-band is used, ISEE-A is capable of returning 16 kbps and ISEE-B 8, the total is a sizeable fraction of the GEOS data return. These are so-called 'high' bit rates which will be used about 20% of the time to look at special features. The 'normal' bit rates are a factor 4 lower.

Special attention is paid to cleanliness, conductive coatings similar to those on GEOS will be used on all three spacecraft, and the ISEE magnetic- and electromagnetic-cleanliness requirements are perhaps more stringent. ISEE-A carries 13 experiments, ISEE-B 8. About 100 internationally known scientists from 30 different institutes are collaborating in this mission, making it a large project.

4. The Orbit

The orbit has a strong impact on correlation with ground-based investigations, and definition of this has given a certain amount of difficulty. The problem was not that of finding a suitable orbit, but one of finding *any* orbit that would give a three-year life without superfreezing in a long eclipse or telemetry dropouts.

The original requirements were to maximise the number of bow-shock crossings (this set the apogee height to 23 R_e) and to make as many bow-shock crossings as possible early in the mission in order to have good measurements whilst the instrument failure risk was low. Presently, IMS representatives asked if we could arrange for A and B to be near the Scandinavian field lines in the northern winter. In the meantime, the weight of the payload had grown to the point where the perigee had to be set at the very low altitude of 280 km in order to reach the 23 R_e apogee. This made the task of finding a launch window for the orbit conditions very hard. Finally, the Investigators Meeting held in Berkeley during December 1974 decided that plasma-sheet crossings were important. This last requirement placed over-riding restrictions on the launch window, and a small series of 10 min windows in October and November was found to be the only viable possibility for 1977. Another series of small windows opens in March of 1978, but there is no possibility of launch in between. So it is no longer a question of choosing a suitable orbit, but of looking to see what we have. The orbit parameters for a launch on 15 October 1977 are listed in Table 2. For this launch, the apogee is in the northern hemisphere; for a 1978 launch it is in the southern. For co-ordinated measurements, the initial line of apsides is the most important quantity. The sketch in Figure 4 is intended to remove any confusion about what is meant by the line of apsides and which way it moves. Looking down on the ecliptic plane from the north, an initial line of apsides of

TABLE 2

Orbit parameters for a launch on 15 October 1977.

Launch time	October - November 1977
Apogee	23 R_e
Perigee	282 km
Initial equatorial inclination	28.3°
Argument of perigee	310°
Initial line of apsides (15 Oct.)	11.30 local time

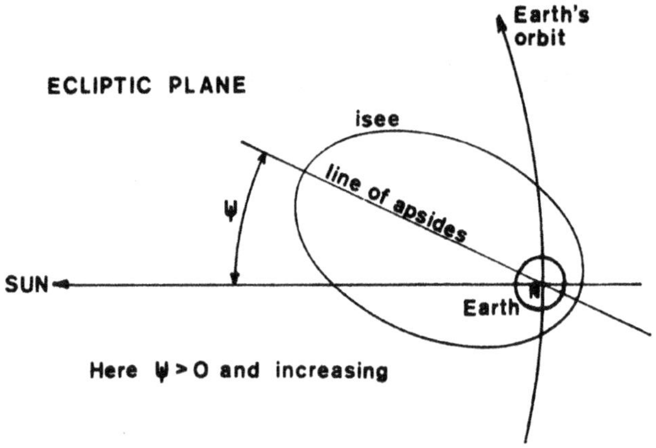

Figure 4. Sketch of position of initial line of apsides for 15 October launch. Projection of the orbit onto the ecliptic plane is as viewed from the north.

11.30 local time means it is 9° clockwise from the Earth-Sun line. It turns out that for a stable three-year life, the line of apsides remains fixed in space, and it is possible to predict precisely where the spacecraft will be at any time in its life even though the exact launch date is not known. The positions of the initial line of apsides are shown in Figure 5, which represents the ecliptic plane of the magnetosphere viewed from the north. Superimposed on this is a circle marked out with the months of the year; the projection of the line of apsides onto the ecliptic will move around this circle according to the date. This diagram has been drawn for a 1977 launch; for a 1978 launch it is necessary to rotate the date circle through 180°. It is immediately possible to predict the times and depths of neutral-sheet crossings, and Table 3 is the result of a number of spot checks that were made at Goddard using the Fairfield modification of the Russell and Brody equation to find the

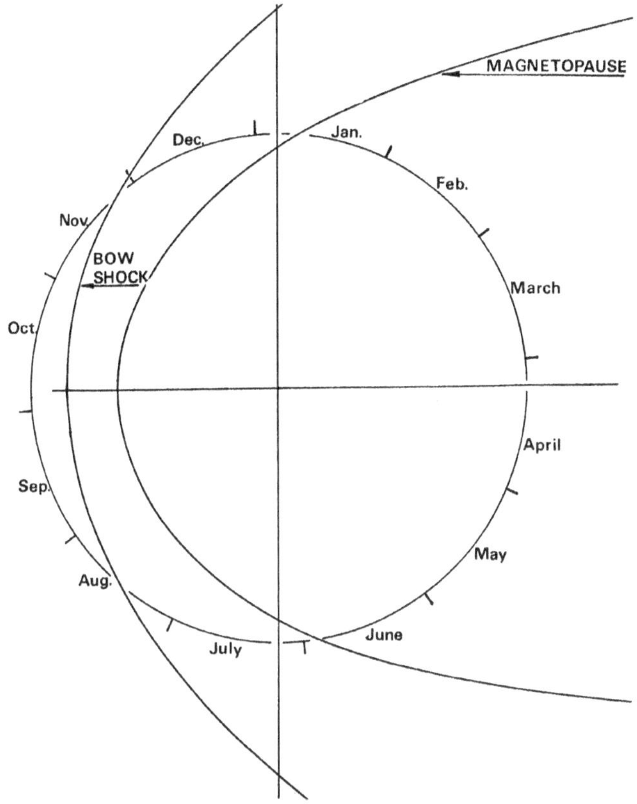

Figure 5. Position of bow shock and magnetopause in the solar ecliptic plane as determined by Fairfield [J. Geophys Res. 76, 6700 (1971)]. Superimposed circle represents predicted positions of line of apsides for any month of year, following launch in Autumn of 1977. For launch in Spring 1978, this circle should be rotated through 180°.

neutral sheet. Two important points emerge from looking at this table. The first is that neutral-sheet crossings begin in February and continue through to June for each of three years of the mission life. This means that the spacecraft will pass through this important region both in the northern and southern winters. The second point is that the depth of the crossings becomes deeper as time progresses; this can be seen from examining the immersion times in the second column from the left. Consequently, there is time for a learning process before the best measurements where, hopefully, measurements would be made with maximum efficiency.

TABLE 3

Neutral-sheet crossing parameters for three-year mission following launch in Autumn 1977. Two successive revolutions have been investigated every 30 days to cover the daily motion of the neutral sheet with respect to the orbit (the orbital period is about 2.5 days).

Date	Time in NS region $\pm 2R_e$	Spacecraft Position in Geocentric Solar Magnetospheric Co-ordinates (R_e) X enter NS	X exit NS	Y enter NS	Y exit NS	Z enter NS	Z exit NS	Minimum distance to NS (R_e)
14 Feb. 78	6	−11.8	−10.0	−19.1	−19.2	2.0	−0.2	0
16 Feb. 78	6	−13.7	−12.9	−17.0	−18.6	2.0	2.4	1.1
15 Mar. 78	0	−17.6*		−13.4*		2.4*		2.5
18 Mar. 78	7	−15.0	−10.0	−12.0	−10.6	3.6	−0.8	0
14 Apr. 78	4	−14.4	−10.0	−4.6	−5.2	4.8	1.9	0
16 Apr. 78	9	−17.7	−10.7	−4.6	−5.0	6.0	1.5	1.1
16 May 78	1	−11.6	−10.0	0.6	0.3	3.8	3.0	1.5
18 May 78	4	−13.9	−10.0	4.1	1.4	4.9	2.9	0
14 Jun. 78	0	−10.0*		7.2*		5.2*		2.6
16 Jun. 78	3	−10.9	−10.0	12.3	10.1	4.0	2.3	0
9 Feb. 79	4	−11.1	−10.5	−18.6	−19.5	2.1	−2.3	0
11 Feb. 79	12	−11.8	−11.3	−14.6	−19.0	1.9	2.6	0
10 Mar. 79	15	−18.1	−15.9	−11.6	−13.6	3.2	1.1	0
13 Mar. 79	5	−17.7	−16.1	−12.7	−12.9	2.6	−1.7	0
11 Apr. 79	18	−21.3	−11.2	−2.9	−4.9	5.0	−1.0	0
14 Apr. 79	5	−18.7	−15.5	−3.6	−4.3	2.5	0.7	0
12 May 79	18	−19.6	−11.7	7.5	0.6	6.4	−0.2	0
13 May 79	5	−18.4	−16.4	6.2	4.2	4.1	2.1	0
9 Jun. 79	5	−13.0	−11.0	10.6	7.1	3.8	−3.9	0
11 Jun. 79	4	−13.2	−12.8	14.8	12.5	2.2	0.2	0
4 Feb. 80	5	−10.3	−10.1	−15.5	−18.0	2.8	−2.5	0
7 Feb. 80	11	−10.1	−10.8	−11.9	−18.0	0.9	2.6	0
4 Mar. 80	23	−16.3	−14.8	−9.8	−14.1	2.7	−2.2	0
7 Mar. 80	6	−18.1	−17.5	−12.2	−13.1	2.0	−2.2	0
3 Apr. 80	6	−21.4	−20.0	−4.5	−5.5	2.2	0.1	0
5 Apr. 80	20	−21.4	−15.8	−2.2	−5.4	4.1	−1.9	0
4 May 80	21	−19.3	−16.9	8.5	2.6	5.4	−0.4	0
7 May 80	5	−20.0	−19.3	7.2	5.8	3.4	1.5	0
3 Jun. 80	4	−14.8	−14.1	12.9	11.0	3.0	−0.6	0
5 Jun. 80	7	−13.9	−14.2	16.4	13.9	2.8	−0.2	0

* Did not enter neutral sheet. Co-ordinates of closest approach are given.

5. The Separation Strategy

One important area where IMS investigators can have an influence is in the planning of the ISEE A and B separation strategy. Almost nothing has been decided on this matter, and one of the reasons is because it is a complex subject, so a few words of explanation are necessary. The idea is that the separation between A and B can be altered so that the distance will be appropriate to the size of feature being measured. The range of distances considered is from approximately zero, which is good for interspacecraft calibration, to a maximum of 10 000 km, although there is some discussion about reducing the maximum to 5000 km. The separation direction will be along the path of the orbit, called the 'true anomaly', and here we come to a limitation of the mission. Since we have only two spacecraft and not a cluster, we can only resolve space and time variations in one dimension, along the separation direction. However, the separation method has been designed so that this direction is generally at right angles to the moving planes of the magnetosphere. This is illustrated in Figure 6, which shows the orbit, the direction of the lines showing the separation direction. Thus, this does not appear to be an important restriction at present, but is has to be kept in mind when separation strategy is considered. Figure 6 also illustrates a second difficulty. The length of the lines represents the separation distance (not to scale) and this can be seen to vary wildly around the orbit. This can lead to confused thinking; for instance, a separation of 1000 km at

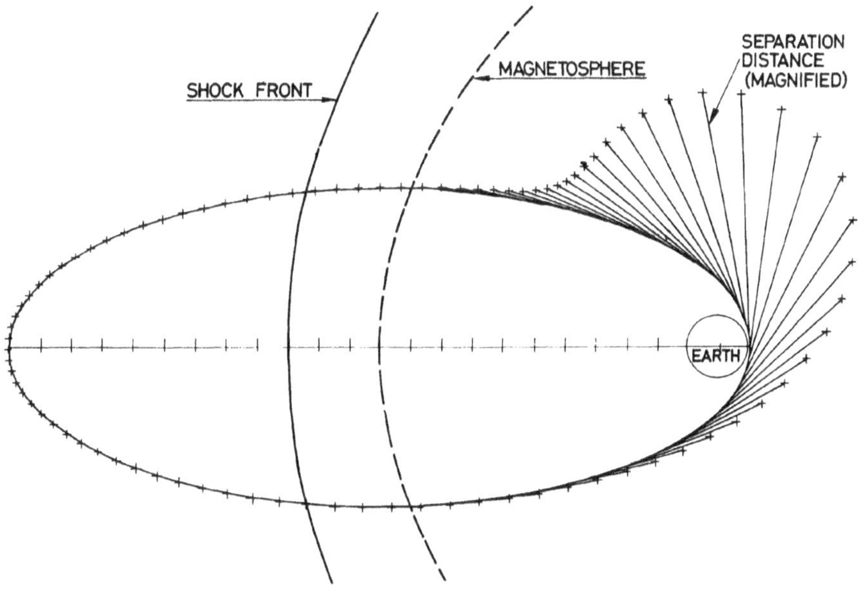

Figure 6. Orbit, with direction of separation represented by directions of radiating lines. Length of line indicates distance of separation (not to scale).

an altitude of 5 R_e means something quite different at apogee. It is possible to construct a nomogram to give a rough guide in this matter (Fig. 7). This makes use of the fact that, although the separation distance varies around the orbit, the separation time remains constant. If a separation distance of 1000 km is required at 10 R_e, a straight line across the nomogram shows the separation time to be 370 s. To find the separation distance at 20 R_e altitude for the same orbit, a straight line through 20 R_e and 370 s shows this to be 390 km. This graph is reasonably correct, but it is intended only as a guide and heavy reliance should not be placed on the figures obtained from it. The curve of the line at low altitudes reflects the distortion due to the difference between the straight-line separation between A & B (which is used in the nomogram) and separation distance along the orbit curve around perigee. This effect is not seen at apogee, where separation distances are much shorter.

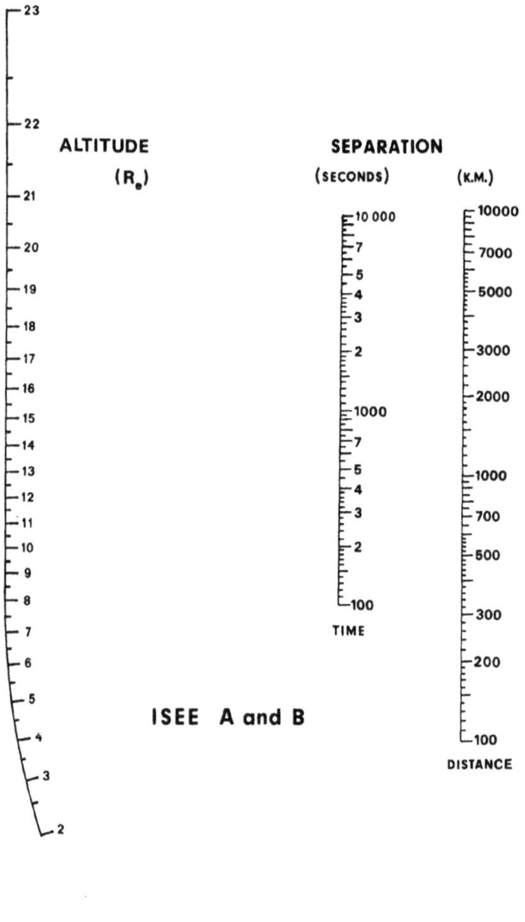

Figure 7. Nomogram for determining separations at different altitudes for any given separation time.

Because of the relatively small amount of gas available (16 m/s) it is proposed that separation variations will be achieved by a drift policy. The general idea is to give the ISEE-B a 'kick' to start it drifting away, and then when maximum separation is reached another 'kick' to reverse the direction. Figure 8 demonstrates the sort of thing that can be done. It illustrates that by using 'kicks' to vary the drift rate, more time can be spent at shorter separation distances, and economies in gas can be made by allowing the spacecraft to drift past each other. This particular cycle is obsolete as it does not fit in with the launch window. However, to have an estimate of gas consumption, if this particular cycle is continued for three years, the amount of gas used is just under 8 m/s, half of the available capacity. To formulate a separation strategy, it will be necessary to identify targets, decide on the separation distance needed for best observations and fit these into a drift envelope of some form. IMS

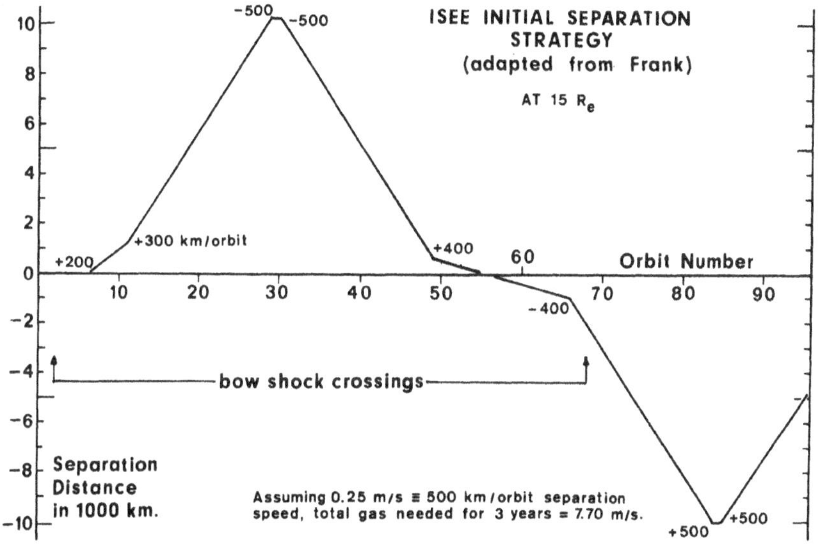

Figure 8. Separation as a function of time for a theoretical separation strategy to illustrate basic principles.

investigators could be of particular help here in identifying targets for ISEE that may shed light on their measurements.

ISEE will be able to make a large contribution to the IMS by virtue of its own 'self-contained' measurements, such as those of the fine structure of the bow shock and magnetopause, analysis of magnetosheath features, measurements of electric fields in the outer magnetosphere, and so on. However, in addition it can add enormously to the power of world-wide IMS measurements, as the following two simple examples demonstrate.

The first is interplanetary plasma measurement. With good observations of solar-wind features, together with measurements by GEOS, other spacecraft and world-wide ground-based networks, it should be possible to establish which interplanetary structures cause what magnetospheric effects and especially to prove solar-wind energy coupling to the magnetosphere. ISEE-C will be sited in the best place for solar-wind measurements and will make these continuously; no other available spacecraft meets these two special conditions. In addition, whenever ISEE-A and B are outside the magnetosphere, they will provide interplanetary plasma data that will be more comprehensive, detailed and analytical than any previously available.

The next points have to do with substorms and the plasma sheet, subjects that still contain the most intransigent problems of the magnetosphere, and the understanding of these phenomena is one of the key steps to the understanding of the dynamics of the magnetosphere. Because substorms involve a large part of the magnetosphere, correlated global measurements will be necessary for any attempt at analysis. The role of ISEE-A and B during these measurements will be to investigate the changing shape of the plasma sheet as the storm progresses and observe the movements of particles and fields inside the sheet during this period. The direction of travel and the pitch-angle and energy spectra of fast plasma flows will be measured and with a chain of other observation posts including GEOS it should be possible to decide if plasma sheet particles convect down to the ionosphere (and the aurora). The last closed field lines will be identified using electrons as tracers. Observation of magnetic and electric fields in the region of the plasma sheet will be made to discover what changes take place during the break-up phase. Other important things to look for are electric-field changes that correlate with the onset, and how the electric field varies at different locations as the substorm progresses.

If the field lines of the narrow, well-defined auroral latitudes are projected back to the plasma sheet, they spread out and the region suitable for investigation becomes relatively large, which should make features easier to see. However, another of the limitations of the ISEE mission is that, in 1978 at least, the time spent in these important regions will be fairly short, and thus it will be crucial to co-ordinate measurements so that the maximum amount of data will be taken during these times. For example, ISEE spacecraft would run at high bit rates, balloons would be at maximum altitudes, rockets would be fired into this time slot, GEOS would be on the right flux tube, etc. With regard to the GEOS placement programme, this is a difficult subject, but there may be some advantages in examining it relative to the ISEE-A and B movements to see if any changes would improve correlation.

Finally, the basis of IMS is intercorrelation of data, and it is very important that investigators should know what data are available and be able to obtain them easily and quickly. The ISEE contribution here will be to have an outline of the measurements available on microfilm and tapes at world Data Center A within (hopefully) two weeks of acquisition.

THE ISEE-C MISSION AND THE IMS

K.-P. Wenzel

Space Science Department, ESTEC
Noordwijk, Netherlands

Abstract

The ISEE-C spacecraft (formerly known as the IME 'Heliocentric') forms part of the joint NASA/ESA 'International Sun-Earth Explorers' mission, which is to make important contributions to the International Magnetospheric Study. The ISEE-C spacecraft will be placed in 1978 in the solar wind upstream of the Earth at about 235 R_E distance, to be unaffected by the presence of the magnetosphere. Besides making important measurements in its own right, this spacecraft will be able to monitor those solar wind features that will convect towards the Earth and that can give rise to processes that occur in or around the magnetosphere.

1. Introduction

ISEE-C (formerly the IME 'Heliocentric') is the third spacecraft in the joint NASA/ESA International Sun-Earth Explorers (ISEE) programme. Whilst the ISEE-A and -B pair (Mother/Daughter) will be in an eccentric Earth orbit investigating phenomena inside the magnetosphere and its boundaries [Durney, 1975], the ISEE-C spacecraft will be placed into a heliocentric orbit and will monitor the interplanetary medium continuously.

ISEE-C is presently scheduled for launch in July 1978 and has a planned operational lifetime of 3 yr. This 'late' launch date has an impact on the International Magnetospheric Study (IMS), since there is no overlap between ISEE-C operation and the presently defined IMS (1976/77). There would be clear advantages in extending the IMS to 1978, or even into 1979, to allow for maximum correlation with the other ISEE and the GEOS spacecraft.

This paper should be read in conjunction with the description of the A and B missions [Durney, 1975]. It summarises how the ISEE-C mission could contribute to the co-ordinated observations during the IMS.

Knott & Battrick (eds.), The Scientific Satellite Programme During the International Magnetospheric Study, 81–85.
All Rights Reserved. Copyright © 1976 by D. Reidel Publishing Company, Dordrecht-Holland.

2. Orbit and Spacecraft

Early in the development of the ISEE programme, a strong need for an interplanetary monitor became apparent. Several types of orbits were discussed, prior to the concept of the heliocentric orbit being chosen. This orbit places the spacecraft near the Sun-Earth line, about 235 R_E (~ 0.01 astronomical units) upstream from the Earth. ISEE-C will thus be continuously in the interplanetary medium, far from the disturbing influence of the Earth's environment (effects of bow shock such as upstream waves and counter-streaming particles), but sufficiently close to be able to maintain a high bit rate (2048 bps) during the entire mission.

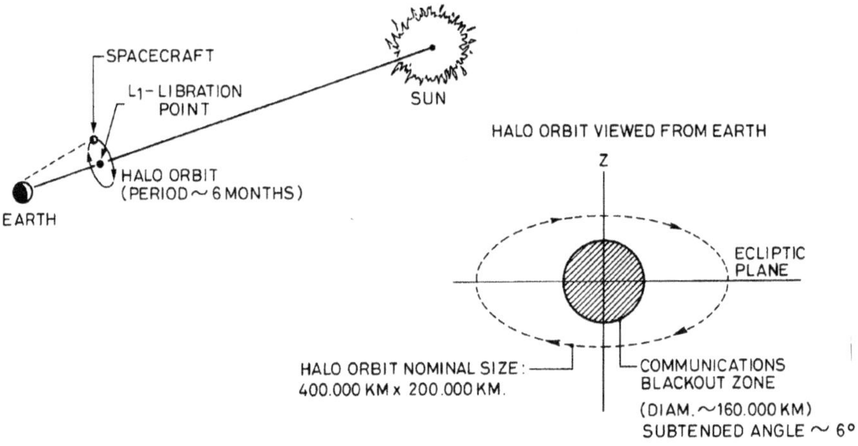

Figure 1. *ISEE-C orbit.*

The spacecraft will be in a halo orbit around the libration point (L_1) between the Earth and the Sun, where the gravitational forces due to the Sun and Earth and the centrifugal force cancel (Fig. 1). The orbital plane of the halo orbit will be perpendicular to the Sun-Earth line. To eliminate most of the solar interference in the spacecraft-to-Earth communication link, the diameter of the halo orbit must be larger than the 'blackout' zone caused by solar radiation. The exact configuration of the halo orbit is still to be determined; in particular, possible interference events such as the increase in solar radio frequency output during solar flares are under study.

Figure 2 shows the ISEE-C spacecraft with its booms in deployed configuration. Its structure is very similar to that of the Mother, but it will have a hydrazine system for attitude control and propulsion. The spin rate will be 20 rpm.

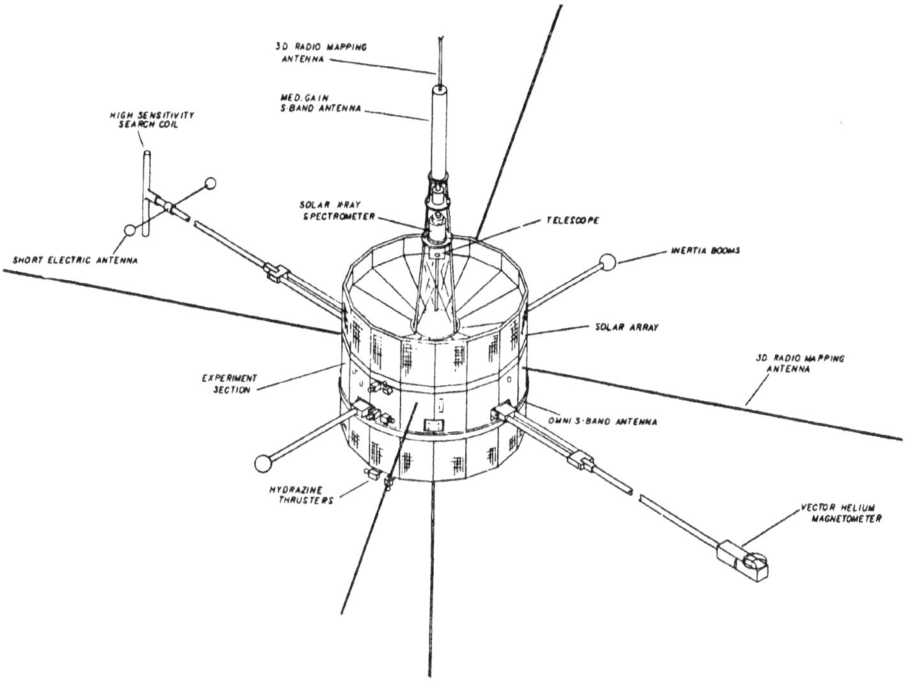

Figure 2. *ISEE-C spacecraft in orbital configuration.*

3. Scientific Aims

The main scientific goals of the ISEE-C mission can be summarised as follows:

(i) to support the magnetospheric satellite missions (ISEE-A and B, GEOS, etc.)
(ii) to study various interplanetary and solar phenomena
(iii) to measure the isotopic composition of solar and galactic cosmic rays
(iv) to serve as a reference in the near-Earth interplanetary medium in support of deep space probes (e.g. MJS).

Mission objectives (i) and (ii) are those of main interest for the IMS, the key to which lies in co-ordinated satellite observations. Once in orbit, the ISEE-C spacecraft will monitor the features of the upstream interplanetary medium about 1 h before the solar wind reaches the magnetosphere. It will thus deliver the interplanetary information needed to interpret simultaneous observations made by the Mother-Daughter behind the bow shock and to study the response of the magnetosphere to interplanetary parameters and their changes. Space/time separation of some interplanetary features will also be possible by comparing Heliocentric and Mother/Daughter observations at times when the latter are outside the bow shock.

4. The Payload

The payload complement (Table 1) reflects the main scientific goals of the ISEE-C mission as listed above. The satellite is properly instrumented to achieve comprehensive and high-time-resolution studies of many features in the interplanetary medium. Most of the 12 experiments carried by ISEE-C measure properties that are of direct interest to the IMS programme; in particular, the vector magnetic field, temperature, density and velocity of the solar-wind plasma, the composition (relative abundance and charge states) of solar-wind ions, the intensity, anisotropy and composition of solar particles over an extremely broad energy range. Solar particles, for example, can be used to study particle entry into the magnetosphere.

Among the various solar and interplanetary phenomena to be investigated on ISEE-C are plasma waves, shocks, solar particle propagation, anisotropies, local acceleration, Forbush decreases, wave-particle interactions, co-rotating regions on the Sun, solar modulation, and so on. All of the latter have been studied before, but can be investigated better both far away from the Earth and by utilising a variety of recent technological advances. This includes conductive coating of the spacecraft as well as implementation of new detector developments.

TABLE 1
ISEE-C payload

Experiment	Measurement range	Principal investigator	Main contributing institutes
Magnetometer	4 ranges: ± 4 to $\pm 146\gamma$	Smith	Jet Propulsion Laboratory
Plasma waves	Electric: 20 Hz - 100 kHz Magnetic: 20 Hz - 10 kHz	Scarf	TRW Systems
Solar-wind plasma	p, e: 5 ev - 7 keV	Bame	Los Alamos
Plasma composition	470 eV/q - 10.5 MeV/q	Ogilvie	GSFC/University of Bern
Protons	0.03 - 1.4 MeV	de Feiter*	Utrecht/ESTEC/Imperial College
Electrons	2 - 800 keV	Anderson	University of California
Solar X-rays	6 - 228 keV		
Radio mapping	type III solar bursts	Steinberg	Meudon/GSFC
Low-energy cosmic rays	5 keV/q - 6 MeV/q	Hovestadt	MPE, Garching/University of Maryland/University of Arizona
Medium-energy cosmic rays	0.5 - 500 MeV/nucl. ($z = 1 - 26$)	von Rosenvinge	GSFC
Heavy isotopes	2 - 200 MeV/nucl. ($z = 3 - 28$)	Stone	California Institute of Technology
High-energy cosmic rays	31 - 495 MeV/nucl.	Heckman	University of California
Cosmic-ray electrons	5 - 400 MeV	Meyer	University of Chicago

* Recently deceased.

Details on the distribution and co-ordination of ISEE data are described in a separate paper [Wilcox, 1975]. A separate ISEE-C data pool tape providing low-time-resolution data from various experiments for intercomparison and indexing is foreseen and will be made available to the IMS.

5. Conclusions

ISEE-C forms an essential part of the International Sun-Earth Explorers programme. It will – during the overlapping operational period – provide the interplanetary support required to achieve many of the aims of the International Magnetospheric Study. In addition, however, it will make important autonomous measurements and will break new scientific ground, especially in the fields of solar and interplanetary phenomena and in the field of cosmic rays.

References

Durney, A.C., Present status of ISEE (A and B) mission, This volume.
Wilcox, J.M., Method, content and format of ISEE data distribution, This volume.

DIAGNOSIS OF SPATIAL PLASMAS WITH A RELAXATION SOUNDER

Jacqueline Etcheto

Centre de Recherche en Physique de
l'Environnement Terrestre et Planétaire (CNET-CNRS),
Issy-les-Moulineaux, France

Abstract

Study of waves with low group velocities has allowed us to make a detailed interpretation of the signals reflected back from a plasma excited by a relaxation sounder. Study of the frequency and amplitude development of these waves in the vicinity of the characteristic frequencies of the medium (gyrofrequency and harmonics, upper hybrid frequency and plasma frequency) leads to determination of such plasma parameters as modulus of magnetic field, electron density, and temperature. Results obtained by this means from three rocket experiments launched under differing ionospheric conditions are presented. Similar experiments will be flown aboard GEOS and ISEE-A and the interpretative techniques described here will enable us to determine the characteristics of the plasma surrounding the spacecraft.

1. Introduction

The purpose of this paper is to make clear to nonspecialists the principle of the relaxation-sounding technique and the capabilities foreseen for the GEOS and ISEE-A sounders. Since the information we hope to obtain on the plasma surrounding the spacecraft should be of interest, as complementary data, for many studies (the most obvious being the generation of natural waves; see Russell, this volume), we would like to familiarise other experimenters with the technique. This paper is in no way an attempt at a complete review of existing knowledge in this field, as such literature already exists [McAfee, 1973].

Knott & Battrick (eds.), The Scientific Satellite Programme During the International Magnetospheric Study, 87–97.
All Rights Reserved. Copyright © 1976 by D. Reidel Publishing Company, Dordrecht-Holland.

2. Description of a Sounder

When a relaxation sounder, or ionosonde, is flown on a spacecraft, a short radio-frequency pulse of duration τ is transmitted and the sounder monitors the returning waves for a much longer period t_r. The central frequency f_T of the transmitted waves is changed in steps, the receiver being tuned to the same frequency band, thus sweeping through a wide frequency range.

Up to now, this type of experiment has only been flown in the ionosphere, working in the megahertz range [Chapman & Warren, 1968; Higel *et al.*, 1972].

3. Relaxation Signal: Measurement of Electron Density and Gyrofrequency

In the general case, the transmitted waves propagate away from the spacecraft and only a short echo is received from the ionosphere. When the frequency of the signal is close to one of the resonance frequencies of the medium, however, the waves propagate in the vicinity of the spacecraft with a very weak group velocity and in a slowly varying medium are reflected back to the satellite. The sounder then receives a relaxation signal lasting for milliseconds, although the signal has been transmitted only for some tens of microseconds.

In the ionosphere, the observed resonances occur at the plasma frequency, at the upper hybrid frequency, at the gyrofrequency and at its harmonics, corresponding to the Bernstein modes of propagation.

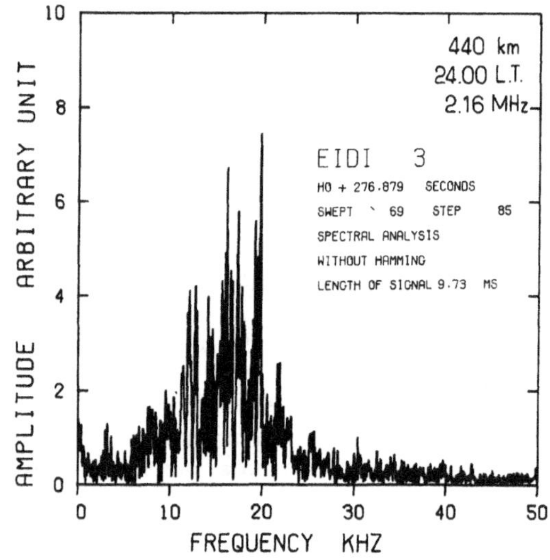

Figure 1. *Frequency spectrum of the relaxation signal at the plasma frequency, measured during the EIDI experiment. Only a 50 kHz bandwidth centred at 2.16 MHz is shown.*

Figure 1 shows the frequency spectrum of a relaxation signal at the plasma frequency measured during the EIDI experiment [Higel, 1975], during which both the envelope of the signal and its waveform were telemetered. This spectrum has been compiled using the signal received in a 50 kHz bandwidth, centred at 2.16 MHz. This spectrum is 10 kHz wide, allowing the plasma frequency to be measured with a 0.5% accuracy, i.e. an accuracy of 1% on the density. Figure 2 represents a density profile measured in this way during one EIDI rocket flight, using only a 50 kHz resolution [Petit *et al.*, 1972].

In the same way, the electron gyrofrequency can be measured. Figure 3 [Bitoun *et al.*, 1975] shows relaxation signals at the harmonics of the gyrofrequency; at the top the spectrum made in the same way as in Figure 1, and below the envelope of the signal in semi-logarithmic co-ordinates. At twice the gyrofrequency, the spectrum of the signal is less than 2 kHz wide, allowing a gyrofrequency of about 1 MHz to be measured with an accuracy of 1%.

We shall see later that the relaxation signal is not exactly at the resonance frequencies, but the offset is less than the above accuracy.

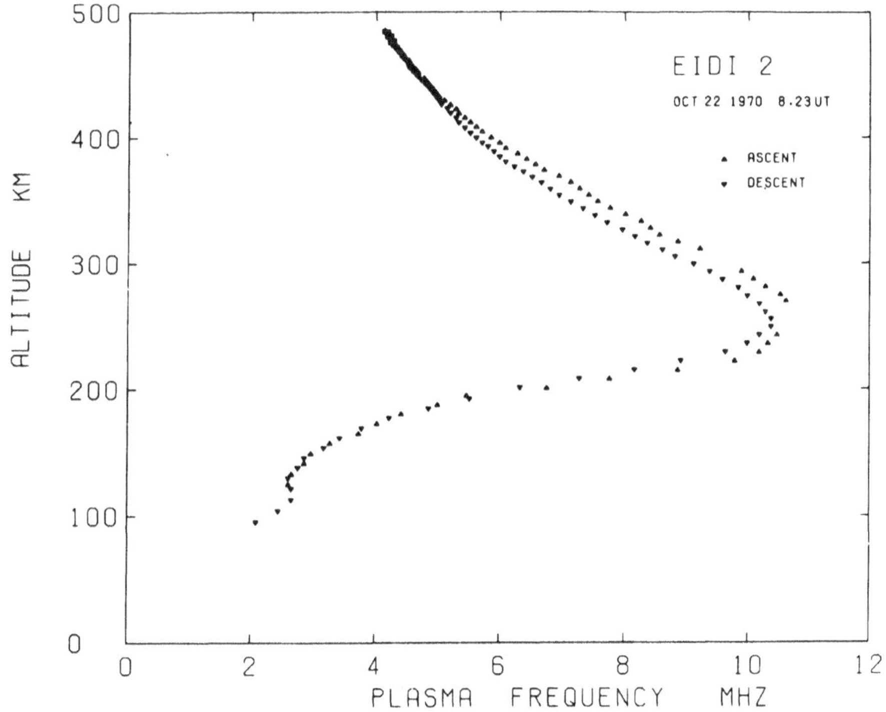

Figure 2. *Profile of the plasma frequency as measured during the EIDI 2 rocket flight. The difference between ascent and descent is due to the horizontal gradient in the electron density.*

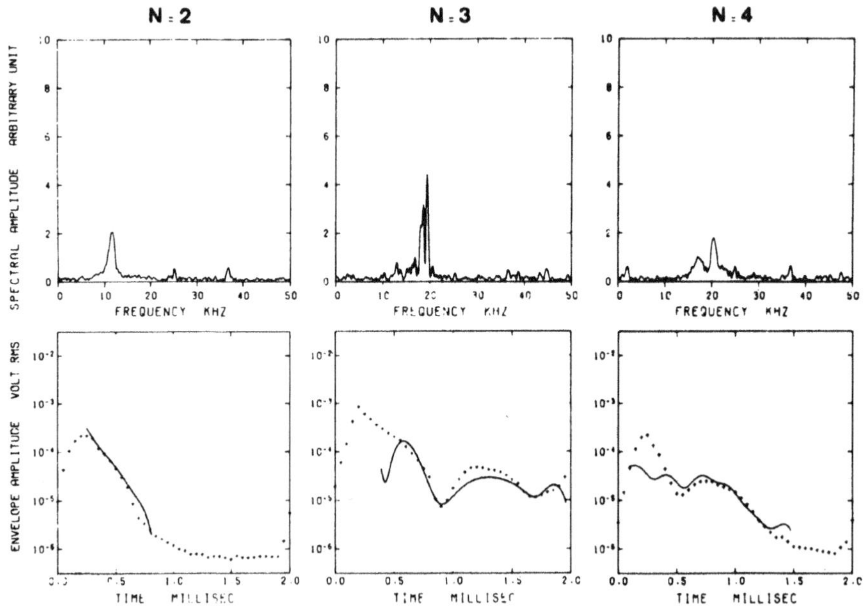

Figure 3. *Resonances at the second, third and fourth harmonics of the gyrofrequency. Above, the spectra of the relaxation signals, computed in a 50 kHz frequency band. The gyrofrequency is about 1 MHz. Below, the signal envelope on a semi-logarithmic scale. The crosses represent data, the solid line theoretical computations.*

4. Oblique Echoes

We shall now examine the physics involved in this phenomenon without doing the computations themselves, which are rather intricate and can be found elsewhere [McAfee 1968, 1969, for the plasma frequency; Aubry *et al.*, 1970, for the upper hybrid frequency; Bitoun *et al.*, 1975, for the harmonics of the gyrofrequency].

Interpretation of the relaxation signals in terms of oblique echoes was first proposed by McAfee [1968].

Near a resonance, the index of the waves, which are electrostatic, varies rapidly with frequency in a given direction and with the propagation angle close to this direction. Figure 4 (bottom) shows the index curves, in polar co-ordinates, for various frequencies normalised to the plasma frequency. In the direction parallel to the magnetic field, the index varies by a factor 2 when the frequency varies by 4×10^{-3}, leading to group velocities of a few tens of kilometres per second. Similarly, the index varies very rapidly with the direction of the wave normal: about a factor 2 in 4°. The rapid variation in the index with frequency leads to weak group velocities, and its dependence on the direction of the wave vector to strong refraction.

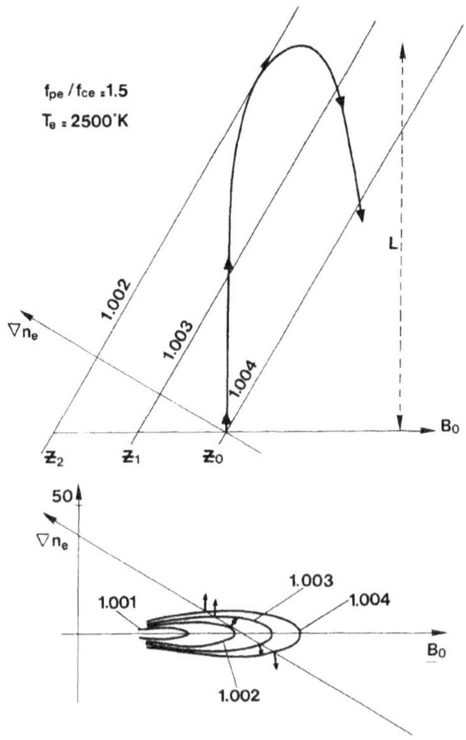

Figure 4. *Below, index curves in polar co-ordinates for various frequency/plasma-frequency ratios. The scale for the index is linear and the point $n = 50$ is indicated on the vertical axis. The arrows indicate the direction of the group velocity at each point of the wave trajectory. Above is the ray path of the waves; L is the maximum elongation of the trajectory.*

In the type of display used in Figure 4, the group velocity of a wave of given frequency, having a wave vector in a given direction, is normal to the corresponding index curve at this point. Assume now that the medium is slightly inhomogeneous, the density gradient being in the direction ∇n_e in the figure. When a wave propagates, its frequency normalised to the local plasma frequency varies, and its index is given successively by different index curves. Due to Snell's law, a wave with a given initial wave vector will refract during its propagation in such a way that the tip of the wave vector will stay on a line parallel to the gradient's direction. At the altitude Z_0 of the transmission, the normalised frequency of the wave is 1.004. The group velocity pictured by the arrow is normal to the index curve.

When it propagates to the altitude Z_1 where the plasma frequency had decreased by 1‰, the group velocity has changed. At the altitude Z_2, where the plasma frequency has again decreased by 1‰, the group velocity is normal to the gradient: the wave is reflected. At the top of Figure 4, the corresponding ray path is drawn.

The spacecraft has moved by this time and receives the reflected wave. The elongation L of the trajectory is usually several hundreds of metres in the upper ionosphere, and should be several tens of kilometres in the magnetosphere.

The mechanism of oblique echoes has been explained here for resonance at the plasma frequency, where it is controlled by the density gradient. The principle is the same at the harmonics of the gyrofrequency or the upper hybrid frequency, but in the first case the magnetic field gradient controls the reflection and in the second both the density and magnetic-field gradients play a part.

The purpose of studying resonances is therefore to find the path followed by the waves, by ray tracing, and to fulfil a 'rendezvous' condition, both in space and time, between the waves and the spacecraft. It has been shown that for all the resonances studied so far, two waves are received simultaneously at each instant. Each of these waves is affected by a Doppler shift $\omega = \omega_0 - \mathbf{k}\cdot\mathbf{V}_s$, where $\mathbf{V}_s$ is the relative velocity between the spacecraft and the medium. Therefore, beats between these two waves are generally observed: the relaxation signal has a fine structure. An example of these beats can be seen in Figure 3 for the resonance at three times the gyrofrequency: the spectrum exhibits two peaks, each corresponding to a wave following a distinct path, and the envelope shows the corresponding amplitude modulation. The beats are not always so clearly seen because, depending upon the geometry of the system (position of the antenna with respect to the magnetic field), each of the waves arriving at the antenna can have a wide spectrum. This aspect has been studied in detail for the plasma resonance by Fleury & Petit [1974].

We shall now establish on which parameters of the medium this fine structure depends, leading to the possibility of a plasma diagnostic other than the electron density and magnetic field.

5. Temperature Dependence

The dispersion relation for the waves depends on the electron temperature and the signal received at each instant is therefore temperature dependent also. We can hope then, by studying the time evolution of the received signal, to obtain information on the electron temperature.

The first attempt was made by Warnock *et al.* [1970], who studied the beat frequency as a function of time delay after transmission for a plasma resonance and deduced an electron temperature with an 8% accuracy. A similar study of the time evolution of the beat frequency has been made by Feldstein & Graff [1972] for plasma and upper hybrid resonances, on several examples. Unfortunately, it has not been possible to determine the temperature from both types of resonances on the same part of the orbit. An example of this work is shown in Figure 5; the solid curves represent the beat frequency as a function of delay for various temperatures. The temperature deduced for the ionosphere is about 2500° K, which is not unreasonable at noon for an altitude of 500 km. Nevertheless, the accuracy of the determination is reduced (to about 20%) by a 'superstructure' in the data: superimposed on the monotonic variation of the beat frequency are oscillations that have not yet been interpreted.

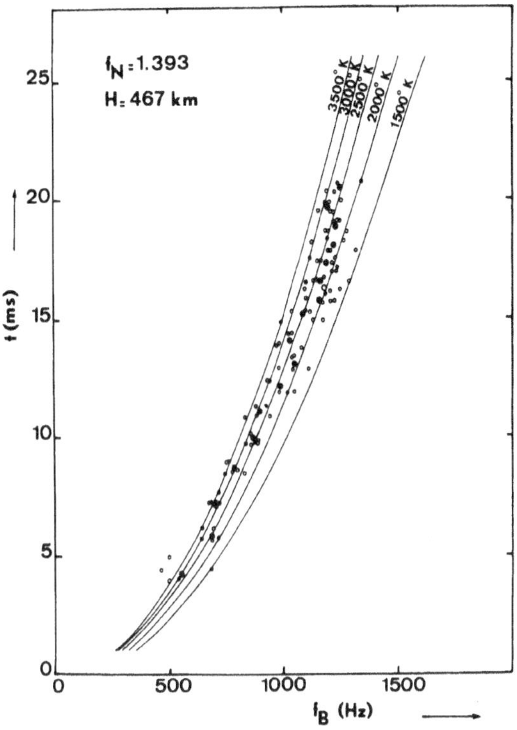

Figure 5. *Beat frequency versus delay for a resonance at the plasma frequency. The open circles are measured values and the points are averages of these measurements. The solid curves represent the theoretical variation for various temperatures.*

McAfee *et al.* [1971] studied, for one resonance at the upper hybrid frequency, the evolution of the frequency received with delay. They compared the theoretical results for various temperatures with measured values. The accuracy of the agreement was of the same order as in the above beats study.

In a similar way, it is possible to study the fine structure of the resonance at the harmonics of the gyrofrequency: our group has paid particular attention to this type of resonance recently. Bitoun *et al.* [1975] have computed, in a very complete way, the signal received at the antenna, and it is possible to make even more sophisticated studies. In Figure 3, a comparison between theoretical results and experiment is shown for three examples of resonances at the harmonics of the gyrofrequency. The solid curves are the computed amplitude of the received signal versus delay, using the best-fitted temperature, and the points represent measured values. The amplitude scale is arbitrary, since the antenna response is unknown and only the time evolution of the signal is meaningful.

6. Spacecraft Velocity Dependence

The resonance signal is not only temperature-dependent, it is sensitive too to the relative velocity of the spacecraft and the medium. No attempt to obtain information on this parameter has been made so far, since in the ionosphere it is the velocity of the satellite in its orbit, which is well known. Conversely, in the outer magnetosphere it is likely to be the drift velocity of the plasma with respect to the satellite, which is a very interesting parameter. We have seen that each received wave is Doppler-shifted by the quantity $\mathbf{k}\cdot\mathbf{V}_S$. The fine structure of the signal (particularly the beats) will therefore be affected by terms containing this quantity.

For the resonance at the plasma frequency, the wave vector is approximately parallel to the magnetic field, while for resonances at the harmonics of the gyrofrequency and at the upper hybrid frequency, it is approximately perpendicular to it. Both components of the drift velocity are therefore hidden in the resonances and we hope to be able to extract them. Moreover, for the resonances at the harmonics of the gyrofrequency, the wave vector (and the electric field of the wave) lies in the plane defined by the magnetic field and the velocity. If we are able to measure the direction of the wave vector, the magnetic field being measured by the magnetometer, we could obtain not only an indication of the modulus of the velocity, but also its direction. Nevertheless, measurement of an electric-field direction using three antennas is known to be difficult and the accuracy of such a velocity determination can surely not be high.

7. Discussion

The problem that has been studied up to now is that of explaining the structure of the relaxation signals in a known medium, the ionosphere. We have now to reverse the problem: given the structure of the relaxation signal, we must guess the nature of the medium. If for the magnetic-field intensity and electron density the answer is obvious (and accuracy better than 1%), it is much more difficult to deduce the medium's other parameters. The density and magnetic-field gradients, electron temperature and spacecraft velocity are intimately linked in the fine structure of the various resonances. Only by a simultaneous study of the fine structure of several resonances can we hope to reach these parameters. So far only a few, very inaccurate, temperature determinations have been made, assuming the gradients and velocity to be known from other means.

Another point is that a relaxation sounder has never been flown in the outer magnetosphere. The plasma there is different from the ionosphere: i.e. diluted, hot, non-Maxwellian. What will be the appearance of the resonances? Will they even be observable? Gaffey *et al.* [1975] have shown that the dispersion relation close to the upper hybrid resonance is modified in a non-Maxwellian plasma: the relaxation signals will therefore surely be modified also. We are presently working on these problems.

8. GEOS and ISEE Sounder Capabilities

The characteristics of the sounders for GEOS and ISEE are summarised in Table 1. The frequency range for GEOS is 77 kHz, while on ISEE we have two frequency ranges, switchable by telecommand: 0-50 kHz for the remote parts of the orbit and 0-300 kHz for use at lower altitudes.

TABLE 1
Characteristics of the sounders for GEOS and ISEE

	GEOS	ISEE A
Frequency range	0-77 kHz	0-50 kHz or 0-300 kHz
Minimum step duration	86 ms	32 ms
Minimum time resolution for a complete sweep	22 s	4 s

We have several possible listening times after transmission, selected by telecommand. The shortest is used to explore a frequency range quickly and to identify the resonance frequencies, thus giving a measurement of the electron density and magnetic field. The longer listening times are to be used to study the fine structure of the relaxation signals. The 'step duration' is the time during which the transmitter and the receiver are tuned to the same fixed frequency. It is the time interval between two successive transmitted pulses. Its minimum value is 86 ms for GEOS and 32 ms for ISEE, leading to a minimum time resolution for a complete sweep of 22 s for GEOS and 4 s for ISEE.

For GEOS, by taking advantage of the stationary position of the spacecraft, we foresee a system for automatic real-time commanding of the experiment. A ground-based minicomputer handles the data and transmits telecommand orders in real time. It identifies the resonances and tracks them in a restricted frequency range in order to improve the time resolution. For instance, with the shortest step duration, if the explored frequency range is restricted to 4.8 kHz, the same frequency is explored every 1.38 s. The difficulty is to find the best compromise for exciting the resonance frequency as often as possible without losing the resonance if its frequency has shifted in the meantime.

9. Conclusions

The relaxation sounders currently used at low altitudes have yet to be flown in the outer magnetosphere and hence the studies discussed are in the nature of an exploratory experiment. Nevertheless, there is a good chance that we can determine the electron density and gyrofrequency. We also hope to obtain some further information about the ambient plasma.

The great advantage of the sounding technique lies in its independence of conditions in the vicinity of the spacecraft (floating potential, photoelectrons, etc.) since the waves propagate for several kilometres from it. Its drawback, besides the uncertainty inherent in any exploratory experiment, is that all the measurements are integrated over the particle distribution.

10. Acknowledgement

The author wishes to thank the members of the CNET 'resonance team' for their many fruitful discussions.

References

Aubry, M., J. Bitoun & P. Graff, Propagation and group velocity in a warm magnetoplasma, *Radio Sci.* **5**, 635–645 (1970).

Bitoun, J., L. Fleury & B. Higel, Theoretical study of gyroresonances with application to rocket experiments, *Radio Sci.*, 1975.

Chapman, J.H. & E.S. Warren, Topside sounding of the earth's ionosphere, *Space Sci. Rev.* **8**, 846–865 (1968).

Feldstein, R. & P. Graff, Daytime and nightime electron temperatures from topside resonances, *J. Geophys. Res.* **77**, 1896–1904 (1972).

Fleury, L. & M. Petit, Antenna dependence of plasma resonance patterns, *J. Geophys. Res.* **79**, 4817–4820 (1974).

Gaffey, J.D. Jr. & R.E. Laquey, Upper hybrid resonance in the magnetosphere, Submitted to *J. Geophys. Res.*, 1975.

Higel, B., J. Bitoun & M. Petit, Rocket observation of two waves beating at 3 f_H resonance, *J. Geophys. Res.* **77**, 6254-6258 (1972).

Higel, B., Non stationnarité des signaux de résonance des plasmas ionosphérique et magnétosphérique, *Ann. Télécom.*, 1975.

McAfee, J.R., Ray trajectories in anisotropic plasma near plasma resonance, *J. Geophys. Res.* **73**, 5577–5584 (1968).

McAfee, J.R., Topside resonances as oblique echoes, *J. Geophys. Res.* **74**, 802–808 (1969).

McAfee, J.R., T.L. Thompson, W. Calvert & J.M. Warnock, Rocket observations of topside resonances, *J. Geophys. Res.* **77**, 5542–5550 (1972).

McAfee, J.R., Electron plasma resonances in the topside ionosphere. Fundamentals of Cosmic Physics, Vol. 1, 71–118, Gordon and Breach Ltd., 1973.

Petit, M., J. Bitoun, R. Feldstein, P. Graff & B. Higel, Ondes de plasma à vitesse de groupe faible, *Ann. Télécom.* **27**, 417–429 (1972).

Warnock, J.M., J.R. McAfee & T.L. Thompson, Electron temperature from topside plasma resonance observations, *J. Geophys. Res.* **75**, 7272–7275 (1970).

DISCUSSION

M. Dryer: Can the sounding technique be used, in principle, to measure N_e, T_e, and relative plasma velocity in the magnetosheath as well as the interplanetary medium?

J. Etcheto: The sounding technique is even more exploratory in the solar wind and magnetosheath than in the outer magnetosphere and it is impossible before launch to be sure of anything. In measuring N_e, the main question is knowing whether or not the Landau damping will allow the plasma resonance to be observed.

In the solar wind, we have computed that for average medium characteristics the relaxation signal should be observed.

In the magnetosheath it is more difficult to estimate, since the irregularities of the density will affect the propagation in an unknown way. In measuring T_e and V_s, I do not think that we shall get any reliable measurement based on the study of the fine structure of the resonances in a turbulent region: it needs a careful determination of the ray trajectory and travelling time of the waves, which will not be possible in such a medium.

In these regions, I should trust more the propagation experiment between mother and daughter, as it is working with electromagnetic waves which are subject to less damping.

D. Jones: The refractive index surfaces shown were labelled as being for a cold plasma. What is the definition of cold, since the surfaces are obviously not those derived from 'cold magnetoionic theory'?

J. Etcheto: There is a misunderstanding. The refractive index surfaces were traced for a hot plasma with a temperature of a few thousands of degrees. What was labelled cold was the resonance cone as computed for a cold plasma and shown for comparison.

G.L. Wrenn: The analysis for electron-temperature determination which you described is very interesting, but the curves you showed did indicate that the temperature resolution or accuracy decreased with increasing temperature. Does this mean that extrapolation to magnetospheric temperatures of say 10^4 or 10^5 °K would be inherently inaccurate or do other factors come into play?

J. Etcheto: What I have shown was not an attempt to determine the temperature, but just an example to show that it has an influence. We usually observe several resonances, each one depending in a different way on the temperature, and several features of each resonance (in particular, the resonance length) depending on the temperature. What we have to do is to put together all the characteristics of the resonances we are able to measure and find, for each type of plasma, which one is the most temperature sensitive. As you pointed out, the decay of the $2f_{ce}$ resonance is not likely to be the good one for GEOS.

We are presently working on this problem and are not able at present to give an estimate for the foreseen accuracy.

PART 3

QUASISTATIC MAGNETOSPHERIC PHENOMENA AND PARTICLES

MAGNETOSPHERIC DC ELECTRIC FIELDS; PRESENT KNOWLEDGE AND OUTSTANDING PROBLEMS TO BE SOLVED DURING THE IMS

F.S. Mozer

Physics Department and Space Sciences Laboratory,
University of California, Berkeley, USA

Abstract

Present knowledge of the relationships between perpendicular and parallel electric fields in the ionosphere and magnetosphere and of their configurations, sources, and variations is summarised with a view to illustrating some important problems that should be addressed by IMS electric-field research programmes. These problems include determination of the source, morphology, and dynamics of parallel electric fields, the role of parallel electric fields and time-varying magnetic fields in decoupling magnetosphere-ionosphere dynamics, the configuration of the neutral line and its effect on the electric-field coupling from the interplanetary medium to the magnetosphere for arbitrary interplanetary magnetic-field directions, the importance of processes other than reconnection in producing the magnetospheric potential field, and the dynamics of polarisation and other processes that produce low-frequency turbulence and small-scale spatial structure in the electric fields.

1. Introduction

A summary of present knowledge and some outstanding problems of magnetospheric DC electric-field research will be undertaken by dividing the magnetosphere into two parts (called the 'ionosphere' and the 'magnetosphere') with a boundary between the two at an altitude of one or a few thousand kilometres. The reasons for this division are:

(a) There is no necessary relationship between ionospheric and magnetospheric electric fields.
(b) The overwhelming fraction of existing data has been obtained in the ionosphere. Hence, some of the presently most valid estimates of the magnetospheric elec-

Knott & Battrick (eds.), The Scientific Satellite Programme During the International Magnetospheric Study, 101–131.
All Rights Reserved. Copyright © 1976 by D. Reidel Publishing Company, Dordrecht-Holland.

tric field are obtained by extrapolating ionospheric data, in spite of item (a) above.

(c) The greatest impact on electric-field research during the IMS years is likely to come from *in-situ* measurements in the magnetosphere proper.

In the following sections, theoretical relationships between ionospheric and magnetospheric electric fields will be derived, some of the highlights of ionospheric electric fields data will be reviewed, and the existing body of magnetospheric data will be discussed and compared, when possible, with ionospheric data. These discussions will focus on some of the outstanding research problems that might be solved by the IMS satellite programme.

2. Relationships between Ionospheric and Magnetospheric Electric Fields

The closed contour ABCD of Figure 1 consists of two neighbouring magnetic field lines (BC and AD) and the segments AB and CD, each of which is perpendicular to the local magnetic field.

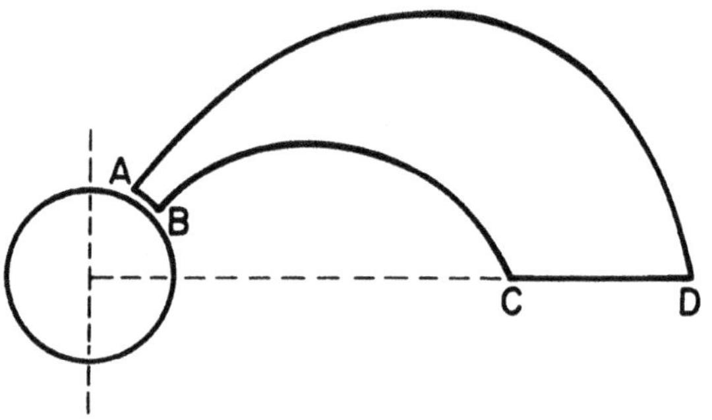

Figure 1. *Geometry of contour associated with surface and line integrals of Equation (2).*

Application of

$$\nabla \times \mathbf{E} = -\frac{1}{c}\frac{\partial \mathbf{B}}{\partial t} \tag{1}$$

to this contour yields:

$$E_e S_{CD} - E_i S_{AB} = -\int_D^A \mathbf{E} \cdot \mathrm{d}\mathbf{l} - \int_B^C \mathbf{E} \cdot \mathrm{d}\mathbf{l} - \frac{1}{c}\frac{\partial}{\partial t}\iint \mathbf{B} \cdot \mathrm{d}\mathbf{A} \tag{2}$$

where E_e and E_i are the meridional components of the perpendicular electric field at the equator and in the ionosphere respectively, S_{CD} is the length of the line segment CD, dl is an element of length along either of the magnetic field lines of interest, and dA is an element of area on the surface bounded by the contour of Figure 1. In the simplest case, the line integrals on the right side of Equation (2) are zero because the parallel electric field is assumed to be zero, and the last term of Equation (2) is zero because time variations of the magnetic field are assumed to be unimportant. For this simple coupling between the ionosphere and magnetosphere, the ratio of meridional perpendicular electric fields in the ionosphere and at the equator is given by the ratio S_{CD}/S_{AB}, which depends only on the magnetic field geometry. For a dipole magnetic field [Mozer, 1970]

$$E_i/E_e = S_{CD}/S_{AB} = 2L\,(L - 3/4)^{1/2} \tag{3}$$

Whereas, for the electric-field components perpendicular to both the meridional plane and the magnetic field line,

$$E_i'/E_e' = L^{3/2} \tag{4}$$

where L is McIlwain's L-shell parameter. The validity of these relationships between equatorial and ionospheric electric fields may be destroyed by the presence of parallel electric fields or time-varying magnetic fields. In such an event, there is no necessary relationship between ionospheric and magnetospheric electric fields and magnetospheric convection and dynamics cannot be obtained from ionospheric electric-field measurements. It is the purpose of the following sections to discuss the relative importance of parallel electric fields and time-varying magnetic fields in the electric-field mapping process.

Parallel electric fields destroy the above simple mapping relationships only when the sum of the line integrals in Equation (2) is comparable to either of the terms on the left side of that equation. The order of magnitude of the largest parallel electric field that preserves the simple coupling relations may be estimated by assuming that the parallel electric field is everywhere constant and that its contribution to Equation (2) comes from the difference of lengths of the neighbouring magnetic field lines. For this case, the condition that the parallel field contribution to Equation (2) be negligible is [Mozer, 1970]

$$E_e >> E_{\|} \tag{5}$$

Since $E_e \simeq 1$ mV/m, a uniform parallel electric field of, say, 0.1 mV/m would not invalidate equatorial electric-field estimates made from ionospheric data. Such a parallel electric field would produce a potential drop of about 5 kV along auroral-zone magnetic field lines. Thus, substantial parallel fields and potential drops might exist with or without affecting the simple coupling between ionospheric and magnetospheric electric fields. One of the important goals of electric-field research during the IMS should be to investigate the extent to which parallel electric fields uncouple ionospheric and magnetospheric convection and dynamics.

The effect of the surface integral in Equation (2) may be estimated from consideration of magnitudes of nonpotential electric fields in the ionosphere and at the equator.

The typically observed upper limit of $\partial B/\partial t \simeq 1$ gamma/s in the auroral-zone ionosphere combines with Equation (1) to require that $\nabla \times \mathrm{E} \simeq 1$ mV/m/1000 km. Since typically observed electric-field variations in the ionosphere are at least an order-of-magnitude larger, the contribution of nonpotential fields in the ionosphere is generally negligible.

These observed ionospheric electric-field variations are reduced by a factor of L^3 when mapped into the equatorial plane according to the simple law of Equation (3), because both the field decrease and scale-size increase vary as $L^{3/2}$. Thus, the observed ~ 10 mV/m/1000 km variation in the ionosphere might be only ~ 0.3 mV/m/1000 km near the equator. A similar variation would be produced from a time-dependent magnetic field having $\partial B/\partial t = 0.03$ gamma/s. Since this value of $\partial B/\partial t$ is comparable to or smaller than that often observed near the equator [Cummings & Coleman, 1968; Block & Carpenter, 1974] nonpotential electric fields are probably significant near the equatorial plane. This conclusion has also been demonstrated by earlier calculations [Wang & Kim, 1972; Block & Carpenter, 1974].

Because nonpotential electric fields are probably important near the equator but not in the ionosphere, there is no required relationship between electric fields in these two regions. An important goal of electric-field research in the next decade will be to measure the extent of magnetosphere-ionosphere decoupling induced by time-varying magnetic fields through simultaneous *in-situ* observations of electric and magnetic fields in the two regions.

3. Significant Results of Ionospheric Electric-Field Research

3.1. THE AVERAGE, PERPENDICULAR, IONOSPHERIC, ELECTRIC FIELD

The average ionospheric electric field is very similar to that predicted before the first measurements were made. These predictions will be briefly reviewed, both in their own right and because they offer a basis for viewing and organising the experimental data.

The concept of an ionospheric electric field and two-cell convection pattern was contained but not stated in analyses of ground-based magnetometer data that led to the D_s current system [Chapman, 1956]. In the same time period, certain classes of convective motions in the magnetosphere were extensively discussed but not related to observations or other magnetospheric processes [Gold, 1959]. It remained for Axford & Hines [1961] to recognise the central role of convection in auroras, particle energisation, and a variety of high-latitude phenomena, and to relate a convection model to the D_s current system and other experimental data. In Figure 2, a view of the ionosphere from above the pole is presented along with solid curves that represent the convection pattern of Axford and Hines. The major features of this pattern are:

(a) an anti-sunward convection across the polar cap
(b) a return flow at lower latitudes
(c) as a result of (a) and (b), there is a flow reversal region near dusk (or dawn) such that the high-latitude flow is opposite to that at low latitudes.

These features have been verified by several thousand hours of data obtained from incoherent scatter radar, balloon, rocket, and satellite observations. The techniques for making such measurements are described elsewhere [Mozer, 1973a]. Some of the earliest or most complete of these results will be used to illustrate this experimental knowledge of polar-cap, auroral, and lower latitude electric fields.

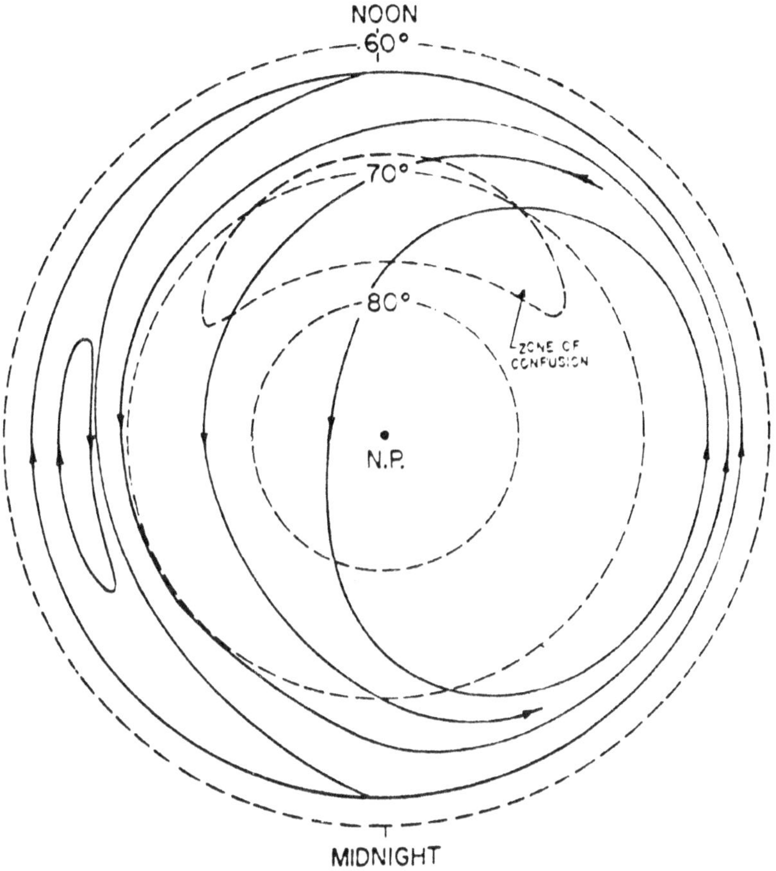

Figure 2. *Pattern of ionospheric convection in a rotating frame of reference as deduced from viscous interaction by Axford & Hines [1961].*

The first extensive observations of polar-cap electric fields were made with the Injun-5 satellite. One example of data obtained during a series of polar-cap traversals is illustrated in Figure 3, in which the one measured component of the plasma convection velocity

$$\mathbf{v} = c\frac{\mathbf{E} \times \mathbf{B}}{B^2} \tag{6}$$

is plotted along the trajectory [Cauffman & Gurnett, 1971]. The principal results of these measurements were that an anti-sunward convective flow was often observed over the poles and it reversed direction in the auroral zone. This field reversal is one of the essential features of the two-cell convection pattern predicted earlier.

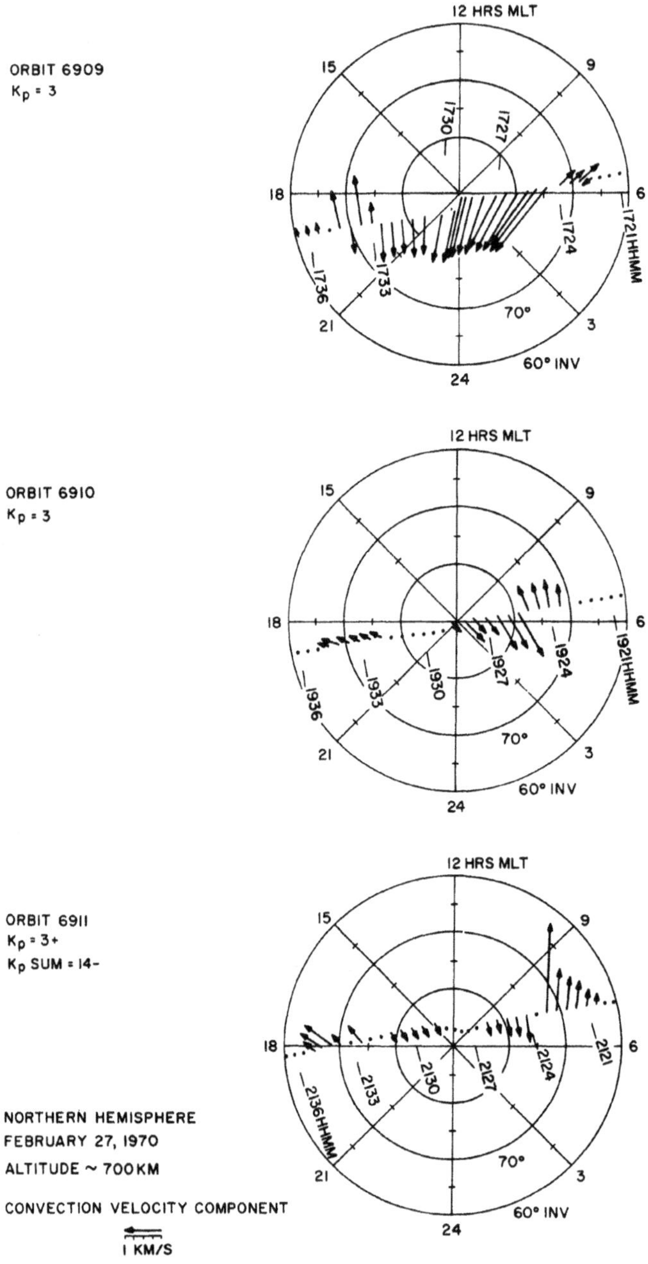

Figure 3. *Pairs of convection velocity reversals observed near dawn and dusk on three successive passes of Injun-5 satellite.*

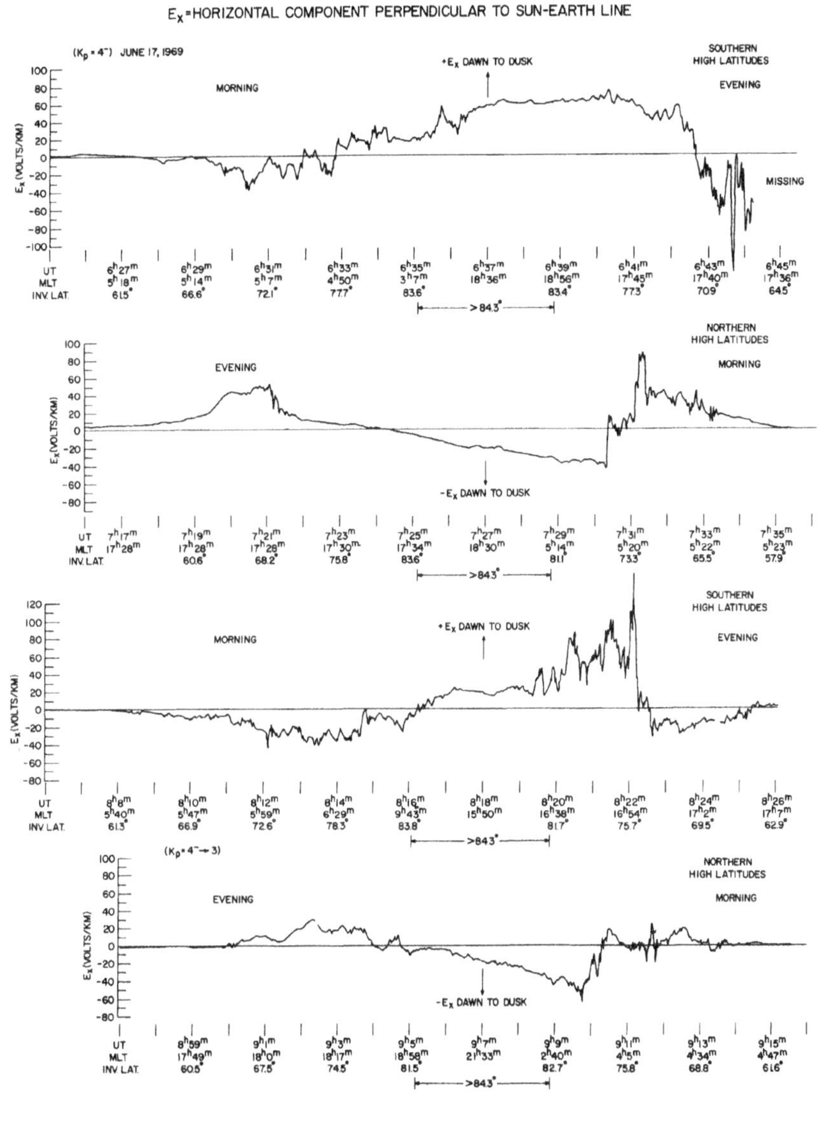

Figure 4. *Dawn-to-dusk electric-field component measured near the dawn-dusk meridional plane by successive polar-cap passes of OGO-6 spacecraft.*

Because of the relatively poor sensitivity of the Injun-5 measurement, it remained for the OGO-5 [Heppner, 1972] and balloon measurements [Mozer *et al.*, 1974] to confirm that a dawn-to-dusk electric field corresponding to anti-sunward convection exists in the polar-cap ionosphere nearly all the time.

An example of several polar-cap passes of OGO-5 measurements of the dawn-dusk electric field component is given in Figure 4 while the hourly averages of more than 200 h of balloon data are plotted as ionospheric electric-field vectors in Figure 5. The lowest latitude data of Figure 5 was obtained at a site that was often outside the polar cap and was, hence, in the field-reversal region discussed above. Such field reversals are also apparent in the OGO data.

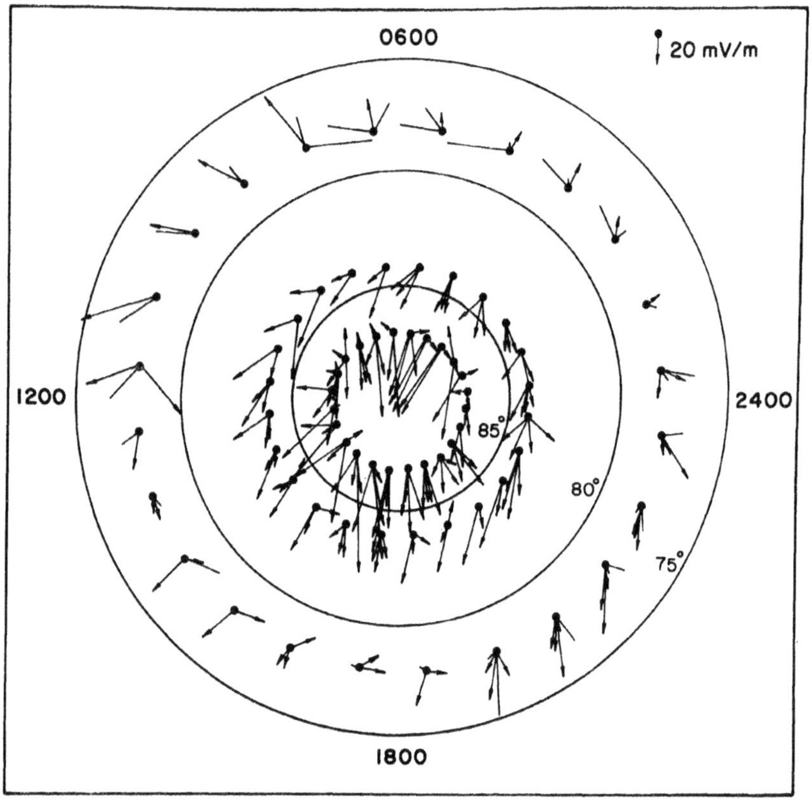

Figure 5. *One-hour average electric-field vectors measured on balloons and plotted in a geomagnetic latitude - magnetic local time nonrotating frame of reference.*

Characteristics of the auroral-zone electric field also include large westward plasma convection (northward electric fields) prior to local midnight and eastward convection (southward electric fields) near and after local midnight. Examples of these convection flows as observed by barium releases [Haerendel, 1972a] and incoherent scatter radar [Doupnik *et al.*, 1972] are given in Figures 6 and 7.

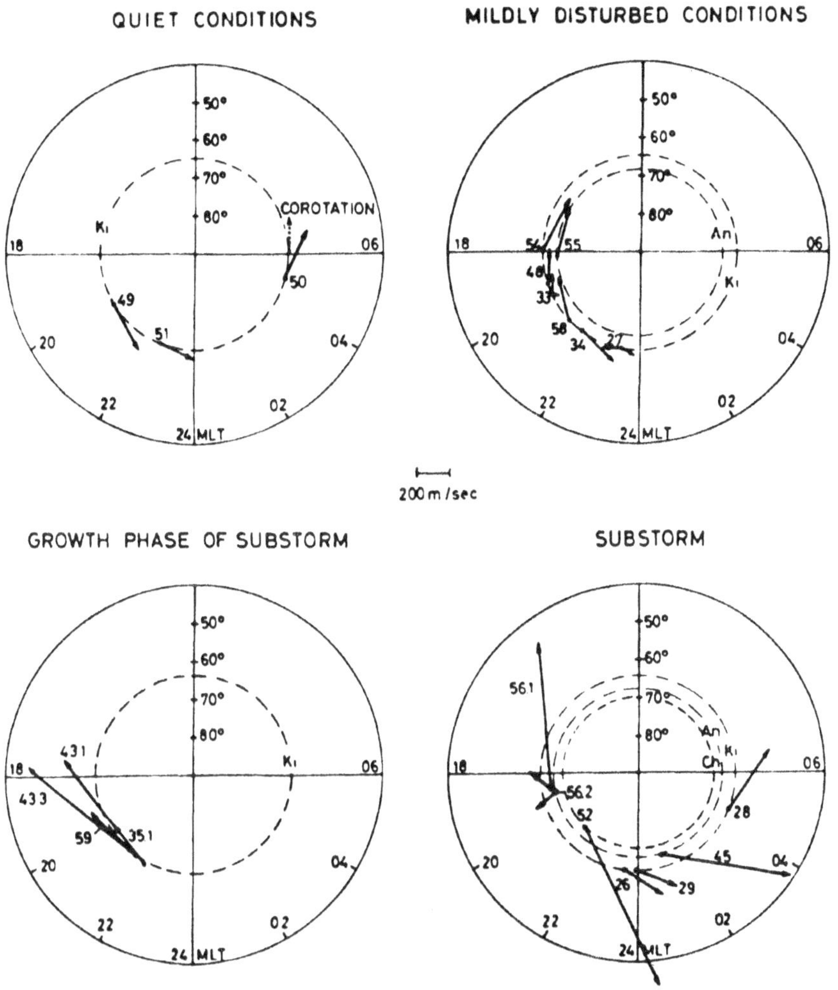

Figure 6. *Mean ionospheric drift velocities for four different conditions of magnetic activity, as determined by barium-cloud releases.*

The most extensive published set of auroral-zone electric field measurements has been obtained on balloons. In Figure 8, 1 h averages of electric-field components measured in a nonrotating frame of reference on 32 balloons are presented [Mozer & Lucht, 1974]. The variability of the electric field is well-illustrated by these data, as is the existence of a predominately northward field near dusk, a southward field near and after local midnight, and a dominance in magnitude of the north-south component over the east-west component.

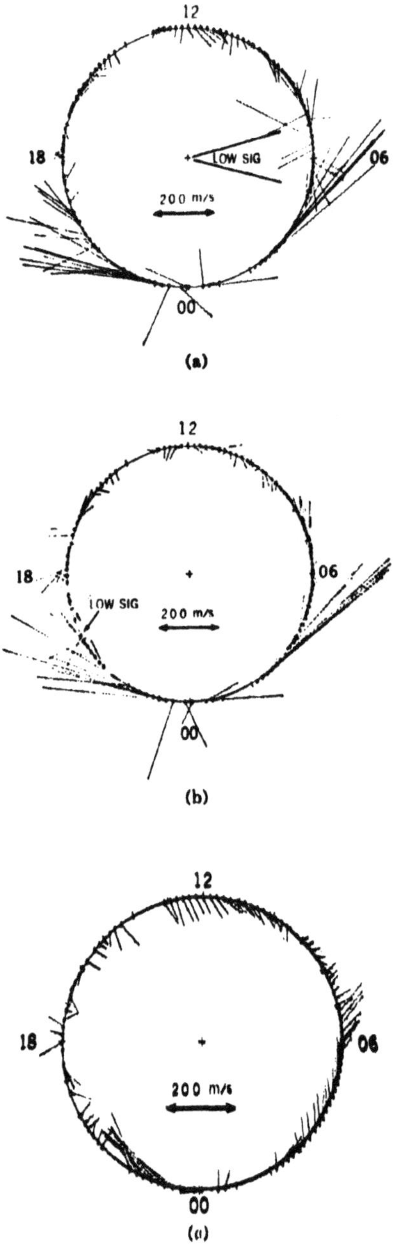

Figure 7. *Incoherent scatter radar measurements of plasma drift velocities at three altitudes: (a) 220 km (b) 163 km, and (c) 106 km.*

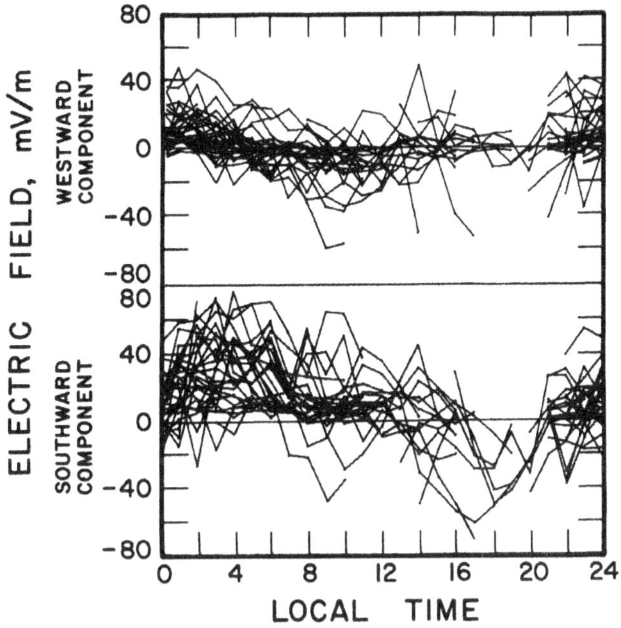

Figure 8. *One-hour averages of electric-field components measured in a nonrotating frame of reference on 32 balloons flown in the auroral zone.*

The data of Figure 8 have been averaged and mapped into the equatorial plane according to generalisations (for the nondipole nature of the magnetic field) of Equations (3) and (4) to produce the equatorial electric-field vectors of Figure 9. The equipotential contours of this figure were obtained from the electric-field data and from assumptions that the field is uniform across the tail at a distance of 15 R_e and that the magnetopause is not an equipotential. Several conclusions may be drawn from this figure:

(a) The average auroral zone convection pattern is sunward, in agreement with the predictions of Axford and Hines.

(b) Despite the dominance of the southward component of the electric field in the ionosphere near local midnight, the equatorial electric field is nearly westward and the equatorial plasma flow is sunward near local midnight on auroral zone field lines.

(c) Conclusions (a) and (b) above depend on the simple mapping relationship between ionospheric and equatorial electric fields that requires parallel electric fields and time-varying magnetic fields to be unimportant in the mapping sense. Hence, these conclusions on the magnetospheric electric field and plasma convection may not often be valid. This criticism of the above analyses can only be tested by further observations.

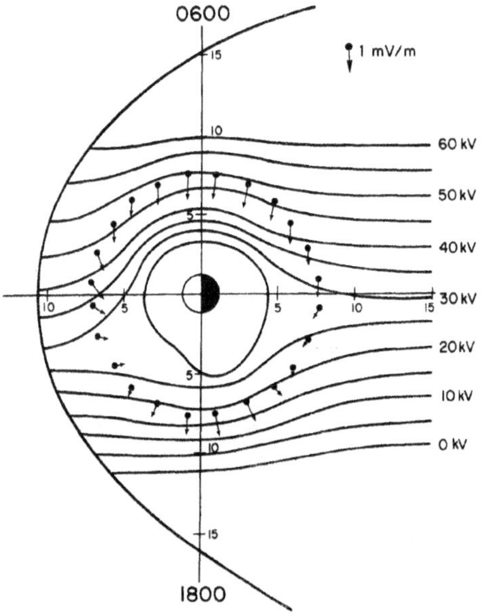

Figure 9. *Hourly averaged electric-field vectors plotted in the equatorial plane of a nonrotating frame of reference, as viewed from above the north pole. The solid curves are electric equipotentials deduced from experimental data assuming that the electric field is uniform across the tail at a tailward distance of 15 R_E from the Earth.*

The mid- and low-latitude ionospheric electric field is generated primarily by corotation and the ionospheric dynamo, as is illustrated by estimates of these and the magnetospheric electric field given in Figure 10 [Mozer, 1973b]. Predictions of the ionospheric dynamo theory are uncertain because knowledge of the global neutral wind pattern is required to estimate the electric field. Thus, it is not yet clear whether the typical low-latitude electric field has a semi-diurnal or diurnal variation. Data showing a semi-diurnal variation of the two components of convection velocity measured with an incoherent scatter radar [Evans, private communication] are given in Figure 11. Convection electric field measurements with barium clouds [Haerendel, 1972b] and by balloons [Mozer, 1973b] suggesting a diurnal variation of the average low-latitude field are given in Figures 12 and 13. This situation, discussed in further detail elsewhere [Mozer, 1973b], certainly requires collection of more experimental data before it can be resolved. Because of the large **v** x **B** electric field induced by satellite motion through the Earth's magnetic field and because of the planned orbits of most of the IMS satellites, it is not likely that satellite observations will contribute to this or other problems associated with the low-latitude ionospheric electric field.

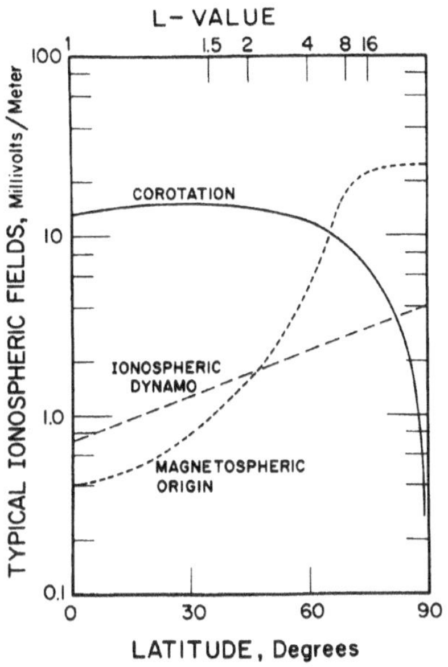

Figure 10. *Typical electric fields observed in a nonrotating frame of reference that arise from corotation, the interaction of tidal neutral winds with the ionosphere, and the interaction of the solar wind with the terrestrial magnetic field.*

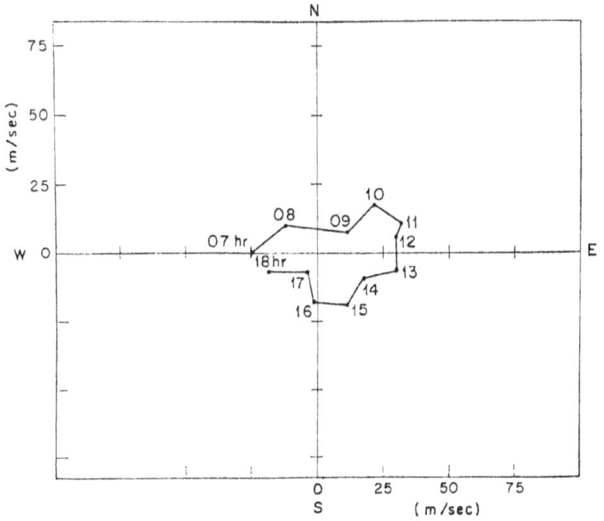

Figure 11. *Average hourly drift velocity vector above the Millstone Hill ionospheric backscatter facility near L = 3.*

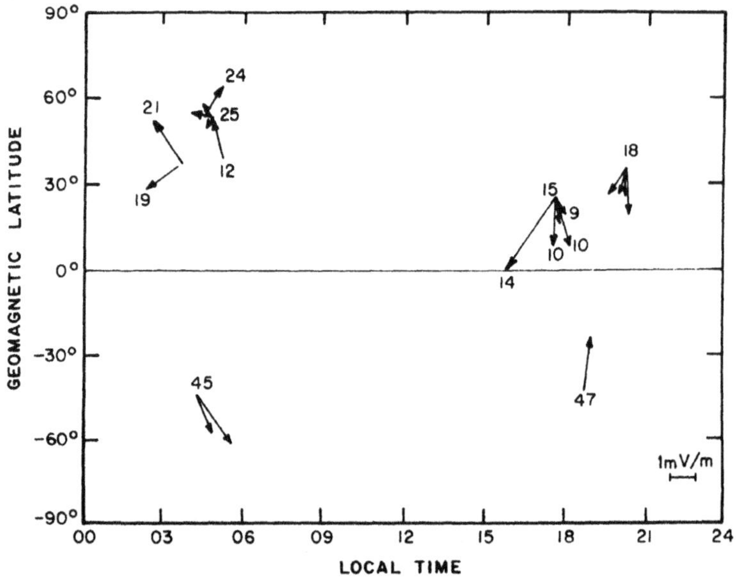

Figure 12. *Examples of electric-field vectors measured by several mid-latitude barium releases.*

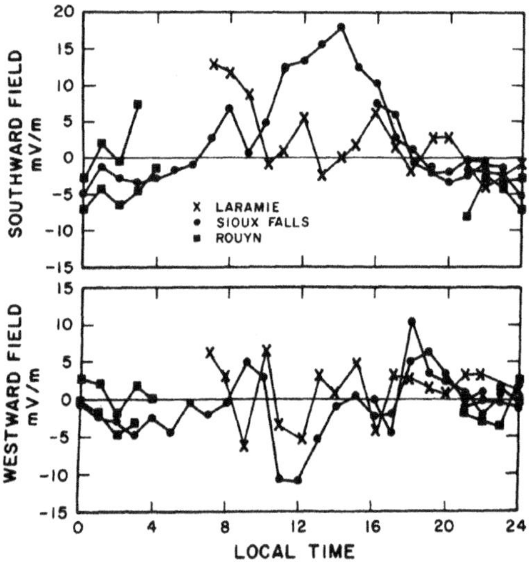

Figure 13. *One-hour averages of ionospheric electric-field components measured in a rotating frame of reference on balloons between L = 2.7 and 3.5.*

3.2. THE SOURCE OF THE PERPENDICULAR IONOSPHERIC ELECTRIC FIELD

The two competing theories of the source of the magnetospheric electric field have been the viscous interaction model of Axford & Hines [1961] and the reconnection model of Dungey [1961]. Because of the overwhelming evidence showing the relationship of a variety of magnetospheric processes to the southward component of the interplanetary magnetic field [references summarised in Gonzalez & Mozer, 1974], there is little doubt but that reconnection plays a significant role in magnetospheric processes. The quantitative extent to which reconnection is understood or dominates other processes can best be studied by correlating interplanetary parameters with polar-cap electric-field data, since the polar-cap electric fields are more likely to map to the ionosphere and to be less affected by local polarisation effects than are the auroral-zone electric fields. Such a study of the OGO-5 data [Heppner, 1972] indicated certain relationships between the polar-cap electric field and the interplanetary magnetic field, but concluded that the observations were not consistent with simple reconnection theories. There has been some dispute over the validity of this conclusion [Russell & Atkinson, 1973].

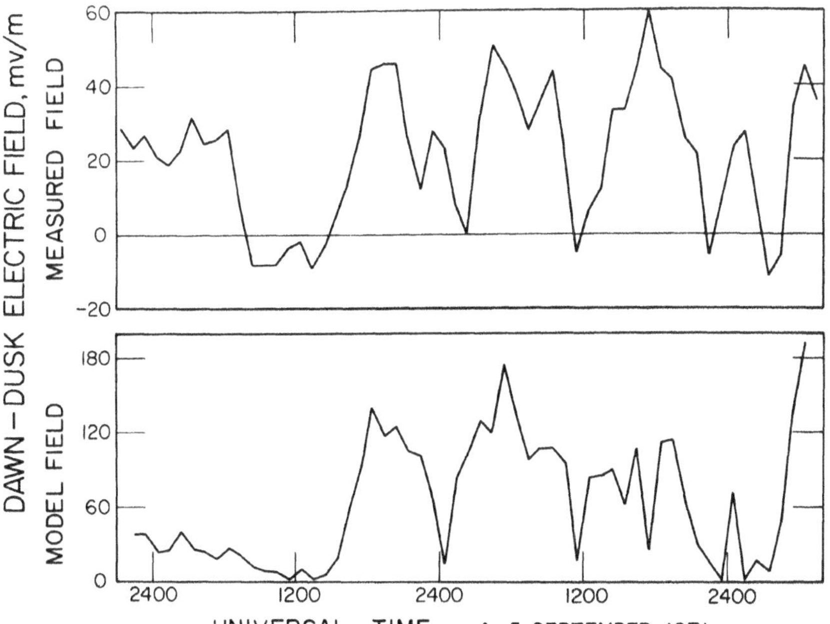

Figure 14. *Hourly averages of the model electric field and the mean of the dawn-to-dusk components of the field measured at Resolute Bay and Thule. Vertical scale of the model field is a factor 3 greater than that of the measured field.*

The polar-cap balloon electric field data of Figure 5 have also been compared with a two-dimensional reconnection model whose inputs utilise interplanetary magnetic-field and solar-wind velocity data [Gonzalez & Mozer, 1974]. The comparison, illustrated in Figure 14, suggests that reconnection accounts for the polar-cap electric-field data. Furthermore, reconnection appears to dominate other processes, since the observed dawn-dusk electric-field component became small or even negative during each of several intervals when the expected reconnection rate was zero.

Further experimental data are required to completely resolve the question of the source of the ionospheric, polar-cap electric field. Such data might also provide quantitative estimates of the orientation of the neutral line, the rate of energy input into the magnetosphere, etc. With an understanding of the field source and with further experimental data, it should be possible to study large and small scale distortions of the magnetospheric electric field associated with nonuniformities in the ionospheric or magnetospheric particle populations. These problems should all be clarified by the IMS research programme.

3.3. PARALLEL ELECTRIC FIELDS IN THE IONOSPHERE

The existence of parallel electric fields greater than the runaway field in the ionosphere (~0.1 mV/m) requires the presence of anomalous resistivity caused by wave-particle interactions, since the parallel currents and power dissipation would otherwise be orders-of-magnitude too large. Anomalous resistivity theories have been applied to the magnetosphere [Kindel & Kennel, 1971; Papadapoulos & Coffey, 1974] with the conclusion that measured parallel currents or energetic particle precipitation should drive instabilities to produce large parallel electric fields (compared to the runaway field) at altitudes of a few thousand kilometres. Direct attempts to measure parallel electric fields have only been made at altitudes below about 500 km. Such measurements by double-probe techniques are difficult because the magnetic-field direction and rocket velocity vector are nearly parallel. Hence, there is no spin periodic variation of the separation distance along the magnetic field line between the pair of probes whose potential difference yields the parallel electric field, and DC offsets due to differences of work functions, sheaths near the probes, etc., influence the measurement [Mozer, 1973a]. Nevertheless, some ten different rockets instrumented by research groups working in five different laboratories have produced measurements of parallel electric fields of $\simeq$ 10 mV/m [Mozer & Bruston, 1967; Mozer & Fahleson, 1970; Kelley *et al.*, 1971; Ungstrup *et al.*, 1972; Kavadas, private communication, 1973; Mahon *et al.*, 1973; Bruston & Fahleson, 1974; Haerendel *et al.*, 1975; Pedersen, private communication, 1975]. An example of one such measurement is given in Figure 15 [Haerendel *et al.*, 1975].

Extensive electric-field measurements have also been made by observing the bulk plasma flow of barium clouds released in the sunlit ionosphere at dusk or dawn. Such single-point measurements cannot directly give parallel field strengths because the parallel flow is related to the parallel electric field through the normal or enhanced parallel conductivity, the plasma expansion and diffusion, etc. [Haerendel *et al.*, 1975]. However, on one occasion, a shaped-charge barium release reached

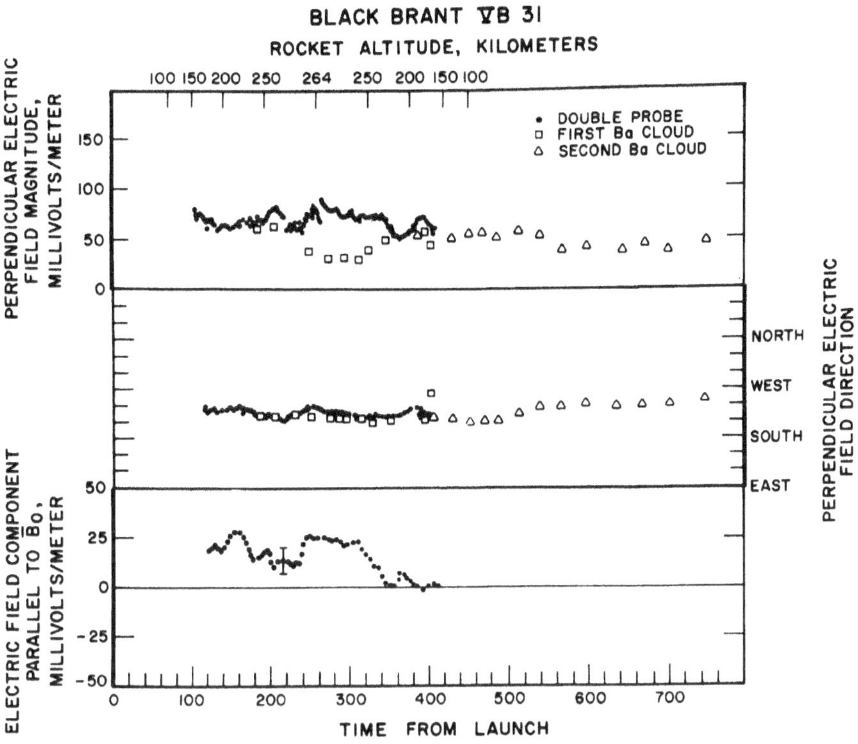

Figure 15. *Electric-field components measured with double probes and barium clouds on a rocket. The barium-cloud data does not yield a parallel electric-field measurement and only the initial data from each cloud should be compared with the double-probe data because each cloud quickly drifted away from the rocket.*

both ends of a magnetic field line at $L = 1.24$, and the conjugate perpendicular flows were observed to be different [Wescott *et al.*, 1974]. This result requires the presence of important parallel electric fields at low altitudes, even deep within the plasmasphere.

The existence of large parallel electric fields anywhere within the magnetosphere is critical to both an improved understanding of the plasma physics of anomalous resistivity and to the magnetospheric physics of ionosphere-magnetosphere coupling and particle acceleration. Hence, the extensive observation of parallel electric fields should be one of the major goals of IMS satellite programmes. The controversies associated with data existing below 500 km altitude and the total lack of observations at higher altitudes only serve to strengthen this conclusion.

Because of the importance of parallel electric fields in the magnetospheric problem, a brief survey of evidence for their existence from particle and other measurements will be given. Perhaps the most direct of this evidence is the observation of predominantly upward-moving electrons [McDiarmid *et al.*, 1961; Heikilla & Mathews, 1964; Cummings *et al.*, 1966; Mathews & Clark, 1968; Ogilvie, 1968; Arnoldy & Choy, 1973; Reasoner & Chappell, 1973; Carlson, 1974] and protons [Mozer & Bruston, 1966a] at altitudes below 500 km. Observations of field-aligned downward-moving electrons [Lampton *et al.*, 1967; Hoffman & Evans, 1968; Holmgren *et al.*, 1970; Arnoldy, 1971; Frank & Gurnett, 1971; O'Brien & Reasoner, 1971; Whalen & McDiarmid, 1972; Maehlum & Moestue, 1973; Paschmann *et al.*, 1972; Bosqued *et al.*, 1974; Arnoldy *et al.*, 1974] and ions [Hultqvist *et al.*, 1971] have also been used to deduce low-altitude acceleration by a large parallel electric field above the rocket or satellite.

Another class of particle observations involves correlations of fluxes and angular distributions of simultaneously measured protons and electrons on rockets [Mozer & Bruston, 1966b; Reme & Bosqued, 1971] from which the existence of low-altitude acceleration mechanisms has been deduced. High-time-resolution studies of the dispersion in arrival times of auroral electrons have also placed the electron production region at relatively low altitudes [Cummings *et al.*, 1966; Lampton, 1967; Mozer, 1968; Johnstone & Davis, 1974].

The existence of nearly monoenergetic precipitation electron fluxes [Albert, 1967; Evans, 1968; Westerlund, 1969; Albert & Lindstrom, 1970; Chase, 1970; Choy *et al.*, 1971; Evans *et al.*, 1972; Arnoldy *et al.*, 1974] has often been cited as evidence for parallel electric fields above the observation altitude. Evans [1974] has recently placed this interpretation on a firmer basis by showing that such parallel fields would reflect upgoing atmospheric secondaries to produce a downcoming energy spectrum that is in excellent agreement with observations over a wide energy interval.

Recent evidence indicates that the Earth is an extremely strong emitter of kilometric electromagnetic radiation during substorm events [Gurnett, 1974]. The frequency spectrum and angular distributions of this radiation as observed with a dipole antenna on IMP-6 suggest that it is emitted at altitudes below 12 000 km and probably as low as 3000 km. It probably arises from conversion of electrostatic wave energy to electromagnetic waves via a wave-wave interaction in a process that is extremely efficient, since as much as 1% of the precipitated electron energy escapes to space as the kilometric radiation.

These conclusions are reinforced by several observations of low-frequency electrostatic turbulence above subpolar latitudes at altitudes below 1000 km [Heppner, 1969; Maynard & Heppner, 1970; Laaspere *et al.*, 1971; Kelley & Mozer, 1972]. This turbulence produces high-latitude spread-F and scintillations of radio stars and satellites [Kelley, 1972]. Under the assumption of low wave frequency before the Doppler shift associated with the satellite motion, these signals are intense enough to cause wave-particle interactions which reduce the parallel conductivity by an amount sufficient to allow parallel electric fields of 10 mV/m at an altitude of 400 km [Kelley & Mozer, 1972].

3.4. IONOSPHERIC ELECTRIC-FIELD VARIATIONS DURING SUBSTORMS

The most complete study of substorm electric-field variations in the ionosphere near midnight has been carried out by averaging balloon data from 19 substorms [Mozer, 1971]. The results, illustrated in Figure 16, indicate that the westward field component increases prior to the substorm onset, but it does not vary during the expansion phase (which begins at $t = 0$ in Fig. 16). On the other hand the southward field component increases greatly during the expansion phase to drive a westward Hall current that accounts for the negative bay. It is not certain that these electric-field variations affect the majority of the magnetospheric plasma since nonpotential and parallel electric fields are liable to be of greatest importance during such active periods. Thus, while the ionospheric electric-field variations during substorms are well-measured, it remains one of the principal tasks of electric-field research to measure magnetospheric electric fields during substorms and to correlate the plasma dynamics, instabilities, etc. with these fields.

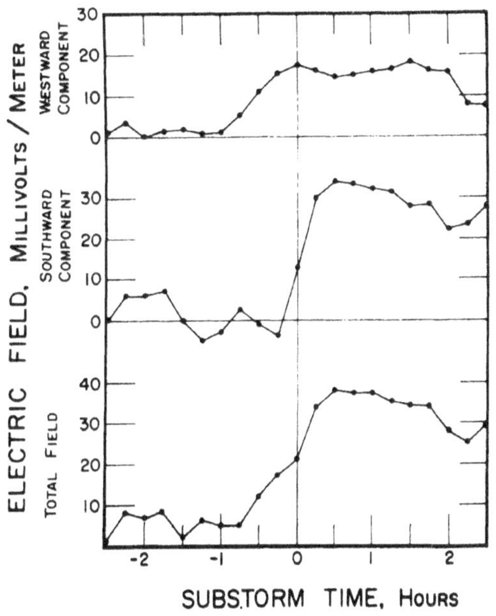

Figure 16. Fifteen-minute averages of electric fields measured on balloons near local midnight during nineteen isolated substorms.

4. Magnetospheric Electric-Field Measurements

Because magnetospheric electric-field data is not as extensive as the ionospheric data, it is not possible to discuss magnetospheric results in terms of their average properties. Instead, several examples of isolated magnetospheric measurements will be presented together with comparisons with ionospheric data when possible, to illustrate the extent of and problems associated with magnetospheric electric-field research.

The most general model of the average equatorial electric field has been obtained from studies of the dispersion in arrival times at the synchronous orbit of different energy plasma particles, since these arrival times are determined by single particle trajectories in the assumed potential fields [McIlwain, 1974]. An example of equatorial potential contours thereby obtained is given in Figure 17. Both the similarities and the differences between this model field and the mapped ionospheric data of Figure 9 are interesting. Whether the differences are real and stem from effects of parallel or nonpotential perpendicular electric fields, or whether they are not real and will disappear as more data is collected, will be decided from *in-situ* measurements obtained during the International Magnetospheric Study.

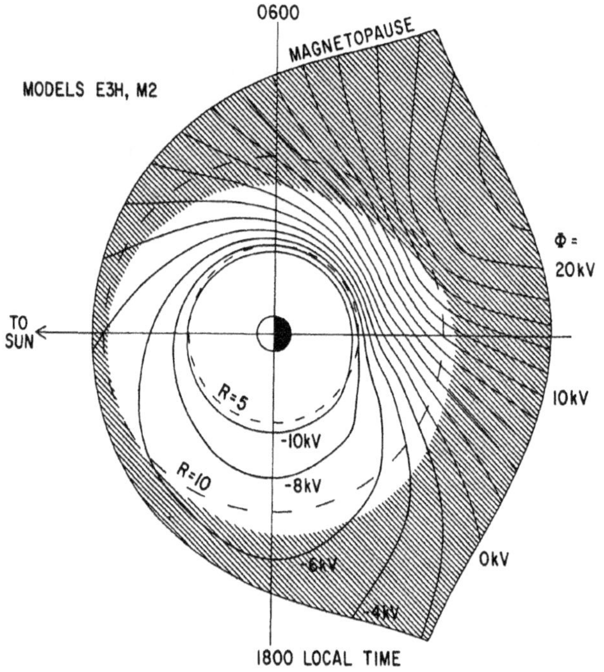

Figure 17. Equipotentials of the equatorial electric field deduced from the energy-dependent dispersion of plasma clouds as observed on the ATS-5 satellite.

An analysis of the frequency dispersion of natural VLF whistler mode waves produced the first measurements of the westward component of the electric field near the equatorial plane [Carpenter & Stone, 1967]. Data for such analyses is obtained predominantly inside or near the plasmasphere where the cold plasma density is sufficiently high to allow duct formation and wave propagation.

A typical example of such a westward equatorial electric-field measurement during a substorm is given in Figure 18 [Carpenter *et al.*, 1972]. The field enhancement before and during the substorm is similar to that observed in the ionosphere (see Fig. 16). The eastward-turning of the electric field following the substorm seems to be a general property of the equatorial but not ionospheric measurements. Whether this is due to decoupling between the equatorial and ionospheric fields, to an L-dependence of the magnetospheric field which manifests itself because the equatorial data is obtained at smaller L-values (see Fig. 10), to shielding by the plasmapause of the magnetospheric electric field [Mozer, 1973b], or to some other cause, must be determined by *in-situ* measurements during the IMS.

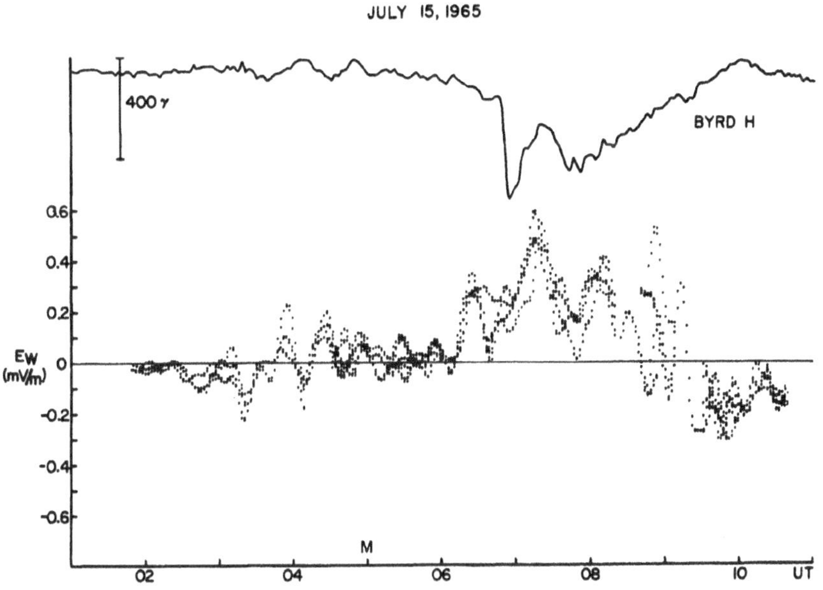

Figure 18. Values of westward electric field near the equatorial plane obtained from analyses of VLF whistler-wave dispersion.

On two occasions, whistler measurements of equatorial fields have been obtained simultaneously with ionospheric field measurements on balloons [Mozer *et al.*, 1973; Mozer *et al.*, 1974]. These two examples are illustrated in Figures 19 and 20, in which the ionospheric and equatorial data are plotted on scales having a ratio given by Equation (4), such that the ionospheric and equatorial curves would be identical if the fields mapped perfectly and the equatorial field were uniform. The agreement between the equatorial and ionospheric data is rather good in view of the possible causes for disagreement enumerated above. Whether such agreement will continue to be obtained or whether certain classes of differences will manifest themselves and lead to an improved physical understanding, are some of the interesting questions to be resolved by future research.

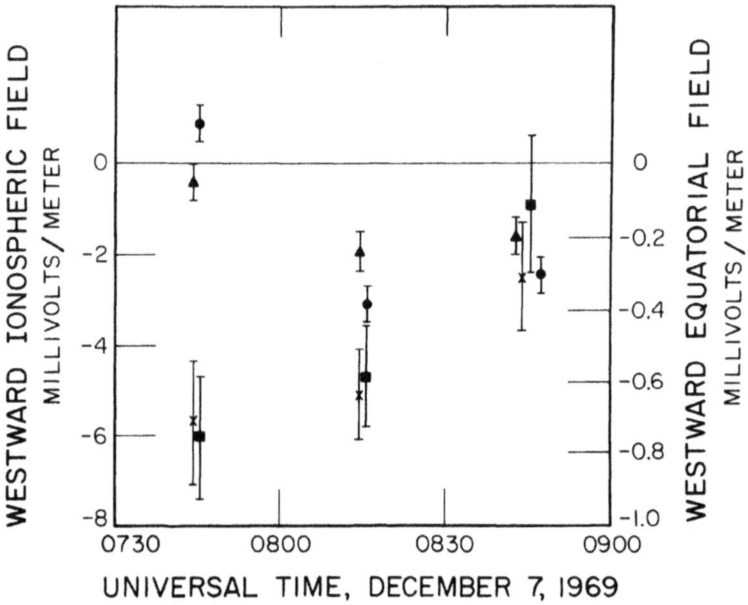

Figure 19. *Thirty-minute averages of the simultaneous balloon and VLF measurements of westward electric fields on 7 December 1969. Circles denote VLF equatorial data at L = 4, triangles VLF equatorial data at L = 2.7, squares northern-ionosphere balloon data, and crosses southern-ionosphere balloon data.*

Figure 20 was obtained in conjunction with a barium release in the deep magnetosphere, which is one of some half-dozen measurements of electric fields above the ionosphere made by observing the motion of a barium cloud in the combined **E** and **B** fields [Haerendel & Lüst, 1970; Adamson *et al.*, 1973; Rieger *et al.*, 1974; Wescott *et al.*, 1974; Haerendel, private communication, 1975]. The cloud associated with the data of Figure 20 was released near 0305 UT near the magnetic equator at a geocentric altitude of about 6 R_e at a time near local midnight. The barium cloud measurements covered the time interval of greatest discrepancy between the balloon and VLF measurements and they agreed with the VLF data [Adamson *et al.*, 1973; Rieger *et al.*, 1974]. On the other hand, the ground magnetometer variations agreed with the sense of the balloon measurement and disagreed with the equatorial data. This example of a pair of selfconsistent ionospheric measurements and a pair of selfconsistent equatorial measurements that disagree with the ionospheric data may be due to decoupling between the ionosphere and magnetosphere. However, further examples are required before this or any other conclusion may be drawn.

Magnetospheric electric-field estimates have also been made by calculating the first

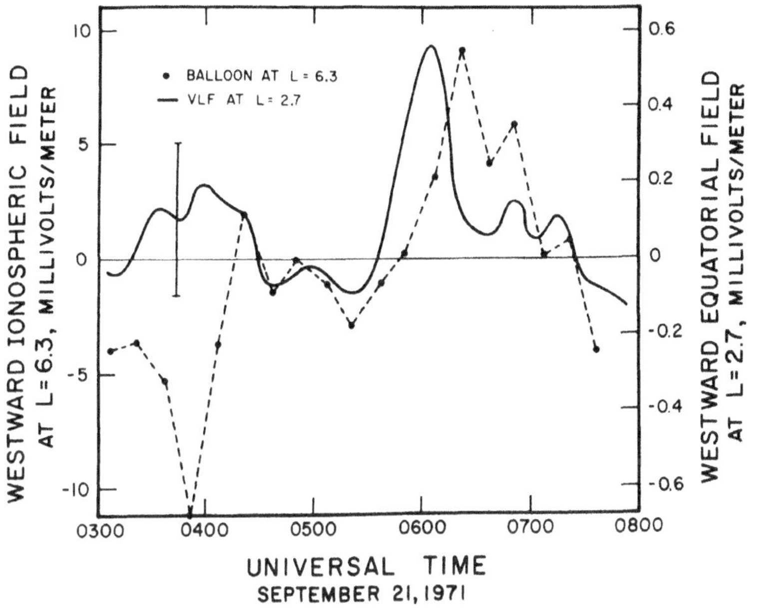

Figure 20. *Comparison of 15 min averages of balloon westward ionospheric field measurements with VLF whistler measurements representing the westward electric field at the equatorial plane at* $L \sim 2.7$.

velocity moment of measured particle distribution functions and obtaining an electric field from this mean velocity and Equation (6) [Freeman, 1968; Freeman *et al.*, 1968; Prakash, 1972; Hones *et al.*, 1972, 1974; Burke & Reasoner, 1973]. In general these estimates have produced fields of a fraction of a millivolt/metre, which is in excellent agreement with earlier described magnetospheric measurements and with extrapolations of the ionospheric data. An example of flow measurements is presented in Figure 21 [Hones, 1975] as obtained from an IMP satellite in the Earth's tail at $18 R_e$. An interesting feature of this and many other examples is that the observed flow is tailward before the expansion phase of a substorm and earthward afterwards. Such a flow variation may be associated with the outward motion of the neutral line during the substorm as the tail field collapses to provide energy for the storm. One problem with much of this data stems from the absence of magnetometers on the Vela satellites and the attendant inability to distinguish between flows parallel and perpendicular to the magnetic field. Since only the perpendicular flow is associated with an electric field by Equation (6), determination of the tail electric-field geometry and complete plasma dynamics must await more complete measurements on IMS satellites.

Another technique for determining electric fields from plasma flow involves observations of particle shadowing by the Moon [Van Allen & Ness, 1969; Anderson &

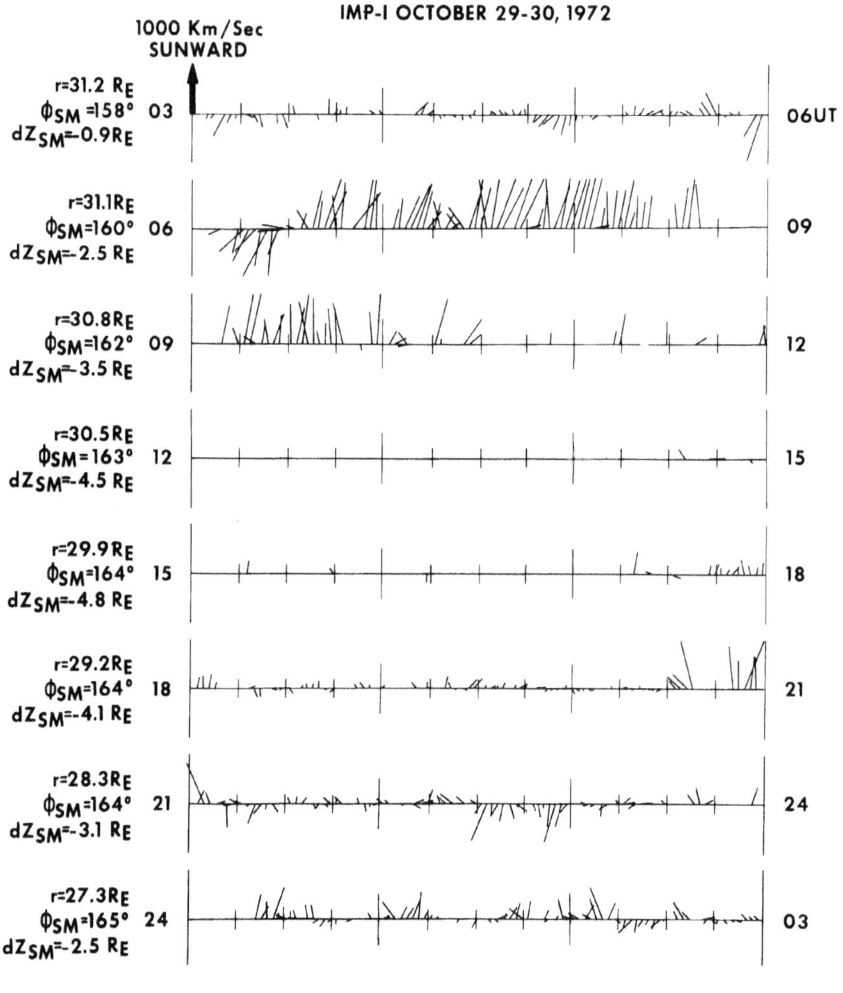

Figure 21. Plasma drift velocities in the tail deduced from low-energy ion measurements on an Imp satellite before and after a substorm that occurred near 0600 UT.

Lin, 1969]. In particular, the electric-field dependance of the return position near the Moon of low-energy solar particles that mirror near the Earth after just missing the Moon on their earthward trajectories may be used to obtain tail electric-field data. Examples of fields deduced by such analyses are presented in Figure 22 [McCoy *et al.*, 1975]. The measured fields generally point in the dawn-to-dusk direction, have magnitudes of a fraction of a millivolt/metre, and produce a cross-tail potential drop of about 40 kV.· These results are in excellent agreement with

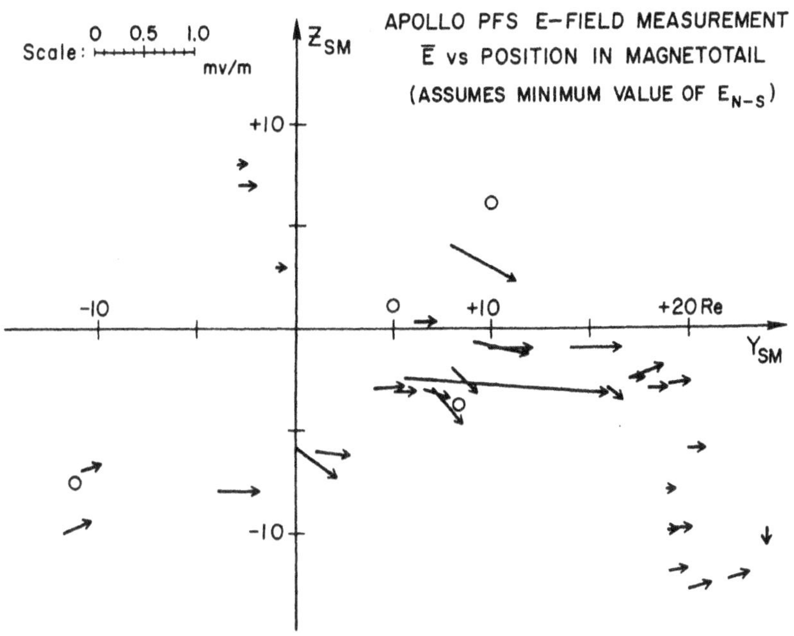

Figure 22. *Electric-field vectors measured in the magnetotail by observing particle shadowing effects of the Moon on a lunar orbiting satellite.*

theoretical expectations and ionospheric data. They should be augmented by further measurements during the IMS which will provide a data base with which to study nonuniformities of the tail field, its variations with interplanetary conditions and substorm activity, relations of the field to particle dynamics, etc.

5. Summary

Some of the outstanding problems of electric-field research to be solved by IMS satellite and other programmes are the determinations of:

1. The existence and morphology of parallel electric fields in the ionosphere and magnetosphere. The instabilities and saturation mechanisms responsible for the anomalous resistivity that produces them. The relation of parallel electric fields to particle acceleration, auroras, etc. The extent and importance of the decoupling of the magnetosphere-ionosphere dynamics by parallel electric fields.
2. The magnitudes and morphologies of nonpotential electric fields in the magnetosphere. Their importance in particle acceleration and other dynamic processes.

The extent of decoupling of the magnetosphere-ionosphere dynamics by non-potential fields.

3. The source of the perpendicular potential electric field in the magnetosphere during active and quiet periods. The extent to which processes other than reconnection play roles. The detailed geometry and physics of the reconnection mechanism, including plasma flow to the reconnection line, geometry of the reconnection line, etc. The extent to which the magnetopause is or is not an electric equipotential.
4. Polarisation effects that distort the magnetospheric electric-field geometry and influence convection and other dynamic processes.
5. The causal relationship between magnetospheric convection and processes such as radial diffusion, the generation of distribution functions that lead to instabilities, particle acceleration, substorms, etc.
6. The relationship of electric fields to plasmaspheric processes. The extent to which the plasmapause is an electric equipotential and to which fields of magnetospheric origin penetrate into the plasmasphere.
7. The average magnetospheric electric field and convection pattern.

6. Acknowledgements

This article was prepared while the author was on sabbatical leave in France. He wishes to thank J.J. Berthelier and the Groupe de Recherche Ionosphérique for their hospitality and assistance during this time.

References

Adamson D., C.L. Fricke, S.A.T. Long, W.F. London & D.C. Ridge, Preliminary analyses of NASA. Optical data obtained in barium ion cloud experiment of September 21, 1971, *J. Geophys. Res.* **78**, 5726 (1973).

Albert, R.D., Energy and flux variations of nearly monoenergetic auroral electrons, *J. Geophys. Res.* **72**, 5811 (1967).

Albert, R.D. & P.J. Lindstrom, Auroral-particle precipitation and trapping caused by electrostatic double layers in the ionosphere, *Science* **170**, 1398 (1970).

Anderson, K.A. & R.P. Lin, Observations of interplanetary field lines in the magnetotail, *J. Geophys. Res.* **74**, 3953 (1969).

Arnoldy, R.L., Rapid fluctuations of energetic auroral particles, *J. Geophys. Res.* **76**, 8258 (1971).

Arnoldy, R.L. & L.W. Choy, Auroral electrons of energy less than 1 keV observed at rocket altitudes, *J. Geophys. Res.* **78**, 2187 (1973).

Arnoldy, R.L., P.B. Lewis & P.O. Isaacson, Field-aligned auroral electron fluxes, *J. Geophys. Res.* **79**, 4208 (1974).

Axford, W.I. & C.O. Hines, A unifying theory of high-latitude geophysical phenomena and geomagnetic storms, *Canad. J. Phys.* **39**, 1433 (1961).

Block, L.P. & D.L. Carpenter, Derivation of magnetospheric electric fields from

whistler data in a dynamic geomagnetic field, *J. Geophys. Res.* **79**, 2783 (1974).
Bosqued, J.M., G. Cardona & H. Rème, Auroral electron fluxes parallel to the geomagnetic field lines, *J. Geophys. Res.* **79**, 98 (1974).
Bruston, P. & U.V. Fahleson, Electric field associated with low altitude particle acceleration in an aurora, submitted to *J. Geophys. Res.*, 1974.
Burke, W.J. & D.L. Reasoner, Observation of plasma flow in the neutral sheet at lunar distance during two magnetic bays, *J. Geophys. Res.* **78**, 6790 (1973).
Carlson, C.W., Rocket measurements of auroral zone low energy charged particles, PhD Thesis, University of California, Berkeley, 1974.
Carpenter, D.L. & K. Stone, Direct detection by a whistler method of the magnetospheric electric field associated with a polar substorm, *Planet. Space Sci.* **15**, 395 (1967).
Carpenter, D.L., Keppler Stone, Jan C., Siren & T.L. Crystal, Magnetospheric electric fields deduced from drifting whistler paths, *J. Geophys. Res.* **77**, 2819 (1972).
Cauffman, D.P. & D.A. Gurnett, Double probe measurements of convection electric fields with the Injun-5 satellite, *J. Geophys. Res.* **76**, 6014 (1971).
Chapman, S., *Vistas in Astron.* **2**, 912 (1956).
Chase, L.M., Energy spectra of auroral zone particles, *J. Geophys. Res.* **75**, 7128 (1970).
Choy, L.W., R.L. Arnoldy, W. Potter, P. Kintner & L.J. Cahill, Jr., Field-aligned particle currents near an auroral arc, *J. Geophys. Res.* **76**, 8279 (1971).
Cummings, W.D., R.E. LaQuay, B.J. O'Brien & M. Walt, Rocket-borne measurements of particle fluxes and auroral light, *J. Geophys. Res.* **71**, 1399 (1966).
Cummings, W.D. & P.J. Coleman, Simultaneous magnetic field variations at the earth's surface and at synchronous equatorial distance, 2. Magnetic storms, *Radio Sci.* **3**, 762 (1968).
Doupnik, J.R., P.M. Banks, M.J. Baron, C.L. Rino & J. Petriceks, Direct measurements of plasma drift velocities at high magnetic latitudes, *J. Geophys. Res.* **77**, 4268 (1972).
Dungey, J.W., Interplanetary magnetic field and the auroral zones, *Phys. Rev. Lett.* **6**, 57 (1961).
Evans, D.S., The observation of near monoenergetic flux of auroral electrons, *J. Geophys. Res.* **73**, 2315 (1968).
Evans, D.S., Precipitating electron fluxes formed by a magnetic field aligned potential difference, *J. Geophys. Res.* **79**, 2853 (1974).
Evans, D.S., B. Maehlum & T. Wede, High latitude observations of field-aligned electron beams (abstract), *EOS Trans. Am. Geophys. Union* **53**, 731 (1972).
Frank, L.A. & D.A. Gurnett, Distributions of plasmas and electric fields over the auroral zones and polar caps, *J. Geophys. Res.* **76**, 6829 (1971).
Freeman, J.W., Observation of flow of low energy ions at synchronous altitude and implications for magnetosphere convection, *J. Geophys. Res.* **73**, 4151 (1968).
Freeman, J.W., C.S. Warren & J.J. Maguire, Plasma flow directions at the magnetopause on January 13 and 14, 1967, *J. Geophys. Res.* **73**, 5719 (1968).
Gold, T., Motions in the magnetosphere of the earth, *J. Geophys. Res.* **64**, 1219 (1959).
Gonzalez, W.D. & F.S. Mozer, A quantitative model for the potential resulting from reconnection with an arbitrary interplanetary magnetic field, *J. Geophys. Res.* **79**, 4186 (1974).

Gurnett, D.A., The earth as a radio source; Terrestrial kilometric radiation, *J. Geophys. Res.* **79**, 4227 (1974).

Haerendel, J. & R. Lüst, Electric fields in the ionosphere and magnetosphere, in Proc. Symposium on Particles and Fields in the Magnetosphere (Ed. B.M. McCormac), pp. 213-228, Springer, New York, 1970.

Haerendel, G., Plasma drifts in the auroral ionosphere derived from barium releases, in Earth's Magnetospheric Processes (Ed. B.M. McCormac) p. 246, D. Reidel, Dordrecht, Netherlands, 1972a.

Haerendel, G., Electric fields and their effects in the ionosphere, in Solar-Terrestrial Physics (Ed. E. Dyer), p. 87, D. Reidel, Dordrecht, Netherlands, 1972b.

Haerendel, G., M.C. Kelley, H. Kappler, U.V. Fahleson & F.S. Mozer, Electric field measurements in a major magnetospheric substorm, for publication in *J. Geophys. Res.*, 1975.

Heikilla, W.J. & D.L. Mathews, Direct observation of low energy electrons in the disturbed ionospheric E-region, *Nature* **202**, 789 (1964).

Hones, E.W. Jr., Recent observations relating to the dynamics and origin of the magnetotail plasma sheet, presented at the Nobel Symposium, Kiruna, Sweden, April 1975.

Heppner, J.P., Magnetospheric convection patterns inferred from high latitude activity, in Atmospheric Emissions (Eds. B.M. McCormac & A. Omholt) p. 251, Van Nostrand Reinhold, New York, 1969.

Heppner, J.P., Polar cap electric field distributions related to the interplanetary magnetic field direction, *J. Geophys. Res.* **77**, 4877 (1972).

Hoffman, R.A. & D.S. Evans, Field aligned electron bursts at high latitudes observed by OGO-4, *J. Geophys. Res.* **73**, 6201 (1968).

Holmgren, L.A., P. Christophersen & W. Riedler, On the pitch angle dependence of auroral electron fluxes in the keV range, *Phys. Norv.* **4**, 85 (1970).

Hones, E.W., Jr., J.R. Asbridge, S.J. Bame, M.D. Montgomery, S. Singer & S.I. Akasofu, Measurements of magnetotail plasma flow made by Vela 4 B, *J. Geophys. Res.* **77**, 5503 (1972).

Hones, E.W. Jr., A.T.Y. Lui, S.J. Bame & S. Singer, Prolonged tailward flow of plasma in the thinned plasma sheet observed at R approximates 18 earth radii during substorms, *J. Geophys. Res.* **79**, 1385 (1974).

Hultqvist, B., H. Borg, W. Riedler & P. Christopherson, Observations of magnetic-field aligned anisotropy for 1 and 6 keV positive ions in the upper ionosphere, *Planet. Space Sci.* **19**, 279 (1971).

Johnstone, A.D. & T.N. Davis, Low-altitude acceleration of auroral electrons during breakup observed by a mother-daughter rocket, *J. Geophys. Res.* **79**, 1416 (1974).

Kelley, M.C., Relationship between electrostatic turbulence and spread-F, *J. Geophys. Res.* **77**, 1327 (1972).

McIlwain, C.E., Substorm injection boundaries, in Magnetospheric Physics (Ed. B.M. McCormac) p. 143, D. Reidel, Holland, 1974.

Kelley, M.C. & F.S. Mozer, A satellite survey of vector electric fields in the ionosphere at frequencies of 10 to 500 Hertz: 1. Isotropic, high-latitude electrostatic emissions, *J. Geophys. Res.* **77**, 4158 (1972).

Kelley, M.C., F.S. Mozer & U.V. Fahleson, Electric fields in the nighttime and daytime auroral zone, *J. Geophys. Res.* **76**, 6054 (1971).

Kindel, J.M. & C.F. Kennel, Topside current instabilities, *J. Geophys. Res.* **76**, 3055 (1971).

Laaspere, T., W.C. Johnson & L.C. Semprebon, Observations of auroral hiss, LHR noise, and other phenomena in the frequency range 20 Hz - 540 kHz on OGO-6, *J. Geophys. Res.* **76**, 4477 (1971).

Lampton, M., Daytime observations of energetic auroral-zone electrons, *J. Geophys. Res.* **72**, 5817 (1967).

Lampton, M.R., R.D. Albert, K.A. Anderson & L.M. Chase, Rocket observations of charged particles in the auroral zone, Paper presented at Birkeland Symposium, Sandefjord, Norway, 1967.

Maehlum, B.N. & H. Moestue, High temporal and spatial resolution observations of low energy electrons by a mother-daughter rocket in the vicinity of two quiescent auroral arcs, *Planet. Space Sci* **21**, 1957 (1973).

Mahon, H.P., M. Smiddy & R.C. Sagalyn, Parallel B electric fields between 120 km and 165 km, *EOS Trans. Am. Geophys. Union* **54**, 395 (1973).

Mathews, D.L. & T.A. Clark, Simultaneous observations of electron fluxes, ionization, and luminosity in an aurora, *Canad. J. Phys.* **46**, 201 (1968).

Maynard, N.C. & J.P. Heppner, Variations in electric fields from polar orbiting satellites, in Particles and Fields in the Magnetosphere (Ed. B.M. McCormac) p. 247, D. Reidel, Dordrecht, Netherlands, 1970.

McDiarmid, I.B., D.C. Rose & E. Budzinski, Direct measurement of charged particles associated with auroral zone radio absorption, *Canad. J. Phys.* **139**, 1888, 1961.

McCoy, J.E., R.P. Lin, R.E. McGuire, L.M. Chase & K.A. Anderson, Magnetotail electric fields observed from lunar orbit, Preprint, 1974.

Mozer, F.S., Rapid variations of auroral particle fluxes, *J. Geophys. Res.* **73**, 999 (1968).

Mozer, F.S., Electric field mapping in the ionosphere at the equatorial plane, *Planet. Space. Sci.* **18**, 259 (1970).

Mozer, F.S., Origin and effects of electric fields during isolated magnetospheric substorms, *J. Geophys. Res.* **76**, 7595 (1971).

Mozer, F.S., Analyses of techniques for measuring DC and AC electric fields in the magnetosphere, *Space Sci. Rev.* **14**, 272 (1973a).

Mozer, F.S., Electric fields and plasma convection in the plasmasphere, *Rev. Geophys. Space Phys.* **11**, 755 (1973b).

Mozer, F.S. & P. Bruston, Observation of the low altitude acceleration of auroral protons, *J. Geophys. Res.* **71**, 2201 (1966a).

Mozer, F.S. & P. Bruston, Auroral-zone proton-electron anticorrelations, proton angular distributions and electric fields, *J. Geophys. Res.* **71**, 4461 (1966b).

Mozer, F.S. & P. Bruston, Electric field measurements in the auroral ionosphere, *J. Geophys. Res.* **72**, 1109 (1967).

Mozer, F.S.. & U.V. Fahleson, Parallel and perpendicular electric fields in an aurora, *Planet. Space Sci* **18**, 1563 (1970).

Mozer, F.S. & D.L. Carpenter, Balloon and VLF whistler measurements of electric fields, equatorial electron density, and precipitating particles during a barium cloud release in the magnetosphere, *J. Geophys. Res.* **78**, 5736 (1973).

Mozer, F.S., W.D. Gonzalez, F. Bogott, M.C. Kelley & S. Schutz, High latitude electric fields and the three-dimensional interaction between the interplanetary and terrestrial magnetic fields, *J. Geophys. Res.* **79**, 56(1974).

Mozer, F.S. & P. Lucht, The average auroral zone electric field, *J. Geophys. Res.* **79**, 1001 (1974).

Mozer, F.S., R. Serlin, D.L. Carpenter & J. Siren, Simultaneous electric field measurements near L = 4 from conjugate balloons and whistlers, *J. Geophys. Res.* **79**, 3215 (1974).

O'Brien, B.J. & D.L. Reasoner, Measurements of highly collimated short-duration bursts of auroral electrons and comparison with existing models, *J. Geophys. Res.* **76**, 8258 (1971).

Ogilvie, K.W., Auroral electron energy spectra, *J. Geophys. Res.* **73**, 2325 (1968).

Papadapoulos, K. & T. Coffey, Anomalous resistivity in the auroral plasma, *J. Geophys. Res.* **79**, 1558 (1974).

Paschmann, G., R.G. Johnson, R.D. Sharp & E.G. Shelley, Angular distribution of auroral electrons in the energy range 0.8 to 16 keV, *J. Geophys. Res.* **77**, 6111 (1972).

Prakash, A., Detection of earthward flow of KeV protons in the geomagnetic tail at lunar distances, *J. Geophys. Res.* **77**, 5633 (1972).

Reasoner, D.L. & C.R. Chappel, Twin payload observations of incident and back-scattered auroral electrons, *J. Geophys. Res.* **78**, 2176 (1973).

Reme, H. & J.M. Bosqued, Evidence near the auroral ionosphere of a parallel electric field deduced from energy and angular distributions of low-energy particles, *J. Geophys. Res.* **76**, 7682 (1971).

Rieger, E., G. Haerendel, H. Kappler, J. Lordl, R. Lust & H. Neuss, Magnetospheric convection derived from barium cloud motions at 5 R_e altitude, Max Planck Institute Report, 1974.

Russell, C.T. & G. Atkinson, Comments on a paper by J.P. Heppner, 'Polar cap electric field distributions related to interplanetary magnetic field direction', *J. Geophys. Res.* **78**, 4001 (1973).

Ungstrup, E., H. Obayashi, I.B. Iverson & A. Bahnsen, Parallel and perpendicular electric fields in a proton aurora, *EOS Trans. Am. Geophys. Union* **53**, 729 (1972).

Van Allen, J.A. & N.F. Ness, Particle shadowing by the moon, *J. Geophys. Res.* **74**, 71 (1969).

Wang, C.S. & J.S. Kim, The effect of an electric field induced by a time-dependent ring current on the particle drift motion, *Planet. Space Sci.* **20**, 2039 (1972).

Wescott, E.M., E.P. Rieger, J.C. Stenbaek-Nielsen, T.N. Davis, H.M. Peck & P.J. Bottoms, L = 1.24 conjugate magnetic field line tracing experiments with barium shaped charges, *J. Geophys. Res.* **79**, 159 (1974).

Westerlund, L.H., The auroral electron spectrum extended to 45 eV, *J. Geophys. Res.* **74**, 351 (1969).

Whalen, B.A. & I.B. McDiarmid, Observations of magnetic-field-aligned auroral-electron precipitation, *J. Geophys. Res.* **77**, 191 (1972).

DISCUSSION

C.T. Russell: (i) The ionosphere may be set into motion from both above and below. What is your estimate of the effects of neutral winds in driving ionospheric convection above the polar cap?
(ii) What percentage of the time is the electric field in the polar cap from dusk to dawn? Is this consistent with your simple model of reconnection?

F. Mozer: (i) Figure 10 of my manuscript, which I did not have time to present in my talk, illustrates the fact that the average electric field produced by the ionospheric dynamo at high latitudes is about an order-of-magnitude smaller than is the electric field of magnetospheric origin. Thus, fields from motions below the ionosphere are generally not of prime importance in the auroral zone and polar cap. An exception to this statement would occur if a high level of magnetospheric convection, lasting for hours, set the high-latitude neutral upper atmosphere into motion, after which the magnetospheric convection stopped. The energy in the neutral atmospheric motion would then continue to drive magnetospheric convection for a few hours until the neutral atmosphere slowed down. This would be an example of magnetospheric electric fields produced by motions below.
(ii) As may be seen from the experimental data of Figure 14, there are several periods of a few hours duration each during the 2½ day interval when the polar-cap electric field was in the dusk-to-dawn direction. These periods all occurred when the reconnection rate was small, as evidenced by the model calculation of Figure 14 whose inputs are the measured interplanetary magnetic-field components and solar-wind velocity. My interpretation of these dusk-to-dawn electric fields is two-fold. First, they offer evidence that all other convection mechanisms are small compared to reconnection because the dawn-to-dusk component of the **E**-field becomes negative when reconnection is turned off. Second, the negativeness may be associated with the relaxation of the magnetosphere toward its lower energy steady state that applies in the absence of reconnection. I do not believe that one can have a dusk-to-dawn polar-cap field in the steady state because the Poynting flux would be out of the magnetosphere into the interplanetary medium and thus the magnetosphere would energise the solar wind.

D. Jones: It was mentioned that there was evidence from kilometric radiation for field-aligned electric fields. Please expand on this statement.

F. Mozer: One of the problems of anomalous resistivity theories is that they require a very large energy deposition in the anomalous resistivity region. From the work of Kindel and Kennel, this region should be in the general area from which the kilometric radiation is being emitted. Thus, the existence of kilometric radiation, which carries off up to 10^9 W of power, helps ease the burdens of those who like to believe in anomalous resistivity.

K.G. Fälthammar: It is worth emphasising that anomalous resistivity may not be the only mechanism for maintaining parallel electric fields. A second possible mechanism (also related to field-aligned currents) is the electric double layer. With this mechanism, unlike the anomalous resistivity, there is no problem of accounting for energy dissipation, because the energy goes primarily into keV particles that deposit this energy in the denser layers of the ionosphere. A third possibility is that of different pitch-angle distributions of positive and negative particles. This requires an essentially collisionless plasma and a mirror-type magnetic field, and therefore would apply in the magnetosphere rather than the ionosphere. This mechanism may maintain a parallel electric field even in the absence of a net current.

F. Mozer: You are certainly correct, and your comments only further emphasise the necessity for co-ordinated parallel electric field, current, and wave measurements.

MAGNETOSPHERIC CONVECTION INDUCED BY INTERPLANETARY MAGNETIC-FIELD VARIATIONS

K. Maezawa

Institute of Space and Aeronautical Science, University of Tokyo, Japan
Presented by T. Obayashi

Abstract

The dependence of the polar-cap magnetic disturbance on the polarity and magnitude of the Z-component of the interplanetary magnetic field is investigated by regression analysis using hourly values. The Svalgaard-Mansurov effect has been eliminated assuming a linear dependence on the Y-component of the interplanetary field. It is shown that as the northward component of the interplanetary magnetic field increases, a characteristic current system appears in the polar cap. This current system is composed of two current vortices in the dayside polar cap, one in the pre-noon sector and the other in the afternoon sector. The current direction is antisunward in the central polar cap, suggesting that the *sunward* plasma convection is induced in the polar cap. Current intensity is strongest at $\phi_M \sim 84°$ around the noon meridian.

We propose that the tail field lines are reconnected with the northward interplanetary field on the polar side of the dayside polar cusp, and as a result plasma convection is induced which is closed within the high-latitude magnetosphere. On the other hand, when the interplanetary magnetic field is directed southward, a transpolar current sheet appears covering the whole polar cap ($\phi_M \geqslant 77.5°$). The characteristics of this transpolar current sheet are as follows: (i) On the dayside, especially around noon, the direction of the current is roughly consistent with the Hall current direction expected from the dawn-to-dusk electric field, while at night it is considerably skewed from the noon-midnight meridian. Skewing of the current direction can be explained by the effect of the currents external to the ionosphere. (ii) Strength of the current is almost linearly dependent on B_z when the interplanetary magnetic field is directed southward ($\theta < -45°$). However, the current intensity is also a function of the magnitude of B_y, this being apparent when $B_z \sim 0$. This indicates that the dayside reconnection rate is a function of $|B_y|$ as well as of B_z, and information is derived about the applicability of the three-dimensional reconnection model in the presence of finite $|B_y|$.

Knott & Battrick (eds.), The Scientific Satellite Programme During the International Magnetospheric Study, 133–166.
All Rights Reserved. Copyright © 1976 by D. Reidel Publishing Company, Dordrecht-Holland.

1. Introduction

In Dungey's reconnection model of the magnetosphere [Dungey, 1961], reconnection of geomagnetic field lines with the southward interplanetary field produces antisunward plasma convection in the high-latitude magnetosphere, and the associated electric field drives an ionospheric Hall current whose direction is sunward in the polar cap and antisunward in the lower latitude zone. According to this model, the magnitude of the magnetospheric convection is expected to be dependent on the southward component of the interplanetary magnetic field. In Fairfield & Cahill [1966] and other earlier studies of the IMF (interplanetary magnetic field) effect on the magnetosphere, the magnetic activity in the auroral zone has been compared with the IMF data. Statistical analyses by Arnoldy [1971] have shown that, of the various IMF parameters, the time integral of the southward component of the interplanetary magnetic field is best correlated with the AE index of the auroral zone magnetic activity, which is a measure of energy released in the auroral zone mainly during the expansion phase of substorms.

It has come to be noted, however, that the occurrence of individual substorms is not necessarily associated with a specific feature of the interplanetary magnetic field [Akasofu *et al.*, 1973b; Kamide & Akasofu, 1974]. Sudden impulses sometimes seem to trigger substorms [Kokubun, 1971; Burch, 1972a]. Furthermore, it has been noted that the substorm expansion phase is often preceded by a magnetic energy increase in the magnetotail. These observations indicate that the substorm expansion phase is not a direct manifestation of interaction between the interplanetary magnetic field and the magnetosphere, but is caused by a release of energy that has been stored in the magnetotail.

Magnetic variations in the polar cap, on the other hand, are expected to be under the direct influence of the convection-driving force, because the polar cap is connected via geomagnetic field lines to the magnetospheric boundary region where the convection driving mechanism operates; we can infer the direction of convective motion in the boundary region from the equivalent current direction of magnetic disturbances in the polar cap. Direct influence of the interplanetary magnetic field on the polar-cap magnetic variations has been noted by Nishida [1968b]. He found that world-wide geomagnetic fluctuations with a time scale of a few hours, designated DP 2 [Nishida, 1968a, b], are often correlated with fluctuations in the B_z (north-south) component of the interplanetary magnetic field with a short (10 – 20 min) delay. The correlation is more prominent in the polar cap, and the current pattern is consistent with Dungey's reconnection model.

It has recently become known [Svalgaard, 1968; Mansurov, 1969; Friis-Christensen *et al.*, 1972; Svalgaard, 1973] that the B_y (azimuthal) component of the interplanetary field also influences the magnetic field in the polar cap. Detailed statistics [Svalgaard, 1973] have shown that the positive B_y component produces counterclockwise zonal current in the polar ionosphere, while the negative B_y component causes clockwise zonal current. This phenomenon (Svalgaard-Mansurov effect) also indicates that the polar-cap current is directly controlled by the interplanetary magnetic field.

In this paper we shall attempt a detailed statistical analysis of the dependence of polar-cap magnetic variations on the interplanetary magnetic field, in order to obtain further information on the reconnection process and the associated magnetospheric convection. In particular, we shall concentrate on the following three problems.

The first and most important point is to clarify the difference in the response of the polar-cap current to the interplanetary magnetic field with different polarities of the B_z component. It has been noted that the intensity of the auroral electrojet is closely related to the magnitude of the *southward* component of the interplanetary magnetic field, but is almost insensitive to the northward component. A decrease in the total magnetic flux of the dayside magnetosphere, as revealed by the dayside magnetopause observations [Aubry *et al.*, 1970; Maezawa, 1974] as well as low-altitude polar cusp observations [Burch, 1972b, 1973], occurs only when the interplanetary magnetic field is southward. While this seems to be consistent with the Dungey model, there is mounting evidence which suggests that the geomagnetic field lines are always connected to the interplanetary ones. The effect of the B_y component of the interplanetary magnetic field can be seen irrespective of the degree of geomagnetic activity. Observation of solar protons and electrons indicates permanent connection between the polar cap and interplanetary magnetic field lines [Morfill & Scholer, 1972, 1973; Fennell, 1973]. Furthermore, Iwasaki [1971] found that when the interplanetary magnetic field is northward, abnormally directed currents (i.e. currents directed almost opposite to the DP 2 or DS current) tend to appear in the polar cap. All these facts suggest that the interplanetary magnetic field interacts with the magnetosphere even when it is directed northward. This contrasts with the aforementioned selective dependence of substorm activity on the southward interplanetary field and suggests that the interaction mechanism in the case of a northward interplanetary magnetic field is different from that in the case of a southward interplanetary magnetic field. We shall clarify this point by separating the cases corresponding to the northward polarity of B_z and obtaining the pattern of current associated with the northward component of the interplanetary field.

The second point is concerned with the absolute direction of convection associated currents. Current directions in the polar cap obtained so far from geomagnetic variations (e.g. those of DS, DP 1 and DP 2 current systems) generally point 20° – 60° westward of the Sun. It has been noted [Heppner *et al.*, 1971] that this direction is not in full agreement with the dawn-to-dusk electric field, which is expected from the reconnection model and has been observed by recent satellite and balloon measurements [Heppner, 1972a, b; Mozer *et al.*, 1973]. However, in order to discuss the absolute direction of current associated with a particular component of the interplanetary magnetic field, we should extract statistically only the part dependent on that component from geomagnetic variations. This will be done in this paper using regression analysis between the interplanetary and geomagnetic data.

The third point is concerned with the functional form of the dependence of the polar-cap current on the interplanetary magnetic field. Analyses of the AE index have shown that it is linearly correlated with the southward component of the interplanetary magnetic field. However, recent theoretical models of magnetic field reconnection suggest that the rate of dayside reconnection should be also dependent on $|B_y|$ [Nishida & Maezawa, 1971; Sonnerup, 1974; Gonzalez & Mozer, 1974]. To examine the applicability of the basic concept of these models, we will investigate how the convection magnitude depends on the magnitudes of the B_y and B_z components of the interplanetary magnetic field, after first eliminating the Svalgaard-Mansurov effect.

2. Method of Analysis

2.1. DATA AND CO-ORDINATE SYSTEMS

Polar-cap magnetic-field data used in this analysis are hourly values from stations at Thule (TH), Resolute Bay (RB), Mould Bay (MLB), and Godhavn (GO) in the northern polar cap. Corrected geomagnetic latitudes for these stations are 87.7°, 84.3°, 80.5°, and 77.6°, respectively [Hakura, 1965]. For the sake of reference, universal times corresponding to the local geographic and geomagnetic noon in summer at each station [Montbriand, 1970] are shown in Table 1. The magnetic local time (MLT) at each station is defined here by referring to this local magnetic noon. Only the summer data (July and August 1966; May, June and August 1968) are analysed because the current intensity in the polar cap is generally higher and the nature of the current easier to recognise during the summer months. Interplanetary magnetic data used are hourly values from the Explorer satellites 28, 33 and 35.

TABLE 1
Geomagnetic and geographic noons at the observatories used in the analysis

	Thule	Resolute Bay	Mould Bay	Godhavn
Geomagnetic Noon (UT)	14.0	20.0	23.0	14.0
Geographic Noon (UT)	16.5	18.5	20.0	15.5

To represent components of magnetic-field perturbation vectors observed at a fixed MLT, three co-ordinate systems are used in the following analysis, depending upon the purpose. The first system, in which the vector magnetic field is represented as (X, Y, Z) is defined locally referring to the direction of the corrected geomagnetic pole seen from each station; the X-axis points to the pole, the Z-axis is vertically downward, and the Y-axis completes the right-handed three-dimensional Cartesian co-ordinate system. In this system, the X-component of magnetic-field perturbation corresponds to a zonal ionospheric equivalent current, while the Y component represents a transpolar equivalent current flowing perpendicular to parallels of latitude. The second system in which the vector magnetic field is represented as (X_s, Y_s, Z) refers to the solar direction; the Z-axis is vertically downward, the X_s-axis lies horizontally in the plane defined by the solar direction and the Z-axis, and the Y_s-axis completes the right-handed Cartesian system. In this system, the X_s component of magnetic field perturbation corresponds to a dawn-to-dusk current, and the Y_s component to a current parallel to the Sun-Earth direction. In the third system, in which the magnetic-field vector is represented as $(H_\parallel, H_\perp, Z)$, the $H_\parallel$-axis is taken to be parallel to the direction of the local horizontal field disturbance induced by the southward interplanetary magnetic field. (This direction will be determined through the regression analysis between the interplanetary and geomagnetic data.) In this system, the $H_{\parallel}$ component represents the magnitude of the current induced by the southward interplanetary magnetic field.

In order to represent the spatial distribution of currents, the magnetic latitude/magnetic local time co-ordinate system is used, which is thought to be most appropriate for the present purpose.

Interplanetary magnetic field is expressed using the solar-magnetospheric co-ordinate system. We denote three components of the interplanetary magnetic field in this system by B_x, B_y and B_z, respectively.

2.2. ELIMINATION OF B_y DEPENDENCE

In order to clarify the dependence of polar-cap magnetic-field variations on the B_y and B_z components of the interplanetary magnetic field, linear regression analysis is utilised. Since the geomagnetic field is found to respond to the interplanetary field with a time delay of the order of 30 min, geomagnetic hourly values centred at the time of 30 min after the original hourly data are obtained by interpolation of the original data and used for the subsequent analysis. The advantages of using regression analysis are as follows:

(i) We can separate a particular type of magnetic-field disturbance from other disturbances superposed on it, provided that they are of statistically independent origin. For example, when we examine the dependence of the polar-cap current on one particular component of the interplanetary field, the effects of all those physical processes which are not related to that particular interplanetary field component are averaged out statistically.

(ii) Linear regression coefficients do not depend on the base line from which the geomagnetic field is scaled. This enables us to determine the absolute direction of the perturbation field related to a specific component of the interplanetary magnetic field.

In particular, the presence of S_q^0 does not affect our result due to (ii) and the B_y-dependent part (Svalgaard-Mansurov effect) can be separated from the effect of B_z due to (i) by virtue of the regression analysis. In some cases, however, mutual correlation is found between B_y and B_z. In such cases, B_y and B_z could not be regarded as statistically independent. Hence in the following we employ a more sophisticated method by taking linear partial-correlation coefficients. (Linear dependence of the Svalgaard-Mansurov effect on B_y has been noted by Friis-Christensen *et al.* [1972]). First, we divide the whole data set into two groups, one for $B_z > 1\gamma$, and the other for $B_z < 1\gamma$. This is done because the degree of B_y dependence may depend on the polarity of the B_z component of the interplanetary magnetic field. (The threshold value of 1γ is chosen in accordance with the division made in the later analysis.) We then calculate partial regression coefficients between B_y and each component of the geomagnetic field ('partial', in the sense that the effect of the mutual relation between B_y and B_z is eliminated using partial correlation analysis.) These partial regression coefficients calculated for each MLT and for each station are shown in Figure 1. The left panel shows the regression coefficients for the geo-

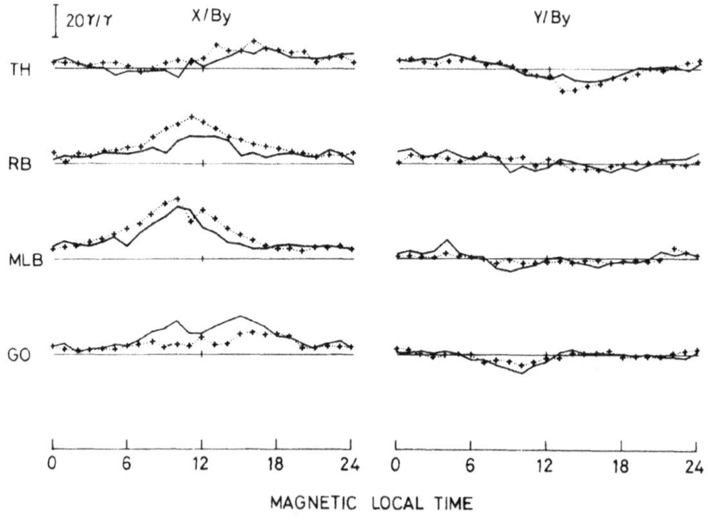

Figure 1. *Linear regression coefficients between B_y (solar magnetospheric y-component of the interplanetary magnetic field) and the geomagnetic horizontal components, plotted against magnetic local time. Dotted lines represent the case $B_z < 1\gamma$.*

magnetic X-component and the right panel for the geomagnetic Y-component. In both panels, dotted lines represent the case for $B_z > 1\gamma$, and solid lines the case for $B_z < 1\gamma$. Results for different stations are shown from top to bottom in order of decreasing latitude. We note the following two tendencies:

(i) For both $B_z > 1\gamma$ and $B_z < 1\gamma$, large values of regression coefficient are seen only for the X-component and for daytime hours. Correlation is peaked at Resolute Bay and Mould Bay around noon.

(ii) Although the forms of MLT dependence of the regression coefficients are roughly the same for the two ranges of B_z, the magnitude of coefficients for X is higher for the $B_z > 1\gamma$ case than for the $B_z < 1\gamma$ case at Thule, Resolute Bay, and Mould Bay, while this tendency is reversed at Godhavn.

Point (i) indicates that the ionospheric current corresponding to the B_y dependence is approximately zonal and is stronger on the sunlit side of the polar cap. It is known that these points characterise the Svalgaard-Mansurov effect [Svalgaard, 1968; Mansurov, 1969; Friis-Christensen *et al.*, 1972; Svalgaard, 1973; Kawasaki *et al.*, 1973]. Point (ii) implies that the zone of strong current on the dayside shifts from higher to lower latitudes as the interplanetary magnetic field becomes more southward [Friis-Christensen & Wilhjelm, 1975]. Using these regression coefficients separately for two different ranges of B_z, we eliminate the Svalgaard-Mansurov effect from the geomagnetic data. The resultant data set is used throughout the following analysis.

3. Effect of the Southward Component of the Interplanetary Magnetic Field

As will be shown in the following analysis, the nature of the solar wind – magnetosphere interaction is different for different polarities of the north-south component of the interplanetary magnetic field. Accordingly, we separate the analysis into two cases, namely $B_z > 1\gamma$ and $B_z < 1\gamma$.

In this section, the case of $B_z < 1\gamma$ is examined. Adequacy of the choice of the threshold value $B_z = 1\gamma$ rather than $B_z = 0\gamma$ will be discussed later.

3.1. CURRENT AND CONVECTION PATTERN

First we analyse the pattern of the current system induced in the polar cap by the southward interplanetary magnetic field. Figure 2 shows linear regression coefficients between B_z and the geomagnetic field components X and Y for the range of $B_z < 1\gamma$. In this figure, the sign of the regression coefficients is taken as positive when X or Y increases with decreasing B_z. It is seen that for both X and Y the regression coefficients show roughly sinusoidal variation as a function of magnetic local time at each station.

If the MLT dependence of the regression coefficients is exactly sinusoidal, it means that the corresponding current has a same, fixed direction at all local times. In particular, if we assume that a uniform dawn-to-dusk electric field is set up in the polar cap by the southward interplanetary field through the reconnection process, and that the ionospheric current is mainly the Hall current, the phase of the sinusoidal wave should be such that at noon the coefficient for X is zero and the coefficient for Y is a maximum. With this expectation in mind, the characteristics of Figure 2 can be summarised as follows:

(i) Within 1 h of the local magnetic noon, the regression coefficient for X vanishes and that for Y takes the maximum value at all stations, although Y at Mould Bay is slightly exceptional.

(ii) However, in dark sectors, the phase of the regression curves is advanced by 3 ~ 5 h with respect to the local magnetic midnight. For example, the minimum values of $Y/(-B_z)$ and the night-time modes of $X/(-B_z)$ occur at 20 – 21 MLT on average.

Point (ii) means that the current direction is not uniform in the polar cap. To visualise this, the spatial distribution of the equivalent current is shown in Figure 3 using magnetic latitude / magnetic local time co-ordinates. In this figure, the directions of current vectors are reversed to give the direction of convection. The length of each arrow is proportional to the magnitude of the magnetic-field perturbation caused by a 1γ reduction in B_z.

It is seen that on the dayside, particularly around the noon meridian, convection flow vectors are distributed roughly symmetrically with respect to the Sun-Earth line, while on the nightside their directions are considerably skewed from the Sun-Earth line. For example, in the noon meridian, the apparent flow direction is within 15° of the Sun-Earth line except at Mould Bay, while at pre-midnight hours, the flow direction is deflected westward by as much as 45° at all stations. It is noted

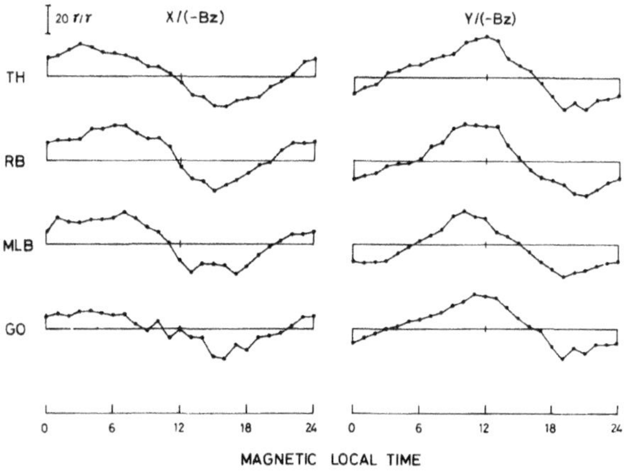

Figure 2. *Linear regression coefficients between B_z and geomagnetic horizontal components X and Y, plotted against magnetic local time at each station for $B_z < 1\gamma$. The left panel is for the X component, and the right for the Y component. The scale shown at the top (left) is common to X and Y.*

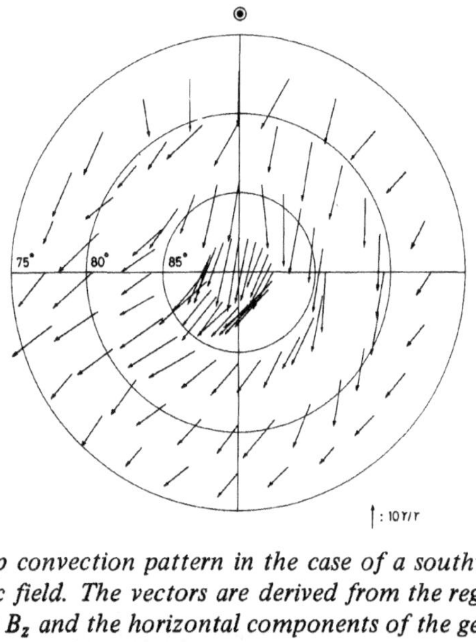

Figure 3. *Polar-cap convection pattern in the case of a southward interplanetary magnetic field. The vectors are derived from the regression coefficients between B_z and the horizontal components of the geomagnetic field for the case $B_z < 1\gamma$. The length of the vector shown at the bottom right corresponds to a 10γ increase in the horizontal geomagnetic field produced by a 1γ decrease in B_z.*

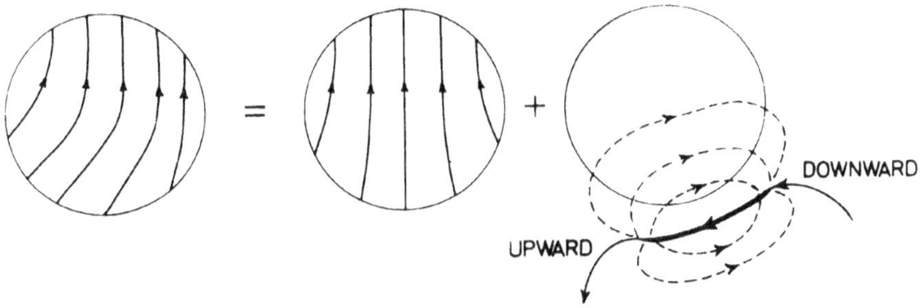

Figure 4. *Interpretation of the current pattern that appears in the case of the southward interplanetary magnetic field. The observed current system (left) is interpreted to be the superposition of the ionospheric Hall current (centre) and the equivalent current of the three-dimensional current system associated with the auroral electrojet (right).*

that the 10° ~ 15° deflection from the Sun-Earth line found around the noon meridian is in agreement with the calculation of the solar-wind flow deflection at the Earth's bow shock [Walters, 1964], when the aberration angle of the solar wind is taken into account.

The above 'bending' of the current direction is a rather unexpected result, since the current direction in the polar cap is often considered to be uniform in both observational and theoretical studies. What causes this skewing of the equivalent current direction?

It is noted in this connection that the direction of the equivalent current in the polar cap derived earlier for various modes of the disturbance field deviates generally westward from the sunward direction. This feature is in apparent contradiction [Heppner *et al.*, 1971] with the idea that the current consists of the ionospheric Hall and Pedersen currents driven by the dawn-to-dusk electric field. This apparent disagreement can be removed [Vasyliunas, 1970; Lyatsky *et al.*, 1974] by taking into account the effect of the field-aligned current, which flows when the ionospheric current is driven by the external electric field. If the ionospheric conductivity is not uniform, the effect of the field-aligned current on the ground magnetic field is not completely offset by that of the ionospheric Pedersen current [Fukushima, 1969]. Assuming a ring-shaped high-conductivity belt along the auroral zone, Vasyliunas [1970] showed that the equivalent current direction in the polar cap is deflected westward at every point in the polar cap, provided that the conductivity distribution has no longitudinal dependence.

In the light of the above theoretical studies, the present result can be interpreted as follows. The skewing of the equivalent current noted on the nightside is probably caused by the effect of the nightside field-aligned current associated with the westward auroral electrojet. In recent theoretical and observational models [Atkinson, 1967; McPherron *et al.*. 1973], the westward electrojet is connected to the tail current via magnetospheric field-aligned currents, which flow from the magnetotail into the ionosphere during post-midnight hours, and flow out of the ionosphere into the tail during pre-midnight hours. Generally speaking, the effect of the field-aligned current connected to a high-conductivity belt in the ionosphere is seen on the ground as though there is an ionospheric return current of the strong electrojet in the belt [Fukushima, 1969]. In the present case, this apparent return current would be directed roughly from 20 h to 8 h, since the centre of the electrojet is usually located a few hours after the midnight meridian. The modification of the equivalent current due to this effect is in good agreement with the observational result, as shown schematically in Figure 4.

On the other hand, the above observational result suggests that the magnetic effect of the field-aligned currents, that necessarily exist to maintain the dawn-to-dusk electric field, is largely masked in the dayside polar cap at the ground level. This can be explained if the conductivity enhancement in the dayside auroral oval is not so great as in the nightside oval, and the magnetic effect of the dayside field-aligned current is largely masked by that of the Pedersen current in the ionosphere.

To support the above contention that the nightside equivalent current in the polar cap is significantly affected by the presence of the electrojet-associated field-aligned currents, cross-correlation analysis between B_z and ground Y is performed. In Figure 5, the left panel shows the cross-correlations at Resolute Bay for night hours (23 – 2 MLT) and the right panel shows those for day hours (11 – 14 MLT). It is seen that the maximum correlation value occurs with the time lag of 30 min – 1 h for the hours of darkness, while for daylight hours the time lag is 0 – 30 min. The time lag of 0 – 30 min is in accordance with the analysis of geomagnetic DP 2 fluctuations [Nishida, 1968b; Nishida & Maezawa, 1971], which are believed to be a

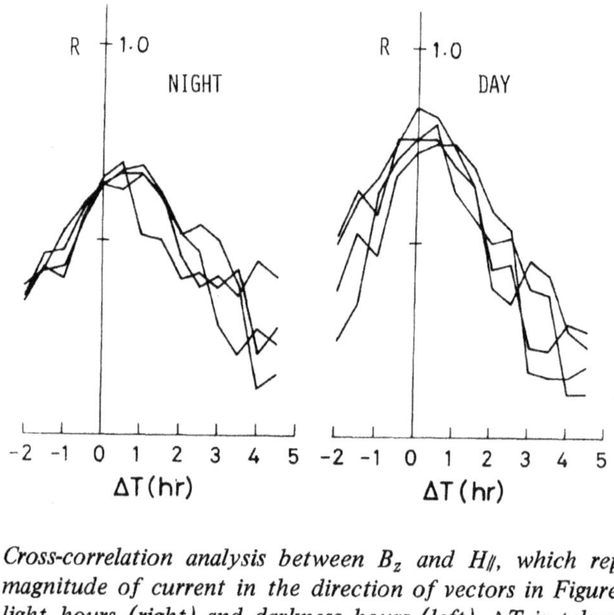

Figure 5. *Cross-correlation analysis between B_z and $H_{//}$, which represents the magnitude of current in the direction of vectors in Figure 3, for daylight hours (right) and darkness hours (left). ΔT is taken as positive when $H_{//}$ lags behind B_z.*

direct manifestation of the convection electric field. On the other hand, the time delay of 30 min ~ 1 h has been obtained in the correlational analysis between the interplanetary B_z and auroral electrojet activity [Arnoldy, 1971; Foster *et al.*, 1971; Kokubun, 1971].

It is thus probable that the electrojet and field-aligned currents significantly affect the polar cap magnetic field in the dark hemisphere. Considering this point, we conclude that the observed current pattern is consistent with the dawn-to-dusk electric field expected from the reconnection model.

3.2. DEPENDENCE OF THE MAGNITUDE OF THE ANTI-SUNWARD CONVECTION ON THE INTERPLANETARY MAGNETIC FIELD

So far, we have implicitly assumed that the polar-cap current depends linearly on the B_z component of the interplanetary field for the range $B_z < 1\gamma$. Here we examine in more detail the dependence of the magnitude of the current on B_z and other components of the interplanetary magnetic field, because the functional form of the dependence is theoretically important. As before, we use the polar-cap magnetic field data adjusted for the Svalgaard-Mansurov effect (the part linearly dependent on B_y). This is because this effect corresponds to a zonal plasma motion, which, on the whole, does not contribute to the estimate of total transpolar convection flow. Furthermore, we can see that the rate of reconnection between the geomagnetic and interplanetary fields should be an even function of B_y. Assume that both the solar-wind flow pattern and the dayside magnetospheric configuration are symmetric with respect to the solar magnetospheric *x-z* plane. In this case, the reconnection rate should not depend on the polarity of B_y because the case with one polarity is the mirror image of the other. Thus, it is expected that the total amount of convection depends only on B_z and the magnitude of B_y. This justifies elimination of the linear dependence on B_y.

The magnitude of convection is scaled by the $H_{\parallel}$ component of the magnetic-field perturbation. The $H_{\parallel}$ axis is defined at each local time in such a way that it is perpendicular to the local direction of the convection shown in Figure 3. Note that a decrease in $H_{\parallel}$ corresponds to an increase in the convection magnitude. No attempt is made to separate the effect of the nightside field-aligned current from that of the ionospheric Hall current, because these two currents seem to be too closely correlated to be separated by the regression analysis.

In the upper panel of Figure 6, $H_{\parallel}$ observed at Thule at local night (2 h MLT) is plotted against $B_{\perp}$ $(=\sqrt{B_y^2+B_z^2})$ for the cases where the latitude angle θ of the interplanetary magnetic field satisfies the condition $\theta < -45°$. In these cases, $B_{\perp}$ is approximately equal to $|B_z|$. It is seen from this panel that the dependence of $H_{\parallel}$ on $B_{\perp}$ is almost linear. Similar analyses have been made for all other stations and other local times. The results confirm that the convection magnitude depends almost linearly on the southward component of the interplanetary magnetic field.

In passing, we should mention the zero level of $H_{\parallel}$ taken in Figure 6. This value is derived from averaging the daily mean magnetic-field values on a five-monthly basis for internationally quiet days in the period analysed. As far as the transpolar convection is concerned, this choice of zero level is thought to be particularly reliable.

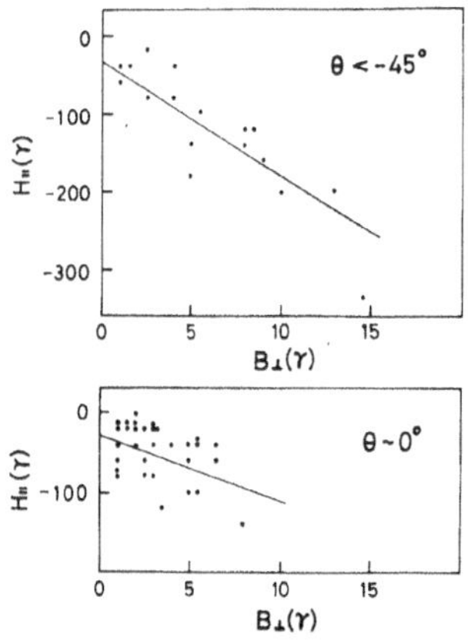

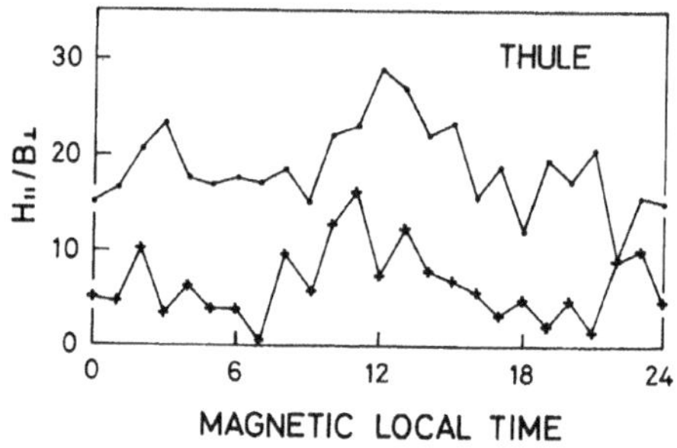

Figure 6. *$H_{/\!/}$ is plotted against $B_\perp$ (= $\sqrt{B_y{}^2+B_z{}^2}$). Upper panel corresponds to the case where the latitude angle θ of the interplanetary magnetic field satisfies $\theta < -45°$, lower panel to the case where $-1\gamma < B_z < 1\gamma$, hence $B_\perp \cong |B_y|$.*

Figure 7. *Linear regression coefficients between $B_\perp$ (= $\sqrt{B_y{}^2+B_z{}^2}$) and $H_{/\!/}$, plotted as a function of magnetic local time at Thule. Dots represent the case $\theta < -45°$, crosses the case $-1\gamma < B_z < 1\gamma$.*

This is because the effect of transpolar current, which appears in Y, should be averaged out when the daily mean is taken (We assume that the daily mean is statistically equivalent to the longitudinal average. Because of the divergence-free nature of equivalent currents, $\oint j_x\, d\ell = 0$, where j_x is the X component of the equivalent current, and $d\ell$ a line element along a circle of constant latitude. This leads to $\oint \Delta Y\, d\ell = 0$, where ΔY is the magnetic perturbation due to j_x).

Now let us examine whether the convection magnitude depends solely on the southward component of the interplanetary magnetic field. For this purpose, we pick the intervals when B_z is nearly zero ($-1\gamma < B_z < 1\gamma$) and plot $H_\|$ against $B_\perp$ at a fixed local time. The result is shown in the lower panel of Figure 6 for the same local time as analysed in the case of $\theta < -45°$. Unexpectedly, $H_\|$ decreases (i.e. convection magnitude increases) with increasing $B_\perp$. This means that even when B_z is zero, the anti-sunward convection in the polar cap is maintained by nonzero B_y. That this dependence on $|B_y|$ is not an incidental or local phenomenon is seen in Figure 7, where the linear regression coefficient between $H_\|$ and $B_\perp$ is plotted as a function of magnetic local time at Thule. Dots and crosses present the case for $\theta < -45°$ and the case for $-1\gamma < B_z < 1\gamma$, respectively. The sign of regression coefficient is taken in such a way that it is positive when the anti-sunward convection is enhanced by an increase in $B_\perp$. The following points are noteworthy:

(i) Regression coefficient between $B_\perp$ and $H_\parallel$ for the case $B_z \sim 0$ $(-1\gamma < B_z < 1\gamma)$ is always positive.
(ii) Magnitude of the regression coefficient is much smaller for the $B_z \sim 0$ case than for the $\theta < -45°$ case.
(iii) Local time dependence of the regression coefficient for the case $B_z \sim 0$ is similar in form to that for the case $\theta < -45°$.

Points (i) and (ii) lend statistical confirmation to the previous result that the convection magnitude depends on $B_\perp$ even when $B_z \sim 0$. It is also noted that regression coefficients are higher during daytime in both cases; this is probably due to the daytime enhancement of ionospheric conductivity.

In summary, the magnitude of anti-sunward convection is found to be a function of $|B_y|$ as well as of B_z. Although B_z-dependence seems to be generally dominant, a weak transpolar convection exists when $B_z \sim 0$, and its magnitude is intensified when $|B_y|$ is enhanced.

3.3. DEPENDENCE OF THE CONVECTION MAGNITUDE ON THE SOLAR-WIND VELOCITY

Murayama & Hakamada [1975] showed that the ratio between the AE index and the southward component of the interplanetary magnetic field is nearly proportional to V_{SW}^2, where V_{SW} is the solar-wind velocity. This suggests that the solar-wind energy input to the magnetosphere depends on the solar-wind velocity, and hence the convection velocity in the polar cap may also depend on the solar-wind velocity. In order to clarify this point, the ratio between $H_\parallel$ and B_z has been calculated for the case $B_z < -1\gamma$. This value is plotted against the solar-wind velocity in Figure 8 for 24 MLT at Resolute Bay. It is evident that this ratio has a significant positive correlation with V_{SW}. Similar plots for other stations and other local times have been performed, and the results support the above conclusion.

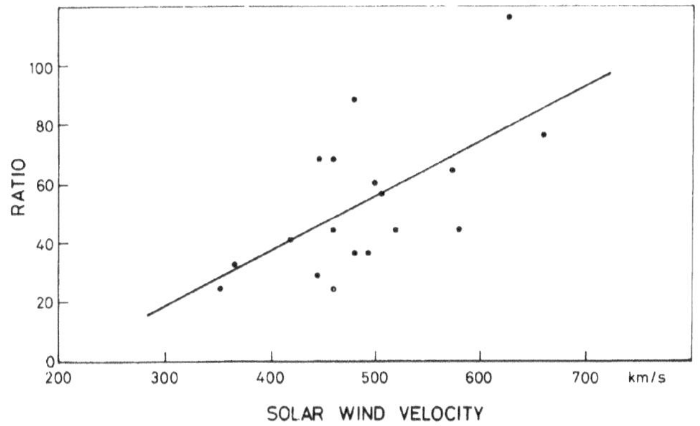

Figure 8. *Ratio between $H_\parallel$ and B_z for the range $B_z < -1\gamma$, plotted against solar-wind velocity.*

Although a linear regression line is shown in the figure for reference, we do not intend to determine the functional form of the dependence, because the scatter in the figure seems to be too large for such a purpose. It is only noted that the regression line, when extrapolated, intersects the abscissa (line $H_{\|} = 0$) for a finite value of the solar-wind velocity.

4. Effect of the Northward Component of the Interplanetary Field

Now let us address ourselves to the question of whether the polar-cap current responds to an intensification of the northward component of the interplanetary magnetic field. Let us first look at the plot of $H_{\|}$ at Resolute Bay versus interplanetary B_z for two local times. The upper panel of Figure 9 corresponds to midnight local time. It is seen that when B_z is negative, $H_{\|}$, which represents the magnitude of convection, responds almost linearly to B_z. Note that $H\|$ does not vanish when $|B_z| = 0$, which confirms that convection still exists when $B_z \sim 0$. When B_z is positive, the dependence of $H\|$ on B_z in the night is much weaker than when B_z is negative. Particularly, when B_z is more than 5γ, $H\|$ becomes almost constant and the magnitude of $H\|$ becomes particularly small ($|H\|| < 20\gamma$). These facts suggest that anti-sunward convection almost ceases when B_z becomes sufficiently large. Note that this interpretation at the same time lends support to our choice of $H_{\|}$ zero level.

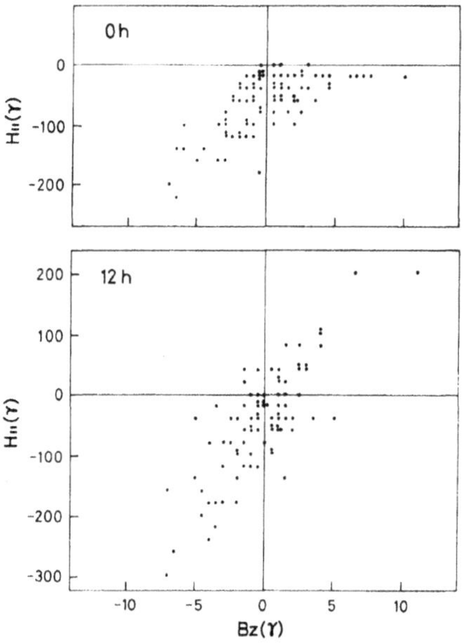

Figure 9. *$H_{\|}$ plotted against B_z for midnight local time (top) and noon local time (bottom) at Resolute Bay.*

Next, let us look at the behaviour of $H_{\|}$ around the noon meridian, as shown in the lower panel of Figure 9. In the negative range of B_z, the dependence of $H_{\|}$ on B_z is similar to the midnight case. However, in the positive B_z range, $H_{\|}$ increases sharply with increasing B_z. This behaviour makes a sharp contrast with the nightside case. Note that the absolute direction of the current for the case of positive B_z is opposite to that for the case of negative B_z. For example, when B_z is 10γ, $H_{\|}$ becomes as much as 200γ. This value is much larger than the ambiguity in the zero level of $H_{\|}$, and shows that the current direction is anti-sunward. It should be also noted that the dependence of $H_{\|}$ on B_z in the positive range seems to be almost linear, the slope being comparable to that in the negative B_z region.

4.1. PATTERN OF CONVECTION DRIVEN BY THE NORTHWARD FIELD

In order to delineate the dependence of the polar-cap current on the northward component of the interplanetary field, we calculate for the range of $B_z > 1\gamma$ regression coefficients between B_z and geomagnetic X and Y for each magnetic local time (MLT). The results are shown in Figure 10 using the same format as in Figure 2, except that the sign of coefficients is such that it is positive when the X or Y component increases with increasing B_z. It is seen that the pattern of MLT dependence of regression coefficients is considerably different from that in the case of $B_z < 1\gamma$. We have seen that the MLT dependence of the regression coefficients in

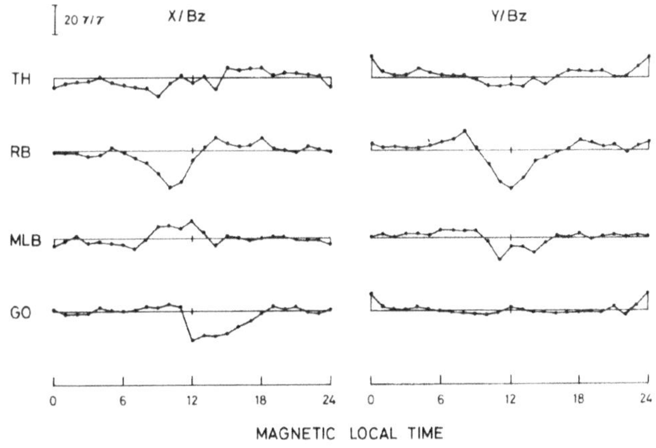

Figure 10. As Figure 2, except that B_z is in the range $B_z > 1\gamma$.

the latter case shows nearly sinusoidal variation, with similar phase and magnitude for the four stations investigated. In the present case, however, the magnitude and form of the variation (particularly in X) are different at different stations, suggesting that the corresponding ionospheric current has a smaller spatial scale than in the case of $B_z < 1\gamma$. More specifically, the following characteristics are noted:

(i) The behaviour of regression coefficients for Y is relatively simple. For all stations except Godhavn, where coefficients are small, they are negative during several hours around noon and positive at other local times. Negative regression coefficients for Y around noon correspond to enhancement of the sunward plasma convection. The magnitude of coefficients is largest at Resolute Bay and decreases at higher and lower latitudes.

(ii) The behaviour of regression coefficients for X is rather complex. A common feature noted at all stations is that large coefficients occur for several hours before and/or after noon. Coefficients are much smaller during the night-time hours. The sign of the coefficients at a fixed local time differs for different latitudes (e.g. compare Resolute Bay and Mould Bay curves at 10 MLT, and Thule and Godhavn curves at 15 MLT). Since a disturbance in X corresponds to an ionospheric current in the zonal direction, the ionospheric current has a rotational nature at these local times.

To visualise these points, the spatial distributions of convection flow vectors determined by these regression coefficients are shown in Figure 11 using the same scale as in Figure 3. It is clear that the mode of the current pattern is different from that in the case $B_z < 1\gamma$. On the dayside, around $\phi_M \sim 84°$, strong sunward convection occurs in association with the increase in the northward component of the interplanetary field. It seems that this sunward convection converges toward a point at approximately $\phi_M = 82°$ on the noon meridian. At lower latitude stations (Mould Bay: $\phi_M = 80.5°$, Godhavn: $\phi_M = 77.6°$) the directions of convection tend to be

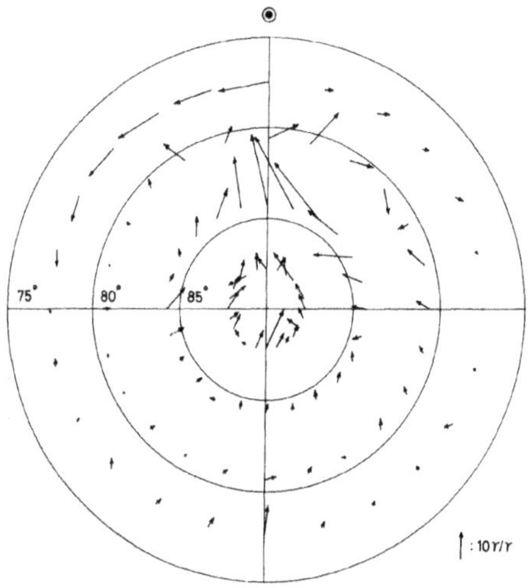

Figure 11. *Polar-cap convection pattern induced by the northward interplanetary magnetic field. The vectors are derived from the regression coefficients for the case $B_z > 1\gamma$. The length of the vector shown at bottom right corresponds to a 10γ increase in the horizontal geomagnetic field produced by a 1γ increase in B_z.*

zonal, flowing away from the noon meridian westward in the pre-noon sector and eastward in the afternoon sector. It is seen that these plasma flow patterns form two plasma-flow vortices, one in the pre-noon sector and the other in the afternoon sector. In the pre-noon sector, the rotational plasma motion is particularly evident from the vector directions at Resolute Bay and Mould Bay. It seems that the vortex on the afternoon side is larger in spatial extent and centred at a slightly lower latitude. At Godhavn (ϕ_M = 77.6°) for example, strong zonal convection away from the subsolar point is particularly clear on the afternoon side, while convection is much weaker on the pre-noon side.

On the nightside, the convection vectors are generally directed sunward at all stations, but their magnitudes are small, particularly at lower latitude stations. In the morning hours, convection velocity generally has a component directed towards the noon-midnight meridian. Thus the convection velocity at the time of positive B_z is not entirely antiparallel to that at the time of negative B_z, even on the nightside. This fact eliminates the possibility that the nightside convection derived as the effect of positive B_z is entirely due to the weakening of the anti-sunward convection that remains when $B_z \sim 0$. The current streamlines are shown in Figure 12. The total current flowing between two consecutive streamlines, induced by a 1γ increase in the northward component of the interplanetary field, is about 4 x 10^3 A.

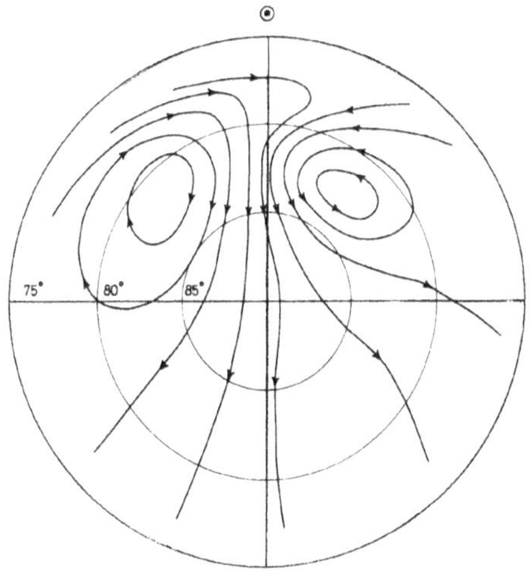

Figure 12. *Equivalent current pattern in the polar cap produced by positive B_z when $B_z > 1\gamma$. The total current induced between two consecutive streamlines by a 1γ increase in B_z is about 4 x 10^3 A.*

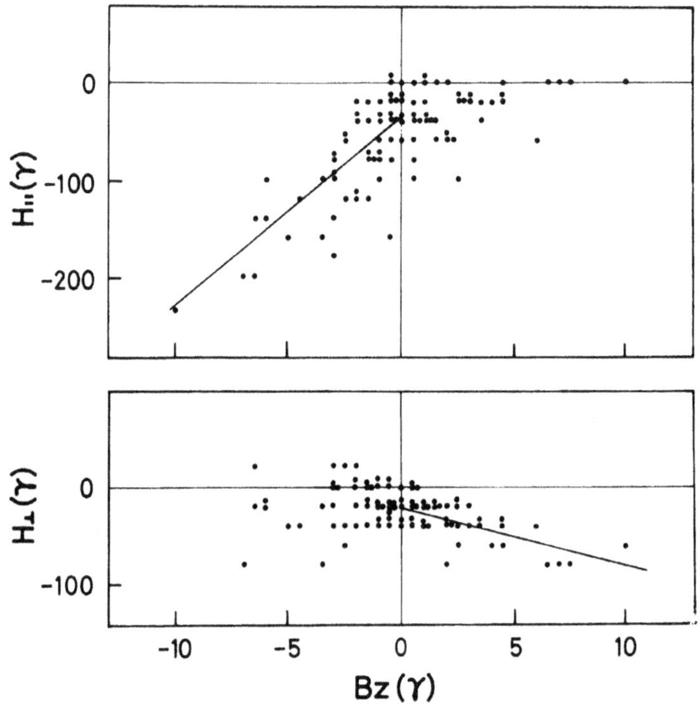

Figure 13. *$H_{/\!/}$ (top) and $H_\perp$ (bottom) plotted against B_z for 6 h MLT at Thule.*

The difference between the dependence of geomagnetic field on northward B_z and that on southward B_z is demonstrated further in Figure 13 where $H_\perp$ and $H_\parallel$ are plotted against B_z for 6 MLT at Thule. As can be seen from comparison of Figures 3 and 11, the convection velocity vectors associated with the northward and southward components of the interplanetary magnetic field are approximately at right angles at this local time at Thule. Hence the effects of the northward and southward interplanetary magnetic fields can be separated here by looking at different components of the geomagnetic field. Indeed, $H_\parallel$, which is the component dependent on the southward field, tends to be insensitive to B_z when B_z is positive, while the dependence of $H_\perp$ on B_z is clear only when B_z is positive, as indicated by the solid line in the lower figure.

In summary, the convection pattern associated with the northward polarity of the interplanetary magnetic field is completely different from that associated with the southward polarity. This implies that there are two distinct modes of the polar-cap convection; one intensified when the interplanetary magnetic field becomes southward, and the other intensified when the interplanetary magnetic field becomes northward. The implications of this conclusion will be discussed in the final section.

5. Background Polar-Cap Current System

We have seen in the previous sections that the interplanetary magnetic field interacts with the magnetosphere irrespective of its direction. While the current driven by this interaction process is our principal interest, it is also important to clarify the nature of the polar ionospheric current which flows irrespective of the state of the interplanetary magnetic field. It has been customary to assume that this 'background' current system represents the daily magnetic variations on very quiet days defined by very low K_p values. At low and middle latitudes, analyses conducted following this line of thought have led to success in extracting the S_q current system, which has been well explained by the dynamo action of the atmospheric wind. However, the situation is more complicated in the polar cap; people have reached different conclusions about the quiet-time polar additional current which is superposed on S_q^0, an extension from the low-latitude S_q.

We should note at this point that the field distribution in the polar cap will not be unique even when K_p is low. As a matter of fact, when B_z is positive, K_p generally tends to be small, reflecting the low subauroral-zone magnetic activity, but the polar-cap field is still activated by the B_y and positive B_z components of the interplanetary magnetic field.

It is thus desirable to minimise the effect of the interplanetary field for extracting the background current pattern which underlies convection-associated time-varying current in the polar cap.

For this purpose, we first choose all time intervals when the hourly values of interplanetary magnetic field satisfy $|B_y| < 2\gamma$, $0 < B_z < 2\gamma$. This range of B_z is chosen to minimise the effect of the anti-sunward and sunward convections deduced in the previous sections. The range of B_z is centred at $B_z = 1\gamma$ because around $B_z = 0$ anti-sunward convection is still present because of nonvanishing $|B_y|$. Furthermore, the magnitude of B_y is restricted to less than 2γ. The range of B_y is taken wider than that of B_z, because a large part of the B_y effect comes from the Svalgaard-Mansurov effect, which has already been subtracted using linear regression coefficients.

We examine the average values of the geomagnetic field perturbation for these selected times. Note that, in this case, to obtain the corresponding current direction, we need a baseline to scale the amount of perturbation. Here we define the baseline in the same manner as the zero level of $H_{\|}$, the validity of which has been discussed earlier. Results are shown in Figure 14 for each of the X and Y components using the scale shown on the top left. For the sake of comparison, the phases of X and Y variations corresponding to S_q^0 are shown at the top (the current responsible for S_q^0 is here assumed to be directed from dusk to dawn).

We first note that the observed Y (right panel) exhibits a quasi-sinusoidal variation, quite similar in phase to that of S_q^0 at each station. The situation is somewhat different for X. Difference in phase from S_q^0 is not so large at Thule and Godhavn, except for a slight advance in phase occurring in the evening sector. At Mould Bay, however, the degree of phase shift is significant, with the minimum X occurring at 15 h MLT. Also, there is a small peak present at 10 h MLT. At Godhavn, the situation is reversed. The minimum value occurs at 9 h MLT in the pre-noon sector, while a peak occurs at 15 h MLT in the afternoon sector. The opposite behaviour of

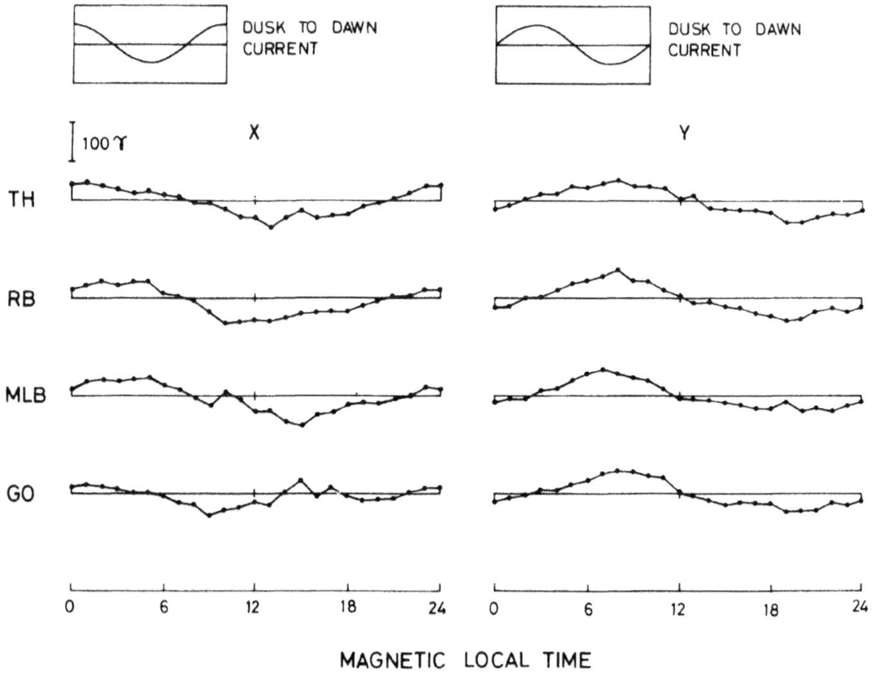

Figure 14. *Geomagnetic components X and Y at each station averaged over intervals when the two conditions* $|B_y| < 2\gamma$ *and* $0 < B_z < 2\gamma$ *hold in interplanetary space.*

X records at these stations suggests that there are current vortices superposed on S_q^0 in the pre-noon and afternoon sectors.

The equivalent flow pattern derived from these records is shown in Figure 15 (to obtain current direction, direction of vectors must be reversed):

(i) On the dayside around the 12 h meridian, the flow direction is nearly from dawn to dusk, in agreement with the S_q^0 current direction adopted in previous studies.

(ii) In the pre-noon and afternoon sectors, flow lines seem to have a kink between the latitudes of Mould Bay and Godhavn. Such kinks are present at 9 – 10 MLT and at 14 – 17 MLT.

(iii) In the dark hemisphere, the direction of flow is at 30° ~ 45° with the dawn-to-dusk direction. The nightside flow is rather uniform when compared to the dayside flow pattern.

Point (i) implies that on the dayside, the anti-sunward convection almost vanishes. Point (ii) suggests that, superposed on S_q^0, a peak and a trough of electric potential exist in the pre-noon and afternoon sectors, respectively, along the 78° ~ 80° latitude zone. Regarding point (iii), it should be noted that the direction of nightside currents resembles that in Figure 4 (right). It was previously suggested that the field-aligned current associated with the auroral zone activity strongly affects the nightside polar cap. Since the auroral zone activity lags behind the southward interplanetary magnetic field, it is quite probable that the effect of the field-aligned current remains in the 'quiet' interval thus defined.

We tentatively conclude that the basic pattern of current in the polar cap would be S_q^0 plus small localised perturbations on the dayside. It should be noted that these

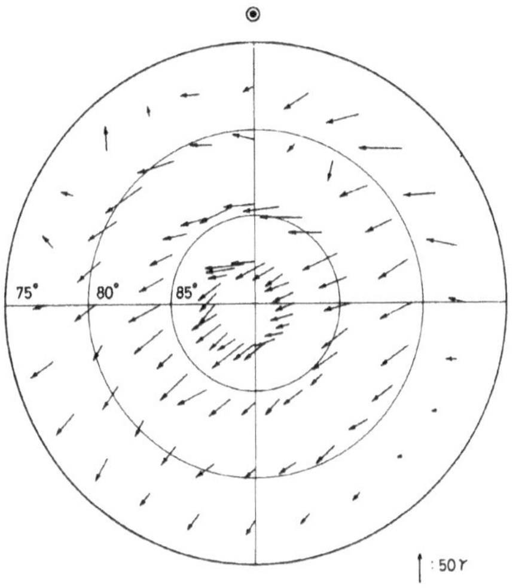

Figure 15. *Convection flow vectors in the polar cap constructed from Figure 14. The length of the vector shown bottom right corresponds to a geomagnetic disturbance of 50γ.*

perturbations would not be the remnant of either the anti-sunward convection or the sunward convection associated with interplanetary B_z, because these pertubations are different in form as well as in spatial extent from either of the above two convection patterns. Rather, the restricted locations of the perturbations suggest that they are caused by the magnetospheric phenomena which are localised in the vicinity of the magnetopause. It may be that these perturbations are caused by the precipitation of solar-wind particles polarised on the magnetopause, as suggested by Kawasaki & Akasofu [1967], though further analysis is needed to confirm this.

6. Summary and Discussion

6.1. SOME REMARKS ON THE METHOD OF ANALYSIS

Before discussing theoretical interpretations of the observations, we will discuss the validity of the assumptions made in the analysis.

(a) Division between the northward and the southward cases

In our analysis of the regression coefficients, we separated the data set into the southward and the northward interplanetary-field cases, and the threshold was taken at $B_z = 1\gamma$. This choice of threshold value is based on the fact that the anti-sunward convection exists when $B_z = 0$ because of the finite $|B_y|$, while the two-cell pattern convection with its direction 'sunward' at the central polar cap is hardly observable when $B_z = 0$. The transition from anti-sunward convection to sunward two-cell pattern convection with increasing B_z can be seen in Figure 16, where we have plotted the dawn-to-dusk component Y_s of the magnetic data at 15 MLT at Godhavn against the north-south component B_z of the interplanetary magnetic field. At this particular location, the flow direction is nearly the same for both anti-sunward and sunward two-cell pattern convection. As a result, Y_s increases with $|B_z|$ for either sign of B_z, so that the distribution of points is V-shaped. Since the change in the slope of the trend in the distribution occurs not at $B_z = 0\gamma$, but at slightly positive B_z, we adopted $B_z = 1\gamma$ as a threshold. This choice was made from inspection of the data from the dayside sector, where the two-cell pattern convection is relatively strong. The threshold value may be different in the night sector, where this type of convection is much weaker in magnitude. Thus on the nightside the direction of the two-cell type convection analysed above may be contaminated by the existence of weak anti-sunward convection even when $B_z > 1\gamma$. However, as noted earlier, differences in the local directions of convection in the early morning hours for two types of convection pattern suggest that two-cell type convection also contributes significantly to the nightside polar-cap record when $B_z > 1\gamma$.

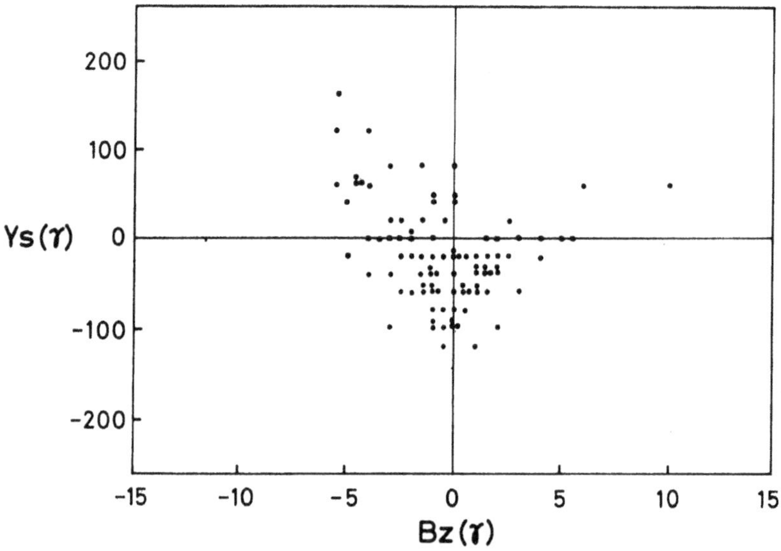

Figure 16. *Y_s (dawn-dusk component of the geomagnetic field) at Godhavn, plotted against interplanetary B_z for 15 h MLT.*

(b) Elimination of the B_y-dependent part

We investigated the magnitude of the anti-sunward convection after eliminating the Svalgaard-Mansurov effect. The validity of this treatment is seen in Figure 17, where the average magnitude of the anti-sunward convection, as expressed by the Y_s component of the field deviation, is plotted along the dawn-dusk meridian for the case $0 < B_z < 2\gamma$ (the dawn side corresponds to the left side of the figure). The top, middle and bottom panels correspond to the different ranges of B_y as shown in the figure. The following points are noteworthy:

(i) When B_y is less than -2γ, the region of strong convection shifts to the dusk side. When B_y is greater than 2γ, the region of strong convection shifts to the dawn side. The situation for negative B_y is roughly the mirror image of that for the positive B_y.
(ii) When the magnitude of B_y is less than 2γ, the above asymmetry almost disappears.

These points are in agreement with the dawn-dusk asymmetry of the convection electric field related to the sign of B_y [Heppner, 1972a, b; Nishida & Obayashi, 1972]. Thus elimination of the B_y-dependent part is almost equivalent to the smoothing out of the asymmetric distribution of the convection flow with respect to the noon-midnight meridian plane, and we can expect that the result obtained locally at each station after eliminating the Svalgaard-Mansurov effect reflects the overall rate of the transpolar convection.

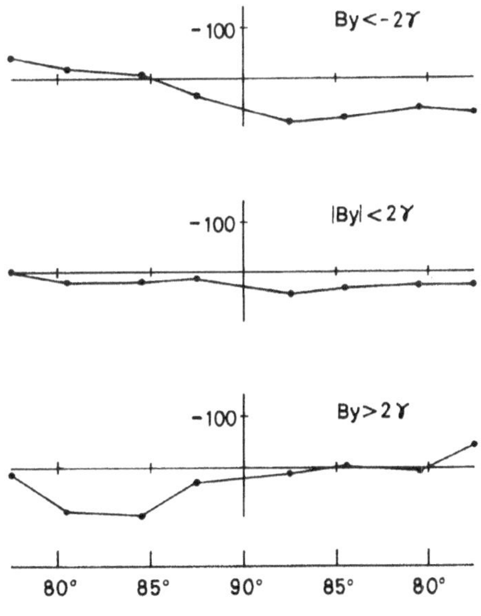

Figure 17. *Y_s, which represents the magnitude of anti-sunward convection, plotted along the dawn-dusk meridian for the cases $B_y < -2\gamma$ (top), $|B_y| < 2\gamma$ (middle), and $B_y > 2\gamma$ (bottom).*

6.2. CONVECTION ASSOCIATED WITH THE SOUTHWARD INTERPLANETARY MAGNETIC FIELD

When the interplanetary magnetic field is southward, a transpolar current sheet covers the entire polar cap ($\phi_M \gtrsim 77.6°$). This current system should be the polar part of DP 1 and DP 2 current systems, for which the effect of the southward interplanetary magnetic field has been found. In view of the time delay obtained in Figure 5, the dominant contribution to the current system on the dayside appears to be DP 2, while on the nightside there seems to be significant contribution from DP 1 (substorm). Newly found characteristics of this transpolar current sheet are:

(i) Direction of the current is consistent with the anti-sunward plasma convection.
(ii) Current intensity depends on B_z almost linearly.
(iii) Current intensity depends also weakly on $|B_y|$.
(iv) Current intensity depends also on the solar-wind velocity.

Points (i) and (ii) support Dungey's reconnection model. In view of this, points (ii) – (iv) are important because they imply that the dayside reconnection rate is a function of B_z, $|B_y|$, and the solar-wind velocity.

We first note that some models [Petschek, 1964; Nishida & Maezawa, 1971; Sonnerup, 1974] predicted that the total magnetic flux reconnected on the dayside magnetosphere in unit time is proportional to B_z^2 when the interplanetary magnetic field is due southward. This is because their discussions are based on the Petschek argument that the maximum value of the speed V_n with which the plasma flows into the reconnection region is proportional to the local Alfvén speed, which is proportional to B_z in the present case. Since the amount of magnetic flux reconnected in unit time per unit length on the reconnection line can be expressed as $V_n B_z$, it is expected that the total amount of magnetic flux reconnected is proportional to B_z^2 when the length of the reconnection line ℓ is fixed.

However, our analysis has shown that the magnitude of the convection is almost a linear function of B_z. The convection speed is also found to depend on the velocity V_{sw} of the solar wind. This suggests that the reconnection rate is not determined by the maximum possible value proposed by Petschek [1964]. It is probable that the reconnection rate is controlled by the efficiency with which reconnected field lines are removed from the reconnection region. In the case of the magnetosphere, this efficiency would be severely controlled by the solar-wind velocity, and it may be justified to suppose V_n is a function of V_{sw}.

Having this point in mind, we consider the case where the magnetic field lines to be reconnected make an arbitrary angle with each other. When V_n is fixed, ϕ, the total magnetic flux reconnected on a unit length of the reconnection line in unit time, can be written simply as

$$\phi \propto B_s \sin \psi \qquad (1)$$

where B_s is the magnitude of the magnetosheath field in front of the dayside magnetopause, ψ the angle that the magnetosheath field makes with the reconnection line. Relation (1) means that the reconnection rate depends on the relative orientations of the magnetosheath field and the reconnection line. When the reconnection line is formed along the equatorial plane, ϕ is proportional to B_z. However, it has been suggested that the reconnection line becomes inclined to the equatorial plane when the interplanetary magnetic field is not due southward [Nishida & Maezawa, 1971; Sonnerup, 1974; Gonzalez & Mozer, 1974]. In particular, assuming that the reconnection line is formed parallel to the average direction of the current inside the reconnection region, Gonzalez & Mozer [1974] obtained

$$\phi \propto B_s \sin \psi \propto \frac{B_s^2 - B_M B_s \cos\theta'}{\sqrt{B_s^2 + B_M^2 - 2B_s B_M \cos\theta'}} \tag{2}$$

where B_M is the geomagnetic field just inside the boundary, and θ' is the angle between the interplanetary and magnetospheric magnetic fields. When it is assumed that B_s around the subsolar reconnection point is proportional to the $B_\perp$ component $(=\sqrt{B_y^2+B_z^2})$ of the interplanetary magnetic field, ϕ given by (2) is proportional to $B_\perp$ when the magnetosheath field is due southward and it increases with $B_\perp$ when $\theta' = 90°$, i.e. when the interplanetary field is parallel to the equatorial plane. Note also that reconnection does not occur (ϕ becomes negative) when $\theta' < \cos^{-1}(B_s/B_M)$, i.e. when the magnetosheath field is sufficiently northward. These points are in excellent agreement with the results obtained in this paper.

To substantiate the agreement between the model and the observations, the correlation coefficient between ϕ and $H_{||}$ can be calculated at each station and compared to that between B_z and $H_{||}$.

Calculation of ϕ necessitates a few assumptions because data on the solar-wind pressure, which influences B_M and B_s, are not always available. Let us denote the magnitude ratio between the magnetosheath magnetic field and the interplanetary field by K. Equation (2) can be written as

$$\phi \propto \frac{K\left(B_\perp^2 - B_\perp \frac{B_M}{K} \cos\theta'\right)}{\sqrt{B_\perp^2 + \frac{B_M^2}{K^2} - 2B_\perp \frac{B_M}{K} \cos\theta'}} \tag{3}$$

TABLE 2

Correlation coefficients between the geomagnetic components $H_{||}$ and the interplanetary magnetic-field parameters that represent the theoretical reconnection rate under particular conditions of the magnetosheath magnetic field

	B_z	F_{20}	F_{15}	F_{10}	$\sin\frac{\theta'}{2}$
Thule	0.71	0.75	0.74	0.73	0.70
Resolute Bay	0.72	0.74	0.74	0.73	0.69
Mould Bay	0.68	0.72	0.71	0.70	0.66
Godhavn	0.66	0.67	0.67	0.65	0.64
Mean value	0.69	0.72	0.72	0.70	0.67

where $B\perp$ and θ' are the magnitude and the co-latitude angle of the interplanetary magnetic field in the solar magnetospheric y-z plane, respectively.

Since ϕ is a function of θ', $B\perp$, K and B_M/K, of which only the former two are known, we proceed by simply assuming K and B_M/K to have a fixed value and calculate correlation coefficients. The result is shown in Table 2 for each station and for various values of B_M/K. (The value of K itself does not affect the correlation values because it is a proportional constant in relation (3)). F_{10}, F_{15} and F_{20} represent the values of ϕ for the cases $B_M/K = 10\gamma$, $B_M/K = 15\gamma$, and $B_M/K = 20\gamma$, respectively. Note that ϕ is proportional to B_z when B_M/K is ∞, and ϕ becomes equal to $\sin \theta'/2$ when $B_M/K = B\perp$, i.e. when the magnitudes of the fields on the magnetosphere and magnetosheath sides of the boundary are the same. It is seen that the correlation values F_{15} and F_{20} are consistently higher than the remaining three variables for all stations. It is particularly important that the correlation values for B_z are always smaller than those for F_{15} and F_{20}. This means that the three-dimensional reconnection model in which the presence of B_y is taken into account is really a step forward in interpreting the observed results concerning dayside reconnection.

6.3. CONVECTION ASSOCIATED WITH THE NORTHWARD INTERPLANETARY MAGNETIC FIELD

When the interplanetary magnetic field is directed northward, a peculiar current system appears in the polar cap. This current system is characterised by two current vortices in the dayside polar cap, one in the pre-noon and the other in the afternoon sector. The current direction is anti-sunward in the central polar cap, suggesting that the *sunward* plasma convection is induced in the polar cap by the northward interplanetary magnetic field.

Dungey [1963] postulated that when the interplanetary magnetic field is northward, it is reconnected with the geomagnetic field on the tail magnetopause. Iwasaki [1972] and Russell [1972] further discussed this possibility and the consequences of such reconnection. Russell stressed in particular that the same interplanetary magnetic field line would not be reconnected with both northern and southern tail field lines at the same time, as assumed by Dungey [1963], so that one end of the reconnected field lines remains open. We here propose a modified version of his model and discuss its consistency with our observations.

In order to see how reconnection is possible on the tail magnetopause, let us briefly examine the necessary conditions for reconnection to occur at a specific point on the magnetopause under the influence of the solar-wind flow. Since the flow energy in the magnetosheath exceeds the magnetic energy, the reconnection process would not be able to alter the magnetosheath flow considerably. Hence we suggest that, for the reconnection to occur at a specific point on the magnetopause, the following conditions should be satisfied by the two-dimensional plasma flow field along the magnetopause:

(i) Plasma flow velocity along the magnetopause has no component perpendicular to the reconnection line.
(ii) The plasma flow field along the boundary is divergent at that point.

Condition (i) is necessary because, if a finite component perpendicular to the reconnection line exists, the reconnection line would be 'blown off' by the solar wind; i.e. as the velocity acquired by the reconnection process is very small the solar-wind flow can carry the reconnection line away with it. Condition (ii) is necessary because if the solar-wind flow is convergent, it opposes the divergent flow induced by the reconnection. In particular, on the noon-midnight meridian, condition (i) means that the reconnection points should also be the stagnation points of the magnetosheath flow.

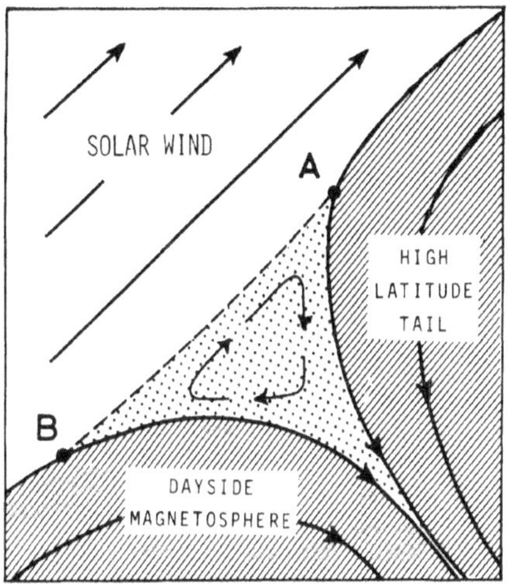

Figure 18. *Postulated flow pattern in and around the polar cusp when reconnection is not taking place. Free stream of the solar-wind plasma makes the trapped plasma in the cusp move in the direction indicated in the figure.*

There are two points which satisfy these requirements on the noon-midnight meridian section of the magnetopause. One is the subsolar point, and the other is a point located just tailward of the dayside polar cusp ('A' in Fig. 18). In the polar-cusp region, the shape of the boundary is such that the existence of the trapped plasma is allowed in the dotted area in the figure [Willis, 1969]. Because of the small but finite viscosity between the plasma trapped in the cusp and the free stream of the solar wind, the trapped plasma will move in the direction indicated by the arrows in the dotted area. The combined flow field of the solar wind and the trapped plasma will then make two stagnation points A and B on opposite sides of the polar cusp. Of these, point A satisfies the condition of diverging flow field along the boundary, while point B does not. Hence only point A and the subsolar point satisfy the necessary conditions to be a reconnection point (when the reconnection process occurs, the flow field is altered a little, so that the plasma will no longer be trapped in the polar cusp. However, point A will remain a stagnation point, because no tangential $J \times B$ force is exerted on the reconnection point).

Let us assume that the reconnection occurs at point A. In this case, as in the case of dayside low-latitude reconnection, the reconnection rate will depend on the direction of the interplanetary magnetic field. Since the magnetospheric field line at point A has a southward component, reconnection is expected to occur predominantly when the interplanetary magnetic field is directed northward.

Two reconnection *lines* will be formed in the real case, one in the northern, and the other in the southern tail magnetopause. The longitudinal extent of these reconnection lines should be rather short, as our observation showed that the sunward convection in the polar cap is limited to the narrow region around the noon-midnight meridian. This implies that only those interplanetary field lines that touch the polar cusp near the noon-midnight meridian can be reconnected to the tail field lines. Thus, in the general case where B_y is finite, an interplanetary magnetic field line will not be reconnected with both northern and southern tail field lines at the same time, so that the reconnected field lines will remain open at the interplanetary end.

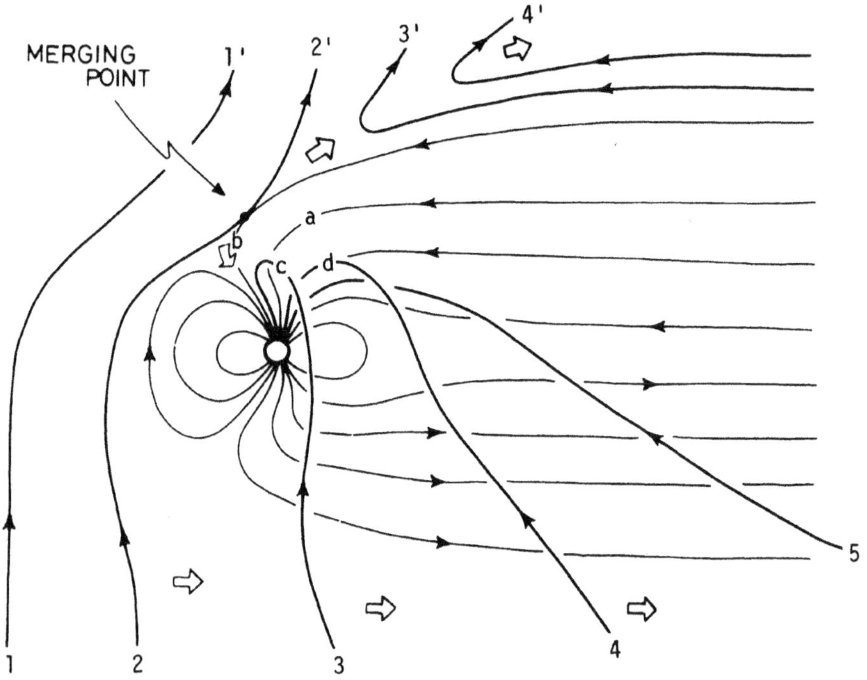

Figure 19. Field and flow pattern around the magnetosphere when the northward interplanetary magnetic field is reconnected with the high-latitude tail field lines. Although not shown in the figure, reconnection also occurs near the southern polar cusp in a symmetrical manner (for details see text).

Let us examine the plasma motion induced by this reconnection process. The $J \times B$ force operating around the northern reconnection line is in the direction of thick arrows shown in Figure 19 near the 'merging point'. The tail part of the reconnected field line is detached from the Earth and convected away with the solar wind. (The motion of field lines is indicated by labelling the location of a particular field line at successive moments by numbers 1′–4′.) The other part of the reconnected field line, which has one end in the polar ionosphere and the other end in interplanetary space, is pulled towards the low-latitude side by the $J \times B$ force. However, since the solar wind quickly convects the interplanetary end of the reconnected field lines around the magnetosphere (this motion being indicated by numbers 1 – 5), the direction of convection on the low-latitude side soon becomes almost parallel to the auroral oval or, more appropriately, parallel to the boundary between closed and open field lines. The field lines eventually enter the boundary region between the tail lobe and the plasma sheet. They are then convected back toward the reconnection point. The motion of the foot of the reconnected field line is indicated in the figure by letters a – d. Although not shown in the figure, the same kind of plasma motion is induced in the southern tail lobe by the interplanetary field lines reconnected on the southern tail lobe.

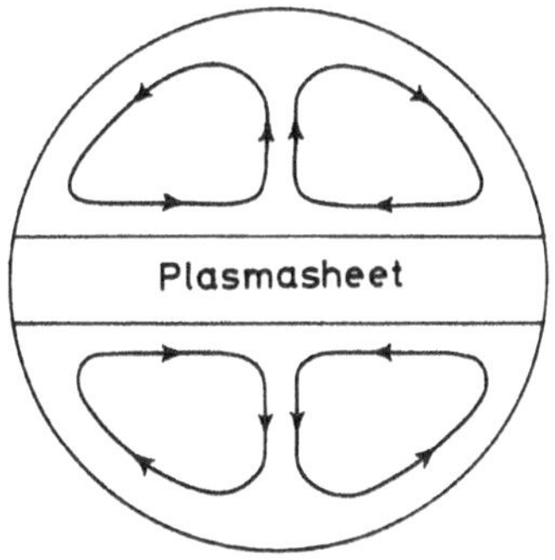

Figure 20. Plasma convection pattern inside the tail viewed toward the Sun at the time of a northward interplanetary magnetic field.

The corresponding convection pattern in the tail section is shown in Figure 20 (as viewed toward the Sun). Let us consider the northern tail lobe. Magnetic field lines at the top of the tail lobe are reconnected to the interplanetary field lines near the polar cusp. Reconnected field lines are pulled downward along the tail surface by the $J \times B$ force. They enter the tail near the tail lobe/plasma sheet boundary and return to the reconnection point by flowing through the tail lobe. When the effect of the $J \times B$ force due to the Y component of the interplanetary magnetic field is neglected, the average convection pattern is symmetric with respect to the midnight meridian plane, and consequently two flow vortices appear in the tail lobe. If this convection pattern is mapped to the ionosphere, it fits well with the observed current pattern with two vortices in the polar cap.

The above reconnection model predicts that the reconnected field lines are convected along the auroral oval when the interplanetary magnetic field is directed northward. This means that along the auroral oval, just poleward of the last closed line, there exists a layer of magnetic field lines which are connected to the magnetosheath. Indeed, some of the observed phenomena related to the open magnetosphere occur along the auroral oval. It is also noted that Burch [1973] found that the poleward boundary of the dayside polar cusp moves appreciably poleward when the interplanetary magnetic field becomes northward, but its equatorward boundary stays in the same position, the total effect being the widening of the polar cusp. This is consistent with the above expectation that when the interplanetary magnetic field is directed northward, the polar-cap field lines adjacent to the auroral oval become open to the magnetosheath, while dayside closed field lines are unaffected. It is also interesting that the field lines convected along the auroral oval will eventually enter the boundary between the plasma sheet and the tail lobe. This means that the magnetosheath particles have direct access to the plasma sheet from dawn and dusk sides. This might be related to the Akasofu *et al.* [1973a] observation that the solar-wind particles flow along the boundary between the plasma sheet and the tail lobe well inside the tail boundary.

The most important conclusion of the above tail reconnection model would be that the reconnection of interplanetary magnetic field lines and the tail lobe field lines does not change the total number of open field lines in the magnetosphere, because the reconnected tail field lines are open from the beginning. This means that although the polar cap field lines are always connected to the interplanetary field lines through either of the two reconnection mechanisms, a change in the total magnetic flux of the tail lobes results only from the dayside low-latitude reconnection, which predominantly occurs when the interplanetary magnetic field is directed southward. It is noteworthy that this conclusion reconciles the disagreement between the observation that the polar-cap field lines are always connected to the interplanetary field lines and the fact that substorm activity shows selective dependence on the southward field.

Finally, we should recall that if B_y is sufficiently large, dayside reconnection can occur even when the interplanetary B_z is slightly positive. This is consistent with the observational fact that substorms sometimes occur when the interplanetary magnetic field is northward. In this connection, Burch [1974] noted that almost all substorms that have been reported by Akasofu *et al.* [1973b] to have occurred during the northward interplanetary magnetic field correspond to the time intervals when the latitude angle θ of the interplanetary magnetic field is within the range $0 < \theta < 30°$. This again is consistent with the present model of reconnection between non-antiparallel field lines.

7. Acknowledgements

The author is deeply indebted to Profs. T. Obayashi and A. Nishida for their kind guidance and stimulating discussions. The author also wishes to thank Profs. N. Fukushima, T. Oguti, T. Tamao and Dr. T. Sato for their useful advice during the present work.

The Explorer 28, 33 and 35 magnetometer data of Dr. N.F. Ness were provided by the National Space Science Data Center through the World Data Center-A for Rockets and Satellites. Ground magnetic data were provided through the World Data Center-C2 for Geomagnetism, Kyoto University. Data processing was performed at the Data Processing Center, Institute of Space and Aeronautical Science, University of Tokyo and at the University of Tokyo Computer Center.

References

Akasofu, S.-I., E.W. Hones, Jr., S.J. Bame, J.R. Asbridge & A.T.Y. Lui, Magnetotail and boundary layer plasmas at a geocentric distance of ~ 18 Re: Vela 5 and 6 observations, *J. Geophys. Res.* **78**, 7257 (1973a).

Akasofu, S.-I., P.D. Perreault, F. Yasuhara & C.-I. Meng, Auroral substorms and the interplanetary magnetic field, *J. Geophys. Res.* **78**, 7490 (1973b).

Arnoldy, R.L., Signature in the interplanetary medium for substorms, *J. Geophys. Res.* **76**, 5189 (1971).

Atkinson, G., The current system of geomagnetic bays, *J. Geophys. Res.* **72**, 6063 (1967).

Aubry, M.P., C.T. Russell & M.G. Kivelson, On inward motion of the magnetopause before a substorm, *J. Geophys. Res.* **75**, 7018 (1970).

Burch, J.L., Preconditions for the triggering of polar magnetic substorms by storm sudden commencements, *J. Geophys. Res.* **77**, 5629 (1972a).

Burch, J.L., Precipitation of low-energy electrons at high latitudes: Effects of interplanetary magnetic field and dipole tilt angle, *J. Geophys. Res.* **77**, 6696 (1972b).

Burch, J.L., Rate of erosion of dayside magnetic flux based on a quantitative study of the dependence of polar cusp latitude on the interplanetary magnetic field, *Radio Science* **8**, 955 (1973).

Burch, J.L., Observation of interactions between interplanetary and geomagnetic fields, *Rev. Geophys. Space Phys.* **12**, 363 (1974).

Dungey, J.W., Interplanetary magnetic field and the auroral zones, *Phys. Rev. Lett.* **6**, 47 (1961).

Dungey, J.W., The structure of the exosphere or adventures in velocity space, in Geophysics (Eds. C. DeWitt, J. Hieblot and A. Lebeau), p. 503, Gordon and Breach, Science Publ., New York, 1963.

Fairfield, D.H. & L.J. Cahill, Jr., Transition region magnetic field and polar magnetic disturbances, *J. Geophys. Res.* **71**, 155 (1966).

Fennell, J.F., Access of solar protons to the earth's polar caps, *J. Geophys. Res.* **78**, 1036 (1973).

Foster, J.C., D.H. Fairfield, K.W. Ogilvie & T.J. Rosenberg, Relationship of interplanetary parameters and occurrence of magnetospheric substorms, *J. Geophys. Res.* **76**, 6971 (1971).

Friis-Christensen, E., K. Lassen, J. Wilhjelm, J.M. Wilcox, W. Gonzalez, & D.S. Colburn, Critical component of the interplanetary magnetic field responsible for large geomagnetic effects in the polar cap, *J. Geophys. Res.* **77**, 3371 (1972).

Friis-Christensen, E. & J. Wilhjelm, Polar cap currents for different directions of the interplanetary magnetic field in the Y-Z plane, *J. Geophys. Res.* **80**, 1248 (1975).

Fukushima, N., Equivalence in ground geomagnetic effect of Chapman-Vestine's and Birkeland-Alfvén's electric current-systems for polar magnetic storms, *Rep. Ionos. Space Res. Japan* **23**, 219 (1969).

Gonzalez, W.D. & F.S. Mozer, A quantitative model for the potential resulting from reconnection with an arbitrary interplanetary magnetic field, *J. Geophys. Res.* **79**, 4186 (1974).

Hakura, Y., Tables and maps of geomagnetic coordinates corrected by the higher order spherical harmonic terms, *Rep. Ionos. Space Res. Japan* **19**, 121 (1965).

Heppner, J.P., Electric fields in the magnetosphere, in Critical Problems of Magnetospheric Physics (Ed. E.R. Dyer), p. 107, IUCSTP Secretariat, c/o National Academy of Sciences, Washington DC, 1972a.

Heppner, J.P., Polar cap electric field distributions related to the interplanetary magnetic field direction, *J. Geophys. Res.* **77**, 4877 (1972b).

Heppner, J.P., J.D. Stolarik & E.M. Wescott, Electric-field measurements and the identification of currents causing magnetic disturbances in the polar cap, *J. Geophys. Res.* **76**, 6028 (1971).

Iwasaki, N., Localized abnormal geomagnetic disturbance near the geomagnetic pole and simultaneous ionospheric variation, *Rep. Ionos. Space Res. Japan* **25**, 163 (1971).

Iwasaki, N., Relation between the interplanetary magnetic field and reverse magnetic disturbances in the polar cap, Proc. 52nd Meeting of the Society of Terrestrial Magnetism and Electricity of Japan, p. 138, 1972 (in Japanese).

Kamide, Y. & S.-I. Akasofu, Latitudinal cross section of the auroral electrojet and its relation to the interplanetary magnetic field polarity, *J. Geophys. Res.* **79**, 3755 (1974).

Kawasaki, K. & S.-I. Akasofu, Polar solar daily geomagnetic variations on exceptionally quiet days, *J. Geophys. Res.* **72**, 5363 (1967).

Kawasaki, K., F. Yasuhara & S.-I. Akasofu, Short-period interplanetary and polar magnetic field variations, *Planet, Space Sci.* **21**, 1743 (1973).

Kakubun, S., Polar substorm and interplanetary magnetic field, *Planet. Space Sci.* **19**, 697 (1971).

Lyatsky, W.B., Yu.P. Maltsev & S.V. Leontyev, Three-dimensional current system in different phases of a substorm, *Planet. Space Sci.* **22**, 1231 (1974).

Maezawa, K., Dependence of the magnetopause position on the southward interplanetary magnetic field, *Planet. Space Sci.* **22**, 1443 (1974).

Mansurov, S.M., New evidence of a relationship between magnetic fields in space and on earth, *Geomag. Aeron.* (Engl. Transl.) **9**, 622 (1969).

McPherron, R.L., C.T. Russell & M.P. Aubry, Satellite studies of magnetospheric substorms on August 15, 1968, 9, Phenomenological model for substorms, *J. Geophys. Res.* **78**, 3131 (1973).

Montbrian, L.E., A simple method for calculating the local time of corrected geomagnetic midnight, *J. Geophys. Res.* **75**, 5634 (1970).

Morfill, G. & M. Scholer, Reconnection of the geomagnetic tail deduced from solar-particle observations, *J. Geophys. Res.* **77**, 4021 (1972).

Morfill, G. & M. Scholer, Study of the magnetosphere using energetic solar particles, *Space Sci. Rev.* **15**, 267 (1973).

Mozer, F.S., F. Bogott, M.C. Kelley & S. Schutz, Polar cap electric fields and magnetospheric convection, Space Sciences Laboratory, Series 13, Issue 96, Univ. of California, 1973.

Murayama, T. & K. Hakamada, Effects of solar wind parameters on the development of magnetospheric substorms, *Planet. Space Sci.* **23**, 75 (1975).

Nishida, A., Geomagnetic D_p 2 fluctuations and associated magnetospheric phenomena, *J. Geophys. Res.* **73**, 1795 (1968a).

Nishida, A., Coherence of geomagnetic DP 2 fluctuations with interplanetary magnetic variations, *J. Geophys. Res.* **73**, 5549 (1968b).

Nishida, A. & K. Maezawa, Two basic modes of interaction between the solar wind and the magnetosphere, *J. Geophys. Res.* **76**, 2254 (1971).

Nishida, A. & T. Obayashi, Magnetospheric convection, in Critical Problems of Magnetospheric Physics (Ed. E.R. Dyer), p. 179, IUCSTP Secretariat, c/o National Academy of Sciences, Washington DC, 1972.

Petschek, H.E., Magnetic field annihilation, in AAS-NASA Symposium on the Physics of Solar Flares (Ed. W.H. Hess), NASA SP-50, p. 425, 1964.

Russell, C.T., The configuration of the magnetosphere, in Critical Problems of Magnetospheric Physics (Ed. E.R. Dyer), p. 1, IUCSTP Secretariat, c/o National Academy of Sciences, Washington DC, 1972.

Sonnerup, B.U.Ö., Magnetopause reconnection rate, *J. Geophys. Res.* **79**, 1546 (1974).

Svalgaard, L., Sector structure of the interplanetary magnetic field and daily variation of geomagnetic field at high latitudes, Geophys. Paper R-6, Dan. Meteorol. Inst., Copenhagen, Denmark, 1968.

Svalgaard, L., Polar cap magnetic variations and their relationship with the interplanetary magnetic sector structure, *J. Geophys. Res.* **78**, 2064 (1973).

Vasyliunas, V.M., Mathematical models of magnetospheric convection and its coupling to the ionosphere, in Particles and Fields in the Magnetosphere (Ed. B.M. McCormac), p. 60, Reidel, Dordrecht, Holland, 1970.

Walters, G.K., Effect of oblique interplanetary magnetic field on shape and behavior of the magnetosphere, *J. Geophys. Res.* **69**, 1769 (1964).

Willis, D.M., The influx of charged particles at the magnetic cusps on the boundary of the magnetosphere, *Planet. Space Sci.* **17**, 339 (1969).

DISCUSSION

F. Mozer: (i) Reconnection occurs whenever it provides the energetically easiest means for the solar wind to flow around the magnetospheric obstacle. Your reconnection picture for $B_Z > 0$ looks as if it might require more energy than if the plasma and magnetic field simply continued to flow around the obstacle, which it has almost cleared. Can you comment on this?
(ii) Can you decide whether the ionospheric flow that you observe when $B_Z > 0$ is a steady-state phenomenon or is the magnetosphere settling down to its lowest energy state after reconnection is abruptly turned off?

T. Obayashi: In answer to question (ii), the ionospheric convection, when $B_Z > 0$, seems to be a steady-state phenomenon, though a more careful time-series analysis is required to confirm this.

W.P. Olson: Can you comment on relative contributions to your currents from the neutral winds and magnetospheric sources? It seems your work might be suggesting that the winds are more important than Mozer has implied. (Mozer gave an excellent summary of the strength of the various sources of **E** as a function of latitude).

T. Obayashi: Yes, I would refer you to Mozer's summary.

MAGNETOPAUSE POSITION AND THE RECONNECTION PROBLEM

V. Formisano

Laboratorio Plasma Spazio
CNR, Frascati, Italy

Abstract

A multisatellite study of the position of the magnetopause has allowed the first statistical study of the constant F^2/k, describing the kind of interaction that exists between the solar-wind plasma and the Earth's magnetosphere. Using HEOS-1 plasma data and OGO-5 observations of the magnetopause, we have tried to study the dependence of the magnetopause position D, after normalisation for geometry and dynamical pressure, on the IMF latitude ϑ and on the solar wind β (plasma pressure/magnetic pressure ratio). Differences are found for the average value of D depending on whether $\vartheta > 0°$, $\vartheta < 0°$ and $\beta > 0.9$ or $\beta < 0.9$.

Using HEOS-1 as a solar-wind monitor (plasma and magnetic field) and OGO-5 as a magnetopause monitor, we have studied the mechanism of magnetic reconnection at the boundary of the magnetosphere by studying the dependence of the position of the magnetopause on some interplanetary solar-wind parameters. The magnetopause observations ($\simeq$45 cases used in this study) were all at high geomagnetic latitude (close to the cusp) and therefore very sensitive to magnetic merging, as this phenomenon should move the cusp equatorward and the magnetopause inward. The magnetopause observations were normalised to the same solar-wind dynamical pressure and a correction was made for geometry (using Faifield's [1971] average magnetopause shape) so as to reduce all the observations to distances along the Sun-Earth line. For different solar-wind parameters, two ranges of values were defined, and for each range the average subsolar magnetopause position was computed. The effect (importance) of that parameter on the magnetopause position (and therefore on the magnetic-field reconnection process) was deduced from the difference between the two values of the average subsolar magnetopause position for the two parameter ranges.

Knott & Battrick (eds.), The Scientific Satellite Programme During the International Magnetospheric Study, 167–168.
All Rights Reserved. Copyright © 1976 by D. Reidel Publishing Company, Dordrecht-Holland.

The results are summarised in Table 1 where, for each parameter, the point separating the low range from the high range of values is given; usually this point coincides with the average value of that parameter in the solar wind. Table 1 gives the average subsolar magnetopause position for the low range (D_2), for the high range (D_2) and the absolute value of the difference $|D_1 - D_2|$, to compare the effects of the different parameters considered.

It appears that not only is the IMF ϑ important, but also the proton number density N_p and the magnetic-field intensity B. Consideration of the parameters B/N, important for the vacuum merging model, and $B/\sqrt{N}$, important for the Petscheck model [see Vasyliunas (1975)], does not allow a clear determination of the correct merging model.

TABLE 1

	IMF ϑ	MMF ϑ	β	Dst	N_p	B	B/N	$B/\sqrt{N}$
Separation point	0°	0°	0.9	-30γ	4.5cm^{-3}	5.6γ	1.3γ	2.6
$\bar{D}_1 R_E$	9.6	9.7	10.0	10.0	9.7	10.3	10.7	10.6
$\bar{D}_2 R_E$	10.3	10.2	9.8	9.9	10.6	9.7	9.7	9.7
ΔR_E	0.7	0.5	0.2	0.1	0.9	0.6	0.6	0.9

References

Fairfield, D.M., Average and unusual locations of the earth's magnetopause and bow shock, *J. Geophys. Res.* **76**, 6700 (1971).

Vasyliunas, V.M., Theoretical models of magnetic field line merging, *Rev. Geophys. & Space Phys.* **13**, 303 (1975).

NEW THEORETICAL ASPECTS OF THE SHAPE OF THE MAGNETOSPHERIC BOUNDARY

F.F. Cap, M.P. Leubner & F.P. Stössel

Österreichische Akademie der Wissenschaften,
University of Innsbruck, Austria

Abstract

On the basis of the magnetopause model developed by Beard and Mead, the dependence of the shape and geometry of the magnetopause, and especially the position of the neutral points, on several effects not included in the old model have been investigated. The approximation that all particles are incident parallel to the Earth-Sun line has been dropped, and an angular distribution for the solar-wind particles introduced, with the intention of taking into account the turbulence of the plasma in the magnetosheath. Under the same assumptions, the effects of various tilt angles of the dipole with respect to the Earth-Sun direction are considered. The alterations caused by a purely southward or northward directed interplanetary magnetic field are calculated and the influence of the magnetic field of the current sheet is analysed. For each of these cases, the pressure-balance equation is given which expresses the equilibrium between kinetic particle and magnetic field pressures. The numerical results for the noon-midnight meridional plane are presented and compared with satellite observations.

1. Introduction

Our investigations are based on the theory developed by Mead & Beard [1960, 1964], Spreiter [1962, 1963] and others who solved the Chapman - Ferraro [1931] problem for the magnetopause. In this theory, the boundary of the magnetosphere is assumed to be of negligible thickness compared to the radius of curvature. The solar-wind particles which move towards the Earth with a speed of some hundred kilometres per second cannot penetrate the magnetopause, but are specularly reflected. In this way they produce a surface current in the boundary layer. This surface current causes a magnetic field which, together with the undisturbed field of the Earth's geomagnetic dipole, constitutes the total field at the boundary.

Knott & Battrick (eds.), The Scientific Satellite Programme During the International Magnetospheric Study, 169–1
All Rights Reserved. Copyright © 1976 by D. Reidel Publishing Company, Dordrecht-Holland.

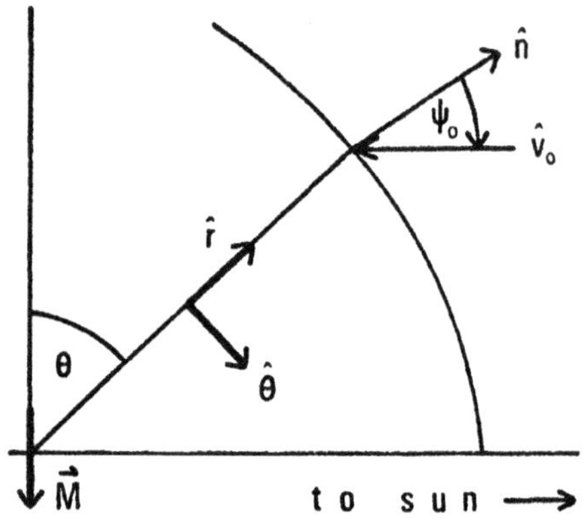

Figure 1. *Co-ordinate system in the noon-midnight meridional plane for the geomagnetic dipole perpendicular to the Earth-Sun direction.*

Figure 1 shows the co-ordinate system in the noon-midnight meridional plane. Polar co-ordinates are used. **M** is the geomagnetic dipole moment, $\mathbf{v}_0$ is the unit vector of the velocity of the incident particles, $\hat{\mathbf{n}}$ the unit normal to the surface and ψ_0 the angle between $\hat{\mathbf{n}}$ and $\hat{\mathbf{v}}_0$, $\cos\psi_0 = -(\hat{\mathbf{n}}\hat{\mathbf{v}}_0)$.

The governing equation of the problem is taken from magnetohydrostatic theory, namely

$$\nabla p = \mathbf{j} \times \mathbf{B}/c \tag{1}$$

where p is the kinetic pressure, **j** the electric current and **B** the magnetic field. Substituting the Maxwellian

$$\text{curl}\,\mathbf{B} = \frac{4\pi}{c}\,\mathbf{j} \tag{2}$$

yields the well-known equation

$$\nabla p = \frac{1}{4\pi}(\mathbf{B}\nabla)\,\mathbf{B} - \frac{1}{8\pi}\nabla\,\mathbf{B}^2 \tag{3}$$

Since the change in the magnetic field in the direction of the field lines is small compared to the change perpendicular to the direction of the magnetic field, the first term on the right-hand side of Equation (3), $(\mathbf{B}\nabla)\mathbf{B}$, which represents the effect of the curvature, can be neglected in a first approximation, as by Beard [1964]. The remaining equation can then be integrated at once with the result

$$p + \frac{B^2}{8\pi} = \text{const.} \tag{4}$$

Further assumptions of the model are that the magnetic field outside the surface vanishes and that there is no plasma inside the magnetopause. This leads directly to Equation (5), the pressure-balance equation:

$$B^2 = 16\pi N_0 m v_0^2 \cos^2 \psi_0 \\ = CN_0 \cos^2 \psi_0 \tag{5}$$

The pressure of the magnetic field is equal to the kinetic pressure of the particle beam, expressed by the statistical term on the right-hand side. N_0 represents the particle number density, m the proton mass (electron pressure can be neglected, because of the much smaller mass of the electrons) and v_0 the velocity.

If **B** is known analytically, Equation (5) would determine the shape and position of the boundary surface. The problem, however, is that the magnetic field due to surface currents cannot be computed until this surface is known. To overcome this difficulty, Beard made the approximation that the magnetic field just inside the boundary is exactly twice the tangential component of the geomagnetic dipole field. This is, in a way, a consequence of the assumption that the external magnetic field vanishes, but it is a still more severe restriction. Equation (5) can now be solved.

2. Distribution Function for the Incident Particles

In Beard's model, all particles move on parallel trajectories and have the same angle of incidence. This assumption is now dropped. It is well known from many satellite observations [e.g. Heppner *et al.*, 1967; Fairfield, 1967; Wolfe & MacKibbin, 1968], and is also clear from the theory [e.g. Kennel & Sagdeev 1967; Spreiter *et al.*, 1966], that the solar-wind plasma becomes turbulent in passing the bow shock. To take account of this turbulence in the magnetosheath, an angular distribution function for the incident particles is introduced [Cap & Leubner, 1974].

A special co-ordinate system is established, which is sketched in Figure 2. The curved line represents the magnetopause. The plane defined by $\hat{\mathbf{N}}$ and $\hat{\mathbf{t}}$ is tangential to the surface. ψ_0 is the main angle of incidence according to Beard. ψ and χ determine the actual direction of incidence for every particle. ψ is measured from $\hat{\mathbf{n}}$ along the meridian plane and χ perpendicular to it.

The particle number density N_0 is divided into two parts

$$N_0 = N_1 + N_2 \tag{6}$$

where N_1 is the number of parallel incident particles and N_2 the number of particles that impinge on the boundary from directions other than ψ_0. These N_2 particles should be subject to an angular distribution function $f(\psi,\chi;\psi_0)$. ψ and χ denote the deviation from the main angle of incidence ψ_0 (angle of incidence for the N_1 particles) for each particle. The particles incident within the differential solid angle dΩ exert a differential kinetic pressure

$$\mathrm{d}p = CN_2 f(\chi,\psi;\psi_0)\cos^2(\hat{\mathbf{n}},\hat{\mathbf{v}})\mathrm{d}\Omega \tag{7}$$

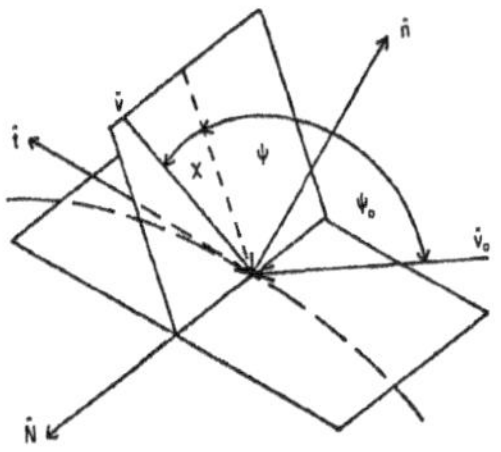

Figure 2. *Special co-ordinate system defining the angles χ and ψ, which describe the deviation of the main angle of incidence ψ_0.*

Integrating and normalising this expression,

$$B^2 = CN_1 \cos^2 \psi_0 + CN_2 \frac{\iint f(\psi,\chi;\psi_0)\cos^2(\hat{n},\hat{v})d\Omega}{\iint f(\psi,\chi;\psi_0)d\Omega} \tag{8}$$

with

$$d\Omega = \cos\chi d\chi d\psi \tag{8a}$$

$$\cos(\hat{n},\hat{v}) = \cos\chi\cos\psi \tag{8b}$$

An essential consideration in the mathematical treatment are the limits of integration. The lower limit is given by the condition that no particles are incident from inside the magnetopause. On the other hand, it is reasonable to assume that no particles strike the boundary from the night side. The integration over ψ therefore runs from $\psi_0 - \pi/2$ to $\pi/2$ (Fig. 3), while χ runs from $-\pi/2$ to $\pi/2$ for every point on the noon-midnight meridian plane.

Since the distribution function should have its maximum value for the direction $\psi = \psi_0$ and $\chi = 0$, a function was chosen of the form

$$f(\psi,\chi;\psi_0) = \cos(\psi_0 - \psi)\cos\chi \tag{9}$$

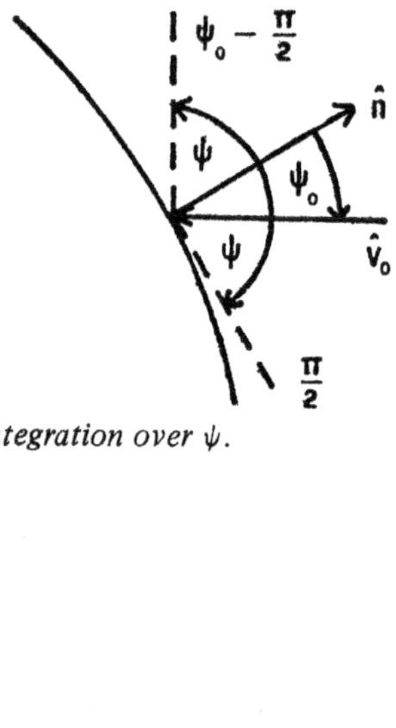

Figure 3. *Limits of integration over ψ.*

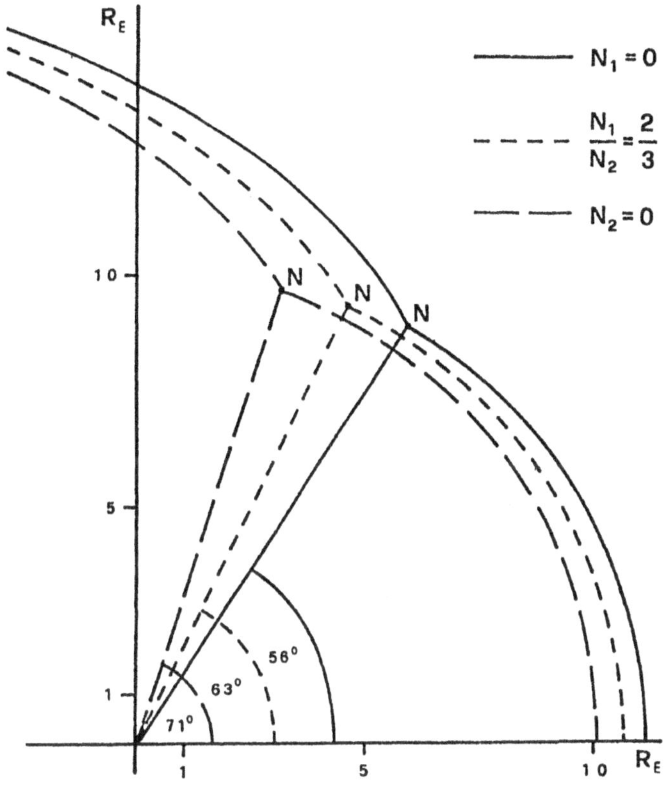

Figure 4. *The magnetopause for the perpendicular dipole for the three cases $N_1 = 0$, $N_1/N_2 = 2/3$ and $N_2 = 0$.*

It is easily seen that f is axially symmetric about the main angle of incidence ψ_0. This choice for f is also justified by satellite measurements [Frank, 1971].

With this distribution function, integration of Equation (8) yields

$$B^2 = CN_1 \cos^2 \psi_0 + \frac{1}{4} CN_2 (1 + \cos\psi_0) \tag{10}$$

Again Beard's approximation that the total field inside the magnetopause is twice the tangential component of the Earth's dipole field is used:

$$B_{total} = 2B_{dt} = 2\ |(\hat{\mathbf{n}} \times \mathbf{B}_d)| \tag{11}$$

$$\cos\psi_0 = -(\hat{\mathbf{n}} \cdot \hat{\mathbf{v}}_0) \tag{12}$$

$$\hat{\mathbf{n}} = (\hat{\mathbf{r}} - \frac{1}{r}\frac{dr}{d\theta}\,\hat{\theta}) \left[1 - \frac{1}{r^2} (\frac{dr}{d\theta})^2 \right]^{-\frac{1}{2}} \qquad \text{(Beard, 1964)} \tag{13}$$

Here $\hat{\mathbf{n}}$ is already 'specialised' for the noon-midnight meridional plane. The surface is symmetric in the meridional plane relative to the azimuthal angle ϕ. Therefore the partial derivatives with respect to ϕ vanish and only the derivatives with respect to θ occur. Hence one obtains an ordinary but nonlinear differential equation by substituting Equations (11), (12) and (13) in Equation (10).

This differential equation describing the magnetopause has the following form:

$$f_4 \left(\frac{dr}{d\theta}\right)^4 + f_3 \left(\frac{dr}{d\theta}\right)^3 + f_2 \left(\frac{dr}{d\theta}\right)^2 + f_1 \left(\frac{dr}{d\theta}\right) + f_0 = 0 \tag{14}$$

The f_t are rather lengthy functions of the independent and the dependent variables θ and r.

For the numerical integration, a predictor-corrector method was used. Figure 4 shows the solutions for the three cases $N_1 = 5, N_2 = 0; N_1 = 2, N_2 = 3$ and $N_1 = 0, N_2 = 5$. As expected, the distance to the subsolar point and also the polar distance to the magnetopause increase, corresponding to the portion of the anisotropic particles in the beam. (The distances are given in Earth radii). The main result, however, is a significant reduction in the latitude of the neutral points. They are too far to the north (about 71°) in Beard's model. For a purely anisotropic particle beam (N_1=0), the neutral points have a latitude of 56°, which is in quite good agreement with several observations [Frank, 1971; Mead & Fairfield, 1975].

3. Tilted Dipole

The condition that the dipole axis is perpendicular to the Earth-Sun direction will now be dropped and the effects of various tilt angles (10°, 20°, 30°) of the dipole relative to the perpendicular position will be considered.

The main difference between this and the last section is already apparent in Figure 5. It is the three-dimensional equivalent to Figure 1, but now the angle λ between the equatorial plane and the velocity vector is introduced. The components of the unit vector $\hat{\mathbf{v}}_0$ can easily be taken from Figure 5, so that $\hat{\mathbf{v}}_0$ reads as follows:

$$\begin{aligned} \hat{\mathbf{v}}_0 &= \hat{\mathbf{r}}\,(-\cos\theta\sin\lambda - \cos\phi\sin\theta\cos\lambda) \\ &\quad + \hat{\theta}\,(\sin\theta\sin\lambda - \cos\phi\cos\theta\cos\lambda) + \hat{\phi}\,(\sin\phi\cos\lambda) \end{aligned} \tag{15}$$

For the pressure-balance equation (10), we need $\cos\psi_0$, which now depends also on λ:

$$\cos\psi_0 = -(\hat{\mathbf{n}}\cdot\hat{\mathbf{v}}_0) \tag{12}$$

The full three-dimensional equations have the form

$$\hat{\mathbf{n}} = a\left(\hat{\mathbf{r}} - \frac{1}{r}\frac{\partial r}{\partial\theta}\,\hat{\boldsymbol{\theta}} - \frac{1}{r\sin\theta}\frac{\partial r}{\partial\phi}\,\hat{\boldsymbol{\phi}}\right) \tag{16}$$

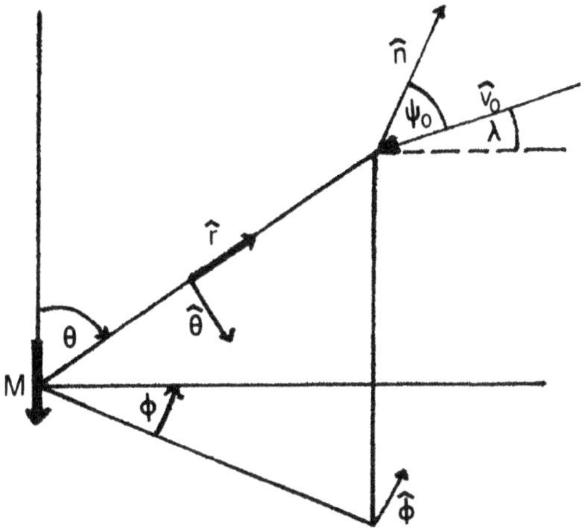

Figure 5. *Three-dimensional co-ordinate system for the tilted dipole; tilt angle λ.*

$$a = \left[1 + \frac{1}{r^2}\left(\frac{\partial r}{\partial \theta}\right)^2 + \frac{1}{r^2 \sin^2\theta}\left(\frac{\partial r}{\partial \theta}\right)^2\right]^{-\frac{1}{2}} \tag{16a}$$

$$(\hat{n}\hat{v}_0) = a\left[-\cos\phi\sin\theta\cos\lambda - \cos\theta\sin\lambda + (\cos\phi\cos\theta\cos\lambda - \sin\theta\sin\lambda)\frac{1}{r}\frac{\partial r}{\partial \theta} - \sin\phi\cos\lambda\frac{1}{r\sin\theta}\frac{\partial r}{\partial \phi}\right] \tag{17}$$

(Beard, 1964)

Here again only the noon-midnight meridional plane is considered, i.e. $\phi=0$ and $\partial/\partial\theta=0$. Substituting Equations (12) and (17) into (10) yields a similar equation to Equation (14), but with modified coefficients f_i. The integration is somewhat more complicated, because the symmetry relative to the equatorial plane is lost. Details and difficulties of the integration are explained in Spreiter & Briggs [1962], who solved the problem for only parallel incident particles. Their curves are plotted (dashed line) in Figure 6 for purposes of comparison. [for higher approx. see Olson, 1969; Choe, Beard & Sullivan, 1973].

The unbroken line in Figure 6 corresponds to the case when all particles are subjected to the distribution function. The dot-dash line between these two cases describes the magnetopause for a beam mixed by the two kinds of particles, namely $N_1 = 2$ and $N_2 = 3$. All distances are given in Earth-radii. Figure 6a again shows the case of the perpendicular dipole. In Figure 6b, the dipole is tilted by 10°. The neutral points no longer lie symmetrically to the equatorial plane. Again the neutral point for the anisotropic particle beam has a lower latitude than that for the isotropic case and the subsolar point moves further from the Earth by about $1R_e$. It is interesting that the northern neutral point for the full distribution function lies nearer to the equatorial plane than that for the untilted dipole. This effect is probably a consequence of the special choice of the distribution function. In nature, this situation occurs for high K_p. The southern neutral point moves away some

(a)

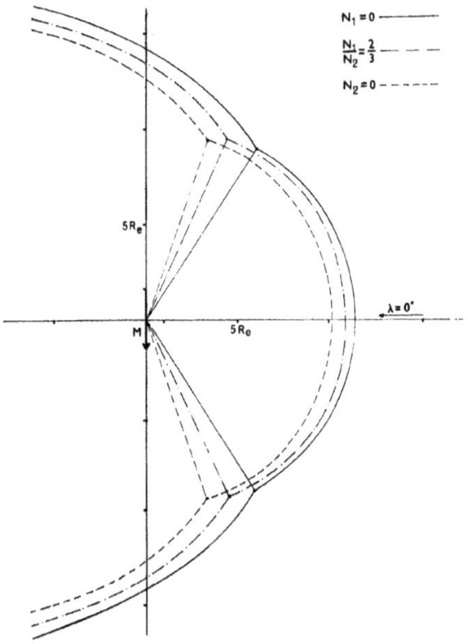

(b)

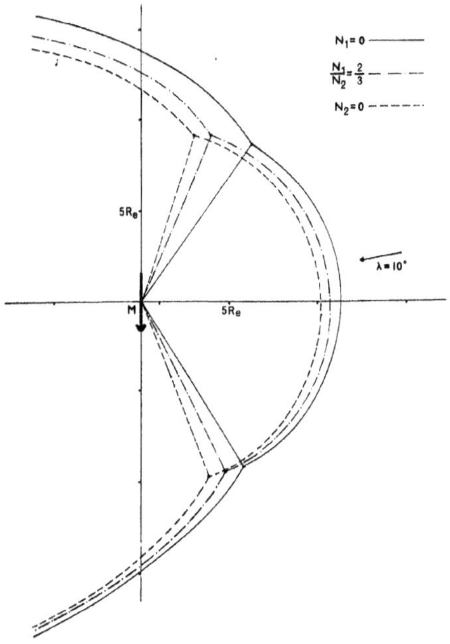

Figure 6 a-d. The noon-midnight meridional plane of the magnetopause for tilt angles of 0°, 10°, 20° and 30°.

(c)

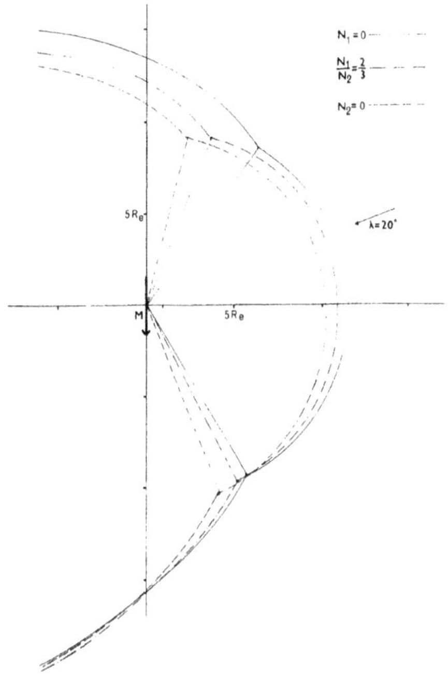
$N_1 = 0$
$\frac{N_1}{N_2} = \frac{2}{3}$
$N_2 = 0$
$\lambda = 20°$
$5R_e$
M
$5R_e$

(d)

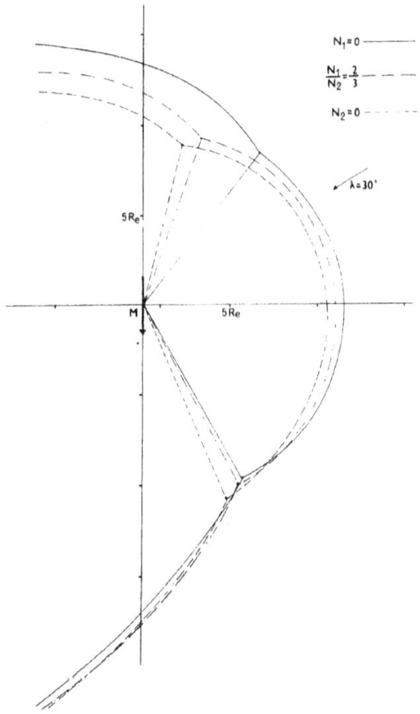
$N_1 = 0$
$\frac{N_1}{N_2} = \frac{2}{3}$
$N_2 = 0$
$\lambda = 30°$
$5R_e$
M
$5R_e$

degrees from the equatorial plane towards the south pole, quite the opposite to the situation for parallel incidence. Also, when the two kinds of particles are mixed, the northern neutral point moves northward relative to the untilted dipole, and so does the southern neutral point.

If the dipole is tilted 20° (Fig. 6c), things are very similar. Again we have the very low northern neutral point for the full distribution function, while the other two northern neutral points are shifted some degrees to the north. Figure 6d shows the solution for a tilt angle of 30°. The northern neutral point for a full distribution function has a latitude of about 51° here (which would be in quite good agreement with measurements for slightly disturbed conditions, i.e. $K_p \sim 3$). However, the southern neutral point is still too far from the equator compared to satellite data, although there is certainly an improvement relative to the curve for parallel incidence. For quiet conditions, the second curve fits the observations best. We are referring especially here to IMP measurements by Mead & Fairfield [1975]. Certainly, the coincidence between theory and data can be improved by varying the ratio N_1/N_2 or using other distribution functions.

4. Tail Magnetic Field

In this section, we treat the problem including a tail field. We are mainly interested in the noon-midnight meridional plane, and especially in the position of neutral points. Because of the symmetry, it is possible to calculate the shape of the magnetopause in this plane in a two-dimensional manner. For a first attempt, we therefore choose the Williams-Mead model of the tail field. Williams & Mead [1965] proposed an infinitely thin current sheet, stretching from $-\infty$ to $+\infty$ in the direction perpendicular to the meridional plane (y-direction in Fig. 7) and from $10\,R_e$ to $40\,R_e$ in

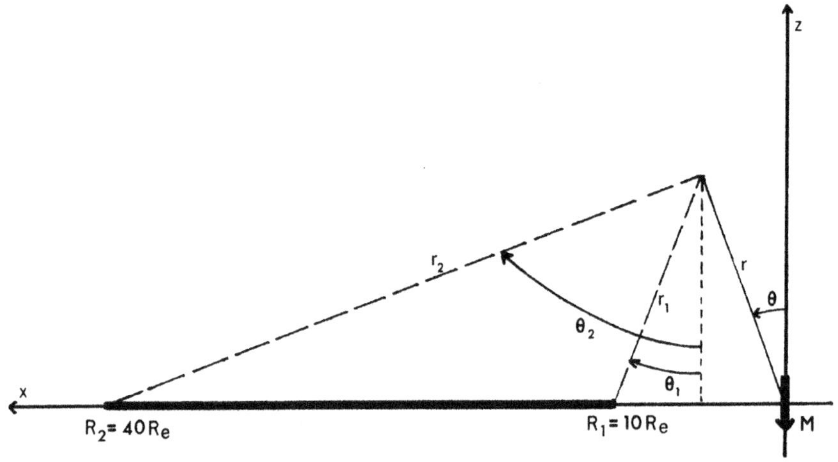

Figure 7. Williams-Mead tail model.

the equatorial plane (x-direction in Fig. 7). The components of the magnetic field induced by such a current strip are

$$B_{tx} = 2j(\theta_2 - \theta_1)$$
$$B_{tz} = 2j \ln \frac{r_1}{r_2} \tag{18}$$

The meanings of the symbols can be seen in Figure 7. The current j is chosen such that the magnetosphere of the tail immediately adjacent to the current sheet is 10γ, respectively 20γ. Written in another form, the components of the field are

$$B_{tx} = 2j(\arc\tan \frac{x - R_2}{z} - \arc\tan \frac{x - R_1}{z})$$
$$B_{tz} = j \ln \frac{(R_1 - x)^2 + z^2}{(R_2 - x)^2 + z^2} \tag{19}$$

For our calculations, these formulae have to be transformed to a polar co-ordinate system:

$$B_{tr} = -2j\, T_1 \sin\theta + j\, T_2 \cos\theta$$
$$B_{t\theta} = +2j\, T_1 \cos\theta - j\, T_2 \sin\theta \tag{20}$$

$$T_1 = \arc\tan \frac{r\sin\theta + R_1}{r\cos\theta} - \arc\tan \frac{r\sin\theta + R_2}{r\cos\theta}$$
$$T_2 = \ln \frac{r^2 + R_1^2 + 2rR_1 \sin\theta}{r^2 + R_2^2 + 2rR_2 \sin\theta} \tag{20a}$$

We again assume that the field outside the magnetopause vanishes. The field inside the magnetopause is the sum of the dipole and the current-sheet field:

$$\mathbf{B}_{d+t} = \mathbf{B}_d + \mathbf{B}_t \tag{21}$$

We use the pressure-balance equation for only parallel incident particles and the perpendicular dipole,

$$B^2 = CN_1 \cos^2 \psi_0 \tag{22}$$

For the field that we have to substitute here, we make an assumption very similar to that of Beard. The magnetic field just inside the boundary surface should be twice the tangential component of the combined field of the dipole and the current sheet.

Then

$$B = 2|\hat{n} \times \mathbf{B}_{d+t}| \tag{23}$$

Using similar procedures for the integration as before, we obtain the results presented in Figure 8. The distance of the subsolar point decreases, while the polar distance of the magnetopause increases. The neutral point is shifted considerably to the south: namely to 56° for 20γ and 64° for 10γ. The most probable value of B_t according to various measurements [e.g. Behannon, 1968] is 20γ.

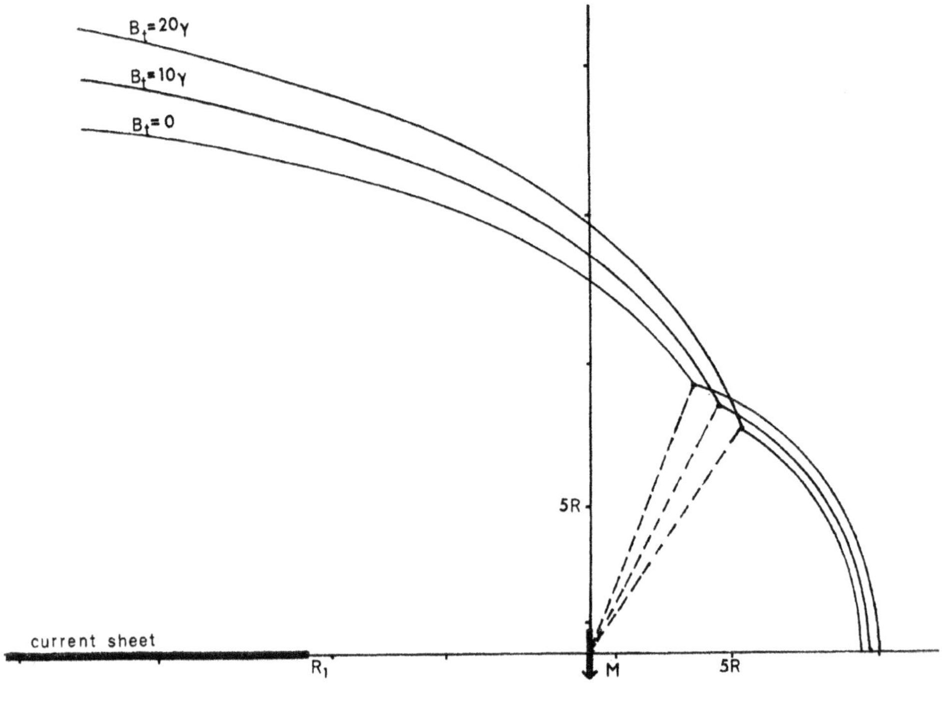

Figure 8. The magnetopause for only parallel incident particles and perpendicular dipole for a tail field B_t of 0, 10, 20 γ.

Thus, the inclusion of a Williams-Mead tail model in a Mead-Beard model of the magnetopause has a very similar effect on the position of the neutral points to introducing an angular distribution function for the incident particles. Because of numerical difficulties, we have not so far succeeded in combining both models. It has also not yet been possible to include the return currents of the current sheet along the tail magnetopause, for example in the form of the two semi-cylinders proposed by Axford *et al.* [1965] and used later by several authors in quite different ways [e.g. Bird & Beard, 1972; Choe & Beard, 1974].

5. Interplanetary Magnetic Field

The last point to be considered is the influence of the interplanetary magnetic field on our pressure-equilibrium model of the magnetopause. IMP 4 [Fairfield, 1971] and OGO-5 [Kivelson & Russel, 1973] observations, for example, have shown that the magnetopause location, and especially the neutral point regions, depend strongly on the north-south component of the interplanetary magnetic field. When the interplanetary magnetic field turns from north to south, the neutral points move equatorward and vice versa.

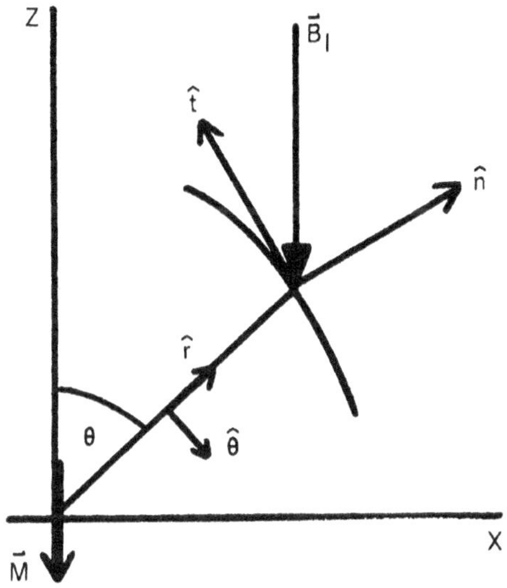

Figure 9. *Co-ordinate system with the interplanetary magnetic field B_I.*

We consider, therefore, in our model an ideal case, the influence of a constant vector $\mathbf{B}_I$ which has only a component parallel to the polar axis (Fig. 9). $\mathbf{B}_I$ is written as

$$\mathbf{B}_I = \pm B_I(-\cos\theta\,\hat{\mathbf{r}} + \sin\theta\,\hat{\boldsymbol{\theta}}) \tag{24}$$

(± indicates the southward or northward interplanetary magnetic field)

Because of the occurrence of an outside magnetic field $\mathbf{B}_I$, the pressure-balance equations have to be written as follows:

$$p + \frac{B^2}{8\pi} = \text{const.} \tag{25}$$

$$p_{in} + \frac{B^2{}_{in}}{8\pi} = p_{out} + \frac{B^2_{out}}{8\pi} \tag{26}$$

$$B_{in}^2 - B_{out}^2 = CN_1 \cos^2\psi_0 + \frac{1}{4}CN_2(1 + \cos\psi_0) \tag{27}$$

$$B_{in} = |\hat{\mathbf{n}} \times (2\mathbf{B}_d + \mathbf{B}_I)| \tag{28}$$

$$B_{out} = |\hat{\mathbf{n}} \times \mathbf{B}_I| \tag{29}$$

where B_{in} and B_{out} are expressions of the magnetic field inside and outside the magnetopause. Also, in the case of an external magnetic field we have to take into consideration only the tangential component (vector product with $\hat{\mathbf{n}}$). This follows immediately from the hydrodynamic equation

$$\nabla p = \frac{1}{c}(\mathbf{j} \times \mathbf{B}) \tag{30}$$

Scalar multiplication by **B** gives

$$(\mathbf{B} \cdot \nabla p) = 0 \tag{31}$$

which implies that the normal component of the total magnetic field vanishes at the surface because p has the same direction as the unit normal $\hat{\mathbf{n}}$.

With the expressions for $\mathbf{B}_d((r,\theta)$, $\mathbf{B}_I(r,\theta)$, $\hat{\mathbf{n}}(r,\theta)$, $\hat{\mathbf{v}}(r,\theta)$ and $\cos\psi_0 = -(\hat{\mathbf{n}}\hat{\mathbf{v}}_0)$we obtain the nonlinear differential equation for the noon-midnight meridian magnetopause

$$\left(\frac{dr}{d\theta}\right)^2 \left[32\frac{M}{r^5}\cos^2\theta\left(\frac{2M}{r^3} \pm B_I\right) - \frac{CN_2}{r^2}\right] + \left(\frac{dr}{d\theta}\right)\left[16\frac{M}{r^4}\sin\theta\cos\theta\left(\frac{4M}{r^3} \mp B_I\right)\right]$$
$$+ \left[16\frac{M}{r^3}\sin^2\theta\left(\frac{M}{r^3} \mp B_I\right) - CN_2\right] = CN_2\left(\sin\theta - \cos\theta\frac{1}{r}\frac{dr}{d\theta}\right)\left[1 + \frac{1}{r_2}\left(\frac{dr}{d\theta}\right)^2\right]^{1/2} \tag{32}$$

The upper (lower) sign describes a southward (northward) directed interplanetary magnetic field. Equation (32) is specialised to $N_1 = 0$ and represents the case in which all particles undergo an angular distribution (see Section 2).

The results of the numerical integration of Equation (32) are shown for two different values of the interplanetary magnetic field ($B_I = 4\gamma$, $B_I = 8\gamma$) in Figure 10 a,b.

Results for Southward Interplanetary Magnetic Field

At low geomagnetic latitude, the magnetopause is compressed; the subsolar distance is reduced. In comparison with the case in which $B_I = 0$, the neutral point location moves equatorward with increasing southward interplanetary magnetic field. The magnetopause distances at high geomagnetic latitudes are increased.

For a northward-directed interplanetary magnetic field, the results are the opposite.

(a)

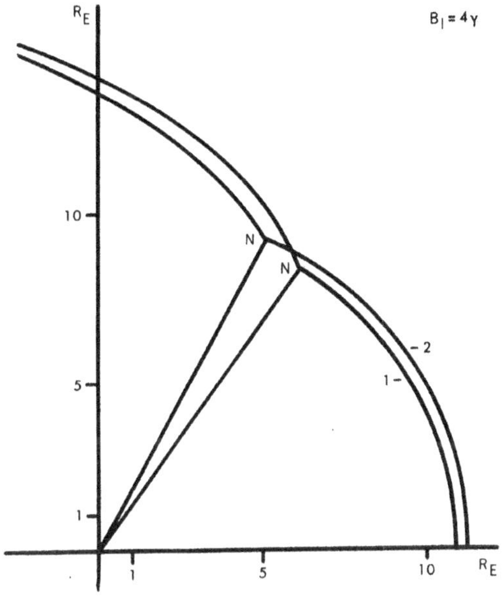

(b)

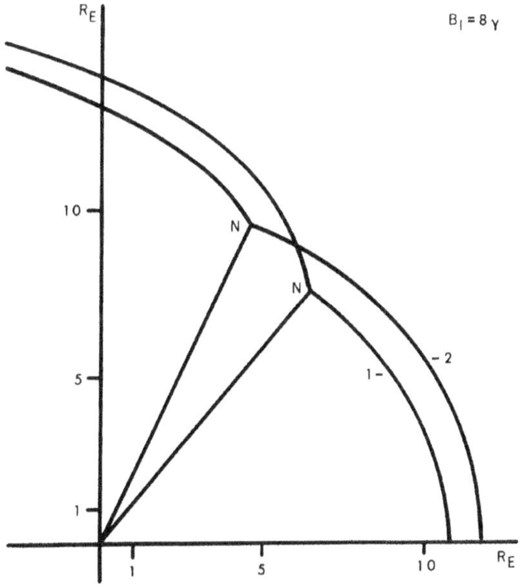

Figure 10 a-b *The magnetopause for purely anisotropic particle beam and perpendicular dipole under the influence of an interplanetary magnetic field B_I of 4 and 8γ. Line 1: B_I southward, 2: B_I northward.*

The discussion of IMP 4 observations by Fairfield [1971] shows that the averaged increase in the subsolar distance when the interplanetary magnetic field turns from south to north is 1.1 R_e. This is the same increase as is obtained from the proposed theoretical pressure-balance model with $B_I = 8\gamma$. The dependence of the neutral point location is highlighted by OGO-5 data [Russel *et al.*, 1971; Kivelson & Russel, 1973]. The average shifting of the neutral points when the interplanetary magnetic field turns is about 8°. The theoretical values obtained with our model are for $B_I = 4_\gamma$ 8° and for $B_I = 8_\gamma$ 15°.

6. Conclusions

We conclude that the qualitative features of the observations agree with those of the theoretical model discussed above using a pressure-balance condition derived from magnetohydrodynamic equations with angular distribution for the incident particles and interplanetary magnetic field.

It has been possible to find some refinements for the pressure-balance model of the magnetopause and to take into account some facts that were neglected in the old model. The shape of the magnetopause in the noon-midnight meridional plane, and especially the position of the neutral points, has been determined considering the turbulence in the magnetosheath, various tilt angles, the magnetic field of the current sheet and the interplanetary magnetic field. To complete the above considerations, it will be necessary to compute the magnetopause including tail field and distribution function; the numerical calculations for this are already in progress. Also, the case for interplanetary magnetic field and parallel incident particles will form a supplement to our investigations.

References

Axford, W.I., H.E. Petschek & G.L. Siscoe, Tail of the magnetosphere, *J. Geophys. Res.* **70**, 1231 - 1236 (1965).

Baker, J.C., D.B. Beard & J.C. Young, Self-consistent method for determining the boundary shape between a plasma and a magnetic field, *Phys. Fluids* **7**, 504 - 510 (1964).

Beard, D.B., The interaction of the terrestrial magnetic field with the solar corpuscular radiation, *J. Geophys. Res.* **65**, 3559 - 3568 (1960).

Beard, D.B., The solar wind geomagnetic field boundary, *Rev. Geophys.* **2**, 335 - 365 (1964).

Behannon, K.W., Mapping of the Earth's bow shock and magnetic tail by Explorer 33, *J. Geophys. Res.* **73**, 907 - 930 (1968).

Bird, M.K. & D.B. Beard, The self-consistent geomagnetic tail under static conditions, *Planet. Space Sci.* **20**, 2057 - 2072 (1972).

Cap, F.F. & M.P. Leubner, A model of the magnetopause using an angular distribution function for the incident particles, *J. Geophys. Res.* **97**, 5304 - 5306 (1974).

Chapman, S. & V.C.A. Ferraro, A new theory of magnetic storms, *Terr. Magnet. Atmos. Elect.* **36**, 77 and 171 (1931).

Choe, J.Y. & D.B. Beard & E.C. Sullivan, Precise calculation of the magnetosphere surface for a tilted dipole, *Planet. Space Sci.* **21**, 485 - 498 (1973).

Choe, J.Y. & D.B. Beard, The near Earth magnetic field of the magnetotail current, *Planet. Space Sci.* **22**, 609 - 615 (1974).

Fairfield, D.H., The ordered magnetic field of the magnetosheath, *J. Geophys. Res.* **72**, 5865 - 5877 (1967).

Fairfield, D.H., Average and unusual locations of the Earth's magnetopause and bow shock, *J. Geophys. Res.* **76**, 6700 - 6716 (1971).

Frank, L.A., Plasma in the Earth's polar magnetosphere, *J. Geophys. Res.* **76**, 5202 - 5219 (1971).

Heppner, J.P., M. Sugiura, T.L. Skillman, B.G. Ledley & M. Campbell, OGO-A magnetic field observations, *J. Geophys. Res.* **72**, 5417 - 5471 (1967).

Kennel, C.F. & R.Z. Sagdeev, Collisionless shock waves in high β-plasma, *J. Geophys. Res.* **72**, 3303 - 3341 (1967).

Kivelson, M.G. & T.C. Russel, Dependence of the polar cusp on the north-south component of the interplanetary magnetic field, *J. Geophys. Res.* **78**, 3761 - 3772 (1973).

Mead, G.D., Deformation of the geomagnetic field by the solar wind, *J. Geophys. Res.* **69**, 1181 - 1195 (1964).

Mead, G.D. & D.B. Beard, Shape of the geomagnetic field solar wind boundary, *J. Geophys. Res.* **69**, 1169 - 1179 (1964).

Mead, G.B. & D.H. Fairfield, A quantitative magnetospheric model derived from spacecraft magnetometer data, *J. Geophys. Res.* **80**, 523 - 534 (1975).

Olson, W.P., The shape of the tilted magnetopause, *J. Geophys. Res.* **74**, 5642 - 5651 (1969).

Russel, T.C., C.R. Chapell, M.D. Montgomery, M. Neugebauer & F.L. Scarf, OGO-5 observations of the polar cusp on November 1, 1968, *J. Geophys. Res.* **76**, 6743 - 6764 (1971).

Spreiter, J.R. & B.R. Briggs, Theoretical determination of the form of the boundary of the solar corpuscular stream produced by interaction with the magnetic dipole field of the Earth, *J. Geophys. Res.* **67**, 37 - 51 (1962).

Spreiter, J.R. & B.J. Hyett, The effect of a uniform external pressure on the boundary of the geomagnetic field in a steady solar wind, *J. Geophys. Res.* **68**, 1631 - 1642 (1963).

Spreiter, J.R., A.L. Summers & A.Y. Alksne, Hydromagnetic flow around the magnetosphere, *Planet. Space Sci.* **19**, 223 - 253 (1966).

Spreiter, J.R., A.Y. Alksne & B. Abraham - Shrauner, Theoretical proton velocity distribution in a flow around the magnetosphere, *Planet. Space Sci.* **14**, 1207 - 1220 (1966).

Williams, D.J. & G.D. Mead, Night side magnetosphere configuration as obtained from trapped electrons at 1100 kilometer, *J. Geophys. Res.* **70**, 3017 - 3029 (1965).

Wolfe, J.H. & D.D. McKibbin, Pioneer 6 observations of a steady-state magnetosheath, *Planet. Space Sci.* **16**, 953 - 969 (1968).

DISCUSSION

G.H. Voigt: (i) Why did you take $B_i = Z \cdot B_t^{dipole}$ inside the magnetopause? This term is valid only if the magnetopause is a plane outer face, and it neglects the effects of a curved boundary.
(ii) Your magnetopause is of tangential discontinuity type with vanishing normal B_n-field components just at the boundary. What happens if a constant interplanetary field is superimposed? Does the IMF ignore the magnetopause (i.e. $B_n^{IMF} \neq 0$ at the boundary)?

F.P. Stössel: (i) This is the usual approximation for the magnetic field at the boundary. It is true that is neglects the effects of the curvature. The curvature is already neglected in the magnetohydrostatic equation by neglecting $\mathbf{B}\nabla)\mathbf{B}$ I think that the approximation is not too bad, because the radius of curvature is very large compared with the thickness of the boundary and so the change in the magnetic field in the direction of the field lines is small compared to the change perpendicular to it.
(ii) The normal component of the interplanetary magnetic field is assumed to be annihilated by the magnetic field of the surface currents in the boundary according to the condition $\mathbf{B}\nabla p = 0$; so that at the magnetopause only the component tangential to the surface remains and only this component contributes to the magnetic pressure.

W.P. Olson: In your talk and in the previous one by Formisano many mechanisms have been suggested for changing the scale size of the magnetosphere. I would like to add to this list the influence of the hydrodynamic behaviour of the magnetosheath plasma. Using pressures obtained by Spreiter's group from supersonic flow past an object simulating the shape of the magnetopause and balancing them against the pressure of the magnetospheric magnetic field yield a magnetopause shape much more like the average shape indicated by recent boundary crossings. The calculated shape is more flared in both the noon meridian (and more blunt near the subsolar point) and the equatorial plane than the shape obtained using the assumption of specular reflection. The differences are as large as 2 or 3 Earth radii sunward of the dawn-dusk plane.

M. Dryer: (to W. Olson, following his comments and slides after Cap *et al.*'s paper). When you used Spreiter's wind-tunnel pressures to obtain your large blunted shape (in the noon-midnight meridian plane), were the test results obtained for *another* shape? If so, it will be necessary to iterate upon the actual required pressure distribution *and* shape in order to get the same results for both the laboratory *and* your analytical results. (I might point out that this philosophy was put forward in our paper of 9-10 years ago: Dryer & Faye-Petersen, *AIAA J.*, 1966).

W.P. Olson: What you say is correct. The computation I performed is clearly not self-consistent. I point out these results to illustrate that the magnetopause position is determined by many factors not included in the specular reflection - pressure balance formalism. In addition to the several mentioned in the last two papers, the use of pressures obtained from a hydrodynamic approach (as done by Spreiter's group) suggests large departures from the classical specular reflection shape. The resulting shape is more flared in both the equatorial regions and the noon-meridian plane at high latitudes.

LOW-ENERGY HEAVY-ION OBSERVATIONS ON THE IMP 7 SATELLITE

C.Y. Fan* & D. Hovestadt
Max-Planck-Institut für Extraterrestrische Physik,
Garching, Germany

G. Gloeckler
Department of Physics and Astronomy,
University of Maryland, USA

Abstract

H, He, and CNO ion pulses have been detected by an electrostatic deflection spectrometer on the IMP 7 satellite in the magnetotail, the magnetosheath, and in the region up-stream from the Earth's bow shock. The spatial distribution of the pulses indicates that the ions are accelerated in the up-stream region and in the near-Earth region of the neutral sheet.

1. Introduction

This paper is concerned with two questions related to the energetic electron and ion pulses frequently observed in the Earth's magnetosheath, magnetotail, and nearby interplanetary space: namely (i) Are these particles related to auroral particles or trapped radiations? and (ii) If not, then where are these particles produced?

2. Instrument

The measurements were made on the IMP 7 satellite with an electrostatic deflection spectrometer (EDS) and a scintillator-PM tube system (ULET). In the EDS system, a particle is first selected according to its energy per charge by means of an electrostatic field, and its energy T and charge Q are then determined separately by an additional measurement of the energy deposit of the particle in an Au-Si solid-state detector. Five detectors were placed in five pre-chosen deflection paths for

---

*Permanent address: Department of Physics, University of Arizona, Tucson, USA.

Knott & Battrick (eds.), The Scientific Satellite Programme During the International Magnetospheric Study, 187–197.
All Rights Reserved. Copyright © 1976 by D. Reidel Publishing Company, Dordrecht-Holland.

measuring positive ions in five different energy/charge intervals simultaneously and three detectors for measuring electrons. Due to the steepness of the energy spectra of these particles, however, only the first three in the positive ion channel, designated P1, P2A and P2B, and the first in the electron channel, designated N1, recorded a sufficient number of particles per 'spike' above their background to yield useful information. The scintillator-PM tube system was used primarily as an anti-coincidence device for a proportional counter telescope designed for measuring low-energy heavy cosmic rays (ULET). Because of its large geometry, it was sensitive enough to register bremsstrahlung produced by electrons of energies < 750 keV or penetrating electrons of energies $\gtrsim 750$ keV bombarding the system. The results obtained thus complement those of the N1 channel. Table 1 lists the characteristics of the rate channels relevant to the present discussion; the details have been published elsewhere [Fan, Gloeckler & Tums, 1971; Hovestadt & Vollmer, 1971].

The IMP 7 satellite was launched on 23 September 1972. It spins, with its axis of rotation perpendicular to the ecliptic plane, with a period of ~ 1.3 s. This orbit has an apogee of 38 R_e, a perigee of 34 R_e, and an inclination of $\sim 28°$ with respect to the ecliptic plane. The ion and electron pulse data obtained with the electrostatic deflection spectrometer and reported in this paper cover the period 23 September 1972 to 10 July 1973; for the electron pulses obtained with the scintillator-PM tube system, this period extends to 20 April 1974.

TABLE 1
Characteristics of the Electrostatic Deflection Spectrometer

Detector	Geometrical Factor (cm^2 sr)	Particle Type	Energy per Charge
P1 (EDS)	0.02	positive ion	125-160 keV/Q
P2A	0.03	positive ion	170-280
P2B	0.03	positive ion	380-650
N1	0.02	electrons	120-190 keV
A (ULET)			

Responds to bremsstrahlung produced by electrons of energies < 750 keV or penetrating electrons of energies ≥ 750 keV bombarding the housing of the anticoincidence cup. The energy threshold is ~ 300 keV.

3. Velocity Anisotropy

The particle pulses in which we are interested are distinguished from solar cosmic rays by their intensity/time profiles; they rise to their peak intensities and then fall back to low levels in times of usually less than 10 min. These particles, in general, flow in an earthward direction. Figures 1a-c show three computer plots of three

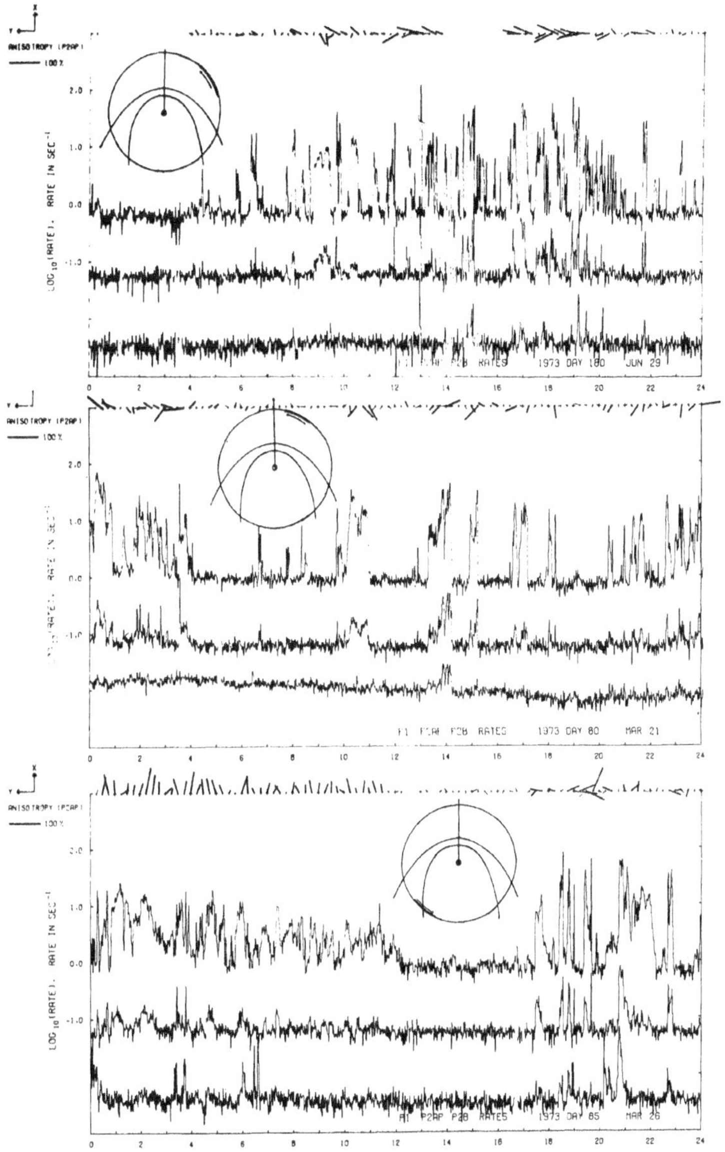

Figure 1a-c. *P1 (125 - 160 keV/Q), P2AP (170 - 280 keV/Q, Z > 2), and P2B (380 - 650 keV/Q) positive ion count rates recorded on (a) 29 June, (b) 21 March, and (c) 26 March 1973. The degrees of velocity anisotropy are exhibited in the top panel and satellite positions are shown as an insert.*

count rates, together with anisotropy information, for ion pulses recorded on 29 June, 21 March and 26 March 1973, respectively. The three rates designated as P1, P2AP, and P2B, are 81.8 s averages for particles in the ranges 125-160 keV/Q, 170-280 keV/Q with the energy threshold set for $Q \geq 2$, and 380-650 keV/Q, respectively. The positions of the satellite on these three days are shown as an insert to each of the figures. Judging from many plots, such as that in Figure 1, we conclude that ion pulses observed in the morning sector (Figs. 1a & 1b) appear to come from the bow shock, while those observed in the evening sector (Fig. 1c) show an anisotropy from the direction of the Earth.

4. Relative Abundance

In order to identify the nature of these particles, we plotted the number of events in each of the three energy ranges 0.125 - 0.160, 0.170 - 0.280, and

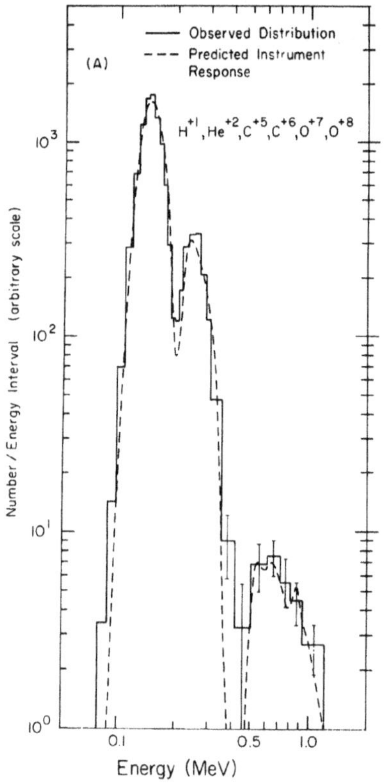

Figure 2. *Particle-distribution histogram for the P1 detector, showing H^{+1}, He^{+2}, and CNO group with charge states in the range +5 to +8.*

0.380 - 0.650 MeV/charge, as a function of their energies. After subtraction of the background, which was measured on board the spacecraft by turning off the high voltage of the EDS deflection plates, the distribution exhibits three distinct peaks, corresponding to particles of charges 1, 2, and a group of heavier ions. Figure 2 shows the distribution, in histogram form, of particles in the 0.125 - 0.160 MeV/Q group recorded on day 287 of 1972. The absence of charge groups 3 and 4 suggests that the three peaks are H^{+1}, He^{+2}, and CNO groups with charge states in the range +5 to +8. We can then determine the relative abundance of H : He: CNO = 9:1:4x10^{-2}. Since the relative abundance is very different from that of radiation-belt particles [Krimigis *et al.*, 1970; Van Allen *et al.*, 1970] and the degrees of ionisation are higher than the equilibrium charge states of ~ 100 keV/Q particles in a cold gaseous medium [Bloom & Sauter, 1971], we conclude that these ions are not particles escaping from the outer radiation belt or the auroral region, but rather solar-wind particles or low-energy 'quiet time' cosmic rays accelerated in the magnetosheath and in the magnetotail [Fan, Gloeckler & Hovestadt, 1975].

5. Morphology

To determine where these particles are accelerated, we have studied the satellite positions where the ion and electron pulses were detected. The following picture has emerged:

(i) Although ion pulses occur rather frequently in nearby interplanetary space (Figs. 1a & b), almost no electron pulses of significant intensity have been detected there.

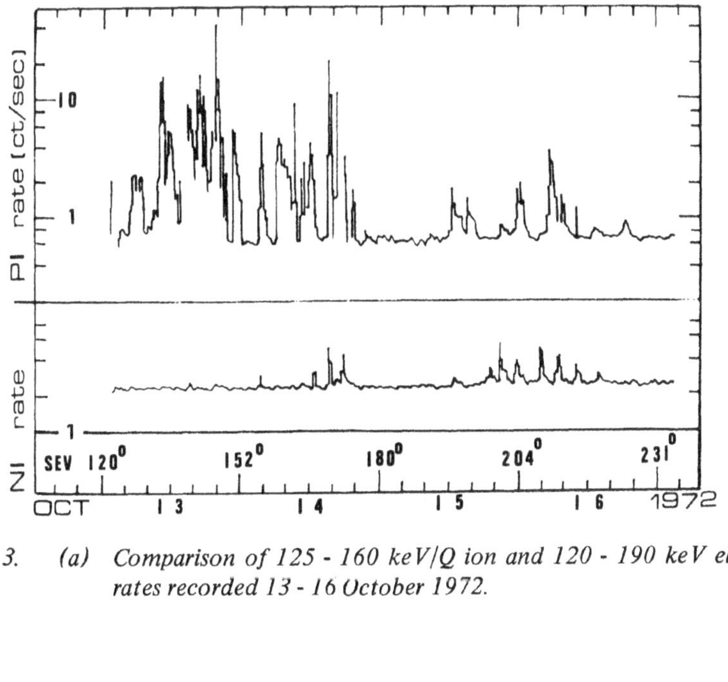

Figure 3. (a) Comparison of 125 - 160 keV/Q ion and 120 - 190 keV electron rates recorded 13 - 16 October 1972.

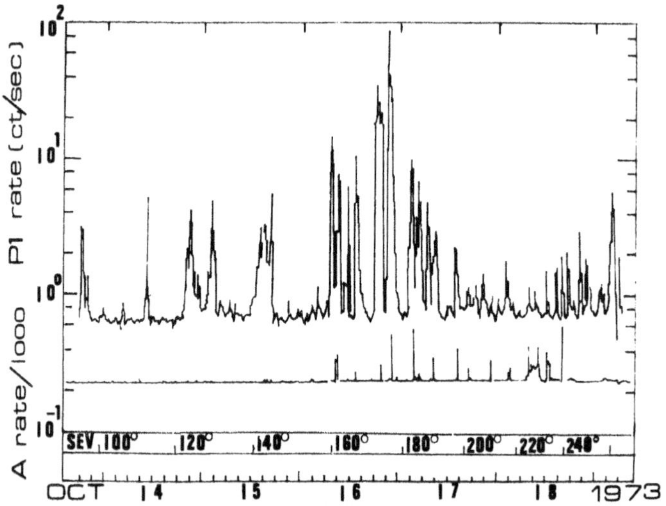

(b) Comparison of 125 - 160 keV/Q ion and E > 300 keV electron rates recorded 14 - 18 October 1973.

(ii) In the magnetosheath and tail, there are electron pulses, but only small numbers compared with those of ions. Furthermore, electron pulses are grouped in a region at larger Sun-Earth-vehicle (SEV) angles than the ion pulses. Comparisons of P1 (125 - 160 keV/Q ions) and N1 (120 - 190 keV electrons), and P1 and A (> 300 keV electrons), plotted against UT with the SEV angles of the satellite indicated, are shown in Figures 3a and b, respectively. It can be seen that on 13 and 14 October, when the SEV angles were 120° - 180°, numerous ion pulses were observed, whereas electron pulses appeared only later on 14 October. On 15 and 16 October, when the SEV angles were 180° - 231°, more electron pulses were detected, whereas ion pulses showed less activity. Several electron pulses appeared when ion counts were barely above the background level. In Figure 3b, again the difference is unmistakable.

To summarise the difference between ion and electron pulse morphology, we plotted the satellite positions projected on to the ecliptic plane where ion pulses were detected in Figure 4a and those where electron pulses were detected in Figure 4b. The sizes of the dots indicate approximately the flux values of these pulses.

6. Discussion

The morning/afternoon asymmetry in the morphology of the ion pulses in the upstream region beyond the bow shock (Fig. 4a) is similar to that of the electron pulses reported by Anderson [1968] and that of the proton pulses in the energy

(a)

(b)

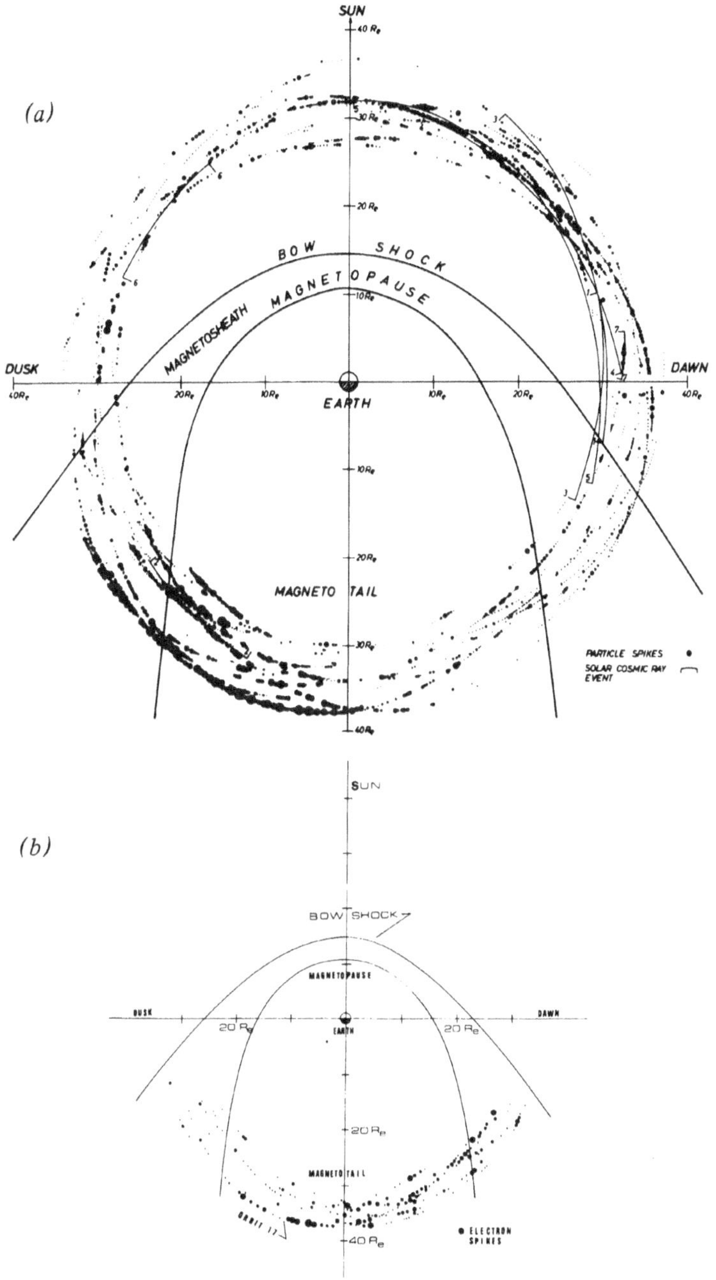

Figure 4. *Morphology of (a) ion pulses and (b) electron pulses projected onto the ecliptic plane. Note the difference in asymmetry with respect to the Sun-Earth line.*

ranges 3 - 10 keV, 30 - 100 keV, and 100 - 1300 keV reported, respectively, by Asbridge *et al.* [1968], Lin *et al.* [1974] and West & Buck [1974]. Such features can be understood if the electrons and the ions are associated with the bow shock. Since, in the upstream region, the particles are guided by the interplanetary magnetic field lines, the motion along the 'garden hose' angle leading away from the bow shock will result naturally in the observed asymmetry.

Regarding the origin of these particles, Lin *et al.* [1974] discussed two alternatives: (i) they are produced at the bow shock by whatever the mechanism may be and then travel upstream large distances, or (ii) they are produced throughout the large upstream interplanetary region, but only on the field lines connected to the bow shock. Based on the fact that the frequency of detecting the proton pulses did not seem to depend on the distance from the bow shock (up to 60 R_e) and there was no difference in the energy spectra measured at near and remote distances, they suggested that the protons were produced throughout the upstream region. The rapid rise and fall in the fluxes observed at the spacecraft arises from the changes in the directions of the field lines that connect or disconnect the spacecraft and the bow shock.

We have re-examined this question in the light of our experimental results and found that if the particles are indeed produced in the upstream region, then it is extremely hard to understand the strong anisotropy exhibited in Figures 1a & b. On the contrary, if the scattering mean free paths of these ions were greater than, say, 30 R_e and the duration of acceleration at the bow shock was longer than the switching on-off time of the lines of force at the spacecraft, we would not expect much attenuation of the ion pulses along the lines of force or any dispersion in the particle energy. We therefore favour the hypothesis that acceleration takes place at the bow shock, rather than in an extended region. On the question of whether the acceleration is actually taking place in front of the bow shock by a Fermi mechanism of the type proposed by Axford & Reid [1963] or the type proposed by Jokipii & Davis [1964], or behind the bow shock [West & Buck, 1974], our experimental results do not allow us to distinguish.

In 1968, Montgomery [1968] reported the morphology of $E > 64$ keV electron pulses observed by the Vela 2 satellite in the magnetotail. A more pronounced asymmetry was detected than is exhibited in Figure 4b. It was thought that this feature was merely a manifestation of the activity of the magnetotail and that the morning side is more active than the evening side. This may still be one of the causes for low-energy electrons, such as the electron pulses (~ 45 keV) reported by Anderson [1965], Montgomery [1968], Meng [1971], and Meng & Anderson [1970, 1971]. However, from the results that we have just presented, the effect of the electric potential across the tail lobe [Van Allen & Ness, 1969; Morfill & Scholer, 1972; Hynds *et al.*, 1974] must play a more important role in producing this observed polarisation.

It has frequently been suggested that the neutral sheet in the magnetotail is a likely site for acceleration of particles [Dungey 1961, 1962, 1963; Levy *et al.*, 1964; Axford *et al.*, 1965; Dessler & Michel, 1966; Alfvén, 1968; Dessler, 1968, 1971]. The reason is that in the neutral sheet an electric field is generated as a result of the annihilation of the magnetic field lines and a zero magnetic field is a required physical condition to allow charged particles to be accelerated along the direction of the electric field without suffering much deflection. Any turbulent residual magnetic field in the region will certainly inhibit the efficiency of the acceleration. This model seems to explain naturally the difference in the frequency of occurrence and the particle fluxes between the ion and electron pulses in the tail region. We should also note that the observation of a 'post shock spike' reported by Gloeckler *et al.* [1974] and Levy *et al.* [1974], also indicated an energy/charge or electric-field acceleration mechanism.

It is instructive to estimate the annihilation rate of magnetic field lines needed to explain the observed ion pulses. Let u be the speed at which the magnetic lines of intensity B are compressed into the neutral sheet of the Earth's tail lobe. If L is the breadth of the tail, the electric potential ϕ induced is then [Axford *et al.*, 1965]

$$\phi = \frac{uBL}{c}$$

Taking the maximum ion energy to be 600 keV according to our observation, $L = 20\ R_e$, and $B = 2\mathrm{x}10^{-4}$ G, we have

$$u = 2.3\mathrm{x}10^{7}\ \mathrm{cm/s}$$

or

$$u = 0.4\ u_A$$

where u_A is the Alfvén velocity for a particle density 0.5 cm^{-1} [Hones *et al.*, 1974].

This rapid rate of annihilation is perhaps another indication of the turbulent state of the accelaration region [Parker, 1973]. Direct measurements of the electric field and the plasma flow in the neutral sheet are now needed to further our understanding of the acceleration mechanism.

References

Electron and Ion Pulses

Anderson, K.A., H.K. Harris & R.J. Paoli, *J. Geophys. Res.* **70**, 1039 (1965).

Anderson, K.A., *J. Geophys. Res.* **70**, 4741 (1965); Proc. 9th Int. Conf. on Cosmic Rays, London, 520, 1966; *J. Geophys. Res.* **74**, 95 (1969).

Anderson, K.A., L.M. Chase, R.P. Lin, R.E. McGuire & J.E. McCoy, Correlated Interplanetary and Magnetospheric Observations, pp. 297-316, Reidel, Dordrecht, Holland 1974.

Armstrong, T.P. & S.M. Krimigis, *J. Geophys. Res.* **73**, 143 (1968).

Asbridge, J.R., S.J. Bame & I.B. Strong, *J. Geophys. Res.* **73**, 5777 (1968).

Bame, S.J., J.R. Asbridge, H.E. Felthauser, R.A. Olson & I.B. Strong, *Phys. Rev. Lett.* **16**, 138 (1966).

Bame, S.J., J.R. Asbridge, H.E. Felthauser, E.W. Hones, Jr. & I.B. Strong, *J. Geophys. Res.* **72**, 113 (1967).

Bame, S.J., The Earth's Particles and Fields (Ed. B.M. McCormac) pp. 317-325, Reinhold, New York, 1969.

Fan, C.Y., G. Gloeckler & J.A. Simpson, *Phys. Rev. Lett.* 13, 149 (1964); *J. Geophys. Res.* 71, 1837 (1966).
Fan, C.Y., G. Gloeckler & D. Hovestadt, *Phys. Rev. Lett.* **34**, 495 (1975).
Fennell, J.F., *J. Geophys. Res.* **75**, 7048 (1970).
Frank, L.A. & J.A. Van Allen, *J. Geophys. Res.* **69**, 4923 (1964).
Frank, L.A., *J. Geophys. Res.* **70**, 1593 (1965).
Gosling, J.T., J.R. Asbridge, S.J. Bame & I.B. Strong, *J. Geophys. Res.* **72**, 101 (1967).
Hones, E.W. Jr., S.-I. Akasofu, S.J. Bame & S. Singer, *J. Geophys. Res.* **77**, 6688 (1972).
Lin, R.P., C.-I. Meng & K.A. Anderson, *J. Geophys. Res.* **79**, 489 (1974).
Meng, C.I. & K.A. Anderson, *J. Geophys. Res.* 75, 1837 (1970); *J. Geophys. Res.* **76**, 873 (1971).
Meng, C.I., *J. Geophys. Res.* **76**, 862 (1971).
Montgomery, M.D., S. Singer, J.P. Conner & E.E. Stogsdill, *Phys. Rev. Lett.* **14**, 209 (1965).
Montgomery, M.D., *J. Geophys. Res.* **73**, 871 (1968).
Murayama, T. & J.A. Simpson, *J. Geophys. Res.* **73**, 891 (1968).
Singer, S., J.P. Conner, W.D. Evans, M.D. Montgomery & E.E. Stogsdill, *Space Research* **5**, 546 (1965).
Vasyliunas, V.M., *J. Geophys. Res.* **73**, 2839 (1968a); *J. Geophys. Res.* **73**, 7519 (1968b).
West, H.I. Jr. & R.M. Buck, *EOS Trans. Am. Geophys. Union* **55**, 693 (1974).

Alfvén, H., *J. Geophys. Res.* **73**, 4379 (1968).
Axford, W.I., H.E. Petschek & G.L. Siscoe, *J. Geophys. Res.* **70**, 1231 (1965).
Axford, W.I. & G.C. Reid, *J. Geophys. Res.* **68** 1793 (1963).
Bloom, S.D., & G.D. Sauter, *Phys. Rev.Lett.* **26**, 607 (1971).
Dessler, A.J., *J. Geophys. Res.* **73**, 209, 1861 (1968); *ibid* **76**, 3174 (1971).
Dessler, A.J. & F.C. Michel, Radiation Trapped in the Earth's Magnetic Field (Ed. B.M. McCormac) p. 447 Reidel, Dordrecht, Holland, 1966.
Dungey, J.W., *Phys. Rev. Lett.* **6**, 47(1961); *J. Phys, Soc. Japan* **17**, Suppl. A-11, 15 (1962); *Planet. Space Sci.* **10**, 233 (1963).
Fan, C.Y., G. Gloeckler & E. Tums, Proc. Int. Conf. on Cosmic Rays, Hobart, **4**, 1602, 1971.
Hones, E.W. Jr., A.T.Y. Lui, S.J. Bame & S. Singer, *J. Geophys. Res.* **79**, 1385 (1974).
Hovestadt, D. & O. Vollmer, Proc. 12th Int. Conf. on Cosmic Rays, Hobart, Tasmania **4**, 1608, 1971.
Hynds, R.J., G.E. Morfill & R. Rampling, *J. Geophys. Res.* **79**, 1332(1974).
Jokipii, J.R. & L. Davis, Jr., *Phys. Rev. Lett.* **13**, 739 (1964).
Krimigis, S.M., P. Verzariu, J.A. Van Allen, T.P. Armstrong, T.A. Fritz & B.A. Randall, *J. Geophys. Res.* **75**, 4210 (1970).
Levy, E.H., F.M. Ipavich & G. Gloeckler, *Geophys. Res. Lett.* **1**, 145 (1974).
Levy, R.H., H.E. Petschek & G.L. Siscoe, *AIAA J.* **2**, 2065 (1964).
Morfill, G. & M. Scholer, *J. Geophys. Res.* 77, 4021 (1972).

Parker, E.N., *Astrophys. J.* **180**, 247 (1973).
Van Allen, J.A., B.A. Randall & S.M. Krimigis, *J. Geophys. Res.* **75**, 6085 (1970).
Van Allen, J.A. & N.F. Ness, *J. Geophys. Res.* **74**, 71 (1969).

DISCUSSION

H.K. Hills: How long does a heavy mass peak persist? With ALSEP/SIDE we have seen magnetosheath data which has a proton peak and then a smaller peak at twice the energy per charge. But the ratio of energies is not always 2; it sometimes shifts to, say 2.5, over a very short time, such that we have been reluctant to conclude that we were seeing heavier ions.

C. Fan: It varies. The durations are in the range from a few seconds (our time resolution limit) to several minutes. The He : H ratio also varies. This is the reason why we believe that these are solar-wind particles accelerated to the observed energies instead of low-energy 'quiet time' cosmic rays, as we suggested in our earlier publication.

M. Dryer: Are you suggesting, then, that the heavy ions can easily penetrate the magnetopause (which, therefore, acts as a filter) into the tail region, where they may then be accelerated?

C. Fan: They are accelerated in the neutral sheet. From the velocity anisotropy measured in the tail, we conclude that the acceleration region must be near the Earth.

H. Balsiger: How do the heavy ions (CNO) behave when the H^+/He^{2+} ratio is changing? Do they vary in the same direction as He^{2+}?

C. Fan: At present, we have not made a systematic study of the variation. Preliminary results seem to indicate that heavier elements and He^{2+} vary in the same way.

K.P. Wenzel: What is the energy range of the electrons in the spikes you showed?

C. Fan: In the N1 channel, the energy range of the electron is 110-190 keV, whereas in the A channel, the energies of the electrons are > 300 keV.

LOW-ENERGY-PROTON REGIME IN THE GEOMAGNETIC TAIL AT LUNAR DISTANCE

H.K. Hills, D.A. Hardy & J.W. Freeman

Rice University, Houston, Texas, USA

Abstract

The Rice University Suprathermal Ion Detector Experiments (SIDE) on the lunar surface have observed a region of low-energy protons in the geomagnetic tail. This region is similar to the 'boundary layer' reported by Hones *et al.* from Vela observations and to the 'plasma mantle' reported by Rosenbauer *et al.* from HEOS-2 in the polar magnetosphere. The plasma is characterised by a proton differential energy spectrum with its peak usually at 50-100 eV, and is clearly a phenomenon distinct from the magnetosheath and the plasma sheet. These protons are seen outside the plasma sheet, but inside the inner edge of the magnetosheath. Continuous observations have been made over 17 R_E in the cross-tail direction and over 12 R_E in the direction perpendicular to the ecliptic. Integral fluxes observed are generally 10^6-10^8 cm^{-2} s^{-1} sr^{-1}, number densities 0.1–5 cm^{-3}, bulk velocity ~100–250 km/s, temperature ~ 0.4–5x10^5°K. The fraction of time these protons are seen while in this region of the tail varies widely from lunation to lunation.
Preliminary multispacecraft comparisons are being investigated, and it may be possible to follow the evolution of this region from near Earth out to lunar distance. There is reasonable expectation that one of the SIDE instruments will still be operating during the IMS.

The data to be presented in this paper come from a satellite which has been mentioned only briefly in this meeting; namely, the Moon. The Suprathermal Ion Detector Experiment, otherwise known from its initials as SIDE, is part of the Apollo Lunar Surface Experiments Package. Three of these instruments were placed on the lunar surface by the astronauts of Apollos 12, 14 and 15. Two of the SIDEs are still operating (more than 5½ years for the Apollo 12 instrument) and it is probable that at least one SIDE will still be operational during the time of the IMS.

Knott & Battrick (eds.), The Scientific Satellite Programme During the International Magnetospheric Study, 199–207.
All Rights Reserved. Copyright © 1976 by D. Reidel Publishing Company, Dordrecht-Holland.

The data to be presented here are derived from that part of the SIDE which measures positive ions in the energy range 10 eV through 3500 eV. This range is covered in 20 channels, which are centred at 10, 20, 30, 50, 70, 100 and 250 eV, and at 250 eV steps from 250 eV through 3500 eV. For further details of the instruments see Hills & Freeman [1971].

Much of the data obtained by SIDE pertains to the Moon itself, but since the Moon traverses the geomagnetic tail once each month, there is also good coverage for magnetospheric data. Observations of a new low-energy plasma regime in the tail are presented here. This new regime lies within the magnetosphere, but outside the plasma sheet, and it is speculated that this plasma is the same as (or closely related to) the 'boundary layer' reported by Hones *et al.* [1972] from Vela data, and also the 'plasma mantle' reported by Rosenbauer *et al.* [1975] from HEOS-2 data. See Hardy, Hills & Freeman [1975] for the first publication of these SIDE results, and Hardy [1974] for a more complete discussion of the data analysis.

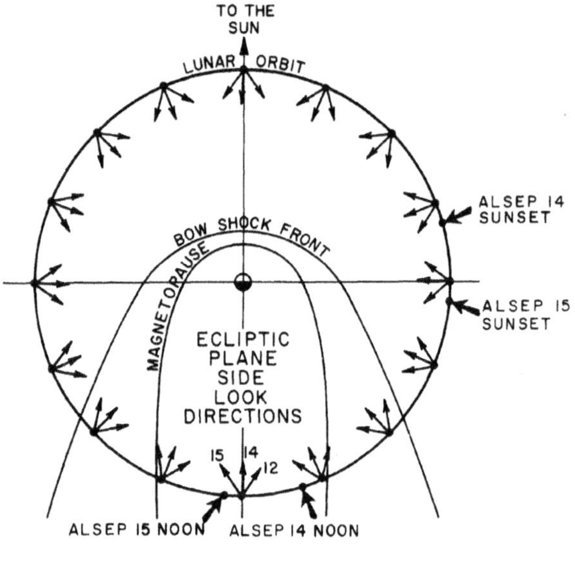

Figure 1. *Look directions of the three SIDE instruments. All look directions lie approximately in the ecliptic plane.*

Figure 2. *Apollo 14 SIDE counting rates for one lunation beginning with local midnight, 1302 GMT, day 253, year 1972. The times at the bottoms of the panels, from left to right, correspond to local times of midnight, sunrise, noon, and sunset. The numbers along the oblique axis denote days after the beginning of the panel. Each spectrum is a 20 min average of the counting rate vs. energy channel. Channels are identified by their central energy in eV.*

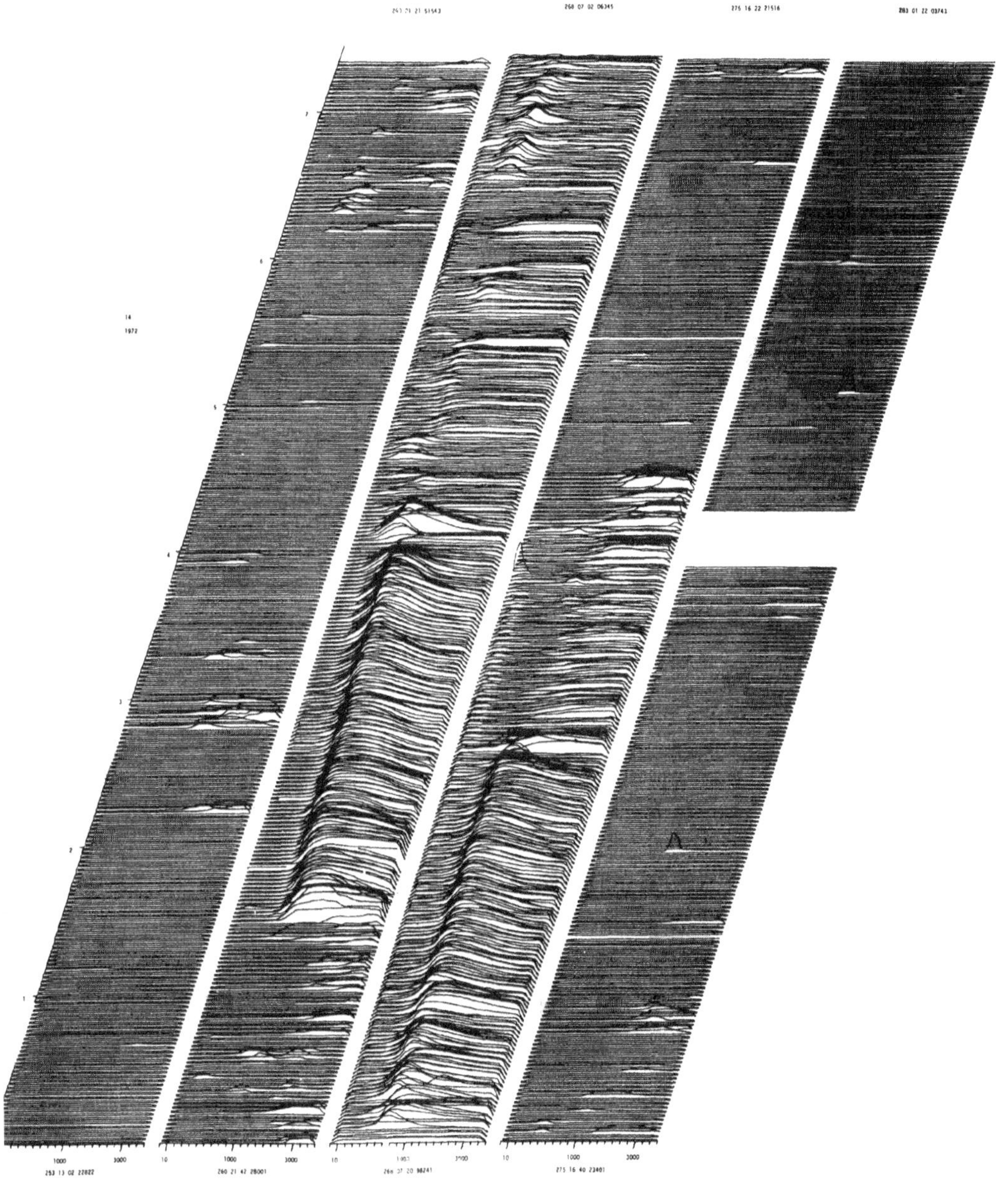

The magnetosheath spectrum, inbound, is seen from 1½ d to ~4 d after sunrise. The low-energy plasma is seen at ~4¾ d, followed closely by the plasma sheet, which is seen intermittently for ~1½ d. The low-energy plasma appears again from ~6½ d after sunrise until noon. The magnetosheath is observed again, outbound, from approximately noon to ~2½ d after noon.

Figure 1 depicts the orbit of the Moon, and shows the look directions of the three SIDEs at various points along the orbit. Note that the Apollo 14 look direction (the central one) is approximately 2° from the Earth (ignoring libration), with the Apollo 12 and 15 instruments looking to either side, spanning a 56° angle. Thus in the magnetosheath there is always one instrument looking approximately into the ion flow along the sheath.

Figure 2 shows one whole lunation of data, i.e. from local midnight at the Apollo side (approximately new Moon) to the next local midnight. Readily distinguishable by their characteristic ion spectra are the magnetosheath, the plasma sheet and the

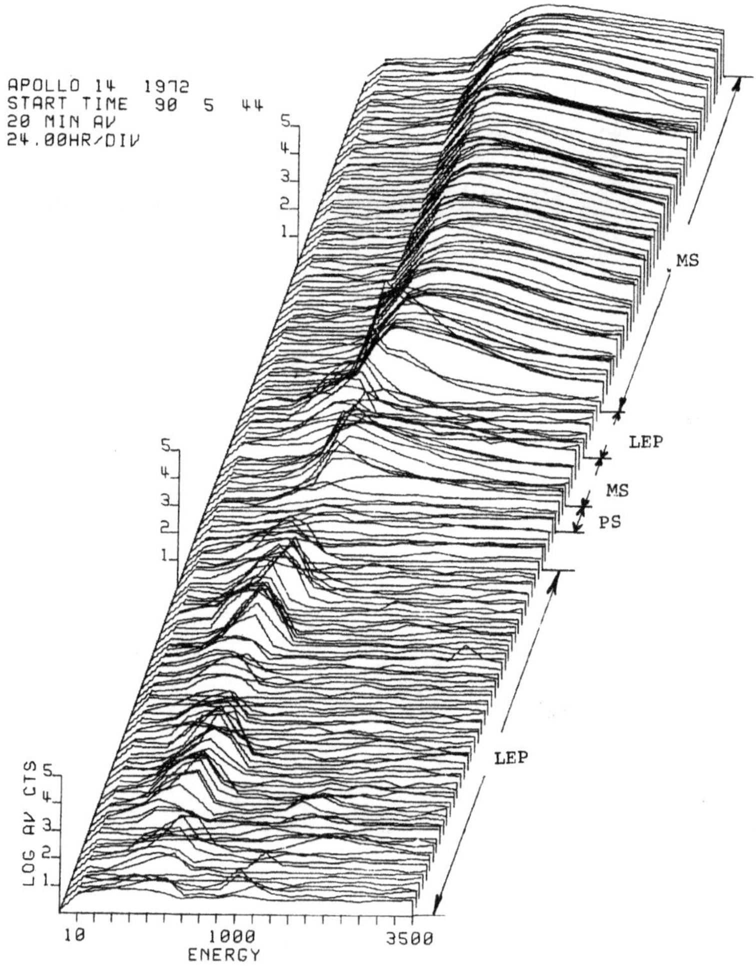

Figure 3. *A small section of a plot similar to Figure 1. The count-rate scales appear at 1 d intervals. See text for identification of the plasma regions.*

low-energy plasma region in the tail lobe. Note that the magnetosheath, plasma sheet, and low-energy plasma spectra are each observed in distinct regions, and are not observed simultaneously. A shorter segment of data is shown in a similar plot in Figure 3, where now the labels identify the spectra as plasma sheet (PS), low-energy plasma (LEP) or magnetosheath (MS). A 24-hour segment of data is shown in Figure 4, with numbers identifying individual spectra which are displayed in Figure 5.

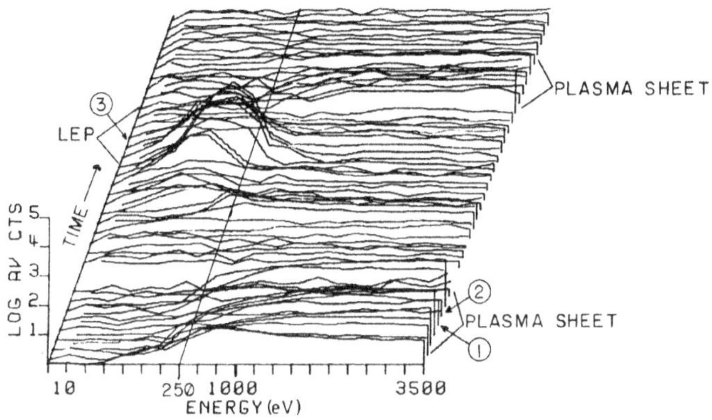

Figure 4. *24 h section of a plot similar to Figure 1. The numbers identify spectra shown in Figure 5.*

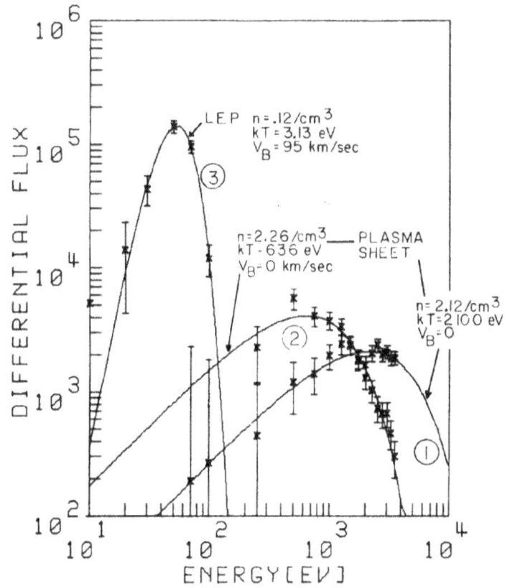

Figure 5. *Individual spectra from Figure 4, with fitted curves and resulting computed plasma parameters.*

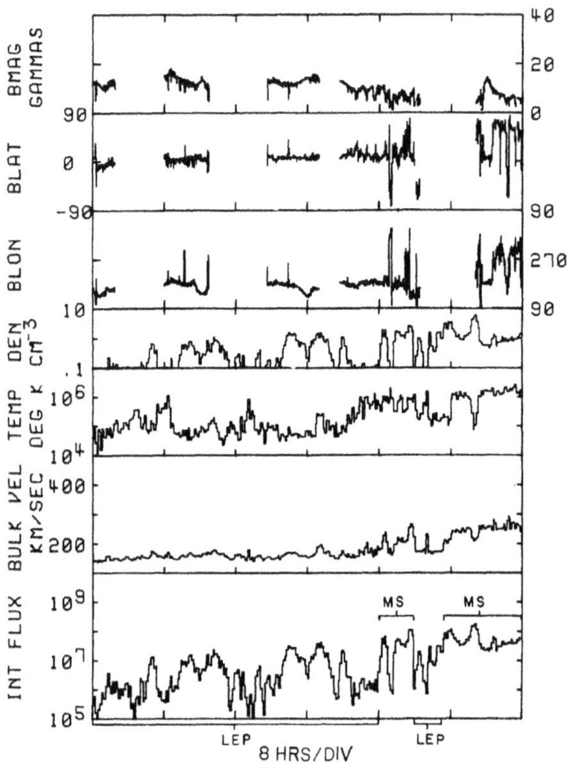

Figure 6. *Plasma parameters computed for a low-energy proton event. Plotted quantities, from bottom to top, are (a) integral flux in the range 10 – 3500 eV; (b) bulk velocity component into the detector; (c) plasma temperature, (d) proton density; (e) longitude (solar equatorial); (f) latitude; and (g) magnitude of the magnetic field at Explorer 35.*

The plasma parameters shown in Figure 5 are obtained by taking the usual moments of the distribution function. The density computation assumes that the flux observed is isotropic over a cone of directions with full angle of 50°. The two plasma-sheet proton spectra shown typify the range of temperatures commonly observed. These plasma-sheet spectra have bulk velocities which are consistent with zero, and have temperatures corresponding to kT of 636 eV and 2100 eV. On the other hand, the low-energy protons in addition to having a differential energy spectrum which is peaked typically at low energies (50 - 250 eV) have a bulk velocity shown here of 95 km/s and a low temperature, with kT = 3.1 eV. The Mass Analyser part of the SIDE has confirmed that these low-energy fluxes are consistent with 100% protons, but there could be a several percent admixture of heavier mass ions.

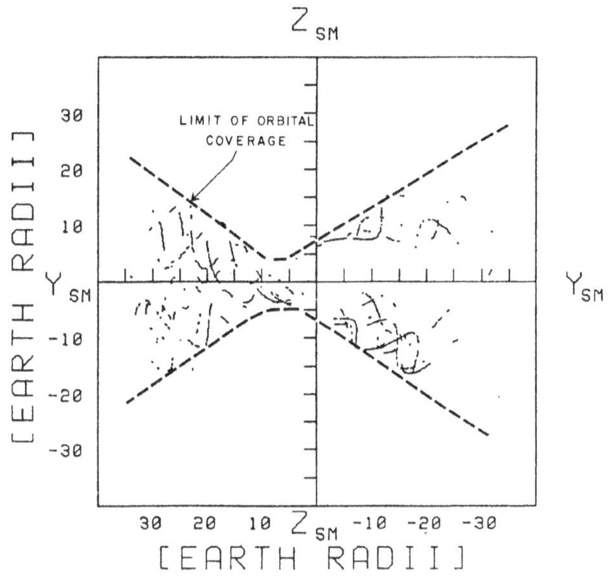

Figure 7. *Portions of the Moon's orbit where the low-energy protons were observed. Orbit coverage is not uniform over the area of the plot.*

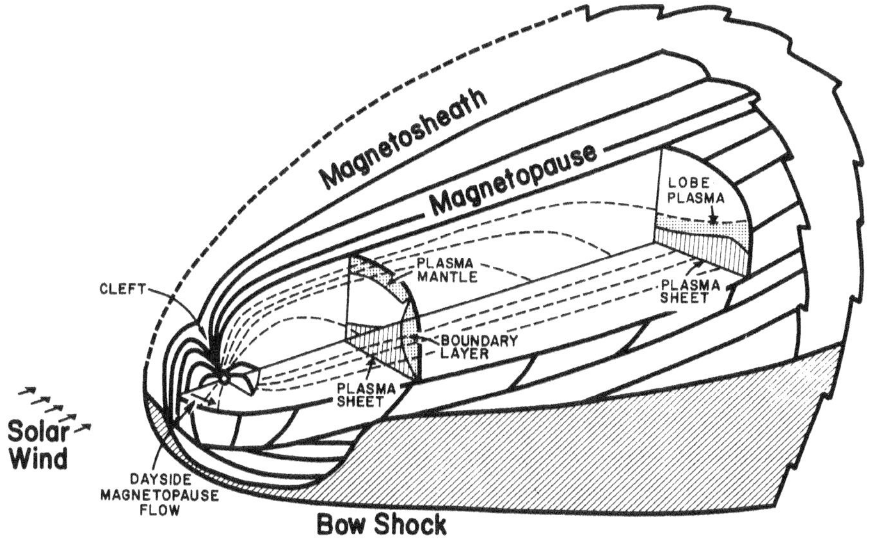

Figure 8. *Structure of the magnetosphere, after Heikkila, showing the low-energy lobe plasma region and related plasma observations.*

Plasma parameters and magnetic-field data are plotted as a function of time in Figure 6, which shows that the LEP regions are encountered in the well-ordered field of the lobe region. Note also the clear changes in the temperature and bulk velocity between the LEP and the magnetosheath regions. The most striking characteristic to use to distinguish the LEP from the other regions, however, is still the differential energy spectrum as seen in Figure 5.

Near the centre of the magnetotail, the orbit of the Moon always takes it relatively near the equatorial plane of the geocentric solar magnetospheric co-ordinate system. Consequently, we cannot say how far the LEP region extends in the $\pm Z_{sm}$ direction near the centre of the tail. However, Figure 7 shows with solid lines those parts of the orbits during which the LEP plasma was encountered, for 1972 and the first quarter of 1973. The dashed lines show the limits of the area traversed by the Moon. It is seen that the low-energy plasma may be found almost everywhere in the tail, when the instrument is not in the plasma sheet. Thus, we have drawn the sketch of the magnetosphere, after Heikkila, as shown in Figure 8. It appears that the low-energy protons observed by SIDE at 60 R_E are the same as the 'boundary layer' observed by Vela [Hones *et al.*, 1972] at 18 R_E and as the 'plasma mantle' observed by HEOS-2 [Rosenbauer *et al.*, 1975] in the polar magnetosphere closer to the Earth, and as a plasma flow observed by ATS-1 [Freeman *et al.*, 1968] near the nose of the magnetosphere. These names, though, seem to be too restrictive to apply to this region as it appears at 60 R_E since it appears to be found everywhere inside and adjacent to the magnetopause. However, the size and nature of the region separating the low-energy plasma from the magnetopause is yet to be clearly determined.

It is speculated that these particles enter near the nose or cusp regions and are convected towards the plasma sheet as they move down the tail, perhaps being accelerated at a neutral line further down the tail. There are several suggested tests of magnetospheric theory made possible by the present data, but these have not yet been carried out. Multi-spacecraft correlations are of great interest in further determination of the nature of the plasma regime.

Acknowledgements

The research has been supported by NASA Contract NAS 9-5911 and by the Atmospheric Science Section, National Science Foundation.

References

Akasofu, S.-I., E.W. Hones, Jr., S.J. Bame, J.R. Asbridge & T.T.Y. Lui, Magnetotail and boundary layer plasma at a geocentric distance of approximately 18 R_e; Vela 5 and 6 observations, *J. Geophys. Res.* **78**, 31, 7257 (1973).

Freeman, J.W., Jr., C.S. Warren & J.J. Maguire, Plasma flow directions at the magnetopause on January 10 and 14, 1969, *J. Geophys. Res.* **73**, 17, 5719 (1968).

Hardy, D.A., Observation of low-energy protons in the geomagnetic tail at lunar distances, MS Thesis, Rice University, Houston, Texas, 1974.

Hardy, D.A., H.K. Hills & J.W. Freeman, A new plasma regime in the distant geomagnetic tail, *J. Geophys. Res. Lett.* 2, 5, 169 (1975).

Hills, H.K. & J.W. Freeman, Jr., Suprathermal Ion Detector Experiment (Lunar Ionosphere Detector) Apollo 14 Preliminary Science Report, NASA SP-272, 1971.

Hones, E.W.,Jr., J.R. Asbridge, S.J. Bame, M.D. Montgomery, S. Singer & S.-I. Akasofu, Measurement of magnetotail plasma flow made with Vela 4B, *J. Geophys. Res.* 77, 28, 5503 (1972).

Rosenbauer, H., H. Grunwaldt, M.D. Montgomery, G. Paschmann & N. Sckopke, HEOS 2 plasma observations in the distant polar magnetosphere: The plasma mantle, *J. Geophys. Res.* (1975).

DISCUSSION

D.A. Bryant: It is possible that the low-energy plasma would map along geomagnetic field lines to the Earth to be located at the poleward boundary of the plasma sheet? If so, it might be the source plasma for the low-energy particles observed on the poleward side of auroral arcs.

H.K. Hills: We have not tried any such mapping of field lines. However, considering the broad region in which we observe these particles at lunar distance, I doubt if they would map down to the area mentioned; but we have not considered this question yet.

C. Fan: Did you correlate your observations with the thinning of the plasma sheet reported by the Los Alamos group?

H.K. Hills: No, we have not made the correlation study.

SOLAR PHENOMENA, WEATHER AND CLIMATE

J.W. King

Appleton Laboratory, Ditton Park, Slough, UK

Abstract

Various ways in which the lower atmosphere is influenced by solar phenomena ranging from short-lived events such as flares to the 11-year and 22-year sunspot cycles are described. Some important consequences for mankind of these 'Sun-weather relationships' are discussed.

1. Introduction

It has been suggested in many papers published in past decades that the weather is influenced by the 11- and 22-year sunspot cycles. The accumulated evidence is now so compelling that it seems no longer possible to deny the existence of strong connections between the weather and radiation changes associated with a whole range of solar phenomena. Research on 'Sun-weather relationships' is expanding in the United Kingdom; this paper reviews some of the relevant results that have been brought to light in the initial stages of this work. What are being sought now are: (a) the mechanisms responsible for the complicated Sun-weather relationships that exist and (b) greater insight into the practical consequences for agriculture, energy consumption and national economies which stem from these relationships.

2. Relationships between the Weather and the Solar Cycle

Figure 1 shows the result of a rainfall analysis published by Xanthakis [1973]. He observed that at high northern latitudes (70° - 80°), the 11-year solar cycle was positively correlated with a 10 cm oscillation in the annual rainfall total; at latitudes between 60°N and 70°N, the opposite behaviour was observed, while at lower

Knott & Battrick (eds.), The Scientific Satellite Programme During the International Magnetospheric Study, 209–222.
All Rights Reserved. Copyright © 1976 by D. Reidel Publishing Company, Dordrecht-Holland.

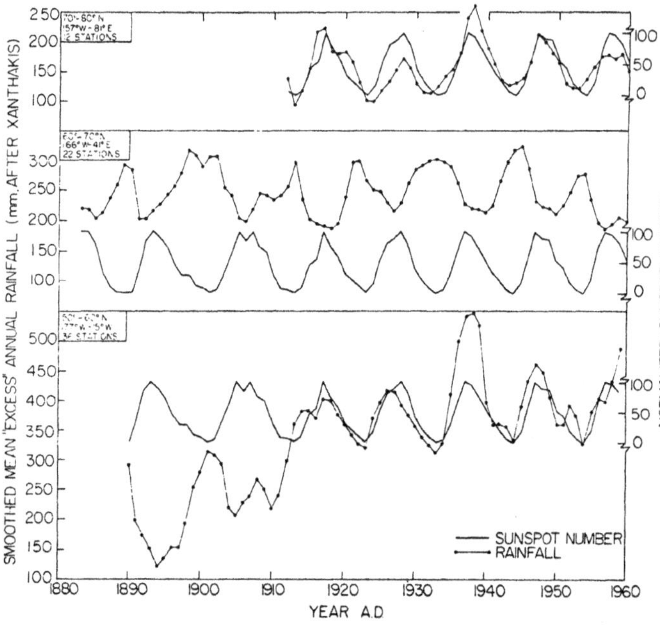

Figure 1. *Smoothed mean 'excess' annual rainfalls in three northern hemisphere zones [after Xanthakis, 1973] compared with the corresponding 'normalised' annual sunspot numbers.*

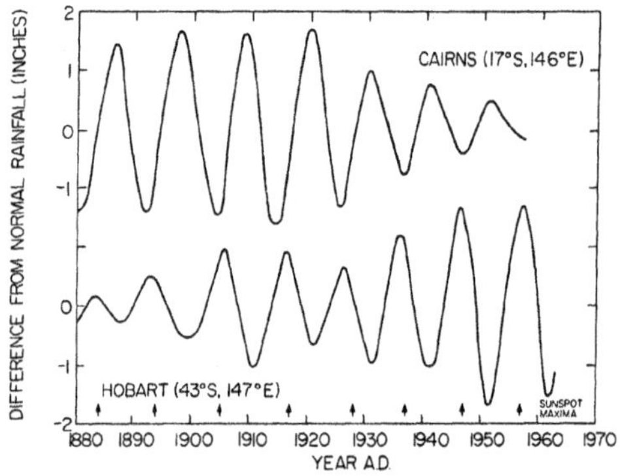

Figure 2. *Curves showing the difference between the actual and normal annual rainfalls at two latitudes in Australia. The curves were obtained by applying an 8-15 yr filter to the annual differences, but it is significant that the extremes of the resulting oscillations occur very close to sunspot minima and maxima [after Bowen, 1974].*

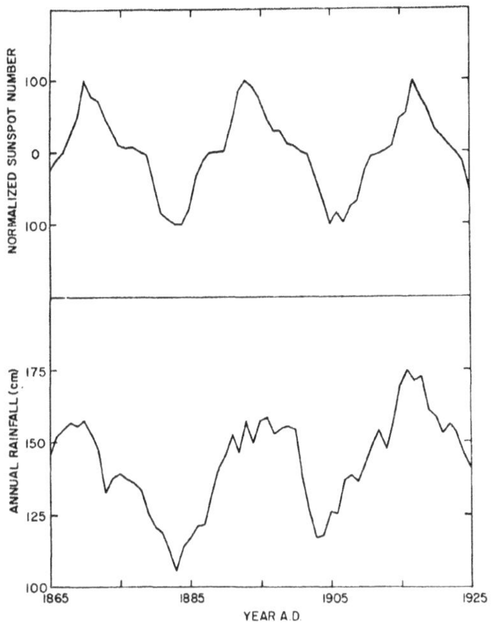

Figure 3. *Upper box: 22-year cycle of normalised annual sunspot numbers. Lower box: Smoothed annual rainfall at Fortaleza in Brazil [after Markham, 1974].*

latitudes still the correlation was negative before about 1915 AD and positive after that. Bowen [1974] reported similar behaviour in the southern hemisphere (Fig. 2) where the phases of the rainfall oscillations associated with the 11-year sunspot cycle are opposite at 17°S and 43°S. The solar-cycle-induced rainfall oscillations referred to range from about 3% to 50% of the normal annual total; a reduction of the rainfall by 25% in each of several years around one of the extremes of the sunspot cycle is obviously of considerable importance.

Figures 3 and 4 show that the annual rainfall totals at Fortaleza in Brazil and at three locations in South Africa were positively correlated with the 'double' sunspot cycle for considerable periods of time. The modulation associated with the double cycle amounted to about 35% of the average annual total at Fortaleza and to about 25% of the average rainfall at the South African stations. Data from Fortaleza are available from 1865 onwards, but after the first 60 yr the relationship between the rainfall and the double cycle changed phase.

Particularly impressive evidence of an association between rainfall and the double sunspot cycle is given in Figure 5, which shows the date within each year from 1844 to 1944 by which one quarter of the year's rainfall at Adelaide (35°S, 139°E) had fallen. This 'quartile date' [Cornish, 1954] oscillates between about April 15 and May 25 in phase with the double sunspot cycle [Whipple, 1936]. Cornish showed

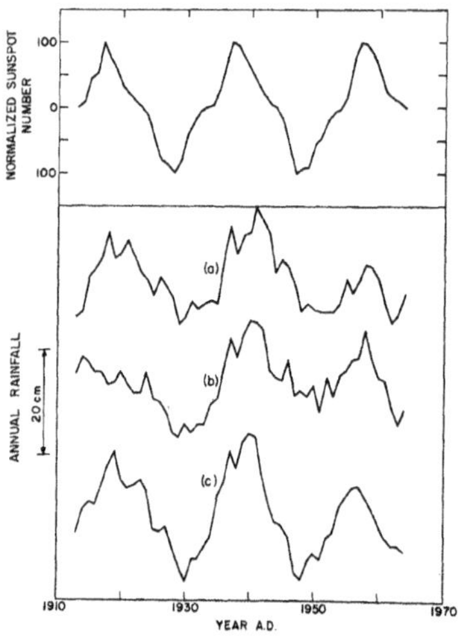

Figure 4. *Upper box: 22-year cycle of normalised annual sunspot numbers. Lower box: Smoothed annual rainfall at three locations in South Africa [after Tyson, 1974]. (a) Rustenburg; (b) Bethal; (c) Dundee.*

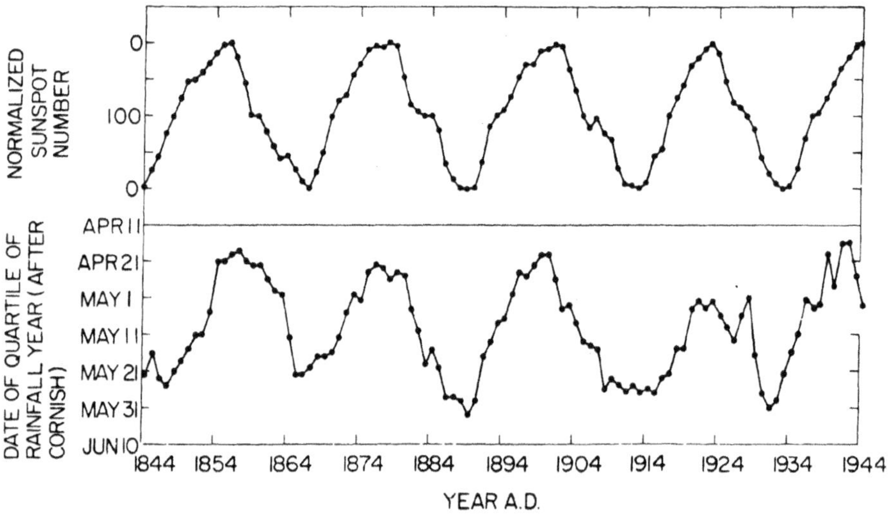

Figure 5. *Upper box: 22-year cycle of normalised annual sunspot numbers. Lower box: 10-year smoothed means of the date by which one quarter of the yearly rainfall fell at Adelaide, Australia [after Cornish, 1954]. The Cornish data points are slightly displaced because he used smoothed means for an even number of years.*

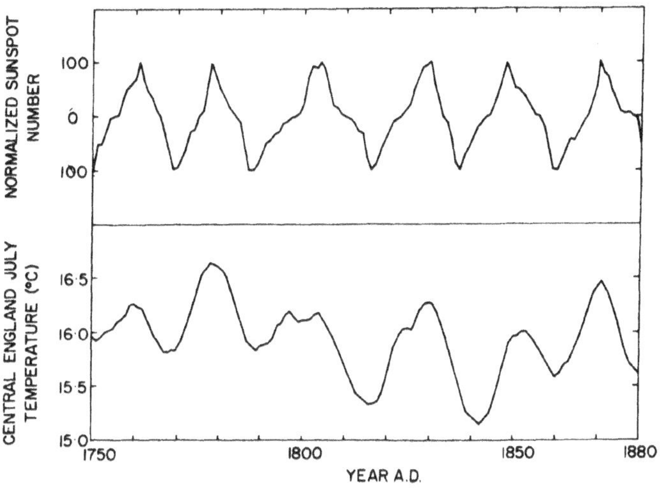

Figure 6. *Upper box: 22-year cycle of normalised annual sunspot number. Lower box: Smoothed central-England July temperatures* [*after Manley, 1974*].

that the dates of the other rainfall 'octiles' exhibited similar oscillations; obviously changes of up to six weeks in the date by which various fractions of the annual rainfall total have fallen are of considerable importance to agriculture.

Another striking illustration of the influence of the double sunspot cycle on the weather has been provided by King *et al.* [1974] who showed (Fig. 6) that during the period 1750-1880 the July temperature in central England exhibited an oscillation of nearly 1°C in phase with the double sunspot cycle. The temperature variation was somewhat anomalous during the years 1840-1855, when the temperature extremes occurred around sunspot minima instead of near sunspot maxima.

The double sunspot cycle appears to influence the climate of the United States in several ways. Thompson [1973] and Roberts [1974] have shown that droughts in various parts of the country occur around every second sunspot minimum; Newman [1965] has shown that winter temperatures in Boston exhibit a 22-yr periodicity and Thompson [1973] has reported that the correlation between the July/August temperature in the corn belt of the United States and the double sunspot cycle is 'remarkable'. Striking evidence relating the double cycle and temperatures in the USA has been provided by Mather [1974].

Figure 7 shows how the incidence of lightning in Great Britain, plotted on an arbitrary scale designed to indicate variations in the number of power failures associated with lightning strikes, varied during four sunspot cycles [Stringfellow, 1974]. The number of events varied by a factor of nearly two in phase with the sunspot cycle. Brooks [1934] first showed that a significant relationship existed between the occurrence of thunderstorms and the annual sunspot numbers. Correlation coefficients up to 0.91 were obtained from long series of data, with the largest

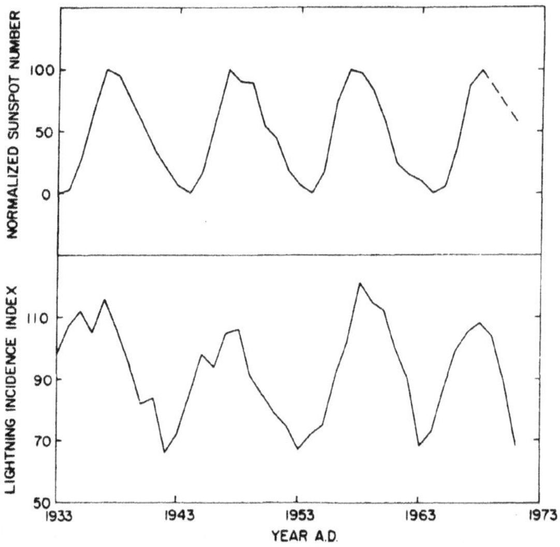

Figure 7. *Upper box: 11-year cycle of normalised annual sunspot numbers. Lower box: 5-year means of the annual lightning incidence index for Great Britain, plotted on a scale proportional to the frequency of lightning strikes on the electricity distribution network* [*after Stringfellow, 1974*].

correlation coefficients relating to high latitudes. Brooks & Carruthers [1953] wrote 'There is a correlation coefficient of + 0.88 between the number of thunderstorms recorded in Siberia and mean annual sunspot relative number. Since it is inconceivable that thunderstorms in Siberia cause sunspots, it is reasonable to assume that sunspots or some other solar phenomenon associated with sunspots cause thunderstorms'. They also established statistically that the observed variation of the frequency of thunderstorms in the West Indies during the sunspot cycle was 'definitely significant'.

3. The Response of the Lower Atmosphere to Short-Lived Solar Phenomena

In recent years several analyses have indicated that short-lived solar phenomena may trigger a response in the lower atmosphere. In order to investigate the tropospheric response to solar flares, Schuurmans [1965] computed the change in the height of the 500 mb level over much of the northern hemisphere during the first 24 h after each of 53 flares; he concluded that the pattern of height changes 'shows a remarkable regularity with a symmetry with respect to the geomagnetic rather than to the geographic pole. The main features suggested by the picture are a rise in

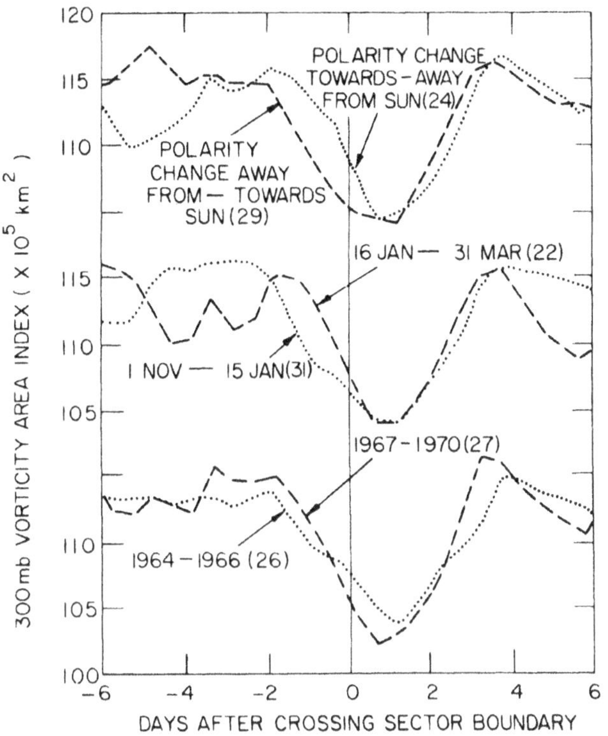

Figure 8. *Average behaviour of the northern hemisphere 'vorticity area index' around the dates of 53 interplanetary magnetic field sector boundary crossings, winter, 1964-1970* [*after Wilcox et al., 1974*].

height in the region of the geomagnetic pole and a wide region of fall in height coinciding quite well with the auroral zone'. Reiter [1973] showed that the influx of stratospheric air to the 3 km level is enhanced on the second or third day after an H_α and/or X-ray flare. He concluded that 'the use of this statistically evident solar-terrestrial relationship would be in order in the practical forecasting of influxes of stratospheric air into the biosphere'. Results such as these show that the circulation of the lower atmosphere is significantly modified after solar flares. The circulation, as measured by the 'vorticity area index' at various heights in the troposphere [Wilcox *et al.*, 1974], also appears to be modified around times when the interplanetary magnetic field sector boundaries cross the Earth. These boundaries originate at the Sun and sweep across the Earth as they rotate during the 27-day solar rotation. Figure 8 shows how the northern hemisphere vorticity area index varied on average during the 53 sector boundary crossings for which dates were known. It appears significant that similar results were obtained irrespective of how the data set was subdivided: three different methods were used (Fig. 8) and in each case the average vorticity area index varied by about 10%.

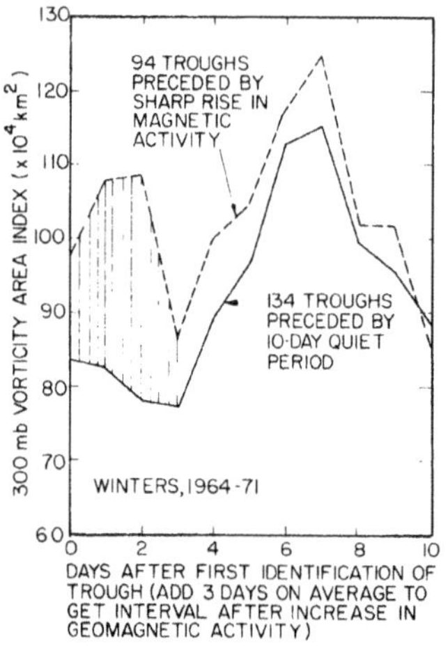

Figure 9. *Average variation of the vorticity area index associated with winter troughs over the northern Pacific and North America during the first 10 days after the troughs were identified. Broken curve: Troughs preceded by magnetic disturbances. Continuous curve: Troughs preceded by 10 magnetically quiet days [after Roberts & Olson, 1973].*

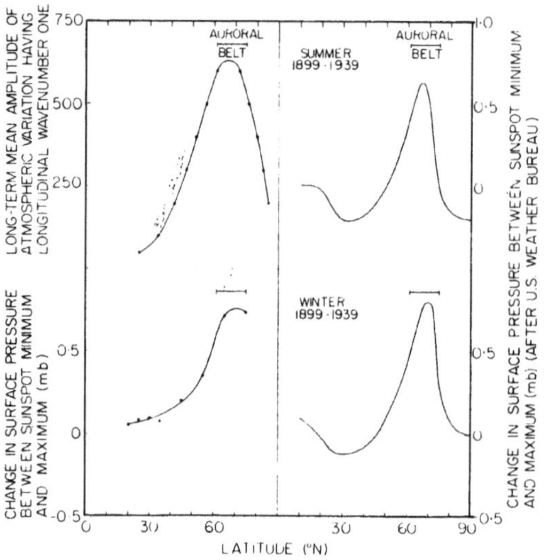

Figure 10. *Evidence suggesting that Sun-weather relationships are most pronounced in the auroral region. Upper and lower right: US Weather Bureau data [Wexler, 1953]. Upper left: From data published by Van Loon et al. [1973]. Lower left: From data published by Miles [1974].*

Solar phenomena tend to recur with periods of the order of 27 days, the synodic period of revolution of the Sun. Panofsky [1967] has shown that 'this period is close to the period of the maximum fluctuations of mean west winds' at upper middle latitudes. Reihl [1956] concluded, after an investigation of the 500 mb circulation along the auroral belt, that 'circulation increases and decreases take place with a period which can be related to the mean solar rotation in the (solar) equatorial zone'. Berkes [1955] has reported the existence of a 27-day periodicity in wind speed, and Egyed [1961] noted a 27-day periodicity in soil temperature.

Many authors, including Sazonov [1965], Mustel [1966], Beynon & Winstanley [1969], Roberts & Olson [1973], Stolov & Shapiro [1974] and Sidorenkov [1974], have reported associations between magnetic activity and lower atmospheric phenomena. For example, Figure 9 shows that the vorticity area index for troughs preceded by a sharp rise in magnetic activity is enhanced by 40% five days after the magnetic disturbance. Stolov & Shapiro [1974] analysed 24 yr of 700 mb heights from 20°N to 70°N and concluded that 'the 700 mb height difference between 20°N and 55°N increases significantly in winter four days following geomagnetic disturbance'.

4. The Role of the Earth's Magnetic Field in Sun-Weather Relationships

Several studies have shown that Sun-weather relationships are most pronounced in the vicinity of the auroral zone. Sarukhanyan & Smirnov [1970], for example, concluded 'the pattern of the amplitude distribution of the 11-year variation of atmospheric pressure shows that the amplitude increases only up to 70°N. Further north the amplitudes decrease. Thus the maximum amplitudes of the 11-year variations are observed in the 65-70°N zone. This zone is known to have the maximum density of solar corpuscular radiation, maximum occurrence of auroras, and maximum frequency of magnetic disturbances'.

Figure 10 shows the latitudinal variation of the change in surface pressure (averaged around the Earth at particular latitudes) between sunspot minimum and sunspot maximum, calculated by the author from data published by Miles [1974]. The pressure variation associated with the 11-year sunspot cycle is particularly large in the auroral belt; unfortunately Miles did not publish data for higher latitudes and it is not possible to use his data to confirm the observation of Sarukhanyan & Smirnov that the effect decreases at latitudes above 70°N; striking confirmation of this observation has, however, been provided by the work of various authors including Clayton [see Wexler, 1953] and is reproduced in Figure 10. The evidence suggesting that the lower atmosphere is influenced by the auroral zone is perhaps not conclusive, but it is certainly sufficient for us to hypothesise that some of the observed Sun-weather relationships are caused by energetic particles that enter the atmosphere at auroral latitudes and, through some unknown mechanism, influence the circulation of the lower atmosphere. If this were the case, it would be expected that the geomagnetic field as a whole would influence the lower atmosphere. Several authors including Sazonov [1974] and King [1974a] have in fact drawn attention to the similarity between meteorological and geomagnetic contour maps. Figure 11,

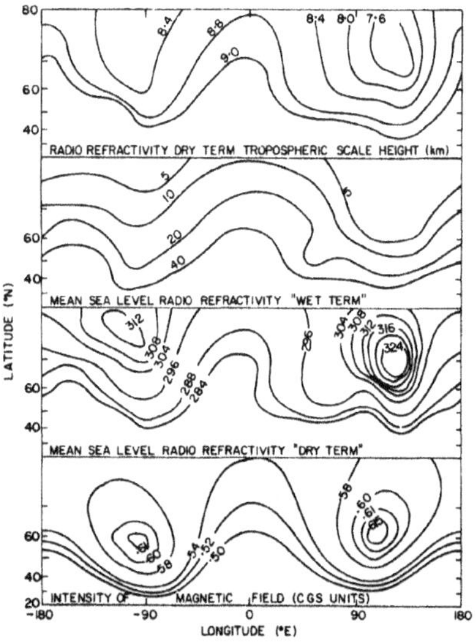

Figure 11. *Comparisons [after King, 1974b] of the spatial variations of three different meteorological parameters with the intensity of the geomagnetic field. The upper three boxes contain maps, published by Bean et al. [1966], which essentially show how the average tropospheric temperature, surface humidity and surface pressure, respectively, vary over most of the northern hemisphere in November. The lowest box contains a map of the intensity of the Earth's magnetic field.*

published by King [1974b], illustrates the remarkable similarity between the spatial variations of three different meteorological parameters and the intensity of the geomagnetic field. Such a high degree of similarity between different parameters is unusual in geophysics and, although it does not prove anything, its implications should not be lightly dismissed, particularly since existing models of the atmosphere do not satisfactorily explain the behaviour observed at high latitudes.

If the morphology of the lower atmosphere is indeed controlled by the Earth's magnetic field in some way, changes of the field which occur as the nondipole component 'rotates', or as the dipole field itself changes, may well be accompanied by climatic variations. Figure 12 shows that variations of temperature and magnetic intensity in the United Kingdom appear to have been associated during this century. These results, and many similar ones reported by Wollin *et al.* [1973, 1974], show that the temporal variations of magnetic intensity and temperature at various places are inversely correlated. Strong negative correlations have occurred not only over decades, but over much longer time spans. Figure 13, for example, shows oxygen isotope data (which provide a measure of temperature) and magnetic intensity values obtained from a single deep-sea core formed during a 500 000 yr period; cold epochs occurred when the magnetic intensity was relatively high and vice versa.

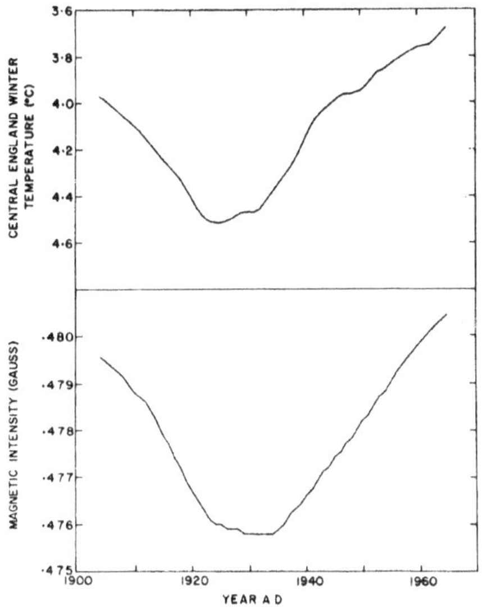

Figure 12. *Upper box: Smoothed central-England winter temperatures [after Wollin et al., 1973]. Lower box: Intensity of the geomagnetic field measured in England [after Wollin et al., 1973].*

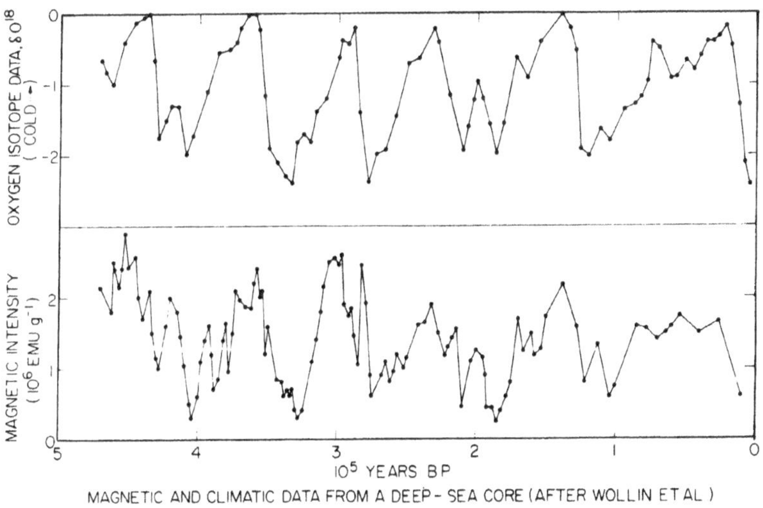

Figure 13. *Magnetic and climatic data obtained from a single deep-sea core [after Wollin et al., 1974]. Upper box: Oxygen isotope data used as an indicator of temperature. Lower box: Intensity of magnetisation of the core.*

5. Mechanisms that may be involved in Sun-Weather Relationships

It has been shown that various empirical Sun-weather relationships exist and that the Earth's magnetic field may play a role in bringing some of these about. It would, therefore, be useful to identify geophysical phenomena that vary during the solar cycle and that are also influenced by the Earth's magnetic field. One such phenomenon is cosmic rays; Ney [1959] drew attention to the fact that at the 500 mb level the amount of ionisation associated with cosmic rays decreases by a factor of about four between sunspot minimum and maximum and he suggested that 'if there is a connection between atmospheric ionisation and thunderstorm activity, for example, a solar-cycle modulation might be observed in climatological data'. The effects of the sunspot-cycle-induced changes of lower-atmosphere conductivity on atmospheric electric fields, cloud formation, temperatures, thunderstorm activity and rainfall do not appear to be clearly known at the present time and the possibility that the solar-cycle variation of cosmic rays may lead to certain Sun-weather relationships cannot be rejected.

Another atmospheric parameter which varies during the sunspot cycle and possibly also during short-lived solar events is the total ozone content. Various authors including Willet & Prohaska [1965], Christie [1973] and Paetzold [1973] have described variations of ozone content that occurred during the sunspot cycle, while Dobson *et al.* [1929] reported that 'there is a small but definite tendency for days with much ozone to be associated with magnetically disturbed conditions'. Weeks *et al.* [1972] concluded that the ozone content is reduced during strong solar proton events. Although it is not known whether such temporal variations of ozone cause or are caused by weather effects, it appears that ozone and weather disturbances are associated.

Apart from cosmic rays and ozone, other mechanisms which may possibly be involved in Sun-weather relationships include:

(a) The influence of energetically weak upper-atmosphere phenomena on lower-atmosphere systems which are close to resonance [Green, 1974].

(b) The solar-cycle variation of the frequency of occurrence of short-lived phenomena; the response of the troposphere to the 11-year sunspot cycle may represent merely an integration of the effects caused by short-lived phenomena such as solar flares or geomagnetic storms which are much more frequent at sunspot maximum than at sunspot minimum.

(c) The effect of water vapour formed from solar-wind protons as suggested by de Turville [1961a, 1961b].

(d) The effects of possible, but as yet unknown, solar-cycle changes in the 'Solar Constant', i.e. in the total electromagnetic energy radiated by the Sun.

(e) Variations of the efficiency of the coupling between the solar wind and the magnetosphere as the direction of the interplanetary magnetic field varies [Svalgaard & Wilcox, 1974] during the sunspot cycle.

(f) The variable reflection by the upper atmosphere of upward-propagating gravity-wave energy [Hines, 1974].

6. Implications of Sun-Weather Relationships

6.1. ENERGY BUDGETING AND ELECTRICITY DISTRIBUTION

Mather [1974] has shown that between 1900 and 1960 in Maryland, Delaware and Virginia, USA, oscillations of temperature averaging 2.5°C (4.5°F) occurred with 'a marked cyclic movement with about 22 yr between successive peaks; these oscillations are in phase with, and there seems little doubt that they are associated with, the 22-year sunspot cycle. A 2.5°C temperature oscillation extending over large areas will obviously affect the demand for energy; it should be noted in this connection that the frequency of lightning strikes affecting electricity supplies (Fig. 7) is strongly correlated with the sunspot cycle in certain parts of the World.

6.2. FOOD PRODUCTION AND FAMINES

Several authors have drawn attention to the influence of the sunspot cycle on agricultural productivity; relationships between the solar cycle and agriculture are hardly surprising when one is reminded, for example, that the phase of the annual rainfall variation can vary by six weeks during the double solar cycle (Fig. 5).

Bray [1965] concluded after a study of forest growth that there appears to be a 'relationship between growing season, climate and solar activity' while King [1973] showed that the length of the growing season estimated for agricultural purposes varies in phase with the sunspot cycle. King *et al.* [1974] used the World wheat production figures for 1949-73 to show how the sunspot cycle influences food production; they concluded that the present low level of World food reserves may be associated with the decline in solar activity during the approach to the 1974-75 sunspot minimum.

6.3. INADVERTENT MAN-MADE WEATHER MODIFICATION

Helliwell & Katsufrakis [1974] have reported that the Canadian electrical power distribution system appears to trigger VLF emissions in the magnetosphere. Such emissions are known to induce electron precipitation and, if particle events do indeed affect the circulation of the troposphere as suggested in Section 3, a real possibility exists that electricity distribution systems may influence the lower atmosphere.

6.4. NATIONAL ECONOMIC SYSTEMS

Several authors have gone so far as to suggest that solar-cycle-induced meteorological effects may be sufficiently important to influence the economies of industrialised nations; a review of this subject was published by Williams [1961]. Current opinion on this matter [Sparkes, 1974] seems still to be that reached by the economist J.M. Keynes [1961] who concluded that the possibility of solar-cycle-induced meteorological phenomena playing a part in harvest fluctuations which in turn affect the trade cycle 'is not to be lightly dismissed'.

7. Conclusions

All the mechanisms referred to in Section 5 require further study in order to determine whether or not they play a role in Sun-weather relationships. It appears unlikely that significant progress will be made in this field until the global morphology of, and particularly the role of the auroral zone in, Sun-weather relationships is known. The coincidence of major international meteorological and geophysical programmes (GARP and the IMS) later in this decade may well stimulate and provide a golden opportunity for collaboration between two groups of scientists who must surely interact more than they do at present – meteorologists and space physicists.

8. Acknowledgment

The author wishes to thank Dr. J.A. Saxton, Director of the Appleton Laboratory, for supporting this work.

References

Full details of the papers referred to can be obtained from the author.

PART 4

PARTICLES AND MAGNETOSPHERIC INSTABILITIES

MULTISATELLITE OBSERVATIONS OF SOLAR PROTONS PENETRATING THE MAGNETOPAUSE

V. Domingo, D.E. Page & K.-P. Wenzel

Space Science Department, ESTEC, Noordwijk, Netherlands

Abstract

Simultaneous observations of low-energy solar protons (1 - 40 MeV) by HEOS-2 in the northern high-latitude magnetotail and by other spacecraft in interplanetary space during the solar proton event of 8 June 1972 are presented. The results indicate that solar protons enter that magnetotail lobe which is not well connected to the interplanetary magnetic field – with respect to solar-particle entry – preferentially close to the Earth ($|X| < 10\ R_E$) rather than farther downtail. The particle density gradients measured in the tail lobe suggest that the particles enter there not only across the polar magnetopause, but also from the neutral sheet.

1. Introduction

The observation of solar proton penetration into the magnetosphere is considered a major tool for the understanding of the gross characteristics of the external magnetosphere, as the protons are tracers of magnetic field structures over much larger dimensions than local observations would allow.

Comparison of fluxes observed over the polar cap regions with fluxes and conditions in the interplanetary medium has been used in a number of cases [e.g. Evans & Stone, 1969; Domingo & Page, 1971; Van Allen *et al.*, 1971; Evans, 1972; Fennell, 1973] to infer, or speculate, whether the magnetosphere has an open or closed structure and how the interplanetary magnetic field connects with the magnetospheric field. Observations of solar-particle fluxes over the polar caps near the Earth indicate the existence of two regions in each polar cap well differentiated in the way in which they appear to be connected to the interplanetary medium: one at low latitudes (near the auroral zones) and the other at high latitudes (above 70 to 80° north or south). The particle fluxes over the low-latitude regions, both north

Knott & Battrick (eds.), The Scientific Satellite Programme During the International Magnetospheric Study, 225–235.
All Rights Reserved. Copyright © 1976 by D. Reidel Publishing Company, Dordrecht-Holland.

and south, usually follow the interplanetary flux levels with only a very short delay (up to a few minutes). On the other hand, the fluxes observed over the central polar caps, or high-latitude regions, may differ markedly for long periods of time (of the order of hours) from the average interplanetary proton flux and may be different from north to south pole.

Morfill & Quenby [1971] have produced, by the integration of particle trajectories in model magnetospheres, a map of particle arrival regions over the polar caps. Each region is related to a particular entry window, specifically to the neutral sheet, the front magnetosphere and the tail magnetosphere. Scholer [1972] has demonstrated that the theoretical mapping of these windows into the polar cap can be confirmed by experimental measurements. A review of the subject can be found in, for example, Morfill & Scholer [1973].

In this paper we discuss the penetration of low-energy (1 - 40 MeV) solar protons into the northern high-latitude magnetotail during a particle event on 8 June 1972. The results indicate that in this case, in which the observed magnetotail lobe was not well (or not directly) connected to the interplanetary magnetic field upstream of the Earth, most of the solar particles penetrated it near the Earth. The anisotropy observations agree with the flux gradient structure across the magnetotail observed for other events [Domingo *et al.*, 1974].

2. Instrumentation

The HEOS-2 proton data used in this study were measured with the main solid-state detector telescope at 90° to the satellite's spin axis [Köhn *et al.*, 1972]. During the period studied, the spin axis was nearly perpendicular to the ecliptic plane. Protons were measured in five energy channels (1-9, 9-12, 12-17, 17-27, 27-36 MeV), the first containing small contributions of electrons of about 0.4 - 0.6 MeV. Proton anisotropy measurements in the ecliptic plane were obtained in four 90° sectors when the count rate exceeded a certain level.

HEOS-2 magnetic field data were kindly supplied by Dr. P.C. Hedgecock of Imperial College, London.

3. Results

3.1. PROTON FLUXES

Figure 1 shows the trajectory of HEOS-2 and the projections of the magnetic-field vector in geocentric solar magnetospheric co-ordinates during the inbound and outbound passages through the high-latitude magnetosphere on 8/9 June 1972. HEOS-2 crossed the magnetopause inbound at $-X_{GSM} \approx 11.5\,R_E$ between 1200 and 1230 UT on 8 June and left the magnetosphere again poleward of the polar cusp at 1830 UT on 9 June.

The arrival of solar protons in near-Earth interplanetary space was measured simultaneously by several spacecraft whose observations were reported for the 'Campaign for Integrated Observations of Solar Flares (CINOF)' [Shea & Smart,

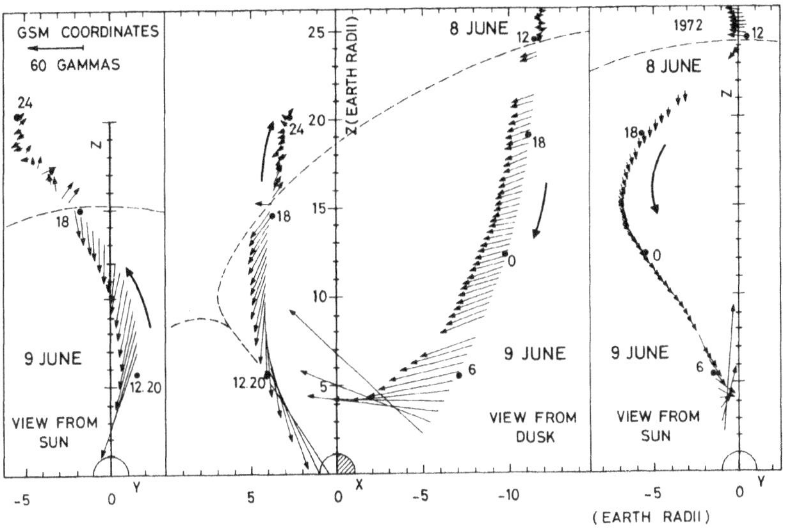

Figure 1. *HEOS-2 position (start of arrows) and magnetic field vector (arrows), projected into the XZ plane (central panel) and the YZ plane (right and left panels) of the geocentric solar magnetospheric (GSM) co-ordinate system from 8 June, 0900 UT to 9 June, 2400 UT, 1972. Every other 1-min average of the magnetic-field projection is plotted. The dashed line is an eyeball' trace of the magnetopause passing through the two observed magnetopause crossings.*

1974]. Observations on Explorer 41, Explorer 43 and Vela 6A indicate a sharp increase in the arrival of solar particles on 8 June between 1400 and 2400 UT, the actual onset time depending on the particle velocity.

Explorer 41 is the only spacecraft known to us in a high-apogee orbit with a proton channel (1 - 10 MeV) that closely corresponds to our low-energy proton channel (1 - 9 MeV). It was actually on an outbound pass of the magnetosphere, crossing the magnetopause at 1730 UT and the bow shock at about 2230 UT. We assume that the magnetosheath fluxes measured on Explorer 41 were representative of the interplanetary fluxes at that time. Comparison of data from Explorer 41 and Explorer 43 – the latter was in the interplanetary medium – justify this assumption.

In Figures 2 and 3 we compare the fluxes measured aboard HEOS-2 and Explorer 41 for two different proton energy channels, namely about 1 - 10 MeV and 10 - 30 MeV. In both cases the level of the Explorer 41 count rate is normalised to the level of the HEOS-2 count rate using long periods when both satellites were in interplanetary space. In both figures the periods during which the spacecraft were inside the magnetosphere are marked by shadowed bars. Except for about 5 h around perigee, HEOS-2 was in the northern lobe of the magnetotail. Throughout

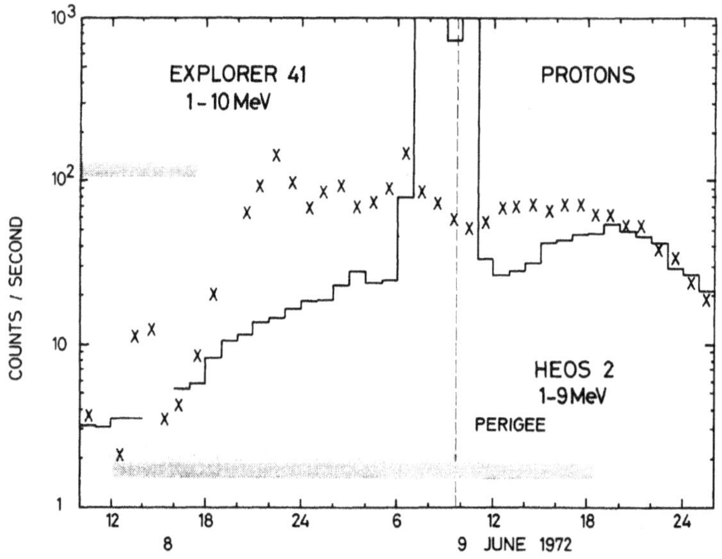

Figure 2. *1 - 9 MeV and 1 - 10 MeV proton count rates measured aboard HEOS-2 and Explorer 41, respectively. Explorer 41 count rates were normalised to the level of the HEOS-2 count rates outside the magnetosphere. Shadowed bars mark the periods when the satellites were inside the magnetosphere.*

this period the magnetospheric flux of 1 - 9 MeV protons (HEOS-2) was substantially lower than the flux in the interplanetary medium (Explorer 41), although the ratio of the magnetospheric to the interplanetary fluxes was monotonously increasing (Fig. 2). The same applies for the 9 - 36 MeV protons, except that in this case the magnetospheric flux reached approximately the same level as the interplanetary flux about 3½ h before HEOS-2 crossed the magnetopause outbound (Fig. 3).

3.2. ANGULAR DISTRIBUTIONS

Figure 4 shows in the top panel the spin-averaged count rates in four proton channels between 1 and 28 MeV together with the maximum and minimum count rates observed in the four angular sectors of each channel during the inbound pass. The extent of the shadowing represents the magnitude of the anisotropy. The bottom panel gives the azimuthal direction of the anisotropy (assuming it to be approximated by a first harmonic distribution) plotted together with the direction of the magnetic field in solar ecliptic co-ordinates. The anisotropy was significant in all channels up to about 0400 UT on 9 June, but thereafter it became marginal (indicated by the fact that the direction appears to be random). The magnetic field in the magnetotail was essentially parallel to the Sun-Earth line. If we assume that

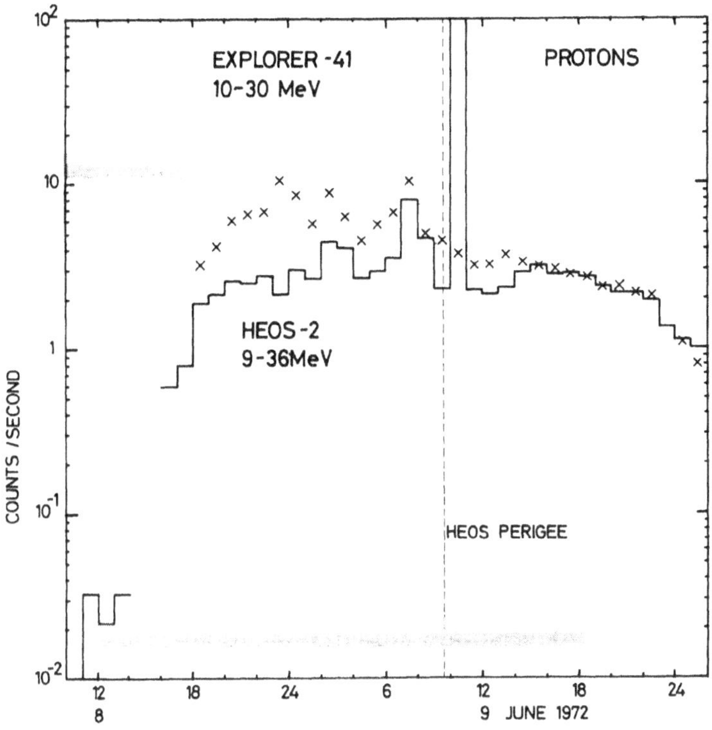

Figure 3. *Same as Figure 2 for 9 - 36 MeV and 10 - 30 MeV protons on HEOS-2 and Explorer 41, respectively.*

the anisotropy vector consisted of two components, one parallel to the magnetic field and the other perpendicular to it, we can see that there was a significant component parallel to the magnetic field lines and pointing toward the Sun (0°), throughout the whole period until about 0400 UT. This means that the dominant flow of particles came from the Sun direction. The perpendicular component, however, changed direction inside the tail around 2100 UT: it first pointed towards dusk, and thereafter towards dawn.

Figure 5 shows the anisotropy of the 1 - 9 MeV channel in the outbound pass through the magnetotail. The anisotropy became significant after about 1430 UT (outside 11 R_E) with a large component perpendicular to the magnetic field. In the last 1½ h inside the magnetosphere (within 2 R_E of the magnetopause crossing) it also had a significant component along the magnetic field lines, with most of the particles moving downtail. In the higher energy channels, the anisotropy was small, but a dominant flow in the direction from the Sun during the last 2 h before crossing the magnetopause could still be observed.

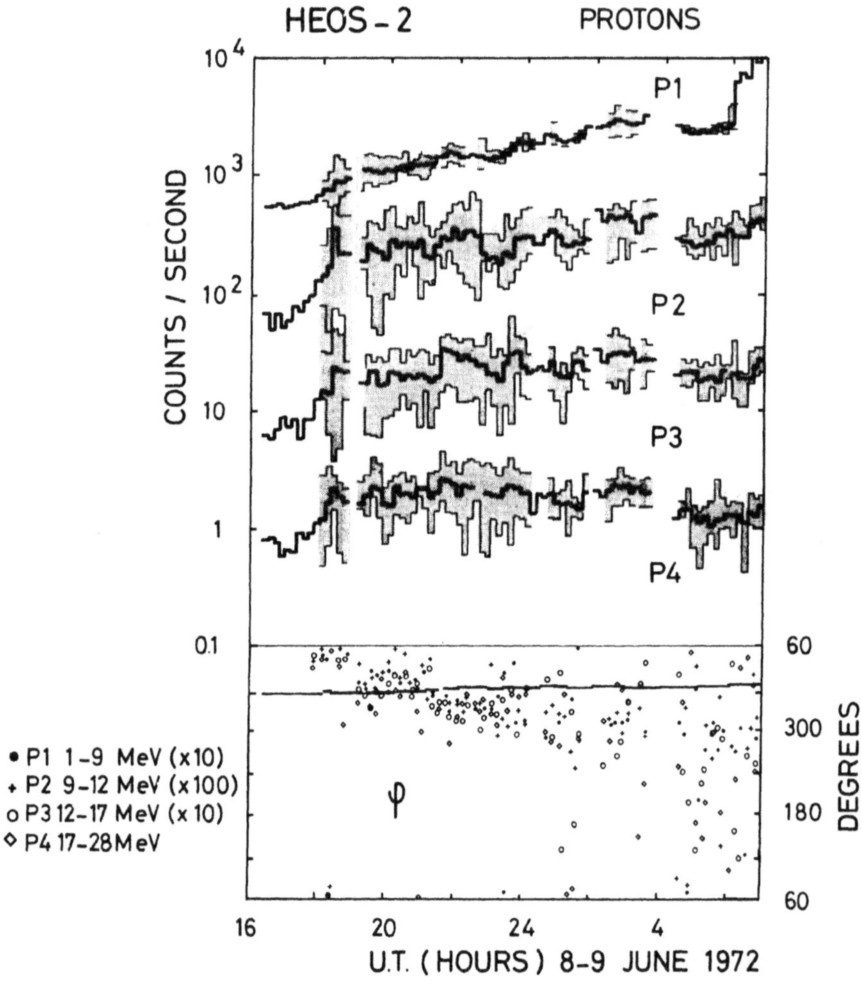

Figure 4. *Upper panel: proton count rates in four channels measured on HEOS-2 during the inbound part of the orbit studied. The upper and lower limits of the shadowing represent the maximum and minimum observed count rates, in the four angular sectors. Lower panel: directions of the local magnetic field (continuous line) and of the anisotropy of the four proton channels, in the ecliptic plane.*

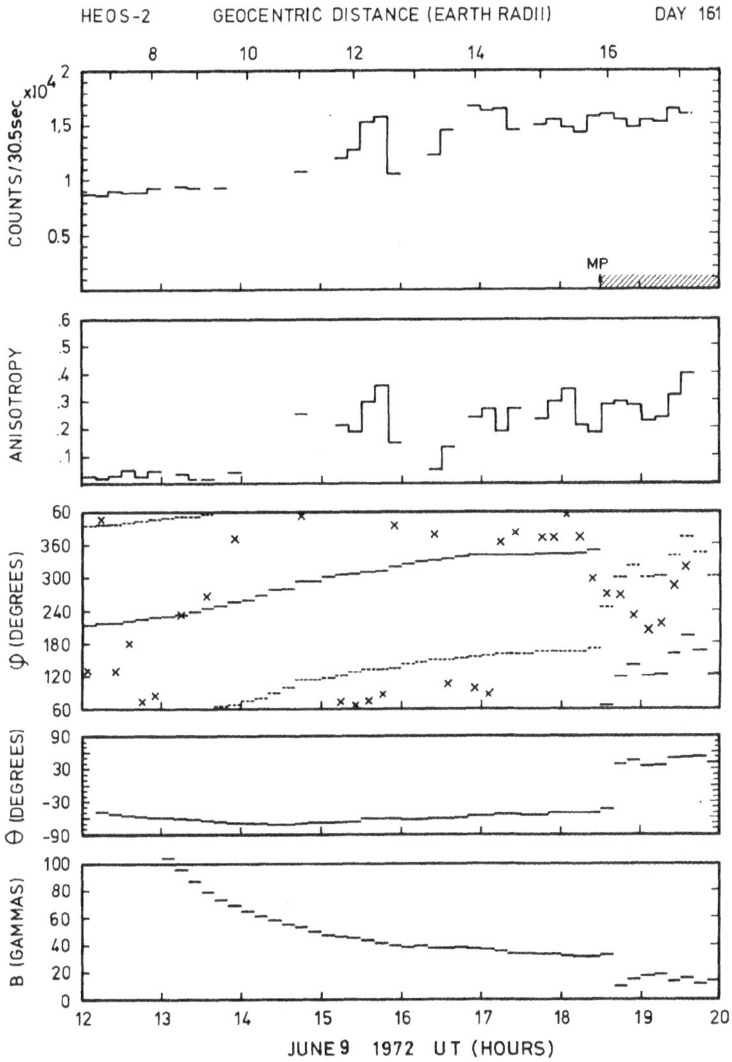

Figure 5. *From top to bottom, versus UT (in hours): (a) the count rate of 1 - 9 MeV protons; (b) anisotropy coefficient (first harmonic); (c) magnetic-field azimuth (continuous line) direction and its opposite, i.e. true azimuth plus 180° (dashed line); direction of the anisotropy of 1 - 9 MeV protons (crosses); (d) latitude of the magnetic-field vector in spacecraft co-ordinates; and (e) intensity of the magnetic field vector during the outbound pass. The magnetopause crossing is indicated (MP).*

4. Discussion

4.1. THE FLUXES

It has been shown [Evans & Stone, 1969; Domingo & Page, 1971; Van Allen *et al.*, 1971; Scholer, 1972] that the low-energy solar proton fluxes observed over the central polar caps of the Earth during a solar proton event are dependent on the interplanetary magnetic field direction, on the angular distribution of the particles in interplanetary space, and/or on the temporal history of the event. In the present case, we have been observing for a long period that part of the magnetotail which is almost certainly connected to the northern central polar cap. The interplanetary magnetic field was directed toward the Sun during most of the period of observation (Explorer 43 [King, 1975]). Therefore, in any model for particle penetration into the magnetosphere, the northern magnetotail lobe was in an unfavourable condition for particle entry. Assuming magnetic reconnection, the conditions were such that the southern magnetotail was connected to interplanetary magnetic field lines upstream of the Earth, and the northern magnetotail to interplanetary field lines downstream of the Earth.

The difference between the flux observed by HEOS-2 in the northern magnetotail lobe and the interplanetary flux may be explained by assuming either that it is due to the time needed for the particles to penetrate or diffuse into the tail at some distant point in the tail, as suggested by Evans & Stone [1969], or that it is due to the fact that HEOS-2 observed only the sunward flowing component of the interplanetary particle flux, as suggested for other events by Fennell [1973].

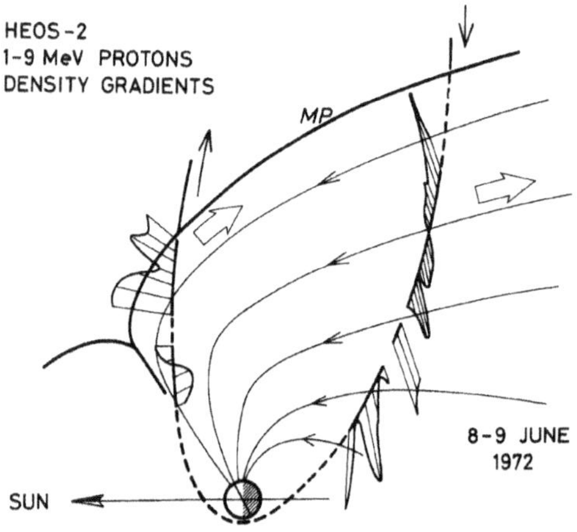

Figure 6. *Schematic of observed proton flow (broad arrows) and inferred proton density gradients along HEOS-2 orbit. The calculated density gradients are plotted along the orbit trajectory in the noon-midnight plane perpendicular to the local magnetic field. Gradients directed towards the magnetopause are plotted to the left of the trajectory, gradients directed towards the neutral sheet to the right.*

4.2. THE ANISOTROPIES OBSERVED INSIDE THE MAGNETOTAIL

The presence of a net flow of particles along the field lines down the tail and away from the Earth observed on both inbound and outbound passes seems to indicate that at this time more protons were penetrating the tail lobe near the Earth than through the downstream tail magnetopause. The anisotropy component at 90° to the magnetic field may be indicative of the existence, at the time of the measurement, of particle density gradients in the tail, as explained by Domingo *et al.* [1974]. On the inbound pass, the perpendicular component of the anisotropy would be consistent with a proton density gradient directed toward the magnetopause in the outer 5 to 8 R_E of the magnetotail (anisotropy directed toward dusk), and directed toward the neutral sheet in the rest (anisotropy toward dawn), i.e. more particles outside the magnetopause and in the neutral sheet than in the tail lobes. On the outbound pass, this anisotropy component would be consistent with a proton density gradient directed toward the magnetopause for the last 4 to 5 R_E inside the magnetopause.

Figure 6 indicates schematically the proton density gradients and flow encountered in the magnetotail. The gradients are inferred from the anisotropy perpendicular to the magnetic field. The flow is along the field away from the Earth. The region of positive gradient toward the magnetopause in the outer high-latitude part of the magnetotail observed on both the inbound and outbound passages can be seen. The reversal of the gradient inside the tail is clearly shown.

Having only one satellite inside the magnetosphere, it is not possible to obtain definite information about the actual routes of penetration of the particles seen flowing away from the Earth. Possible routes are: the front magnetopause and the polar cusp, the neutral sheet and the pseudo-trapping region, the southern polar lobe of the magnetotail and the pseudo-trapping region. The directional reversal of the gradients observed in the magnetotail seems to point to different penetration routes being operative: the magnetopause or polar cusp for the outer magnetotail, and the neutral sheet and/or pseudo-trapping region for the inner.

5. Conclusions

Observation of 1 - 30 MeV protons in the magnetotail following the solar event of 8 June 1972 suggests the following:

(i) When the solar protons penetrated the magnetotail lobe, which was badly connected to the interplanetary field, they did so close to the Earth ($|X| < 10\ R_E$), rather than far downtail. At no time was flow towards the Earth observed in the tail.

(ii) We again interpret the observed cross-field anisotropies as being due to particle density gradients in the tail [Domingo *et al.*, 1974].

(iii) The inferred particle density gradient and its reversal near the centre of the tail lobe suggest that particle entry to the outer high-latitude tail takes place through the front magnetopause, while entry to the inner part of the tail is primarily through the neutral sheet.

6. Acknowledgements

We particularly wish to thank Dr. P. Hedgecock, Imperial College, London, whose data were essential to this study. Thanks are also due to Drs. D.F. Smart and P.A. Shea, Air Force Cambridge Research Laboratories, Bedford, Massachusetts, for providing us with numerical values of data published in the CINOF report, and to Dr. D.J. Williams, Space Environment Laboratory, NOAA, Boulder, Colorado and the World Data Center-A, Rockets and Satellites, GSFC, Greenbelt, Maryland for providing several data sets used as supporting data.

References

Domingo, V. & D.E. Page, North-south asymmetry of solar particle fluxes in polar cap regions, *J. Geophys. Res.* **76**, 8159 (1971).

Domingo, V., D.E. Page & K.-P. Wenzel, Measurements of solar protons in the near earth magnetotail, in Correlated Interplanetary and Magnetospheric Observations (Ed. D.E. Page), p. 507, Reidel, Holland, 1974.

Evans, L.C., Magnetospheric access of solar particles and the configuration of the distant geomagnetic field, Thesis, California Institute of Technology, Pasadena, 1972.

Evans, L.C. & E.C. Stone, Access of solar protons into the polar cap: a persistent north-south asymmetry, *J. Geophys. Res.* **74**, 5127 (1969).

Fennell, J.F., Access of solar protons to the Earth's polar caps, *J. Geophys. Res.* **78**, 1036 (1973).

King, J.H., Interplanetary magnetic field data book, NSSDC, NASA, Goddard Space Flight Center, Greenbelt, Maryland, 1975.

Köhn, D., D.E. Page, B.G. Taylor & K.-P. Wenzel, Interplanetary experiment (S 204) for the HEOS-2 mission, *ELDO/ESRO Sci Techn. Rev.* **4**, 19 (1972).

Morfill, G.E. & J.J. Quenby, The entry of solar protons over the polar caps, *Planet. Space Sci.* **19**, 1541 (1971).

Morfill, G. & M. Scholer, Study of the magnetosphere using energetic solar particles, *Space Sci. Rev.* **15**, 267 (1973).

Scholer, M., Polar-cap structures of solar protons observed during the passage of interplanetary discontinuities, *J. Geophys. Res.* **77**, 2762 (1972).

Shea, M.A. & D.F. Smart (Eds.), Compilation of solar particle and interplanetary measurements acquired during the campaign for integrated observations of solar flares (CINOF), Air Force Cambridge Research Laboratories, Special Report 177, Bedford, Massachusetts, 1974.

Van Allen, J.A., J.F. Fennell & N.F. Ness, Asymmetric access of energetic solar protons to the earth's north and south polar caps, *J. Geophys. Res.* **76**, 4262 (1971).

DISCUSSION

C.T. Russell: I would like to suggest that you avoid the term 'non-reconnecting polar cap' because an open tail must be connected to the interplanetary field in both lobes. Connection on the dayside demands that the same interplanetary field line connects to both lobes. Further, the slowing down of the plasma in the magnetosheath wraps field around the magnetosphere, so that 'northward' field will be antiparallel to the tail field in *both* lobes. The principal difference is that one lobe is connected to the Sun and the other lobe to the interstellar medium.

V. Domingo: I agree with your semantic consideration. I only used the terms 'connecting and non connecting' or 'well connecting or badly connecting' magnetotail lobe, or polar cap, to refer to the phenomenologically observed fact that, often, during the early phase of a solar proton event one observes over one of the polar caps a lower level particle flux than over the other, and to the fact that this feature has been related in many instances to the direction of the interplanetary magnetic field (Van Allen & Ness, Domingo & Page, Fennell, Evans & Stone).

W.P. Olson: We might avoid some of our communication difficulties if we merely describe the observations, which here Domingo has clearly described: the signature of cosmic-ray fluxes over the two polar caps is different and the difference is somehow related to the direction of the interplanetary magnetic field. Other groups (CIT and MDAC) have reported observations in support of the work you report.

Have you considered the possibility that some of your delay times might be explained in terms of 'trapping'? That is, some particles may move around the inner magnetosphere several times before reaching the observation point?

V. Domingo: Thank you for the comment. As regards the trapping consideration, we have not yet made any quantitative study, but we are going to receive fine time resolution data from experiments aboard Explorers 41 and 43 which were in the interplanatary medium for at least part of the time of interest and we will try to correlate the flux structures observed inside and outside the magnetosphere for that purpose.

R.J. Hynds: (i) What were the dimensions of the particle gradients observed near the magnetopause?

(ii) What was the inclination of the magnetic field with respect to the ecliptic plane during the event?

V. Domingo: (i) On the outbound pass (day 9 June) the region over which we observe the outward particle gradient extends for about three Earth radii near the magnetopause. On the inbound pass (day 8 June), the experiment switches itself into angular mode (i.e. becomes able to measure anisotropies) when it is at about 5 R_e from the magnetopause, and the particle density gradient is already there. It remains directed toward the magnetopause until the spacecraft is about 7 R_e away from it. In all this I assume that the shape and distance of the magnetopause remain constant.

(ii) It was near-parallel to the ecliptic planes throughout the period of interest, with a dominantly small southward component between 19 hours on 8 June and 13 hours on 9 June. For the rest of the time, it had a dominantly northward component.

TRAPPED PARTICLES AND WAVES, AND WHAT CAN BE LEARNED FROM MULTISATELLITE EXPERIMENTS

L.R. Lyons

NOAA, Boulder, Colorado, USA

Abstract

Calculations concerning the pitch-angle diffusion resulting from resonant wave-particle interactions can lead to definitive predictions of equatorial pitch-angle distributions and rates of particle loss as a function of particle energy and L-value. Thus, given simultaneous high-altitude measurements of pitch-angle distributions and low-altitude measurements of precipitating fluxes as a function of energy and L, the importance of proposed wave-particle interactions can be verified or discarded. Since many wave-particle phenomena occur over large spatial and temporal scales, exact simultaneity in longitude and time is not necessary. Simultaneous low and high altitude (preferably nearly equatorial) particle measurements could thus greatly increase our understanding of trapped particles and their effects on the ionosphere. Furthermore, given a verified pitch-angle diffusion mechanism and simultaneous low- and high-altitude measurements, accurate low- to high-altitude mappings of field lines and magnetospheric boundaries (such as the plasmapause) could be obtained.

1. Introduction

Pitch-angle diffusion driven by resonant interactions with plasma waves is an important loss mechanism for trapped radiation-belt particles, and the resulting precipitation of particles can have important effects on the ionosphere. Thus determining and understanding the dominant wave-particle interactions affecting trapped particles is necessary not only for understanding the structure of the radiation belts, but also for understanding the coupling between the magnetosphere and the ionosphere.

Knott & Battrick (eds.), The Scientific Satellite Programme During the International Magnetospheric Study, 237–260.
All Rights Reserved. Copyright © 1976 by D. Reidel Publishing Company, Dordrecht-Holland.

The most ideal approach for studying magnetospheric wave-particle interactions would be to self-consistently treat the wave generation and the resulting particle diffusion, and then compare the resulting predictions of wave and particle distributions with satellite observations. However, such an approach is generally not feasible, since it would be necessary to treat the entire inhomogeneous plasma as a whole and to include all source and loss processes for both particles and waves.

An alternative approach for studying these interactions is to calculate diffusion coefficients for particles resonant with a specified wave distribution [Roberts, 1969; Kennel, 1969]. This requires independent knowledge of the distribution of wave energy from observations and/or theoretical analysis. However, the approach yields definitive predictions of particle pitch-angle distributions and loss rates as a function of particle energy and L-value. These predictions can then be directly compared with satellite observations. If the predictions and observations agree for some population of trapped particles, then one can conclude that the dominant pitch-angle diffusion mechanism has been identified for those particles. Furthermore, if the specified wave distribution was obtained from theoretical analysis, it is likely that the predicted wave distribution exists.

In general, pitch-angle distributions for particles resonant with a band of waves are controlled by the distribution of wave energy with frequency, wave normal angle, and geomagnetic latitude. The total wave amplitude is required to calculate particle loss rates. Thus pitch-angle distributions obtained near the equator are needed to determine whether a suggested distribution of waves is the dominant cause of pitch-angle diffusion for a given population of particles. Low-altitude measurements of precipitated particles are needed in addition to the high-altitude measurements to obtain actual particle loss rates and to map high-altitude regions of pitch-angle diffusion to particle precipitation patterns in the ionosphere.

In Section 2, we discuss general concepts of resonant wave-particle interactions and how they may be applied to satellite observations. Examples of how these concepts have been succesfully applied to specific proposed interactions are described in Section 3, along with suggestions for other applications of these concepts. Comments on the use of data from planned and proposed multisatellite missions are given in Section 4.

2. General Concepts

Given a spectrum of waves, it is possible to obtain considerable useful information without performing any complicated calculations. This is accomplished by simply using the resonance condition

$$\omega - k_\parallel v_\parallel = -n\Omega; \quad n = 0, \pm 1, \pm 2, \pm 3, \ldots$$

and the wave-dispersion relation to determine which particles resonate with the waves. Here ω is the wave frequency, Ω is the gyrofrequency of the particle species under consideration, and $k_\parallel$ and $v_\parallel$ are the components of the wave vector and the particle velocity along the direction of the magnetic field. The cyclotron harmonic resonances are given by $n = \pm 1, \pm 2, \ldots$ and the Landau resonance by $n = 0$.

For simplicity, assume $\omega << \Omega$. Then the resonant parallel velocity for each cyclotron harmonic resonance $v_{\|,n} = n\Omega/k_{\|}$ and the resonant parallel velocity for the Landau resonance $v_{\|,0} = \omega/k_{\|}$. Waves generally propagate both up and down field lines so that resonance at each harmonic occurs for particles with parallel velocities of both $v_{\|,n}$ and $-v_{\|,n}$.

Figure 1 illustrates the regions of cyclotron resonance in velocity space for waves distributed over some band of $k_{\|}$'s. Since we are assuming $\omega << \Omega$ for all frequencies in the wave distribution, the $n = 0$ resonance occurs at much smaller values of $v_{\|}$ than do the cyclotron resonances. Resonance at each cyclotron harmonic occurs over a band of $v_{\|}$'s, and no cyclotron interactions occur for $v_{\|}$ less than a minimum value $v_{\|,min}$. That no cyclotron resonances occur for $v_{\|}$ less than a $v_{\|,min}$ does not depend on the assumption $\omega << \Omega$.

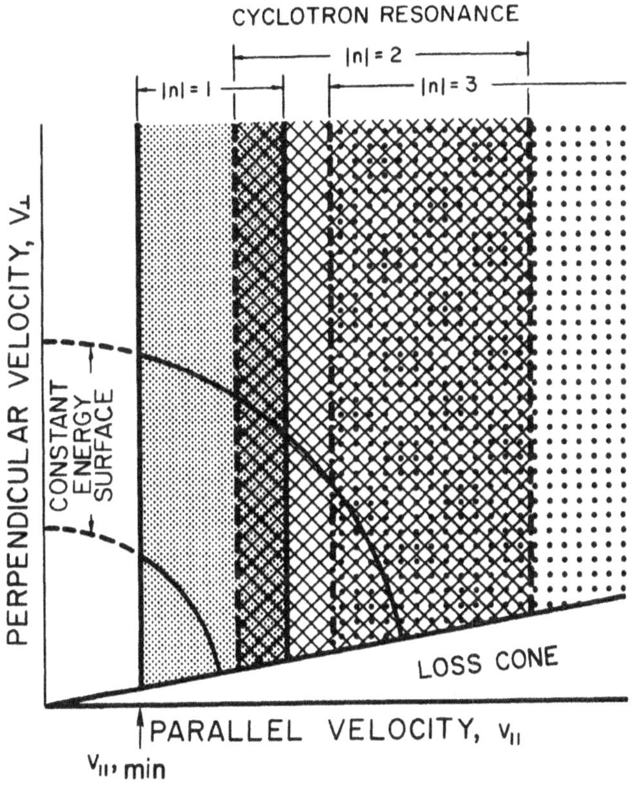

Figure 1. *Regions in the $(v_\perp, v_\|)$-plane of resonance at the $|n| = 1, 2,$ and 3 cyclotron harmonics are indicated by different shadings. Wave energy is assumed to be distributed over a band of $k_\|$'s, with $\omega << \Omega$ for all frequencies in the wave distribution. The minimum parallel velocity for cyclotron resonance $v_{\|,min}$ is indicated.*

The simple picture shown in Figure 1 is modified somewhat because the geomagnetic field is not homogeneous, and wave distributions are generally distributed over some range of geomagnetic latitudes. As a particle moves away from the equator along its bounce trajectory, the increasing geomagnetic field strength generally causes both the particle pitch angle and the resonant parallel velocities $v_{\|,n}$ to increase. Therefore, including the off-equatorial interactions, cyclotron resonance occurs for all equatorial values of $v_{\|}$ greater than the equatorial value of $v_{\|,min}$ and does not occur for lower equatorial $v_{\|}$'s.

Thus in examining particle data obtained near the equator for evidence of resonant wave-particle interactions, one should look for the effects of diffusion at values of $E_{\|} = \frac{1}{2}mv_{\|}^2$ greater than a minimum $E_{\|,min}$. As shall be seen in the examples discussed later, particle pitch-angle distributions measured near the equator as a function of energy can show marked effects resulting from cyclotron resonance occurring at $E_{\|} \geq E_{\|,min}$ but not at $E_{\|} < E_{\|,min}$. Such observations have given strong support to specific wave-particle interactions which have been proposed as the dominant cause of pitch-angle diffusion for specific populations of trapped particles.

Calculations of expected particle pitch-angle distributions and lifetimes can be performed for any mode of weak plasma turbulence by applying quasi-linear theory [Kennel & Engelman, 1966]. General expressions for pitch angle as well as energy diffusion coefficients are given by Lyons [1974a, b], and a technique for averaging over the particle bounce trajectory is given by Lyons *et al.* [1972]. Also, expressions for calculating pitch-angle distributions and lifetimes from the bounce-orbit-average diffusion coefficients are available [Kennel, 1969; Lyons, 1973].

Calculated pitch-angle distributions can be compared directly with measurements. However, it is generally meaningless to compare calculated particle lifetimes to lifetimes obtained from measured particle decay rates in order to test specific loss mechanisms (either wave-particle interactions or other loss mechanisms such as Coulomb collision or charge exchange). This is because other processes such as radial diffusion generally act on time scales comparable to particle lifetimes. (e.g. in equilibrium source and loss processes exactly balance, but the loss rate is not zero). Removing the effects of radial diffusion from data is extremely difficult, since accurate radial diffusion coefficients are needed as well as the first and second derivatives of particle fluxes as a function of L at constant first adiabatic invariant.

However, low-altitude measurements of precipitating particles as a function of pitch angle and energy could yield a direct measurement of the precipitation loss rate as a function of energy. Even if complete pitch-angle distributions are not available, measurements of the energy spectrum of precipitated particles at a fixed pitch angle, together with simultaneous measurements of the equatorial particle distribution, could yield the variation of the particle lifetime with energy.

Using low- and high-altitude satellite measurements to obtain particle loss rates requires that the measurements be taken from the same L-shell. This requires that field lines be traced from the equatorial plane to the ionosphere. While such tracing is nearly impossible using a single satellite, such field-line mapping should be possible from particle measurements at low and high altitudes when the effects of resonant wave-particle interactions are identifiable. Since pitch-angle diffusion at a given L-shell only occurs for parallel energies $E_{\|} \geq E_{\|,min}$, precipitation from pitch-angle diffusion should only occur for energies $E \geq E_{\|,min}$. Thus, since $E_{\|,min}$ generally varies as a function of L because of the L-dependence of the geomagnetic field, mappings of field lines should result from the separate identification of $E_{\|,min}$ versus L at high and low altitudes.

The degree of simultaneity required in longitude and time between the high- and low-altitude measurements must be considered and will depend upon the specific wave phenomena under consideration. The distribution of wave energy with frequency cannot change appreciably between the times and longitudes of the low- and high-altitude measurements, since it is necessary that $E_{\|,min}$ be the same. However, the wave amplitude need not be constant in order to determine $E_{\|,min}$ or to determine the variation of particle lifetime with energy. Certainly wave and pitch-angle distribution measurements on mother-daughter types of satellite missions would greatly assist in determining limits on the time scales over which changes occur.

3. Examples of Effects of Wave-Particle Interactions in Particle Data

3.1. SLOT-REGION ELECTRON LOSSES

A well-understood example where resonant wave-particle interactions have striking effects on particle pitch-angle distributions is the case of electron interactions with the plasmaspheric whistler-mode hiss. Hiss is observed nearly continuously throughout the plasmasphere over a band of frequencies several hundred Hertz wide centred at several hundred Hertz [Thorne *et al.*, 1973]. Lyons *et al.* [1972] calculated pitch-angle diffusion coefficients for electrons resonant with the observed distribution of plasmaspheric hiss. Using the calculated diffusion coefficients, predictions of electron pitch-angle distributions and precipitation rates were obtained.

Examples of the calculated pitch-angle diffusion coefficients and equatorial pitch-angle distributions at $L = 4$ are shown in Figure 2 for 20 keV, 200 keV, and 2 MeV electrons. Cyclotron resonance occurs for equatorial pitch angles up to $\sim 55°$ at 20 keV. Cyclotron resonance extends to increasing pitch angles with increasing electron energy, since cyclotron resonance occurs for all $E_\| \geq E_{\|,min}$ independent of the electron energy. Landau resonance occurs only at pitch angles near 90°, but off-equatorial interactions cause the range of equatorial pitch angles affected by Landau resonance to overlap the pitch-angle range of cyclotron resonance. Note that the pitch-angle diffusion rates are relatively low at values of $E_\|$ just below $E_{\|,min}$. This slow diffusion manifests itself in the equatorial pitch-angle distributions as a region of large slope. Thus the equatorial distributions develop bumps surrounding 90° pitch angles, and this bump decreases in size and pitch angle extent with increasing electron energy. Since $E_{\|,min} \sim B^2/N \sim L^{-2}$ (for an equatorial electron density $N \sim L^{-4}$), these bumps are also expected to decrease with increasing L-value for a constant electron energy.

Such pitch-angle distributions, with the expected variation with energy and L, were first observed on OGO-5 by West *et al.* [1973]. Good examples of their observations are shown by Lyons *et al.* [1972]. Subsequent measurements from

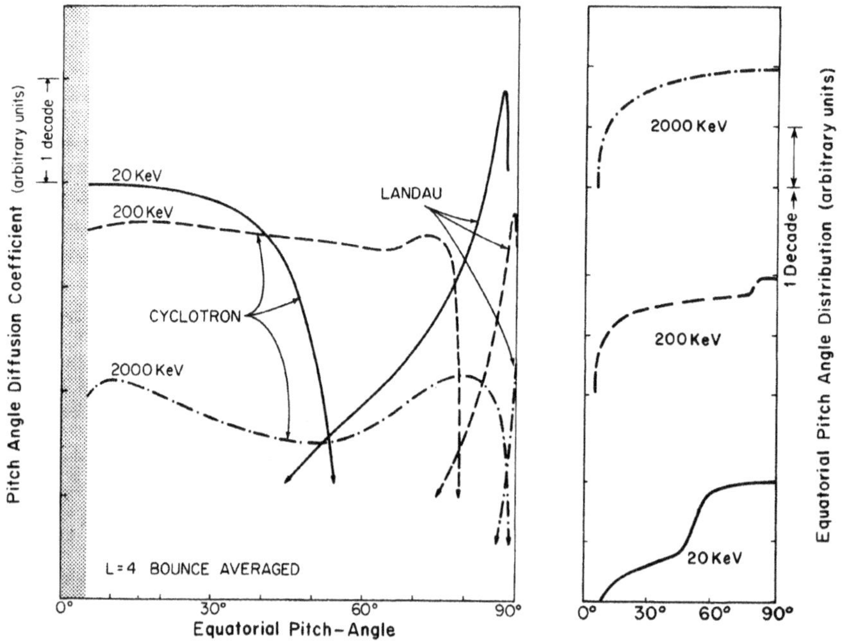

Figure 2. *Bounce-orbit averaged cyclotron and Landau resonant pitch-angle diffusion coefficients as a function of equatorial pitch angle at L = 4 for 20, 200, and 2000 keV electrons. For each energy, the line coding used for the Landau resonant diffusion coefficients corresponds to that for the sum of the cyclotron-harmonic resonances, where the particle energy is indicated. Equatorial pitch-angle distributions, calculated using the diffusion coefficients shown in the figure, are displayed on the right-hand side [from Lyons et al., 1972].*

Explorer 45 (S^3) have shown that such pitch-angle distributions are a general feature of the quiet-time plasmasphere [Lyons & Williams, 1975a]. Electron pitch-angle distributions observed on Explorer 45 are shown in Figure 3 every 0.2 in L between $L = 2$ and $L = 5$ for four energy channels. Dashed lines are the observations and solid lines are the theoretically calculated pitch-angle distributions. Slopes as steep as those of the theoretical distributions cannot be observed because of the 11° detector aperture and the 11.25° angular scan during each measurement; however, the pitch-angle region of increased slope does show in the data. Note the reduction in size and pitch-angle extent of the observed and theoretical bumps surrounding 90° with increasing electron energy and L.

Since $E_{\|,min}$ increases with decreasing L, cyclotron resonance does not occur within the inner zone for the electron energies observed by Explorer 45. Thus, Coulomb collisions have been included in the calculations and the rounding of the

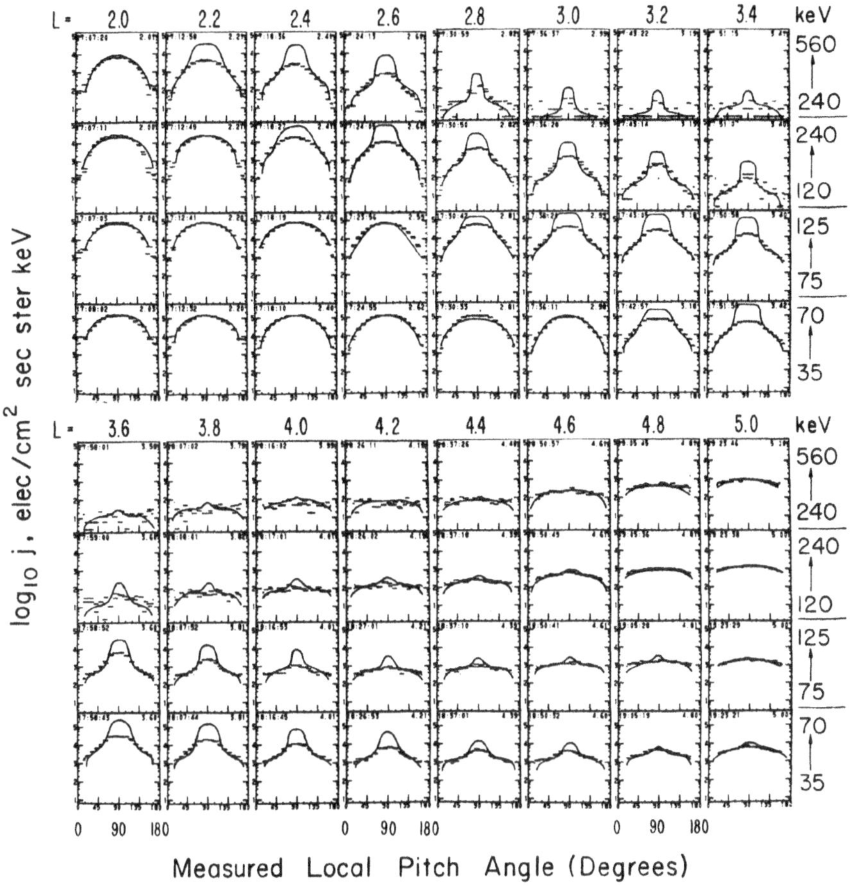

Figure 3. *Equatorial pitch-angle distributions observed on 15 December 1971 (orbit 95) and those predicted by Lyons et al. [1972] to result from resonant interactions with the plasmaspheric whistler-mode wave band. Distributions are shown every 0.2 in L from L = 2 to L = 5, and the four energies are stacked vertically at each L with the lowest energy at the bottom. The dashes give the measured electron flux with their horizontal extent indicating the pitch-angle scan for each measurement. The solid curves show the theoretically predicted pitch-angle distribution for the geometric mean energy of each interval. Diffusion from Coulomb collisions has been added to the theoretical calculations to account for the rounding of the pitch-angle distributions within the inner zone. The vertical positionings of the theoretical distributions are arbitrary on a logarithmic scale and have thus been adjusted to best illustrate the comparison with the observations [from Lyons & Williams, 1975a].*

calculated pitch-angle distributions within the inner zone occurs when Coulomb collisions become the dominant pitch-angle diffusion mechanism. The observed pitch-angle distributions become rounded within the inner zone, with the bump surrounding 90° disappearing at the same L-values as predicted by the calculations ($L \lesssim 2$ for 240–560 keV electrons, and $L \lesssim 3$ for 35–70 keV). It thus appears that Coulomb collisions become the dominant pitch-angle mechanism within the inner zone.

The agreement between the shapes of the calculated and observed pitch-angle distributions, and the agreement between their variations with energy and L strongly supports the contention that resonant interactions with plasmaspheric hiss is the dominant loss mechanism for radiation-belt electrons within the slot-region and outer plasmasphere, with Coulomb losses becoming dominant within the inner zone. Note that the changes in size of the bumps surrounding 90° pitch angles, as a function of energy and L, simply result from cyclotron resonance occurring only for $E_{\|} \geq E_{\|, min}$ with $E_{\|, min}$ decreasing with increasing L.

Examples of Explorer 45 observations of electron pitch-angle distributions obtained just preceding, during, and following the large geomagnetic storm of 17 December 1971 [Lyons & Williams, 1975b] are shown in Figure 4. The observations show that the storm injection results in pitch-angle distributions greatly distorted from their quiet-time shape with no evidence of the bumps surrounding 90°. Following the storm, the distributions return to their pre-storm shape over a period of several days and maintain that shape as the electron fluxes decay back to their quiet-time values. These observations imply that resonant interactions with plasmaspheric hiss become the dominant pitch-angle diffusion mechanism for slot-region electrons almost immediately following storm-time injections.

For the case of slot-region electrons, equatorial measurements have demonstrated rather conclusively that the dominant loss mechanism has been identified and evaluated properly. Unfortunately, simultaneous measurements at low and high altitudes have not been reported which would allow $E_{\|, min}$ to be mapped as a function of L both near the equator and near the ionosphere. However, Imhof *et al.* [1974] have reported low-altitude satellite measurements of precipitating electron energy spectra within the slot region. The observed spectra show peaks which vary with L in a manner consistent with cyclotron resonance with the plasmaspheric hiss. It appears that the slot-region offers an excellent opportunity to check whether simultaneous high- and low-altitude measurements can be used to map precipitation patterns as a function of energy and L from the equatorial magnetosphere to the ionosphere.

Given realistic particle loss rates as a function of energy and L, and an idea as to what the particle source rates should be, one can attempt to balance sources and losses to obtain the structure of trapped particles as a function of energy and L. If the resulting structure agrees with observations, then it is possible to conclude that the sources are reasonably well understood. Lyons & Thorne [1973] balanced the calculated electron loss rates with radial diffusion to obtain an equilibrium structure for radiation-belt electrons throughout the plasmasphere. Radial diffusion was assumed to be driven by substorm-associated fluctuations of the magnetospheric convection electric field as modelled by Cornwall [1968]. The results agreed well with the quiet-time structure observed by Pfitzer *et al.* [1966]. Figure 5 shows a

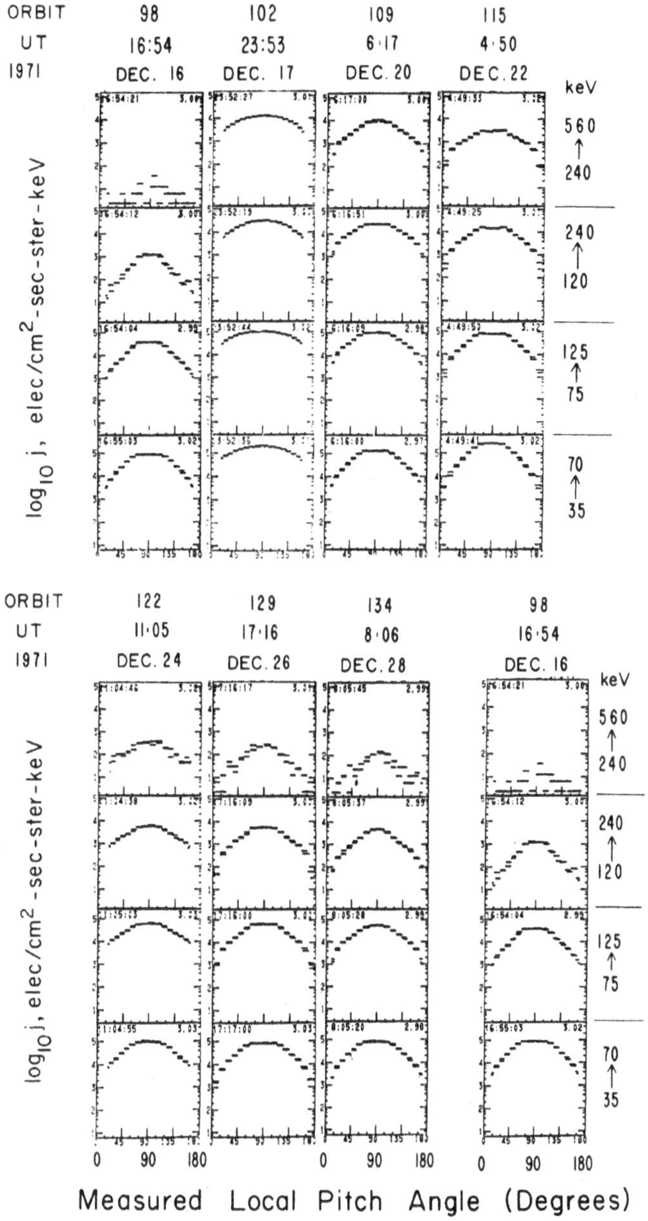

Figure 4. *Observed pitch-angle distributions at $L = 3$ for the periods preceding, during, and following the storm of 17 December 1971.. Pre-storm distributions are repeated at end of figure to indicate return to pre-storm shapes and intensities [from Lyons & Williams, 1975b].*

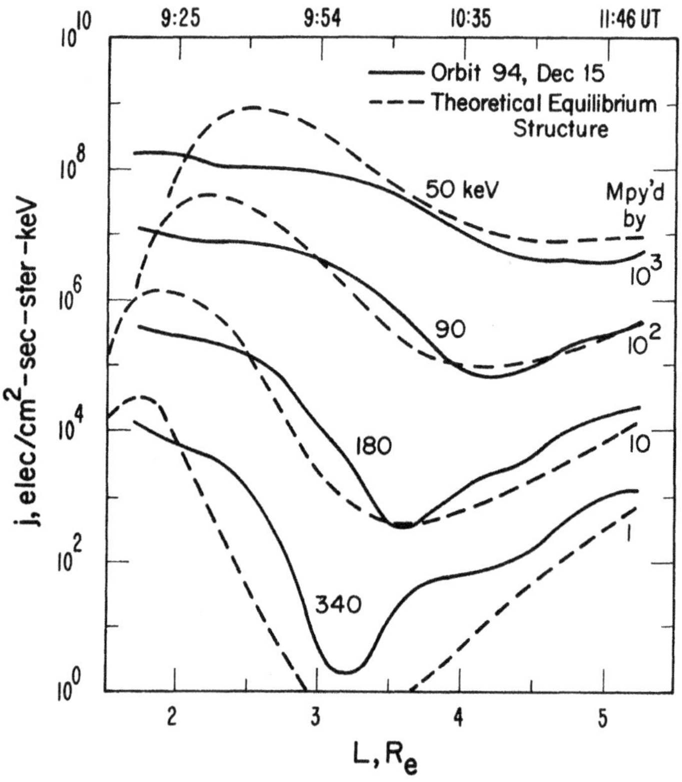

Figure 5. *Comparison between the radial profiles observed on the outbound portion of orbit 94 on 15 December 1971 (solid lines) and the theoretical equilibrium radial profiles (dashed lines) as obtained from the analysis of Lyons & Thorne [1973] using B_w =10mγ, E = 0.05 mV/m, and the observed mean outer-zone energy spectrum. The theoretical profiles are for the indicated energies, which are approximately the geometric means of the four Explorer 45 energy channels. The 180, 90 and 50 keV fluxes have been multiplied by 10^1, 10^2, and 10^3, respectively [from Lyons & Williams, 1975a].*

Figure 6. *Equatorial proton pitch-angle distributions observed on Explorer 45 orbit 103 inbound, ∿ 16 h after the minimum D_{st} of the 17 December 1971 storm main phase. Distributions are shown every 0.4 in L from L = 3 to L = 5, and selected proton energy channels are stacked vertically at each L. No data suffering from the saturation problem with the solid-state detector [Williams et al., 1973] are shown. Elevated fluxes at pitch 90–180° for energies ≤ 14 keV are due to reflected sunlight. Ticks are at constant values of E‖ for each L, with the chosen value of E‖ at each L being equal to the energy of one of the Explorer 45 channels.*

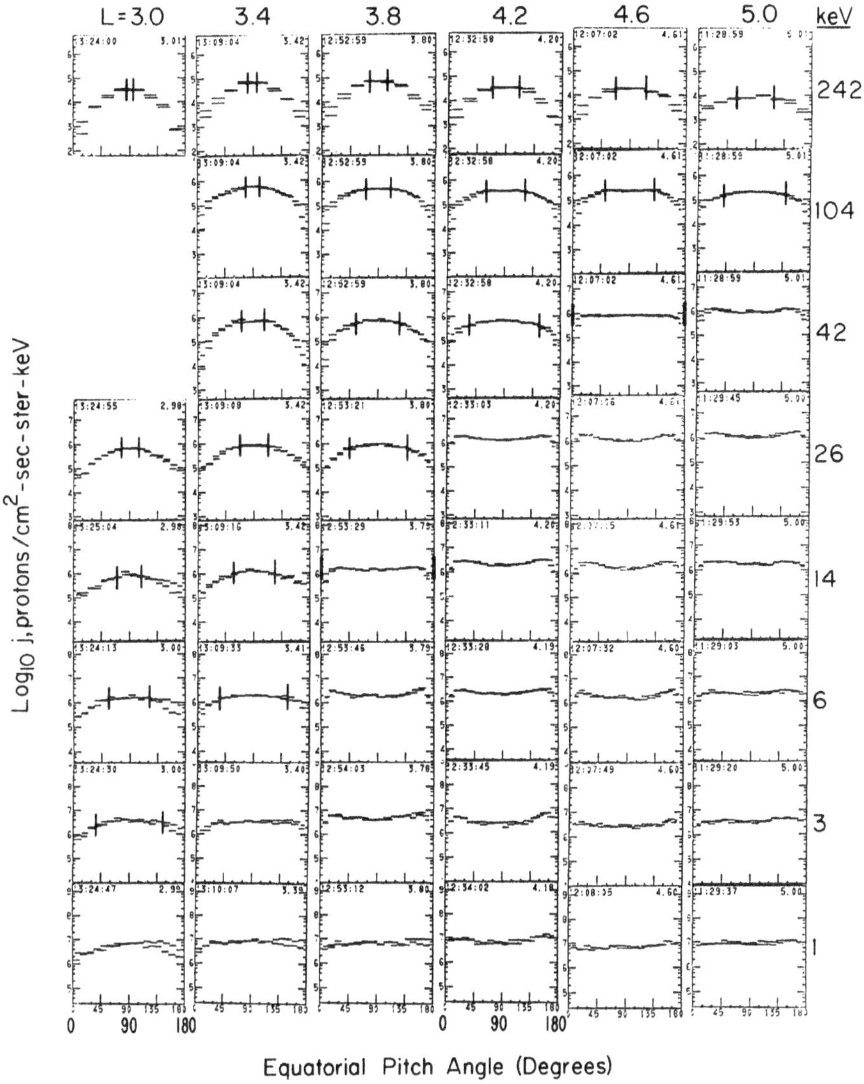
L=3.0
3.4
3.8
4.2
4.6
5.0
keV
242
104
42
26
14
6
3
1
Log10 j, protons/cm2-sec-ster-keV
0
90
180
Equatorial Pitch Angle (Degrees)

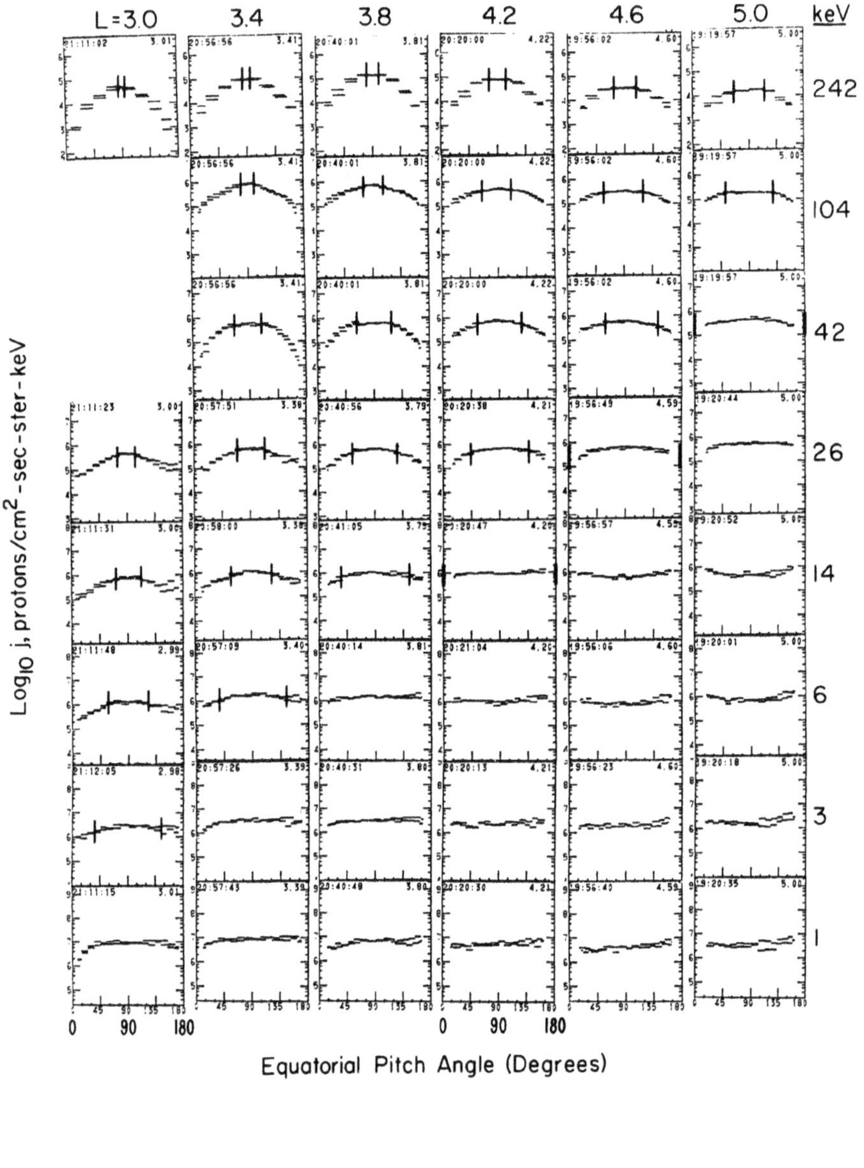

Figure 7. *Same as Figure 6, except for Explorer 45 orbit 104 inbound, ∿ 24 h after the minimum D_{st} of the storm main phase.*

recent comparison [Lyons & Williams, 1975a] between the calculated equilibrium structure and Explorer 45 observations of quiet-time electron fluxes versus L. The calculations reproduce the overall two-zone radial structure and its variations with electron energy reasonably well, with the values of the fluxes being correct to about an order of magnitude. This accuracy is consistent with our knowledge of radial diffusion rates and the amplitude of the hiss throughout the plasmasphere. Thus, equatorial particle measurements indicate that the dominant source and loss processes have been identified for radiation-belt electrons within the plasmasphere.

3.2. RING-CURRENT LOSSES NEAR THE PLASMAPAUSE

Williams & Lyons [1974a, b] have performed a detailed analysis of Explorer 45 measurements of the pitch-angle distributions of ring-current protons obtained during the recovery phase of the large geomagnetic storm of 17 December 1971. Of particular interest for that study is the prediction by Cornwall *et al.* [1970] of the growth of ion cyclotron waves as the hot ring-current protons interact with the cold plasmaspheric plasma. Wave amplification is expected to be sufficient to cause significant pitch-angle diffusion of protons, thus resulting in an important loss process for ring-current protons.

Examples of Explorer 45 measurements of the equatorial pitch-angle distributions of ring-current protons obtained near local midnight during the storm recovery phase are shown in Figures 6 and 7. The observations in Figures 6 and 7 were obtained $\sim$ 16 h (orbit 103) and $\sim$ 24 h (orbit 104), respectively, after the minimum D_{st} of the storm main phase. In each figure, observations from selected Explorer 45 proton energy channels are shown every 0.4 in L from $L = 3.0$ to 5.0.

Note in both figures the transition from nearly isotropic pitch-angle distributions at the lower energies to rounded pitch-angle distributions, peaked at 90° pitch angle, at the higher energies. This transition occurs at higher proton energies with increasing L-value. A careful examination of the nearly isotropic distributions shows that when the pitch-angle scans reach the loss cone, significant intensity decreases are seen implying a nearly empty loss cone [Williams & Lyons, 1974a]. These isotropic distributions with an empty loss cone are indicative of a stable trapped particle population in a region of negligible losses from pitch-angle diffusion. The isotropy must have originated either from an isotropic injection or from strong pitch-angle diffusion during the storm main phase.

Figure 8 shows pitch-angle distributions from orbit 103 for two sample L-values (3.4 and 4.6). Every available energy channel is shown over the range of proton energies for which the transition to rounded distribution occurs. Specifically note the vertical ticks indicated on the pitch-angle distributions. For each L, the equatorial pitch angle at which the ticks are placed gives the same value of $E_{\|}$. That value of $E_{\|}$ is equal to the energy of the Explorer 45 channel with ticks shown at 0° and 180°. Notice that to within the accuracy of the pitch-angle measurements (the angular scan per measurement is 11.25° for $E \leq 104$ keV and 22.5° for $E = 164$ and 242 keV, and the detector aperture is 11°), the pitch-angle distributions are nearly isotropic between the ticks, i.e. for parallel energies less than a given value. At larger parallel energies (pitch angles approaching 0° and 180°), the pitch-angle distribu-

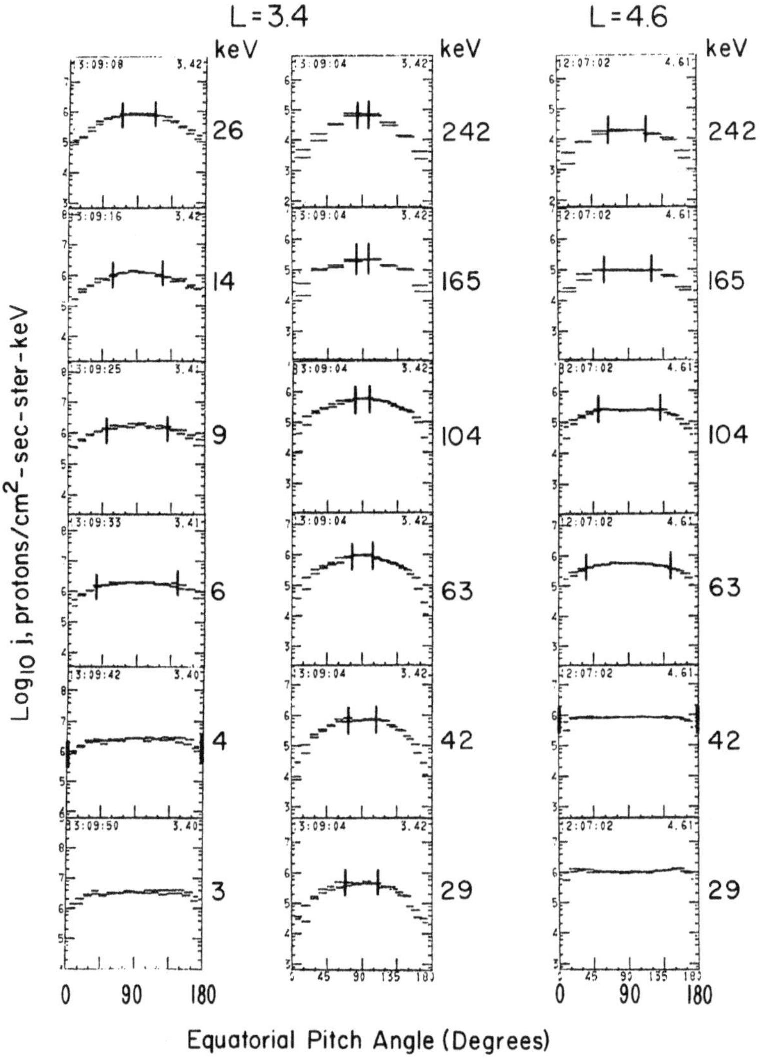

Figure 8. *Pitch-angle distributions from orbit 103 inbound for L = 3.4 and 4.6. Every available energy channel is shown over the range of proton energies for which the transition from flat (with an empty loss cone) to rounded (peaked at 90° pitch angle) distributions occurs. Vertical ticks shown on each pitch-angle distribution are at constant values of $E_\parallel$ for each L. Note that to within the accuracy of the measurements, the pitch-angle distributions are nearly isotropic between the ticks and rounded outside the ticks.*

tions are rounded. Ticks at constant values of $E_{\parallel}$ for each L are also plotted on the pitch-angle distributions of Figures 6 and 7. It can be seen that, for all L-values shown, the ticks quite accurately separate regions of nearly isotropic pitch-angle distributions at lower values of $E_{\parallel}$ from regions of rounded distributions at higher values of $E_{\parallel}$.

Pitch-angle distributions with fluxes monotonically increasing towards increasing pitch angles are expected under conditions of pitch-angle diffusion into the loss cone, except in the limit of strong diffusion. Since the pitch-angle distributions are rounded only at values of $E_{\parallel}$ greater than a minimum value at each L, it is reasonable to conclude that pitch-angle diffusion is being driven by resonant interactions with plasma waves. Applying a resonant-energy analysis, Williams & Lyons [1974a] concluded that the diffusion is driven by resonant interactions with ion-cyclotron waves. Calculations currently in progress of ion-cyclotron wave growth rates, using the measured proton pitch-angle distributions and energy spectra, indicate that ion-cyclotron waves can be significantly amplified by the ring-current protons in the vicinity of the plasmapause [Joselyn & Lyons, in prep.] as predicted by Cornwall *et al.* [1970]. Furthermore, the unstable wave frequencies give values of $E_{\parallel, min}$ consistent with the Explorer 45 observations when realistic cold plasma densities are assumed for the plasmapause region.

Since $E_{\parallel, min}$ varies throughout the plasmapause region (from $L \approx 3$ to $L \approx 5$), simultaneous low-altitude and nearly equatorial measurements should allow a mapping of field lines from the equator to the ionosphere during storm recovery phase. Thus the storm-time inflation of geomagnetic field lines could be measured, and the plasmapause could be mapped from the equator to the ionosphere.

Cornwall *et al.* [1971] predicted that ion-cyclotron wave energy generated in the region of the plasmapause is transmitted along field lines leading to the formation of SAR arcs in the ionosphere. Simultaneous Explorer 45 measurements of the equatorial proton ring current and Fritz Peak observation of SAR arcs have recently been reported [Williams *et al.*, 1975]. The observations indicated that sufficient energy is lost from the observed pitch-angle distribution transitions of the equatorial ring current to account for SAR arcs. Furthermore, the radial structures of observed peaks in the proton loses agree with the multiple SAR arc structure observed on the ground. Order-of-magnitude estimates of the energy-loss time scales indicate that the ring-current energy loss rates are greater than what is required to maintain the observed SAR arcs (Fig. 9).

Simultaneous ring-current and SAR-arc measurements were only available from one satellite pass. More observations will be required to show that the observed agreement in structure and energy requirements is generally true. Furthermore, single-satellite measurements allow only a rough estimate of the proton loss rates as a function of energy and L. More accurate measurements of the time scales over which the pitch-angle distributions transform from flat to rounded are needed, and they should be obtainable from a mother-daughter type of satellite mission.

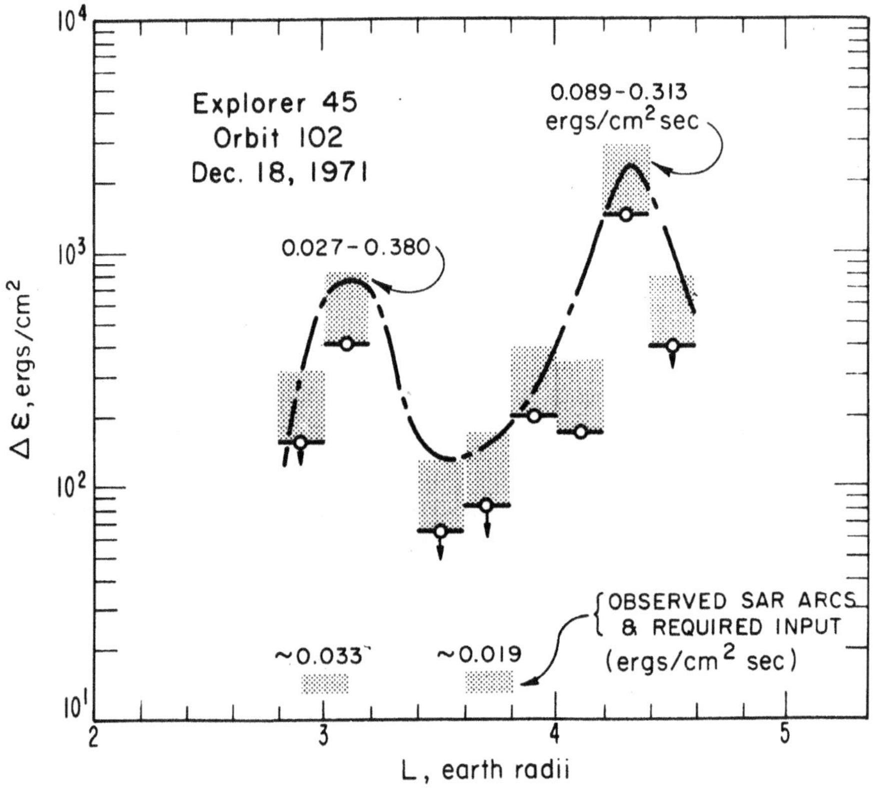

Figure 9. *Energy loss from the ring current and simultaneous SAR-arc observations. Shaded rectangles give energy-loss estimates obtained from the observed Explorer 45 flat to rounded transitions of the quatorial proton pitch-angle distributions. Downward pointing arrows indicate upper limits. Order-of-magnitude estimates of the energy-loss time scales give the estimates of the ring-current energy loss rates given at the two peaks of the energy loss versus L curve. Simultaneously observed (from Fritz Peak) SAR-arc locations and energy input required to maintain the observed arcs are also shown. The spatial correspondence between the ring-current loss peaks and the SAR arcs is good provided the storm-time expansion of the geomagnetic field is considered [from Williams et al., 1975].*

3.3. PRECIPITATION OUTSIDE THE PLASMAPAUSE

Satellite measurements of particle energy spectra and pitch-angle distributions should also help identify the causes of pitch-angle diffusion outside the plasmapause on auroral field lines.

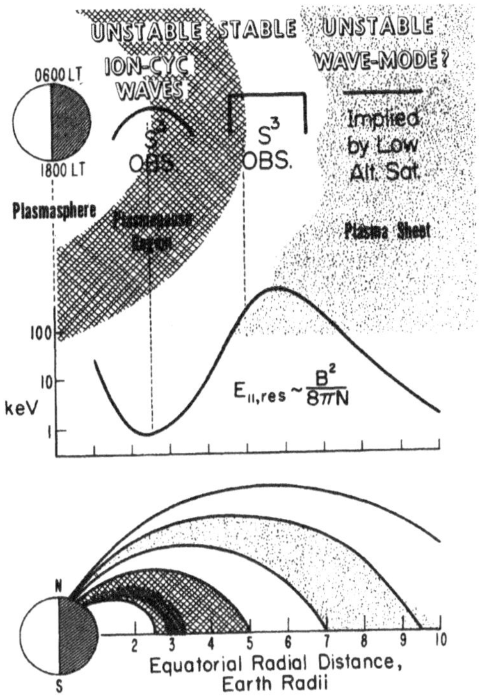

Figure 10. *Schematic of geomagnetic storm recovery phase showing hot ring-current plasma behaviour in the geomagnetic field and its interaction with the cold plasmaspheric plasma. The top panel shows an equatorial projection and the bottom panel shows a midnight-meridian projection. Midnight-meridian features are projected to low altitudes to indicate expected low-altitude latitudinal structure. A semiquantative resonant energy versus altitude plot which used Explorer 45 results is shown for reference. Moderate pitch-angle diffusion occurs for the hot ring-current plasma in the plasmapause due to the amplification of ion-cyclotron waves. Each energy of the ring-current plasma begins its interaction in the plasmapause region at the appropriate value of* $B^2/8\pi N$. *Heavier shading in the plasmapause region is used to indicate that energy flow to the ionosphere from the hot ring-current plasma may be peaked at preferential locations and will depend on the plasmapause shape and hot plasma energy spectrum. More than one peak in the energy flow is possible. Above the plasmapause region the hot ring-current plasma is stably trapped with negligible or no losses due to pitch-angle scattering. Above this region, and above Explorer 45* (S^3) *apogee during recovery phase, there is the region of proton auroral precipitation exhibiting strong turbulence and a full loss cone as implied by low-altitude measurements [from Williams et al., 1975].*

A variety of low-altitude observations [e.g. Amundsen *et al.*, 1972; Bernstein *et al.*, 1970; Mizera, 1974] show intense proton precipitation outside the plasmapause. The spatial relation of this auroral precipitation to the precipitation expected in the plasmapause region during storm recovery phase is illustrated schematically in Figure 10. Note the predicted mappings of the plasmapause region to low altitudes and the proton auroral precipitation to the equatorial plane. It is reasonable to expect the proton auroral precipitation to result from pitch-angle diffusion, since the precipitation is widespread and does not display field-aligned pitch-angle distributions. A suggested mechanism for this precipitation is pitch-angle diffusion driven by resonant interactions with ion loss-cone waves [Coroniti *et al.*, 1972; Rönmark, 1975]. However, high-altitude measurements of the pitch-angle distributions of these protons have yet to be reported. Such measurements will be necessary to determine whether resonant wave-particle interactions in the magnetosphere are responsible for auroral proton precipitation, and if so, to determine the responsible wave mode. Simultaneous low- and high-altitude measurements will be needed to determine precipitation rates and to map high-altitude regions of pitch-angle diffusion to precipitation patterns in the auroral ionosphere.

Similarly, pitch-angle diffusion is a likely cause of auroral electron precipitation at energies above 0.1–1.0 keV. This refers only to widespread, structureless precipitation that is not field-aligned, i.e. to diffuse auroral electron precipitation. Kennel *et al.* [1970] reported OGO-5 observations of intense electrostatic waves on auroral field lines with the dominant emissions being at a frequency near 1.5 times the electron gyrofrequency. They suggested that these waves may be responsible for the pitch-angle diffusion of auroral region electrons. Using the wave vectors predicted theoretically by Young *et al.* [1972], we find that $E_{\|, min}$ for cyclotron resonance with these waves is of the order of 0.1 to 1 keV, as required to explain diffuse electron precipitation.

Lyons [1974c] calculated pitch-angle diffusion coefficients for electrons resonant with these electrostatic waves. The results showed that electrons of energies up to a few keV should be put on strong pitch-angle diffusion throughout the loss cone by waves of typically observed amplitudes. Thus precipitating electrons should have isotropic pitch-angle distributions at energies up to a few keV, with a gradually increasing anisotropy at higher energies. Isotropy should not be observed at equatorial pitch angles approaching 90° for energies below a few keV, since resonance only occurs at $E_{\|} \leq E_{\|, min}$, so that no diffusion will occur at high pitch angles. Figure 11 shows sounding-rocket observations of electron energy spectra at two local pitch angles within the loss cone (76° and 42°) obtained during a post-break-up aurora [from Whalen & McDiarmid, 1973]. Using the observed 76° energy spectra and the calculated diffusion coefficients, Lyons [1974c] calculated energy spectra for the 42° pitch-angle particles given by the dashed lines in Figure 10. Notice that the calculations reproduce the observed pitch-angle anisotropy as a function of electron energy.

This suggests that diffuse auroral electron precipitation may, at least at times, result from interactions with the electrostatic waves at 1.5 times the electron gyrofrequency. However, observations of pitch-angle distributions as a function of energy and high and low altitudes will be needed to verify this suggestion, and to determine if such interactions are the dominant cause of the precipitation.

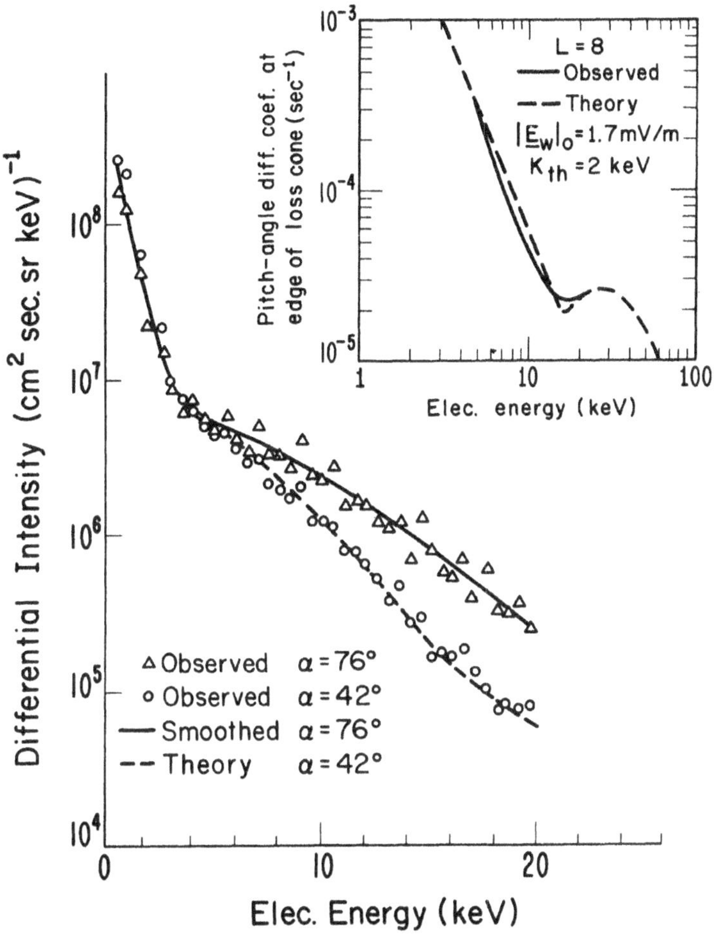

Figure 11. *Sounding-rocket observations of electron energy spectra obtained within the loss cone at two pitch angles ($\alpha = 76°$ and $\alpha = 42°$) during a post-breakup aurora are shown [taken from Whalen & McDiarmid, 1973]. In the upper part of the figure, the pitch-angle diffusion coefficient at the edge of the loss cone is shown versus electron energy as obtained from the anisotropy of the observations (solid line) and as obtained from the theoretical calculations for a wave amplitude $|E_w|_0 = 1.7\ mV/m$ and a thermal energy $K_{th} = 2\ keV$ (dashed line). Also shown, in the main part of the figure, is the smooth line through the $\alpha = 76°$ energy spectrum (solid line) and the theoretical energy spectrum at $\alpha = 42°$ (dashed line) obtained using the theoretical diffusion coefficients and the smoothed $\alpha = 76°$ spectrum [from Lyons, 1974c].*

4. Conclusions

Equatorial measurements of the pitch-angle distributions of trapped particles as a function of energy have been extremely valuable in identifying the effects of specific wave-particle interactions. Such measurements have clearly shown the minimum parallel energy for cyclotron resonance $E_{\|, min}$ as a constant independent of total particle energies, and the measurements have traced the variation of $E_{\|, min}$ with L. The identification of $E_{\|, min}$ throughout the particle data has strongly indicated that the dominant loss mechanisms have been identified for radiation-belt electrons within the plasmasphere and for storm-time ring-current protons in the plasmapause region. Future measurements of equatorial pitch-angle distributions should allow other important wave-particle interactions to be identified. Of particular interest would be the causes of electron and proton pitch-angle diffusion on auroral field lines.

Simultaneous low-altitude measurements would allow direct measurements of particle lifetimes as a function of energy. Furthermore, by mapping $E_{\|, min}$ as a function of L at high and low altitudes, pitch-angle diffusion regions in the magnetosphere could be traced directly to precipitation patterns in the ionosphere. Thus closed magnetic field lines, along with magnetospheric boundaries such as the plasmapause, could be traced from the equator to the ionosphere.

Low-altitude to high-altitude mappings of the effects of wave-particle interactions should be most successful for widespread phenomena that do not vary rapidly with time. Insight into more highly structured phenomena may result from mother-daughter satellite missions.

Field-aligned pitch-angle distributions are frequently observed at low altitudes in auroral regions. Pitch-angle diffusion cannot account for such distributions, and parallel (to the geomagnetic field) electric fields are often invoked as a possible cause [e.g. Evans, 1974 and references therein]. Recently reported ATS-6 observations show intense beams of electrons travelling parallel to the magnetic field [McIlwain, 1975]. It is tempting to associate such beams with parallel electric fields. If simultaneous measurements of electron and proton pitch-angle distributions at low and high altitudes are all consistent with the existence of a parallel electric field, then information on the location and intensity of such electric fields may result.

The planned Electrodynamics Explorer (EE) mission consisting of two polar orbiting satellites, one at low altitudes and one at high altitudes, will certainly be useful for studying the problems discussed here. The fact that both satellites will be near the same longitude will greatly facilitate low to high altitude comparisons. However, equatorial measurements will only be available from a single L-value. Wave-particle interactions generally result in particle distributions with well-defined variations as a function of L, and different interactions occur in different regions. Furthermore, the radial distance of such phenomena varies with the level of geomagnetic activity. Thus an equatorial satellite with a highly elliptical orbit similar to Explorer 45 (S^3) would be of great value in studying wave-particle interactions throughout the magnetosphere and in studying their variations with geomagnetic activity. A proposal has been submitted to NASA for such a satellite (S^3-B) with an apogee of $\sim 7.5\ R_E$ and a perigee of several hundred kilometres. If launched in conjunction with the EE mission, such a satellite would significantly enhance the scientific results obtainable from the mission.

Certainly any future high-altitude particle measurements should include complete pitch-angle distributions. Complete pitch-angle distributions would also be valuable from low-altitude satellites; however, at least a two- or three-point pitch-angle distribution is a necessity.

5. Acknowledgements

Much of the theoretical analysis presented here has resulted from discussion and collaborative work with R.M. Thorne and C.F. Kennel, and much of the Explorer 45 data interpretation has resulted from similar interactions with D.J. Williams. This work was supported in part by NASA Contract S-50028.

References

Amundsen, R., F. Söraas, H.R. Lindalen & K. Aarsnes, Pitch angle distributions of 100- to 300-keV protons measured by the Esro 1B satellite, *J. Geophys. Res.* 77, 556 (1972).

Bernstein, W.B., B. Hultqvist & H. Borg, Some implications of low altitude observations of isotropic precipitation of ring current protons beyond the plasmapause, *Planet. Space Sci.* 22, 767 (1974).

Cornwall, J.M., Diffusion processes influenced by conjugate-point wave phenomena, *Radio Sci.* 3, 740 (1968).

Cornwall, J.M., F.V. Coroniti & R.M. Thorne, Turbulent loss of ring current protons, *J. Geophys. Res.* **75**, 4699 (1970).

Cornwall, J.M., F.V. Coroniti & R.M. Thorne, A unified theory for SAR arc formation at the plasmapause, *J. Geophys. Res.* **76**, 4428 (1971).

Coroniti, F.V., R.W. Fredricks & R. White, Instability of ring current protons beyond the plasmapause during injection events, *J. Geophys. Res.* **77**, 6243 (1972).

Evans, D.S., Precipitating electron fluxes formed by a magnetic field aligned potential difference, *J. Geophys. Res.* **79**, 2853 (1974).

Imhof, W.L., E.E. Gaines & J.B. Reagan, Evidence for the resonance precipitation of energetic electrons from the slot region of the radiation belts, *J. Geophys. Res.* **79**, 3141 (1974).

Kennel, C.F., Consequences of a magnetospheric plasma, *Rev. Geophys.* **7**, 379 (1969).

Kennel, C.F. & F. Engelmann, Velocity space diffusion from weak plasma turbulence in a magnetic field, *Phys. Fluids* **9**, 2377 (1966).

Kennel, C.F., F.L. Scarf, R.W. Fredricks, J.H. McGehee & F.V. Coroniti, VLF electric field observations in the magnetosphere, *J. Geophys. Res.* **75**, 6136 (1970).

Lyons, L.R., Comments on pitch-angle diffusion in the radiation belts, *J. Geophys. Res.* **78**, 6793 (1973).

Lyons, L.R., General relations for particle diffusion in pitch angle and energy, *J. Plasma Phys.* 12, 45 (1974a).

Lyons, L.R., Pitch angle and energy diffusion coefficients from resonant interactions with ion-cyclotron and whistler waves, *J. Plasma Phys.* 12, 417 (1974b).

Lyons, L.R., Electron diffusion driven by magnetospheric electrostatic waves, *J. Geophys. Res.* **79**, 575 (1974c).

Lyons, L.R. & R.M. Thorne, Equilibrium structure of radiation belt electrons, *J. Geophys. Res.* **78**, 2142 (1973).

Lyons, L.R., R.M. Thorne & C.F. Kennel, Pitch angle diffusion of radiation belt electrons within the plasmasphere, *J. Geophys. Res.* **77**, 3455 (1972).

Lyons, L.R. & D.J. Williams, The quiet time structure of energetic (35–560 keV) radiation belt electrons, *J. Geophys. Res.* **80**, 943 (1975a).

Lyons, L.R. & D.J. Williams, The storm and post-storm evolution of energetic (35–560 keV) radiation belt electrons, 1975b (in press).

McIlwain, C.E., Auroral electron beams near the magnetic equator, Nobel Symposium Proceedings, Plenum Pub. Co. Ltd., London, 1975.

Mizera, P.F., Observations of precipitating protons with ring current energies, *J. Geophys. Res.* **79**, 581 (1974).

Pfitzer, K., S. Kane & J.R. Winckler, The spectra and intensity of electrons in the radiation belts, *Space Res.* **6**, 702 (1966).

Roberts, C.S., Pitch-angle diffusion of electrons in the magnetosphere, *Rev. Geophys.* **7**, 305 (1969).

Rönmark, K., Pitch angle diffusion of ring current protons caused by electrostatic ion loss-cone waves, *Planet. Space Sci.* 1975 (in press).

Thorne, R.M., E.J. Smith, R.K. Burton & R.E. Holzer, Plasmaspheric hiss, *J. Geophys. Res.* **78**, 1581 (1973).

West,H.F., Jr., R.M. Buck & J.R. Walton, Electron pitch angle distributions throughout the magnetosphere as observed on OGO-5, *J. Geophys. Res.* 78, 1064 (1973).

Whalen, B.A. & I.B. McDiarmid, Pitch angle diffusion of low-energy auroral electrons, *J. Geophys. Res.* **78**, 1608 (1973).

Williams, D.J., T.A. Fritz & A. Konradi, Observations of proton spectra ($1.0 \leq Ep \leq 300$ keV) and fluxes at the plasmapause, *J. Geophys. Res.* **78**, 4751, 1973.

Williams, D.J., G. Hernandez & L.R. Lyons, Simultaneous observations of the proton ring current and stable auroral red arcs, Submitted to *J. Geophys. Res.* 1975.

Williams, D.J. & L.R. Lyons, The proton ring current and its interaction with the plasmapause: storm recovery phase, *J. Geophys. Res.* **79**, 4195 (1974a).

Williams, D.J. & L.R. Lyons, Further aspects of the proton ring current interaction with the plasmapause: main and recovery phases, *J. Geophys. Res.* **79**, 4791 (1974b).

Young, T.S.T., J.D. Callen & J.E. McCune, High-frequency electrostatic waves in the magnetosphere, *J. Geophys. Res.* **78**, 1082 (1973).

DISCUSSION

M.J. Rycroft: What do you consider to be the roles played by natural whistlers, harmonics of the power lines and manmade VLF signals in precipitating energetic electrons?

L.R. Lyons: While they may cause very localised regions of enhanced precipitation, their effects on the overall trapped particle population should be negligible.

M.J. Rycroft: Are there any significant longitudinal variations of trapped particle fluxes or loss rates, at any particular L-value $\gtrsim 5$?

L.R. Lyons: I do not believe there are significant longitudinal variations in trapped-particle fluxes within the plasmasphere, except possibly near the plasmapause in association with substorms. Longitudinal variations in wave amplitudes, which I believe exist for plasmaspheric hiss, will cause corresponding variations in loss rates.

K.D. Cole: Some recent calculations I have made on Coulomb exchange of energy between ring-current protons and magnetospheric thermal electrons suggest strongly that this mechanism can supply most of the additional heat needs of the mid-latitude ionosphere (including the SAR arc) during storms. Studies of the rapid movement of the SAR arc also support this view (to be reported at AGU, June 1975). Granted the existence of pitch-angle diffusion of protons by generation of ion-cyclotron waves, what is the current state of observation of the intensity of these waves and what calculations (as distinct from indirect estimates) of the energy flow from protons exist at present?

C. Kennel: If the pretty story that Larry Lyons has concocted about the minimum parallel energy above which pitch-angle diffusion is observed continues to hold, it seems that pitch-angle diffusion does indeed occur on SAR arc field lines. Similarly, no-one in this room dares deny Coulomb collisions, which Keith Cole has just told us can supply enough energy for the arcs. Since the two energy estimates for Coulomb and turbulent scattering are comparable, it seems to me that we are no longer in an 'either-or' controversy, but in a fuzzier, but more realistic, theoretical domain. Both mechanisms act; which dominates on a given day depends upon the details of the proton energy distribution in the ring current. Our experiments (and theories!) are not yet sufficiently accurate to sort out how much energy goes into electrons from collisions and how much from waves on a given day.

L.R. Lyons: I agree with Charlie Kennel's comment.

G.Rostoker: How do you assign the position of the tick marks on the pitch-angle distributions which separate flat distributions from rounded ones?

L.R. Lyons: At each L, the tick marks on the distributions are at the same value of $E_{||}$ for all particle energies. The chosen value of $E_{||}$, at each L, is an adjustable parameter. The value selected best separates nearly isotropic distributions at lower value of $E_{||}$ from rounded distributions at higher values.

J. Etcheto: What is your definition of $E_{||}$ minimum?

L.R. Lyons: Wave energy is assumed to be distributed over some band of parallel wave numbers, $k_{||}$. Resonance at each cyclotron harmonic will occur over a band of parallel electron energies $E_{||}$, with that band being determined by the resonance condition and the wave dispersion relation. Including all cyclotron harmonics, and wave energy distributed over some range of geomagnetic latitudes, cyclotron resonance will occur for all equatorial values of $E_{||}$ greater than some minimum value $E_{||,min}$.

K.D. Cole: Observations of N_e and T_e in the topside of the ionosphere imply the existence of a heat source in the magnetosphere which supplies not only the SAR-arc (which is observable above background 6300 Å emission) but also a belt of high T_e about 1500-2000 km broad. How does the ion-cyclotron mechanism for exchange of energy between ring current and plasmaspheric particle operate to supply the implied heat flux over this wide belt, i.e. both inside and outside the plasmasphere?

L.R.Lyons: The Explorer 45 observations clearly show that the ion-cyclotron wavemechanism operates throughout the entire plasmapause region, which covers a radial distance range of more than 2 R_e in the equatorial plane. It is not yet possible to trace the plasmapause region from the equatorial plane to the ionosphere in order to determine whether the ionospheric heating occurs on exactly the same field line as the ion-cyclotron waves.

S. De Forest: Did you try to de-convolve the 22° angular resolution of your detector? Do you ever see enhancements in the 'loss cone' of low-energy ions?

L.R. Lyons: We have not tried to de-convolve the data, nor have we seen any enhancements in the loss cone. While S^3 measures the pitch at the beginning and end of each measurement, we never know precisely how close to 0° pitch angle the satellite scanned during any specific particle measurement. Thus I don't think we could de-convolve the data to obtain any detailed pitch angle distributions within the loss cone.

MAGNETOSPHERIC DYNAMICS AND WAVE-PARTICLE INTERACTIONS

Frederick L. Scarf
Space Sciences Department,
Systems Group of TRW Inc., Redondo Beach, California, USA

Christopher T. Russell
Institute of Geophysics & Planetary Physics,
University of California, Los Angeles, USA

Abstract

It has been demonstrated that two general classes of wave-particle interactions are of great importance for magnetospheric dynamics. Electromagnetic and electrostatic plasma instabilities give rise to relatively narrow-banded spontaneous emissions (e.g. ELF hiss, chorus, three-halves noise, ion cyclotron and ion-plasma-frequency turbulence) that can scatter trapped particles into the loss cone, leading to modified pitch-angle distributions, stable trapping limits, diffuse aurora, proton precipitation events, etc. The current-driven plasma instabilities give rise to impulsive ion acoustic or Buneman mode turbulence that provides very effective energy transfer (via the anomalous conductivity mechanism) at the bow shock and in regions where strong field-aligned currents are observed. We review these interactions and identify significant open questions that must be investigated during the IMS.

1. Introduction

It is now almost ten years since Kennel & Petschek [1966] proposed the first comprehensive theory in which certain plasma instabilities and associated wave-particle interactions control the dynamics of part of the Earth's magnetosphere. During the past decade, many local satellite observations and intensive analysis programmes have conclusively verified the basic general premise that the Earth's magnetosphere is governed by plasma-physics phenomena. However, it has been determined that at the Earth the important wave-particle interactions cover a much broader range of wave modes and particle energies than envisaged in the original Kennel & Petschek paper, which dealt primarily with whistler-mode turbulence (electromagnetic waves having $f < f_c^e$, where f_c^e is the electron gyrofrequency) and

Knott & Battrick (eds.), The Scientific Satellite Programme During the International Magnetospheric Study, 261–280.
All Rights Reserved. Copyright © 1976 by D. Reidel Publishing Company, Dordrecht-Holland.

moderate energy ($E \gtrsim 40$ keV) electrons. We now know that electrostatic waves driven by currents or thermal anisotropies interact very strongly with electrons at all energies, and in general there are corresponding plasma instabilities associated with protons.

Some of the very important dynamical phenomena *known* to be dominated by local plasma instabilities and wave-particle interactions at the Earth include: (i) the development of limiting fluxes for stable trapping of energetic particles; (ii) the mechanisms for ring-current decay and precipitation of electrons to form the diffuse aurora; (iii) energy transfer and heating at the collisionless bow shock and the field-aligned current regions within the magnetosphere; (iv) the mechanisms that provide microscopic coherence and lead to enormous electromagnetic radiation levels from regions above auroral arcs. It is also widely accepted that wave-particle interactions provide the basis for dissipation in field-line merging regions, lead to anomalous electrical conductivity effects that allow parallel electric fields to develop, and give rise to viscous interactions that are significant in the formation of the geomagnetic tail and overall plasma convection pattern. Plasma instabilities are also invoked to explain the formation of SAR arcs and the electron slot, and the scattering of particles onto closed field lines from the open polar cleft. Finally, large-scale dynamical questions involving ionosphere-magnetosphere coupling and variations in line-tying ultimately involve the local plasma conductivity, and its variations because of plasma turbulence effects.

In the remaining sections of this paper we review some observations and analyses of wave-particle interactions in the magnetosphere and we identify significant problem areas that must be examined in detail during the IMS.

2. Magnetospheric Plasma Distribution Functions

It has always been recognised that some plasma distribution functions with 'free energy' sources capable of yielding growing plasma waves have to be present in the magnetosphere. It was anticipated long ago that some kind of two-stream instability would play an important role at the solar-wind/magnetosphere interface. It was also noted at an early stage that the existence of a loss-cone implies a thermal anisotropy that can yield wave amplification.

The enormous amount of detailed information on local particle characteristics gathered in the past decade shows that many mechanisms operate to produce a variety of non-Maxwellian and unstable plasma distributions in the Earth's ionised environment. Within the magnetosphere, thermal anisotropies with $T_\perp > T_\parallel$ develop because of: (i) inward plasma diffusion and convection with conservation of magnetic moment; (ii) betatron acceleration effects; and, (iii) selective pitch-angle scattering into the loss cone. The observed magnetospheric distributions with $T_\parallel > T_\perp$ are probably associated with inward motion and conservation of the longitudinal action invariant, while similar distributions in the solar wind are thought to be connected with conservation of μ, as the interplanetary field strength in the expanding plasma continuously declines. Other significant thermal anisotropies involve heat flux and higher order moments, and these perturbations can be associated with parallel electric fields, currents, particle beams, and heat conduction.

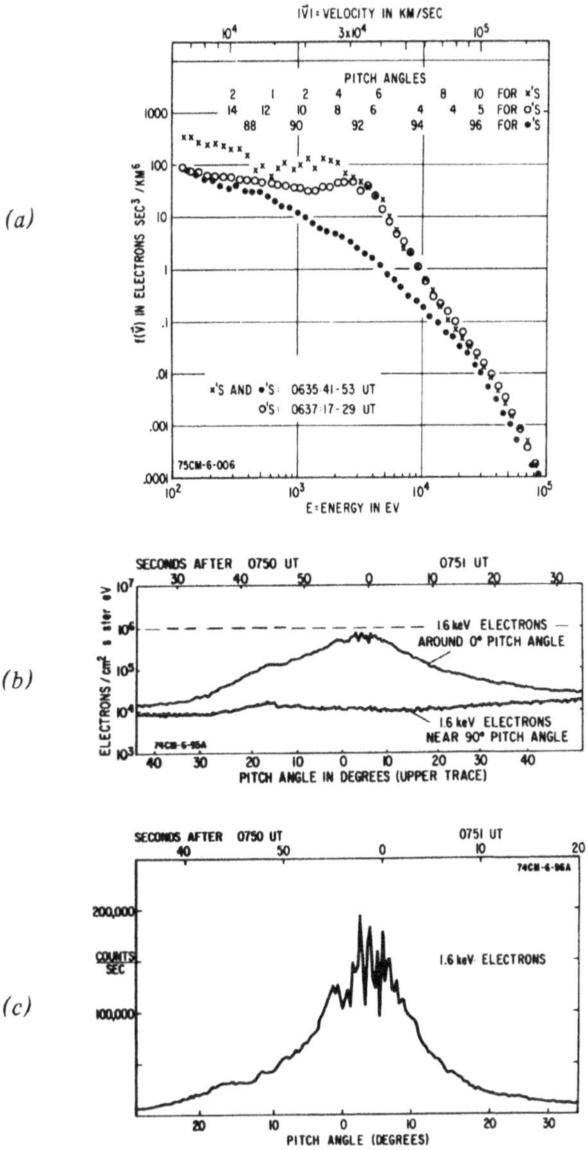

Figure 1. *(a). ATS-6 measurements of the electron distribution function, shortly after a substorm on 20 July 1974.*
(b). Time and pitch-angle dependence of 1-6 keV electrons measured on ATS-6 during a long dwell at fixed energy on 3 July 1974.
(c). Linear replot of the upper trace in Figure 1(b) that shows the large fluctations near zero degree pitch angles.

In recent years it has also become clear that the natural plasmas are locally non-Maxwellian in the sense that there are significant peaks and dips in the energy distributions. These fluctuations arise because: (i) the solar wind and the ionosphere are more or less independent sources of warm and cool magnetospheric plasma; (ii) varying gradient drifts selectively remove trapped particles in restricted energy ranges from the magnetosphere and they leave residual quasi-trapped distributions with deep flux ripples; (iii) collisionless acceleration processes apparently provide enhanced fluxes at certain energies; and (iv) resistive dissipation and heat conduction (and perhaps runaway) contribute to the observed nonthermal tail populations.

The application of the two-stream instability in magnetospheric physics has also been greatly generalised in recent years. Suprathermal flows are detected within the magnetosphere, and it is known that the equivalent electron-proton drift speed in the solar wind and magnetosheath has a significant contribution associated with electron heat conduction. Current-driven instabilities are found to be important at field-merging regions in the magnetosheath, as well as at the bow shock, and the field-aligned currents associated with pressure gradients in the magnetosphere, the cusp boundary and the high-latitude ionosphere can involve streaming instabilities.

Some unusual examples of extreme non-Maxwellian plasma distributions at synchronous orbit are described in Mc Ilwain's [1975] recent paper on the ATS-6 measurements, and Figure 1 shows one example of a very complex distribution function observed by the ATS-6 plasma probe. ATS-6 does not have an on-board plasma-wave investigation to provide correlative data on plasma instabilities, but it now seems clear that natural magnetospheric processes provide enough variety in distribution functions to allow growth of almost any conceivable plasma-wave mode.

3. Cyclotron Resonance Interactions

Early suggestions by Dungey [1963] and Cornwall [1964] that electromagnetic whistler-mode turbulence can lead to particle precipitation were soon followed by the previously cited selfconsistent theory of Kennel & Petschek [1966] which is based on use of the natural gyroresonance instability mechanism for the wave growth (amplification associated with trapped-particle $T_\perp > T_\parallel$ pitch-angle distributions) and the concept of turbulent pitch-angle diffusion arising from the local wave-particle interactions. The Kennel-Petschek theory predicted overall stable trapping limits and precipitation patterns that agreed fairly well with available observations, and this pioneering effort has since stimulated an enormous amount of more detailed analytical activity. In a recent review, Fredericks [1975] described how improved selfconsistent theories for electron-whistler and ion-cyclotron-mode turbulence were developed; these ideas were applied to analyse lightning whistler amplification processes, numerical simulation studies were performed, and the basic concepts were used in attempts to explain localised plasmapause phenomena as varied as ring-current decay, SAR arc formation and the development of the energetic electron 'slot'.

In fact, it is difficult to verify details of the generalised Kennel-Petschek theory for a number of reasons: (i) whistler-mode waves propagate readily in the dispersive magnetospheric plasma, and they are reflected and refracted so that the identification of the source region and the initial wave spectrum is a complex and frequently uncertain undertaking; (ii) the particles that interact resonantly with these waves have relatively high energies (more than tens of kilovolts) and the ions, in particular, drift in and out of the interaction regions; (iii) there are generally significant local sources of resonant particles, and these sources mask the effects of the loss mechanisms.

Only a few aspects of the whistler-mode instability theory have been *directly* verified in some isolated studies. Rosenberg *et. al.* [1970] analysed the arrival characteristics of bursts of precipitating $E \simeq$ 30-100 keV electrons and propagating chorus emissions at Siple Station, Antarctica, and they provided convincing evidence that the chorus interaction with trapped particles near the geomagnetic equator gives rise to pitch-angle scattering that leads to the precipitation. In another area, Kivelson & Russell [1973] showed directly that enhanced whistler-mode turbulence develops in a region with a cold density enhancement, where the theoretical resonant electron energy (proportional to $B^2/8\pi N$) decreases so that more electrons are available to feed the wave growth. However, Kivelson & Russell did not find evidence of enhanced pitch-angle scattering during these events.

During a large substorm event when betatron acceleration produced a sudden local increase in $T_\perp/T_\parallel$, Kivelson *et al.* [1973] showed conclusively that the anisotropy enhancement was associated with chorus growth. Some relevant measurements are displayed in Figure 2, adapted, in part, from an OGO-5 report by Scarf *et al.* [1973a]. The region to the right of the vertical dashed line marks the onset of a $T_\perp > T_\parallel$ distribution for 79 keV electrons, and it can be seen that this change (associated with a change in field magnitude and orientation) was correlated with an increase in 467 Hz search coil output (central panel) representing whistler-mode chorus. The chorus onset at 0732:15 to 0732:40 UT is indeed very well correlated with the anisotropy enhancement, and since the measured plasma parameters give a resonant electron energy ($B^2/8\pi N$) range near 30-90 keV at this time, it may be concluded that the electromagnetic noise enhancement is directly associated with development of a suitable anisotropy in the resonant electron population. It should be noted, however, that this gyro-resonant interaction does *not* lead to anything approaching strong diffusion or near isotropy. During the preceding period with essentially isotropic distributions the chorus was absent, but large-amplitude electrostatic waves with $f \simeq 3 f_c^e/2$ were detected (top panel); we shall discuss this point in more detail below.

The most convincing verifications of the Kennel-Petschek ideas on radiation-belt structure involve analytical treatments based on use of generally measured wave-turbulence spectral characteristics. In a series of papers, Lyons *et al.* [1972] and Lyons & Thorne [1973] assumed a distribution of oblique waves with amplitude and spectral shape that match many measurements of ELF hiss within the plasmapause, and they examined the effects of these waves on the trapped particles in detail. These authors derived energy-dependent and L-dependent pitch-angle distribution lifetimes and equilibrium distribution functions that agree very well with the complex profiles measured on OGO-5 and S^3-A within the plasmasphere. Recently, there have also been significant analytical advances in treating the other half of the problem concerning the origin of the observed gyro-resonant emissions. Etcheto *et al.* [1973] studied the problem of wave-generation and particle diffusion

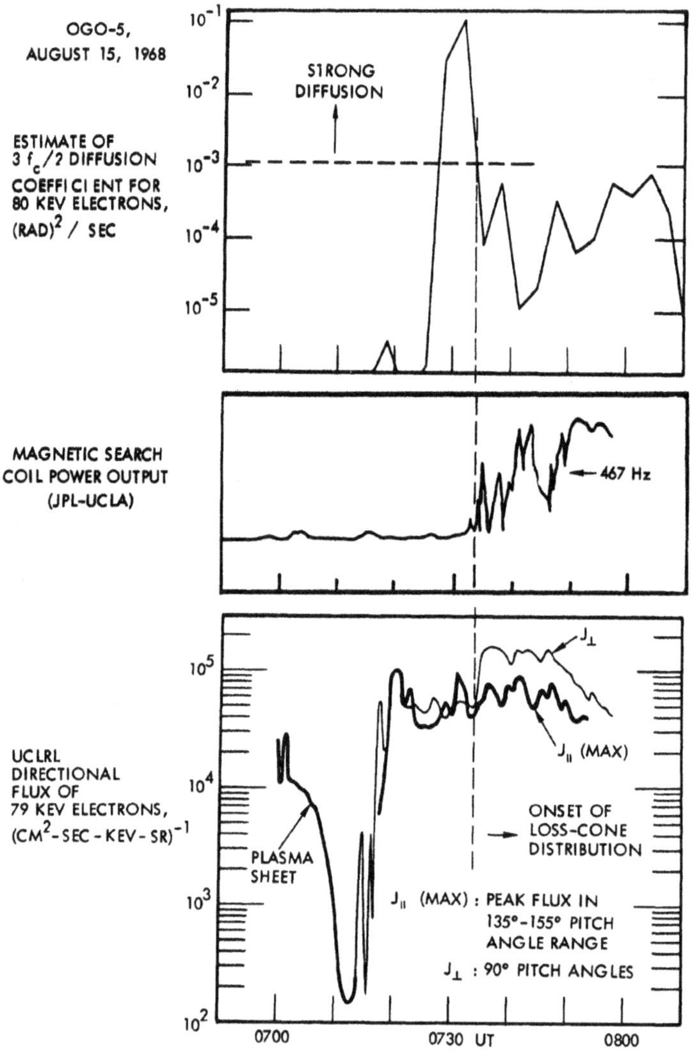

Figure 2. *Wave-particle interaction phenomena detected when OGO-5 was at L = 8, 0040 to 0100 local time, and magnetic latitude near zero degrees during a substorm expansion event. The top panel shows an effective 80 keV electron diffusion coefficient for 3/2 f_c^e waves based on use of Equation (1) with the measured wave amplitude and frequency profile. The central panel gives the output of the magnetic search coil, and the 80 keV electron data in the bottom panel show that at 0733 UT a marked change from near isotropy to a normal loss-cone distribution is associated with the increase in whistler-mode electromagnetic noise.*

selfconsistently, although they did confine attention to parallel propagating modes. These authors used iterative techniques to derive a wave turbulence spectrum that agrees fairly well with the measured ELF hiss spectrum, and they also clarified concepts of the general stable trapping limit and the effects of cold density variations. These semi-analytical papers taken together provide very convincing evidence that the inner radiation belt trapped electron characteristics are controlled by an internal whistler-mode instability associated with the thermal anisotropy.

A very elegant parallel theory of ion-cyclotron turbulence has been developed and applied to explain ring-current decay and proton precipitation. Cornwall *et al.* [1970] first noted that the sudden decrease in resonant energy at the plasmapause should lead to wave growth, greatly enhanced pitch-angle scattering, and rapid ring-current decay. In subsequent reports, associated theories were developed to relate the presumed ion-cyclotron turbulence at the plasmapause to SAR arc generation and formation of an electron slot (see review by Fredricks [1975]). These predictions are in good general agreement with some S^3-A energetic particle observations in the sense that during the geomagnetic storm recovery phase the inner edge of the proton ring current decays just within the plasmapause in a manner consistent with expectations for a moderate ion-cyclotron instability that does not yield strong diffusion [Williams & Lyons, 1974a, b]. However, it is evident that many strong proton precipitation events are not connected with plasmapause phenomena [Bernstein *et al.*, 1974] and other positive ion instabilities must be operative in the plasma sheet.

At present, the proton precipitation problem has some unexplained features. The S^3-A observations show that near the plasmapause there are abrupt changes in pitch-angle distributions for proton energies that appear to be well correlated with the marginal stability resonant energy conditions of the ion-cyclotron turbulence model. However, the formalism of Etcheto *et al.* [1973] indicates that stable trapping limits should be associated with resonant energies for maximum growth rates, rather than conditions for marginal stability. To date, the consequences of this change have not been examined in detail.

Moreover, there has not been unambiguous detection of the expected large-amplitude ($\Delta B \simeq 1\gamma$) ion-cyclotron waves near the plasmapause during storm recovery phases. Large-amplitude ($\Delta B \simeq 3$-4γ), low-frequency electromagnetic waves have been observed in detached cold density regions beyond the plasmapause [Kivelson & Russell, 1973], but these waves do not have the correct frequency or polarisation to represent ion-cyclotron modes. Taylor *et al.* [1974] detected only a few strong low-frequency electromagnetic events on S^3-A, and Gurnett reported (private communication) that only once has any 1.7 Hz signal been detected near the plasmapause with the high-sensitivity Hawkeye-1 search coil.

We conclude that it is now known with certainty that gyro-resonant interactions play an important role in magnetospheric dynamics, but only for trapped electrons in the plasmasphere has it been demonstrated that these interactions completely control the dynamics in the sense that the waves are present throughout the region, with spectral characteristics that strongly modify the local distributions everywhere. Beyond the plasmapause, chorus emissions are known to precipitate electrons, and apparently these waves do control the stable trapping limit; however, the emissions are sporadic and have local time distributions, so that the wave-particle interactions are probably dominant only in the sense of averages over universal and local time.

However, there is still no complete theory to account for whistler-mode emissions observed at the magnetopause, or the surprisingly strong (up to 200 milligamma amplitude) whistler-mode turbulence detected on IMP-7 in the distant plasma sheet at frequencies above the lower hybrid resonance [Scarf *et al.*, 1974]. Comprehensive explanations of chorus fine structure and whistler-mode triggered emissions are also not yet at hand.

4. Narrow-Band Electrostatic Instabilities

The most intense waves detected in the magnetosphere with frequencies related to the local electron gyrofrequency, f_c^e, are the $(n + 1/2) f_c^e$ emissions first discussed by Kennel *et al.* [1970]. Subsequent analyses by Fredricks & Scarf [1973] and Scarf *et al.* [1973a] suggested that these strong emissions (which very frequently have amplitudes as high as 10 mV/m) are substorm related, and Gurnett & Shaw [1973] provided valuable additional information on these emissions. The most common observation of this mode is at $f \simeq 3 f_c^e/2$, and the broadband *E*-field and *B*-field dynamic spectra on the left-hand side of Figure 3 illustrate that the mode is electrostatic. The IMP-6 data displayed on the right-hand side of Figure 3 show that this mode, which is detected beyond the plasmapause, can be excited all the way out to the magnetopause.

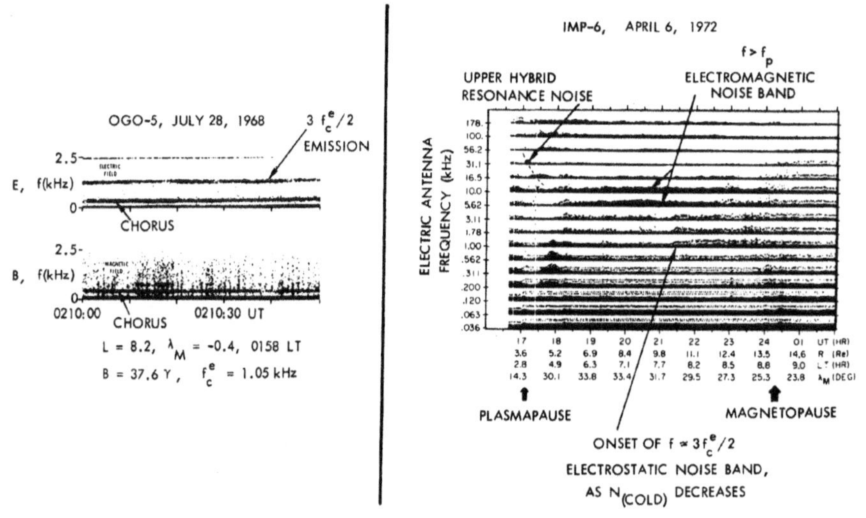

Figure 3. *The data on the left-hand side of the panel show that the $3 f_c^e/2$ emission is electrostatic, in contrast to the chorus signal that is detected with both E and B sensors. The IMP-6 data (right-side) illustrate that the $3 f_c^e$ emissions can be detected well away from the magnetic equator in hot plasma regions extending out to the magnetopause.*

Occasionally, wave amplitudes greater than 10 mV/m are measured during very disturbed periods. During the 15 August of 1968 substorm, the peak wave amplitude was approximately 100 mV/m, and the effective pitch-angle diffusion coefficient shown in the top panel of Figure 2 was estimated using Scarf *et al.* [1973b],

$$D\,(electrostatic) \simeq 10^{-2}\ (eE/p)^2/f \qquad (1)$$

where E is the wave amplitude, f is the frequency, and p is the particle momentum. It was suggested that the unusually intense wave-particle interactions before 0733 UT (to the left of the vertical dashed line in Fig. 2) might yield strong diffusion and near-isotropy for the 80 keV electrons. A much more definitive analysis of the effects of $(n + 1/2)\ f_c^e$ emissions on magnetospheric particle distributions was carried out later by Lyons [1974] who demonstrated that the commonly observed turbulence amplitude of 1-10 mV/m is sufficient to put electrons with energies of several kilovolts in strong diffusion. Lyons' calculations also verified that 100 mV/m waves would indeed lead to strong diffusion for 100 keV electrons, as indicated by the data of Figure 2. In addition, Lyons showed that some detailed low-altitude (rocket) measurements of 1-20 keV electron energy spectra in the loss cone can be very well explained on the assumption that moderate amplitude (a few mV) $3f_c^e/2$ waves produce strong diffusion near the equator. It now seems certain that the $(n + 1/2)\ f_c^e$ modes play a very important role in providing the source of high-latitude electron precipitation fluxes, especially during geomagnetically active periods, and we may regard this interaction as the ultimate origin of the diffuse aurora.

The precise mechanism for generation of the $(n + 1/2)\ f_c^e$ mode is presently somewhat obscure. The recent review by Fredricks [1975] contains fairly detailed discussions of linear stability analyses carried out by Fredricks, Young, Karpman and others, and comparison with a wave-wave interaction theory proposed by Oya. There is little direct information on the specific local plasma distribution characteristics that can be associated with this mode. On OGO-5, no plasma information at all was available (the probe failed soon after launch) and complete IMP-6 results are not yet available. However, Gurnett & Frank [1974] reported that the $(n + 1/2)$ f_c^e electrostatic emissions appear to be quenched whenever the density of very low energy thermal plasma exceeds a small fraction (in the order of 5%) of the suprathermal plasma density. This suggests that the generation mechanism for these modes involves a hump in the energy distribution as well as a thermal anisotropy.

Some very recent research gives important additional information on the $3\,f_c^e/2$ emissions. Abdalla & Kennel [1975] found numerical solutions of the dispersion relations for a 'subtracted' ring-type distribution function, and they noted that zero group velocity solutions existed for $f \approx 3\,f_c^e/2$, $5\,f_c^e/2 < f$ (upper hybrid resonance), with detailed spectral location dependent on the ratio of hot to cold plasma particles. This theory suggests that a continuum is unstable, but that absolute instabilities develop only near the half-harmonic resonances. In another area, Bernstein *et al.* [1975] recently generated very intense electrostatic half-harmonic waves in an extremely large laboratory vacuum chamber, and they identified the generation mechanism in terms of a streaming instability. Since the ATS-6 plasma probe detects streaming [McIlwain, 1975] in regions where OGO-5 and IMP-6 detect

$3f_c^e/2$ noise, these laboratory experiments may provide an important clue to the origin of the magnetospheric half-harmonic bursts.

A different kind of electrostatic instability was recently discussed in connection with the problem of ring-current decay and proton precipitation. Coroniti *et al.* [1972] showed that for an initial ring-current proton distribution peaked near 90° with an almost empty loss cone, there will be sufficiently rapid growth of electrostatic waves with $\omega \approx \omega_p^i$ to provide significant pitch-angle scattering and strong precipitation. Coroniti *et al.* noted that the marginal stability state could have enough residual anisotropy to provide some electromagnetic ion-cyclotron turbulence, but the model calculations indicate that the strongest waves are electrostatic, with $\omega \approx (0.1-0.5)\,\omega_p^i$. The calculations also indicate that strong proton precipitation can readily develop *beyond* the plasmapause in good agreement with the recent results reported by Bernstein *et al.* [1974].

The OGO-5 electric-field experiment had no wideband measurements below 1 kHz, and poor bandpass channel coverage below 560 Hz, but on S^3-A Anderson & Gurnett [1973] reported on the infrequent detection of 20–500 Hz electrostatic wave turbulence just beyond the plasmapause. In addition, it appears that on Hawkeye-1, moderate amplitude ($\approx$ 1 mV/M) electrostatic waves with $f \approx$ 50 Hz are commonly detected beyond the plasmapause (D. Gurnett, private communication). Thus, it now seems that some proton precipitation events may be associated with an important new high-β, low-frequency, wave-particle interaction effect that requires much more study in the future.

5. Current-Driven Instabilities

Currents strong enough to drive plasma instabilities are clearly detected at the bow shock and along the polar-cusp and plasma-sheet boundaries. During storms and substorms, the high-latitude, field-aligned magnetospheric current systems appear to drive especially strong plasma instabilities in the region extending from the topside ionosphere (say 1000 to 2000 km altitude) to a geocentric distance of 2–4 R_e. These phenomena appear to be of major importance for magnetospheric physics because current-driven instabilities produce very important physical effects such as dissipation at the bow shock, anomalous resistivity, and a possible microscopic mechanism for field-line merging.

The most extensive observations in space are related to studies of the collisionless bow shock upstream from the magnetopause, and Figure 4 shows some typical features of the wave-particle-field interaction phenomena detected during early OGO-5 and IMP-6 crossings of the bow shock. We first direct attention to the left side of Figure 4 and discuss the various panels in order, starting from the top.

The magnetic field profile at the shock frequently exhibits ULF magnetic turbulence that can, depending on the interplanetary field orientation, spread into the entire upstream and downstream region. The data displayed here (for a low-β, low-Mach-number laminar shock) are relatively clean, however, and a clear signature of a strong, fairly well-defined current layer parallel to the shock surface is shown in the region centred around 2248:45 UT. It is apparent that the enhanced VLF electric field turbulence shown in the next panel is associated with this current. At

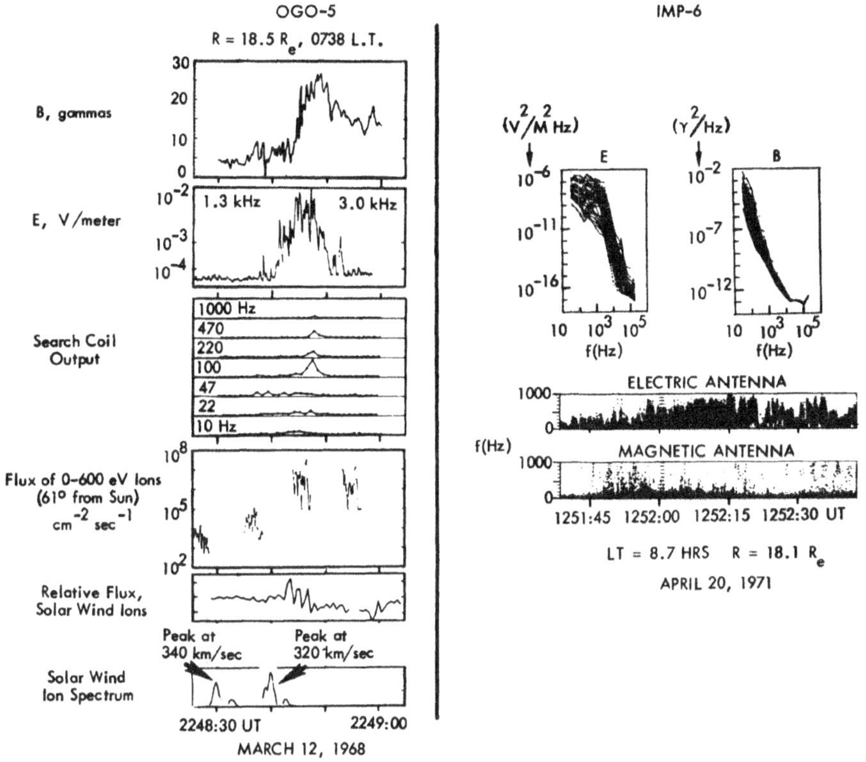

Figure 4. *Typical wave-particle interaction phenomena detected at quasi-laminar shocks. The early OGO-5 observations (left-side) indicate that electrostatic instabilities provide shock dissipation, although whistler-mode turbulence also develops. The characteristic features of the E and B signals at the shock are indicated in the IMP-6 frequency-time spectrogram and power spectra on the right side.*

this time the upstream proton density was 15.1 cm^{-3} [Fredricks *et al.*, 1970], and the proton plasma frequency was 0.82 kHz, so that the waves detected in the 1.3 kHz and 3.0 kHz channels had frequencies somewhat higher than the upstream f_p^+-value in the spacecraft frame of reference.

The search coil panel shows little activity in the 1000 Hz channel, indicating that the kilohertz electric fields involve detection of electrostatic waves; this OGO-5 conclusion of Fredricks *et al.* [1970] was recently verified in great detail by Rodriguez & Gurnett [1975] on the basis of IMP-6 measurements, as shown in the right-hand side of Figure 4. However, it can be seen that electromagnetic waves were detected at lower frequencies; the comprehensive IMP-6 capabilities have been used by Rodriguez & Gurnett to demonstrate that the E to B ratio for the waves with $f < f_c^e$ is consistent with expectations for whistler-mode turbulence.

The 0–600 eV ion measurements shown in Figure 4 were derived from the OGO-5 light-ion spectrometer pointing 61° away from the Sun, and the data thus provide relative information on thermalisation and deflection of solar-wind flow. The lowest panels contain measurements from the OGO-5 ion plasma probe pointing directly toward the Sun. Frequently, as discussed by Neugebauer [1970], it is possible to detect a shift toward lower streaming speed just before the shock, indicating that charge-separation electric fields start to decelerate the upstream ions; typically, as shown here, there are also flux oscillations at the shock, and the local flux spikes may be 2–4 times the upstream flux value.

The left-hand side of Figure 4 clearly shows that the solar-wind thermalisation, deflection and flux compression started well before the whistler-mode turbulence was detected, and it is commonly agreed that electrostatic wave interactions provide the dominant dissipation mechanism for this type of laminar shock.

The right-hand side of Figure 4 shows some other aspects of the typical bow-shock electric-field turbulence. The data from the IMP-6 wideband telemetry link provided the *E*-field and *B*-field dynamic spectra in the lower panels, and these *f-t* diagrams illustrate the basic spectral characteristic of the *E*-field turbulence associated with the current-driven instabilities measured in space; in every case the observed electrostatic waves consist of a number of impulsive or rapidly rising noise bursts covering a relatively broad range in frequency.

In the initial discussions of these bow shock measurements, Fredricks *et al.* [1970], interpreted the VLF electric field turbulence in terms of ion waves associated with a two-stream instability. The disparity between the measured wave frequencies and the anticipated range of 0 to f_p^i (based on the upstream density value) was noted and it was speculated that Doppler shifts play important roles for ion sound waves. In fact, the flux spikes observed within shock structures represent density enhancements, and the local values of f_p^i may exceed the upstream values by factors of 2 or more. However, there does appear to be some problem in explaining the observed *E*-field frequency range in terms of simple linear ion sound waves, since Rodriguez & Gurnett [1975] show that the noise spectral density customarily remains quite high up to about 2 kHz (as illustrated in the upper right panels of Fig. 4), while f_p^i *(nominal)* ≅ 500-700 Hz in the plasma rest frame.

Another difficulty with the ion acoustic two-stream instability is related to observation of limited ranges for $\theta = T_e/T_i$ within the shock structure. Although Montgomery *et al.* [1970] did report a local θ-value as high as 9.3 for one shock crossing, the ions generally appear to be heated much more readily than anticipated in the earlier papers, so that on the basis of calculations by Fried & Gould [1961], electron-proton drift speeds many times higher than the ion thermal speed are needed to provide an ion sound instability. For these reasons, other bow-shock-instability theories were recently discussed. As summarised by Greenstadt & Fredricks [1974], various authors have invoked ion-ion streaming and cyclotron drift instabilities in order to explain the *E*-field shock turbulence, but there is little direct evidence in favour of these interpretations.

Mode identification can be achieved if precise polarisation measurements are made but, to date, no really conclusive measurements *within* the shock are available. OGO-5 was stabilised, and instantaneous polarisation was not measured, while IMP-6 and 8 both spin slowly in comparison with the time needed to transit through a

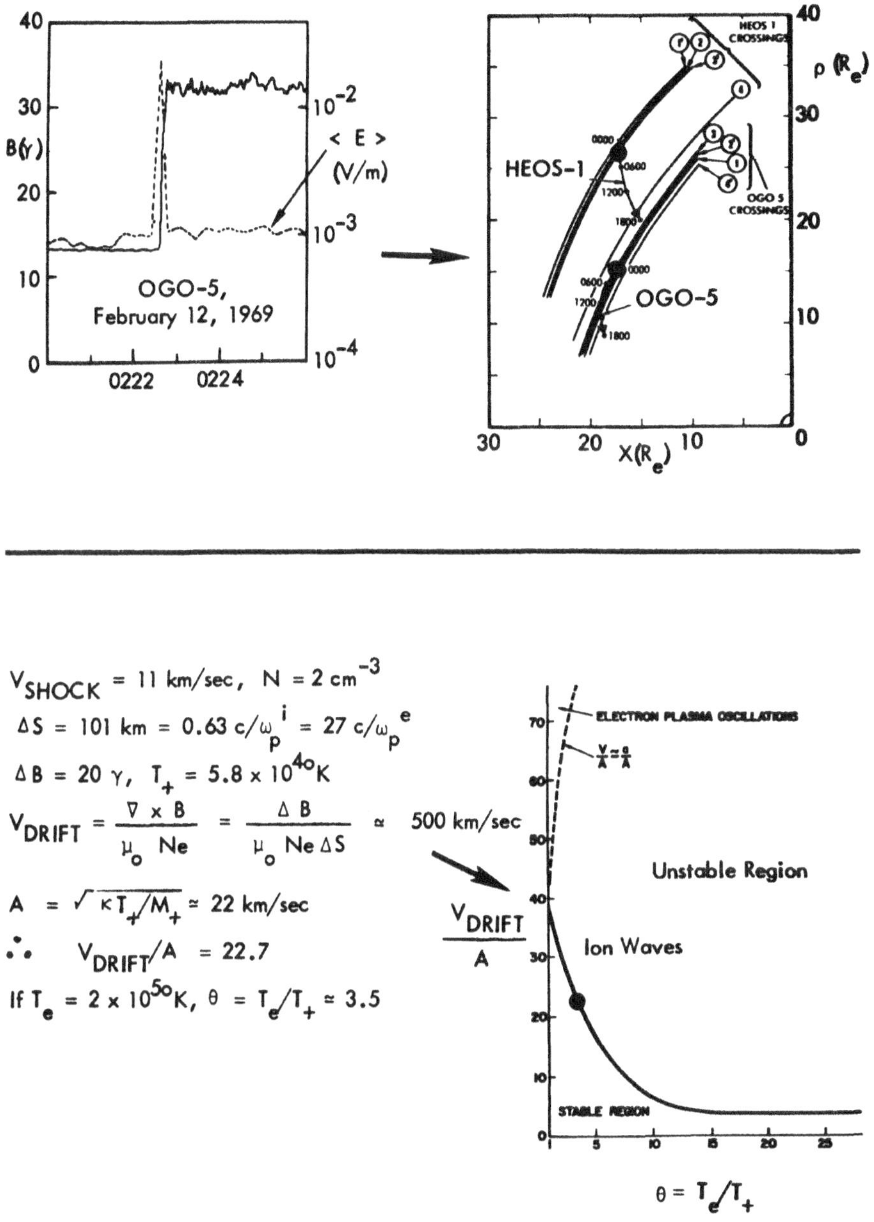

Figure 5. *Instability analysis for the 12 February 1969 laminar bow shock crossing. For this shock HEOS-1 and OGO-5 observations were used to evaluate the rest frame shock current layer thickness.*

laminar shock. However, incomplete polarisation studies reported by Rodriguez & Gurnett [1975] are consistent with detection of longitudinal electrostatic oscillations aligned parallel to the local *B*-field.

One very important problem connected with analysis of the bow-shock plasma instability concerns the motion of the shock. Since it is necessary to deduce the actual current profile from the field-variation measurements, it is also necessary to determine the true shock thickness in the shock rest frame, and this requires use of more than one spacecraft. In a few cases, co-ordinated multispacecraft bow-shock observations have been made [Greenstadt *et al.*, 1975], and Figure 5 contains some OGO-5 data, along with an outline of some preliminary analysis, for one bow-shock crossing on 12 February 1969. The upper-left panel shows a very sharp magnetic gradient and a wideband (1 - 22 kHz) electric-field profile that has a well-confined maximum centred within the *B*-field shock ramp. At this time the upstream density was 2 cm^{-3}, the wind speed was 418 km/s, T_+ was 5.8×10^5 °K the Alfvén Mach number was near 2, and the interplanetary field angle with respect to the shock normal was near 80°.

The upper-right panel in Figure 5 shows the relative locations of HEOS-1 and OGO-5 on 12 February 1969. Greenstadt *et al.* [1975] were able to determine the bow-shock motions for a series of crossings by comparing data from the two spacecraft and for this case they found $V = 11$ km/s, so that the shock thickness and current magnitude could be evaluated, as indicated in the lower left side of Figure 5.

This analysis leads to a surprisingly high value for *V (drift)/A* within the shock. Precise local measurements of T_e were not available on OGO-5, but even if a nominal value of 2×10^5 °K is used, the lower-right panel shows that the ion-wave possibility, based on the Fried-Gould [1961] calculations, cannot be ruled out. If more realistic non-Maxwellian shock electron distribution functions having $T_e \approx 6\text{-}7 \times 10^5$ °K [Montgomery *et al.*, 1970] are considered, along with the effect of electron heat conduction [Forslund, 1970], it is even easier to understand how ion sound waves can be driven by currents in the shock layer. Of course, real advances in this area of bow-shock plasma instability analysis cannot be carried out on the basis of chance multispacecraft observations, but the pending ISEE mission is designed to provide precise information for this type of study during the IMS.

Field-aligned currents are also of considerable importance within the magnetosphere, and the OGO-5 experimenters described the detection of field-aligned currents together with strong electrostatic turbulence in the polar cusp during a large storm [Fredricks *et al.*, 1973; Scarf *et al.*, 1972], and near the low-altitude dayside cusp boundary [Scarf *et al.*, 1975] and inner edge of the nightside plasma sheet [Scarf *et al.*, 1973b] during substorms. Figure 6 shows some of the characteristic wave-particle-field relations observed within a weak high-latitude field-aligned current region in the magnetosphere near midnight. It can be seen that a few strong impulsive rising *E*-field noise bursts, quite similar to the spectra shown in Figure 4, are detected even near this weak field-aligned current. However, as in the case of the bow shock, there can be no unambiguous estimate of the magnetospheric current density based on measurements from a single moving spacecraft, and a complete analysis of this type of important plasma instability must be deferred until observations from a suitable multispacecraft mission are available. Such measurements should be available during the IMS.

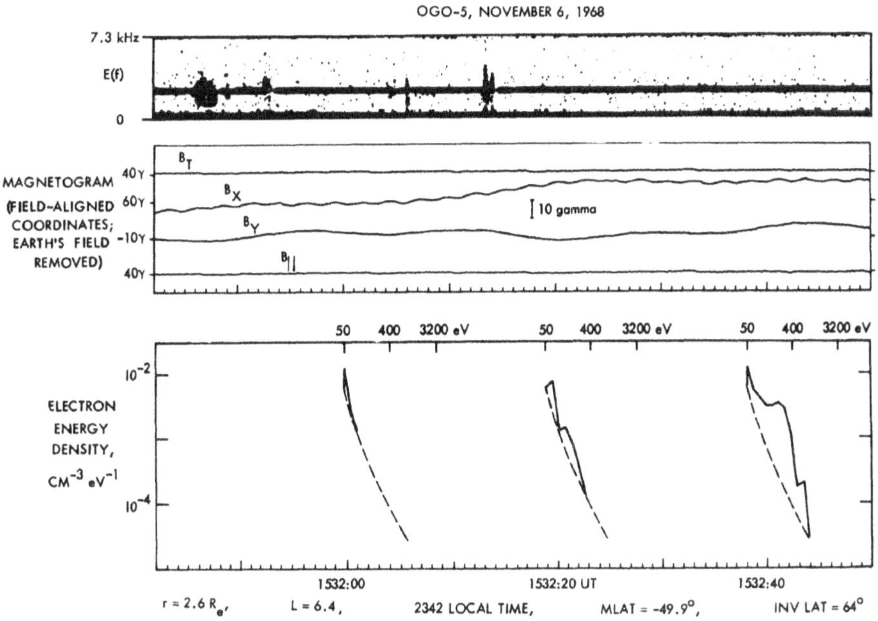

Figure 6. *An example of a weak field-aligned current and sporadic electrostatic noise bursts detected near the inner edge of the plasma sheet during a relatively quiet interval. The small oscillations in the B_X-profile are associated with boom oscillations.*

The plasma turbulence associated with a current-driven instability produces changes in the distribution functions, and the process can be described in terms of an anomalous or turbulent resistivity. As discussed by Kadomtsev [1965], Sagdeev [1967], Biskamp [1973], Papadopolous & Coffey [1974], and others, the effective collision frequency ν_{eff} should be given by

$$\nu_{eff} \simeq \sqrt{\frac{\pi}{2}}\, \omega_p^e \,(\epsilon_0 E^2/2N\kappa T_e) \tag{2}$$

and, in principle, the anomalous resistivity can be evaluated directly by measuring the local plasma-wave turbulence amplitude and the electron moments. For instance, with the OGO-5 shock of Figure 5, ν_{eff} turns out to be approximately 30-60 collisions per second using Equation (2) with $T_e = 2 \times 10^5$ °K, and the anomalous conductivity, $\sigma = N_e^2/m\nu_{eff}$, can account for the shock dissipation.

Recently Fredricks *et al.* [1973] used wave and particle measurements from OGO-5 to argue that during storms anomalous resistivity near the low-altitude boundary of the Earth's polar cusp allows large-scale parallel electric fields to develop. (It should be noted that the polar cusp at higher altitudes during more quiescent geomagnetic conditions has no significant electrostatic noise. For instance, OGO-5 encountered the cusp at 5 R_e near 1300 local time on 14 September 1969 and the E-field noise levels were near background, although field-aligned currents and ULF magnetic oscillations were detected.) In some cases, other aspects of the local measurements are consistent with an anomalous resistivity interpretation. For instance, electron spectra of the type plotted in the bottom panel of Figure 6 frequently show a depletion in low-energy population and appearance of suprathermal particles as the spacecraft encounters the current system and its associated turbulence; these changes could be explained by resistive heating associated with the wave-particle interactions. However, there are still a fair number of important unanswered questions concerning this microscopic process, particularly as it relates to the substorm observations of field-aligned current phenomena and the role of the plasma instability in regulating line-tying and magnetosphere-ionosphere coupling. One of the biggest gaps in this area involves the measurement of the complete particle distribution function in the appropriate unstable current-carrying high-latitude region; McIlwain's [1975] description of the very complex equatorial distributions detected on closed auroral field lines with ATS-6 opens the possibility that extremely non-Maxwellian distributions are present all along auroral field-lines, and local acceleration, streaming, runaway, and similar complex phenomena may well be operative here. Finally, the most significant unanswered general question concerning current-driven instabilities involves the role of this interaction in providing a dissipation mechanism for field-line merging, a process expected to be of great significance at the magnetopause and in the geomagnetic tail.

6. Nonlinear Processes

The conventional discussion of a plasma instability in space involves essentially a linear dispersion relation and the concept of a single growing mode. However, the observations show that such simple treatments are generally not applicable for natural signals, and there is now much evidence that strong wave-wave interactions develop in the magnetosphere. For instance, Coroniti & Kennel [1970] noted that chorus amplitudes are frequently modulated during micropulsation events, and they attempted to explain this in terms of a micropulsation modulation of the particle distribution that affects the instantaneous chorus growth rate. Helliwell [1974] recently summarised many of the other complex wave-wave interactions that involve a variety of magnetospheric signals propagating in the whistler mode, and he presented some of the current nonlinear ideas concerning feedback effects. There are a number of other prominent cases where mode coupling effects are probably quite important; for instance, the current-driven instabilities discussed in the previous section clearly involve coupling of ULF electromagnetic pulses and higher frequency electrostatic waves. This coupling persists even when large-amplitude MHD waves are found in the upstream solar wind [Scarf *et al.*, 1970; Fredricks *et al.*, 1972], and the nonlinear wave-wave interactions can lead to dissipation so that the magnetic pulses steepen into collisionless shocks.

Other important wave-wave interactions are assumed to be operative because the two modes are generally observed together. The whistler-mode turbulence at the shock ramp and at magnetosheath field gradients ('lion roars') appears to involve current-driven modifications of electromagnetic growth mechanisms, and a similar mechanism may be associated with the generation of the strong whistler-mode noise detected with IMP-7 near the neutral sheet in the distant tail during substorms [Scarf *et al.*, 1974]. Since the source of this plasma wave noise is presently unknown, it is also not known how this turbulence affects the plasma sheet dynamics.

The most striking example of a nonlinear interaction in the Earth's plasma environment is the wave-wave coupling that apparently leads to the remarkably intense kilometre-wave radiation from the regions above auroral arcs. Gurnett [1974] showed that at peak intensity the total power radiated is about 10^9 W. This is approximately 1% of the maximum energy dissipated by the precipitating auroral particles. It appears highly unlikely that any incoherent radiation mechanism can account for the tremendous power radiated above the auroral arcs, and it seems to be necessary to invoke mode coupling in which local plasma waves organise the radiating particles to provide a high degree of coherence; similar wave-wave interactions may well be operative at Jupiter.

7. Outlook for The IMS

During the past decade it has become clear that plasma instabilities and wave-particle interactions play a central role in the magnetosphere. Much progress has been made in identifying specific instabilities and in verifying that magnetospheric particles interact strongly with local plasma turbulence. However, certain very important problems (such as the merging dissipation problem and the question of the acceleration of O_2^+ ions to kilovolt energies) are still essentially completely open, with respect to local observations. In dealing with phenomena involving runaway and the inter-relation of field-aligned currents, parallel electric fields and anomalous conductivity, the observational status is still highly preliminary. It seems clear that much better measurements of plasma distribution functions are needed to study field-aligned beams, plasma convection in the magnetosphere, flow in the turbulent magnetosheath, the temporal development of suprathermal particle bumps and distribution function plateaus, and many other topics. In all cases, these measurements must be made on spacecraft with electrostatic specifications that are well-defined so that sheath effects can be identified and removed.

The other advances that must be made during the IMS involve multispacecraft missions. It is known that unambiguous analysis of magnetospheric boundary phenomena at the bow shock, the magnetopause, the plasmapause and the edges of the plasma sheet requires a capability for separating space and time variations. To a large extent, the ISEE programme should provide suitable observations of this nature; however, it must be recognised that ISEE will only yield local information on low-latitude phenomena. It is also known that extremely important local wave-particle interaction phenomena develop at low altitudes (say $r \lesssim 304\, R_e$) on auroral and polar field lines. These processes are to be studied on the Electrodynamic Explorer (EE) mission. It is now evident that this mission is as important for IMS research as any of the others, and it is also evident that a multisatellite version of EE is required to separate space and time variations in the auroral zones. Indeed, although the magnitude of the local current at the bow shock or magnetopause can frequently be deduced from local B-field measurements using symmetry arguments, this analysis cannot be performed within the high-latitude magnetosphere. Thus, in a sense it is more important to have a multispacecraft mission in the EE orbit than in the region covered by ISEE.

8. Acknowledgments

We are indebted to Drs. D.A. Gurnett and C.E. McIlwain for allowing us to describe their recent observations, and we thank M. Abdalla, W. Bernstein, F.V. Coroniti, C.F. Kennel, R.W. Fredricks and D. Williams for helpful discussions. This work was supported by the National Aeronautics and Space Administration under Contract NASW 2659 (TRW) and Grant NGR 05-007-004 (UCLA).

References

Abdalla, M. & C.F. Kennel, This volume, 1975.

Anderson, R.R. & D.A. Gurnett, *J. Geophys. Res.* **78**, 4786 (1973).

Bernstein, W., B. Hultqvist & H. Berg, *Planet. Space Sci.* **22**, 767 (1974).

Berstein, W., H. Leinbach, H. Cohen, P.S. Wilson, T.N. Davis, T. Hallinan, B. Baker, J. Martz & R. Ziemke, *J. Geophys. Res.* **80** (1975).

Biskamp, D., *Nucl. Fusion* **13**, 719 (1973).

Cornwall, J.M., *J. Geophys. Res.* **69**, 1251 (1964).

Cornwall, J.M., F.V. Coroniti & R.M. Thorne, *J. Geophys. Res.* **75**, 4699 (1970).

Coroniti, F.V. & C.F. Kennel, *J. Geophys. Res.* **75**, 1279 (1970).

Coroniti, F.V., R.W. Fredricks & R. White, *J. Geophys. Res.* **77**, 6243 (1972).

Dungey, J.W., *Plan. Space Sci.* **11**, 591 (1963).

Etcheto, J., R. Gendrin, J. Solomon & A. Roux, *J. Geophys. Res.* **78**, 8150 (1973).

Forslund, D.W., *J. Geophys. Res.* **75**, 17 (1970).

Fredricks, R.W., Wave-Particle Interactions in the Outer Magnetosphere, Magnetospheres of Earth and Jupiter (Ed. V. Formisano), D. Reidel, Dordrecht-Holland, 1975.

Fredricks, R.W., G.M. Crook, C.F. Kennel, I.M. Green, F.L. Scarf, P.J. Coleman & C.T. Russell, *J. Geophys. Res.* **75**, 3751 (1970).

Fredricks, R.W., F.L. Scarf, C.T. Russell & M. Neugebauer, *J. Geophys. Res.* **77** 3598 (1972).

Fredricks, R.W. & F.L. Scarf, *J. Geophys. Res.* **78**, 310 (1973).

Fredricks, R.W., F.L. Scarf & C.T. Russell, *J. Geophys. Res.* **78**, 2133 (1973).

Fried, B.D. & R.W. Gould, *Phys. Fluids* **4**, 139 (1961).

Greenstadt, E.W. & R.W. Fredricks, Magnetospheric Physics (Ed. B.M. McCormac), D. Reidel Publishing Company, Dordrecht, Holland, 1974.

Greenstadt, E.W., C.T. Russell, F.L. Scarf, V. Formisano & M. Neugebauer, *J. Geophys. Res.* **80**, 502 (1975).

Gurnett, D.A. & R.R. Shaw, *J. Geophys. Res.* **78**, 8136 (1973).

Gurnett, D.A., *J. Geophys. Res.* **79**, 4227 (1974).

Gurnett, D.A. & L.A. Frank, *J. Geophys. Res.* **79**, 2355 (1974).

Helliwell, R.A., *Space Sci. Rev.* **15**, 781 (1974).

Kadomtsev, B.B., Plasma Turbulence, Acad. Press, London, NY, 1965.

Kennel, C.F. & H. Petschek, *J. Geophys. Res.* **71**, 1 (1966).
Kennel, C.F., F.L. Scarf, R.W. Fredricks, J.H. McGehee & F.V. Coroniti, *J. Geophys. Res.* **75**, 6136 (1970).
Kivelson, M.G. & C.T. Russell, *Radio Sci.* **8**, 1035 (1973).
Kivelson, M.G., T.A. Farley & M.P. Aubry, *J. Geophys. Res.* **78**, 3079 (1973).
Lyons, L.R., R.M. Thorne & C.F. Kennel, *J. Geophys. Res.* **77**, 3455 (1972).
Lyons, L.R. & R.M. Thorne, *J. Geophys. Res.* **78**, 2142 (1973).
Lyons, L.R., *J. Geophys. Res.* **79** 575 (1974).
McIlwain, C.E., Physics of the Hot Plasma in the Magnetosphere (Ed. B. Hulqvist), Plenum Press, London, 1975.
Montgomery, M.D., J.R. Asbridge & S.J. Bame, *J. Geophys. Res.* **75**, 1217 (1970).
Neugebauer, M., *J. Geophys. Res.* **75**, 717 (1970).
Papadopoulos, K. & T. Coffey, *J. Geophys. Res.* **79**, 1558 (1974).
Rodriguez, P. & D.A. Gurnett, *J. Geophys. Res.* **80**, 19 (1975).
Rosenberg, T.J., R.A. Helliwell & J.P. Katsufrakis, *J. Geophys. Res.* **76**, 8445 (1971).
Sagdeev, R.Z., Proc. Symp. Applied Math., **18**, 281 (1967).
Scarf, F.L., R.W. Fredricks, L.A. Frank, C.T. Russell, P.J. Coleman, Jr. & M. Neugebauer, *J. Geophys. Res.* **75**, 7316 (1970).
Scarf, F.L., R.W. Fredricks, I.M. Green & C.T. Russell, *J. Geophys. Res.* **77**, 2274 (1972).
Scarf, F.L., R.W. Fredricks, C.F. Kennel & F.V. Coroniti, *J. Geophys. Res.* **78**, 3119 (1973a).
Scarf, F.L., R.W. Fredricks, C.T. Russell, M. Kivelson, M. Neugebauer & C.R. Chappell, *J. Geophys. Res.* **78**, 2150 (1973b).
Scarf, F.L., L.A. Frank, K. Ackerson & R. Lepping, *Geophys. Res. Lett.* **1**, 189 (1974).
Scarf, F.L., R.W. Fredricks, C.T. Russell, M. Neugebauer, M. Kivelson & C.R. Chappell, *J. Geophys. Res.* **80**, (1975).
Taylor, W.L., B. Parady & L.J. Cahill, Jr., Univ. of Minnesota, Technical Report, June 1974.
Williams, D.J. & L.R. Lyons, *J. Geophys. Res.* **79**, 4195 (1974a).
Williams, D.J. & L.R. Lyons, *J. Geophys. Res.* **79**, 4791 (1974b).

DISCUSSION

K. Knott: In one of your slides you showed black patches indicating anomalous resistivity. Could you comment on the location and extent of regions with anomalous resistivity?

C.T. Russell: There is a paucity of data in the magnetosphere in the middle altitude range (2-5 R_e geocentric) above the auroral oval. In the few passes of OGO-5 through this region, we saw electrostatic noise over most of this altitude range when we encountered field-aligned currents. Theoretically, one expects anomalous resistivity in these currents above 1000 km altitude, but I cannot remember an upper limit on the region of generation.

D. Jones: It is not necessary to invoke plasma wave/electromagnetic wave coupling for the production of kilometric radiation. Kilometric radiation may be explained on the basis of e.m. wave/e.m. wave coupling. In the polar ionosphere, it can be shown that UHR noise becomes coherent. Secondly, it can be very efficiently coupled to the O-wave, since the isoionic contours are perpendicular to the magnetic field.

K.D. Cole: To achieve the conditions for anomalous resistivity you must have ω_p and ω_c comparable, and you must go to altitudes above 1000 km. How are these conditions compatible?

C. Kennel: There is a regime in the upper ionosphere where these frequencies are comparable to within a factor 3.

F. Mozer: I would like to make a comment that one should view anomalous resistivity with caution because the theoretical situation is not at all clear. For example, the question of whether parallel anomalous resistivity can exist in the steady state may not have been resolved because it seems that one cannot get enough waves to produce resonance with a substantial fraction of the distribution function in the one-dimensional case associated with a strong magnetic field.

Secondly a comment with regard to the altitude region over which one might expect anomalous resistivity. Although the existing instability calculations suggest that the resistive region should be above about 1000 km, one can show from general principles such as conservation of momentum that anomalous resistivity enhancements of many orders of magnitude cannot be ruled out at any altitude, even below 500 km.

C. Kennel: I think the plasma physicists have isolated a number of individual effects which could give anomalous resistivity. The trouble is that there has never been a calculation either for the laboratory or the magnetosphere which takes into account enough things, like heat conductivities, to make a definite comparison between theory and experiment possible. Effectively, the anomalous resistance, in my opinion, is sensitive to boundary conditions.

MODE COUPLING OF CERENKOV RADIATION AS A SOURCE OF NOISE ABOVE THE PLASMA FREQUENCY

Dyfrig Jones

Space Science Department, ESTEC, Noordwijk, Netherlands

Abstract

Spacecraft have observed a relatively weak nonthermal continuum radiation which apparently has its source in the outer radiation zone. Upper hybrid resonance noise is also observed in the magnetosphere and it has been suggested that it is produced as Cerenkov radiation. A mechanism is proposed whereby the nonthermal continuum derives its energy from the UHR noise in a process involving electromagnetic mode coupling. On this basis, the low-frequency cut-offs of the continuum are identified with refractive index poles.

1. Introduction

Observations of radio noise by rockets and satellites in the ionosphere and magnetosphere show that enhanced radiation can exist in the frequency range extending from near the local electron plasma frequency f_p to the upper hybrid resonance (UHR) frequency $f_{UHR} = (f_p{}^2 + f_g{}^2)^{1/2}$ where f_g is the electron gyrofrequency. Reports have been given by Walsh *et al.* [1964] using rocket data, Bauer & Stone [1968] using ATS 3 data, Harvey [1968] using Ariel 2 data, Gregory [1969] using Ariel 3 data, Barrington & Hartz [1969] using Alouette 1 and 2 data, Hartz [1969, 1970] and Barrington *et al.* [1971] using Alouette 2 data, Muldrew [1970] using ISIS 1 data, Shaw & Gurnett [1972] and Mosier *et al.* [1973], using IMP 6 data.

Recent observations by spacecraft outside the plasmapause show what appears to be a relatively weak nonthermal continuum radiation with a low-frequency cut-off near the plasma frequency [Brown, 1973; Frankel, 1973; Gurnett & Shaw, 1973; Gurnett, 1974; Gurnett & Frank, 1975]. The radiation has no sharp high-frequency cut-off but a decreasing intensity with increasing frequency up to above 100 kHz,

Knott & Battrick (eds.), The Scientific Satellite Programme During the International Magnetospheric Study, 281–291.
All Rights Reserved. Copyright © 1976 by D. Reidel Publishing Company, Dordrecht-Holland.

where it disappears below the cosmic noise level. The continuum seems to consist of two components: one that is permanently trapped between the higher plasma densities of the plasmasphere and magnetosheath, and an escaping component that propagates freely away from the Earth at frequencies above the magnetosheath plasma frequency.

The purpose of this paper is to illustrate that the two radiations, UHR noise and the nonthermal continuum, which hitherto have been treated as separate entities with separate mechanisms postulated for their existence, may in fact be produced by a common source.

2. Observations

Illustrated in Figure 1 is a period of data observed by the Goddard Space Flight Center radio-astronomy experiment on IMP 6 [Mosier *et al.*, 1973]. The relative noise

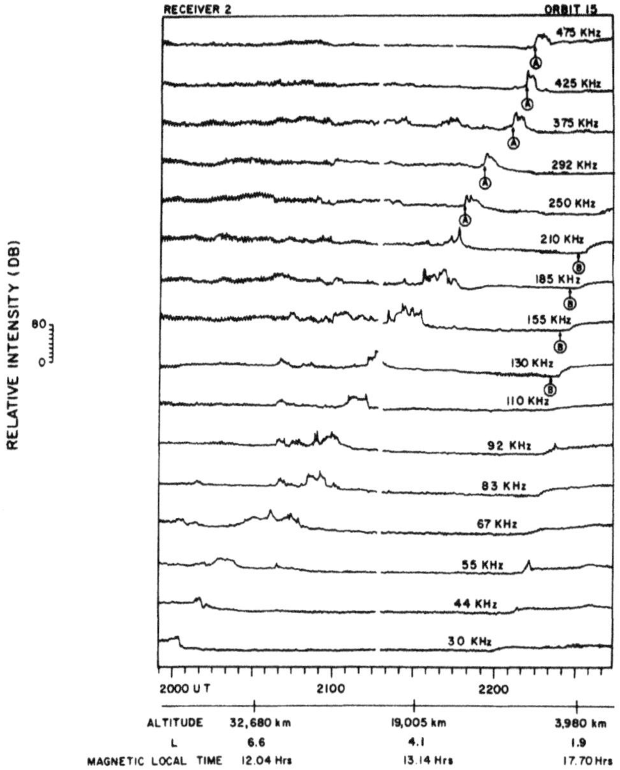

Figure 1. *Output of IMP-6 GSFC radio astronomy receiver (14 May 1971) showing bands of noise observed at different frequencies. Points A designate onset of the noise band when the spacecraft is within the plasmasphere. At points B, the wave frequency is equal to the local electron gyrofrequency. [Mosier et al., 1973].*

intensity is displayed as a function of time for a number of observing frequencies. Also indicated in Figure 1 are the spacecraft position parameters of altitude, geomagnetic shell parameter L, and magnetic local time. A band of noise is observed in the 425 kHz receiver channel, beginning at 22 h 10 m 20 s UT and ending at 22 h 14 m 10 s UT. Similar noise bands precede the 425 kHz band at lower frequencies and succeed it at higher frequencies. The noise band is less well defined outside the plasmasphere. The abrupt increase in intensity at the leading edge of the noise band (designated by point A in Fig. 1), of the order 25-35dB, is identified as being at the local UHR frequency in the plasma. To investigate the detailed characteristics of the noise bands, it is necessary to consider the zeros and infinities in the refractive-index equation and these are shown diagrammatically in the CMA diagram in Figure 2.

In Figure 2, the parameters X and Y are defined as in magneto-ionic theory, $X = f_p{}^2/f^2$ and $Y = f_g/f$ [Ratcliffe, 1962], whereas the parameters R, L, P and S are as defined in plasma physics [Stix, 1962]:

$$R \equiv 1 - [f_p^2 \;/\; f(f + f_g)]$$

$$L \equiv 1 - [f_p^2 \;/\; f(f - f_g)]$$

$$P \equiv 1 - [f_p^2 / f^2]$$

$$S \equiv \tfrac{1}{2}(R + L)$$

The line R=0 in Figure 2 represents the first extraordinary wave (X-wave) cut-off (X=1-Y) and propagation of this wave is possible to the left of the line. The

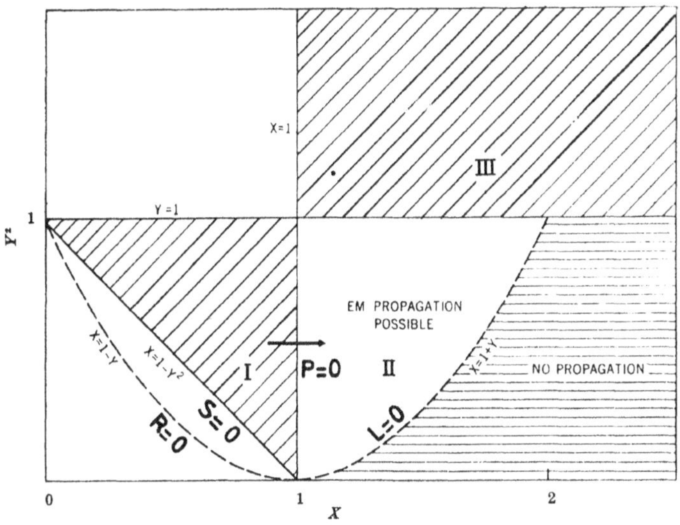

Figure 2. Schematic CMA diagram showing regions where electromagnetic waves can propagate. [*Bauer & Stone, 1968*].

line S=0 indicates the upper hybrid resonance (X=1-Y^2) and propagation of the X-wave is possible to the right of this line; there is a stop-band for the X-wave between R=0 and S=0. The vertical line P=0 is the ordinary wave (O-wave) cut-off (X=1) where the wave frequency equals the focal plasma frequency; the O-wave can propagate in regions to the left of this line. Finally, the line L=0 is the second extraordinary wave cut-off (X=1+Y), beyond which no propagation is possible for wave frequencies above the gyrofrequency, i.e. for $Y < 1$.

Assuming that the sharp increase designated A in Figure 1 is at the local UHR frequency and computing the gyrofrequency from the harmonic expansion for the geomagnetic field, it is possible to identify in the data the various cut-offs defined above. This identification is shown in Figure 3 for the higher frequency channels, where the points labelled R, S, P and L indicate the zero value of those parameters. Little effect is seen at the first X-wave cut-off R=0, but a small increase in intensity is visible even before the upper-hybrid resonance S=0 is reached. Following the sharp increase of about 30 dB at S=0, the noise intensity then decreases by about 20 dB, reaching a null at the plasma frequency P=0. The noise intensity shows considerable variation between P=0 and the second X-wave cut-off L=0, with spacecraft spin modulation superimposed.

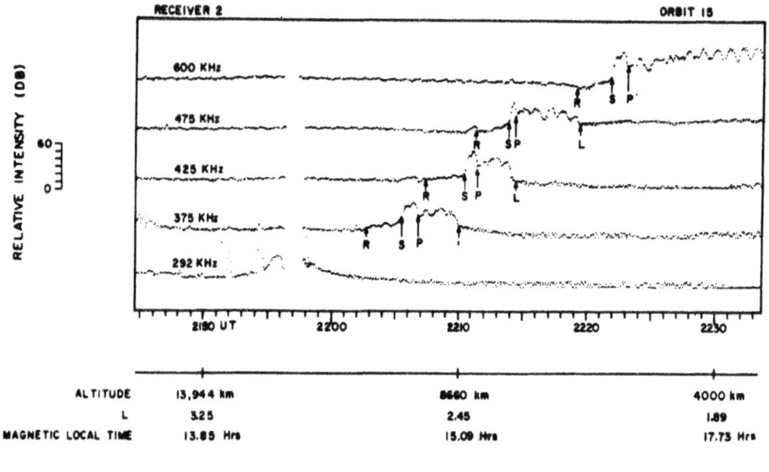

Figure 3. Expanded time scale of Figure 1 showing zero values of plasma parameters R, S, P and L. [Mosier et al., 1973].

It is reasonable to assume that the noise is produced between the upper-hybrid resonance and the plasma frequency, since it is most intense in this region. Hence the noise will be in the X-mode (right-hand polarised) and on this basis two anomalies in Figure 3 become apparent. The first is the low-intensity noise between R=0 and P=0, which as indicated earlier, is a stop-band for the X-wave. The second anomaly is the abrupt decrease at P=0, which is not a characteristic frequency for the X-wave, but is a cut-off only for the O-wave. These features will be considered later.

Cerenkov emission has been suggested as the mechanism responsible for UHR noise. For Cerenkov emission to occur, the particles involved must have velocities greater than the wave phase-velocity in the medium. Hence large refractive indices are required for the production of large powers in the UHR noise frequency range by low-energy electrons. Large refractive indices do exist in this frequency range, as is illustrated in Figure 4, which is a plot of (refractive index $n)^2$ versus X when Y is less than unity, and assuming no collisions [Budden, 1966]. It is clear that Cerenkov radiation produced in the region where $n^2 > 1$, i.e. where $1\text{-}Y^2<X<1$, is free to

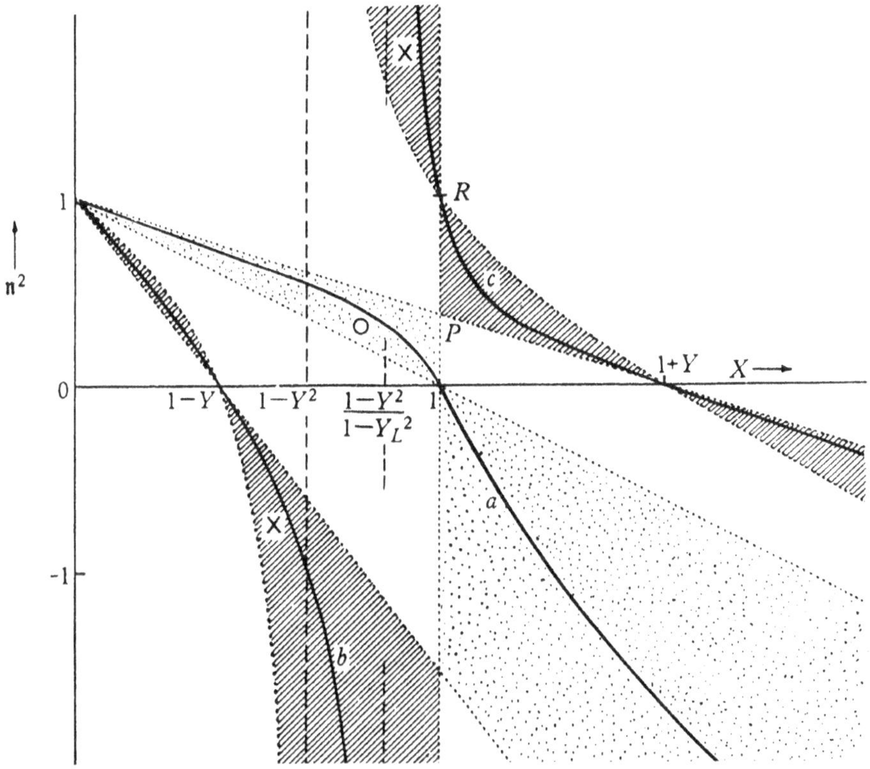

Figure 4. Variation of the square of the refractive index (n^2) with X for intermediate inclination of the Earth's magnetic field, when Y<1. Electron collisions and temperature effects are neglected. [Budden, 1966].

propagate past the level $X=1$ into the region $1<X<1+Y$; the cut-off at $X=1+Y$ is the condition for reflection. The full line labelled c is for wave-normal angle θ with respect to the magnetic field intermediate to zero and $\pi/2$; the closer θ becomes to zero, the closer this line becomes to point P, and this point is important as will be shown later. Calculations show that the Cerenkov radiation from typical magnetospheric electron spectra is more than adequate to provide the power observed in UHR noise [Taylor, 1973].

Returning to the nonthermal continuum observed by spacecraft outside the plasmapause, direction-finding measurements indicate that the radiation is generated in a region extending from 0400 h to 1400 h local time over a broad latitude range [Kurth *et al.*, 1975]. A separate source appears to be situated in the morning magnetosheath. Figure 5 is a spectrogram obtained on IMP 6. It is seen that the radiation has two distinct lower cut-offs which were identified previously as being the R=0 cut-off at ~ 6.35 kHz and the P=0 cut-off at ~ 5.3 kHz. Bands of noise are apparent at higher frequencies. The spin modulation nulls in the spectrogram between 5.3 kHz and 6.35 kHz apparently indicates that the electric field of the radiation is polarised predominantly parallel to the geomagnetic field.

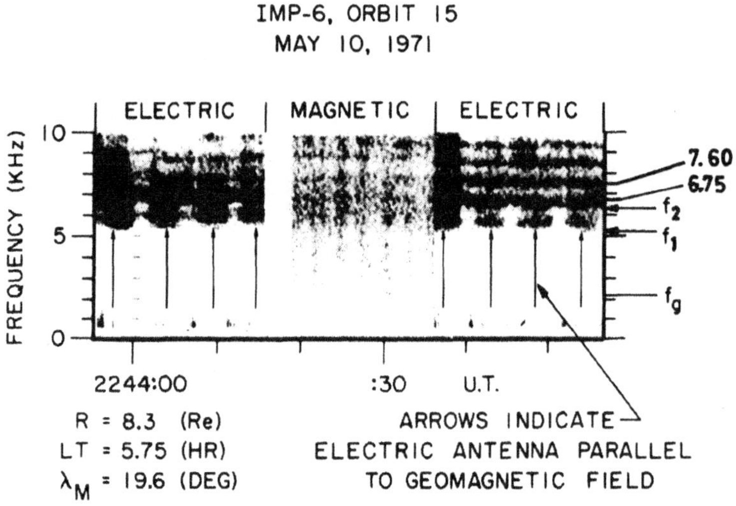

Figure 5. Spectrogram of trapped $f>f_p$ noise observed by IMP-6 University of Iowa plasma-wave experiment. Occurrence of harmonic bands in both electric and magnetic field spectrum is seen. [Gurnett & Shaw, 1973].

Mechanisms that have been proposed as a source of nonthermal continuum radiation are gyrosynchroton radiation by 1 MeV particles [Frankel, 1973] and electrostatic instabilities at $(n+½)f_g$, where n is an integer [Gurnett & Shaw, 1973]. It is suggested that the latter couple in some way to electromagnetic waves and the range of f_g encountered in the magnetosphere would then provide the continuum. A Cerenkov process has not been invoked, presumably because it can be operative only on the X-wave which, as has been shown, cannot propagate very far before encountering complete absorption at the upper-hybrid resonance.

The purpose of this paper is to propose that the observed nonthermal radiation is produced originally as UHR noise by the Cerenkov mechanism, i.e. as X-wave, and that it is subsequently coupled to the O-wave at the level where its wave frequency is close to the plasma frequency, i.e. at X=1, or P=0.

3. Theory

The theory of coupled wave equations [Budden, 1966] yields a coupling parameter which is given by the expression

$$\psi = \Psi \epsilon$$

where

$$\epsilon = X' + iZ'$$

and

$$\Psi = \frac{-\frac{1}{4}\, i\, Y_T^2\, Y_L}{(1-X-iZ)^2\, Y_L^2 + \frac{1}{4}\, Y_T^4}$$

where $Z = \nu/\omega$, ν being the angular collision frequency and ω the angular wave

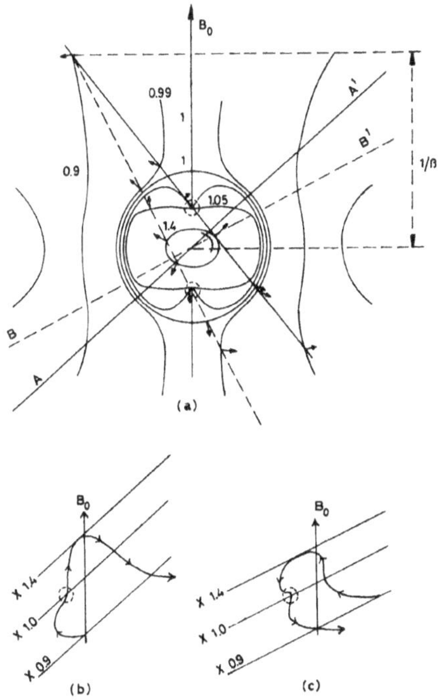

Figure 6. (a) Refractive index surfaces of Z-mode radiation for fixed Y but varying X. Two possible orientations of isoionic surfaces AA′ and BB′ are shown.
(b) Ray path corresponding to isoionic surface AA′.
(c) Ray path corresponding to isoionic surface BB′.

frequency; the primes on X and Z illustrate that the spatial gradients are taken. The parameters Y_L and Y_T are the respective components of Y parallel and perpendicular to the wave normal. It is seen that ψ can become very large if simultaneously $X \simeq 1$ and $Z \approx \frac{1}{2}|Y_T^2/Y_L|$ which are the conditions for 'critical coupling'. In the outer magnetosphere, Z computed on the basis of Coulomb collisions is small and hence critical coupling is achieved if $Y_L >> Y_T$, i.e. nearly longitudinal propagation, at $X \simeq 1$; this is satisfied when curve c in Figure 4 becomes close to point P, resulting in the production of the O-wave at $X \simeq 1$. The O-wave immediately encounters $X=1$ and is reflected, to propagate away freely to lower density regions unaffected by the UHR, which is relevant only to the X-wave.

In order to understand the coupling process in more detail, it is necessary to consider the refractive-index surfaces of the X-wave which are shown in Figure 6a (Budden, 1966). The numbers on the curves are values of X, and the diagram is for $Y = 0.5$. A particle moving with velocity $c\beta$ along the magnetic field B_0 will produce Cerenkov radiation at wave normal angle θ when passing through the region where $X = 0.9$ for example. In theory, the refractive index goes to infinity for small θ and particles of all energies can produce Cerenkov radiation. In practice, the refractive index is limited by collisional and/or temperature effects, and this provides a lower limit to the exciting particle energies. If the iso-ionic contours are parallel to AA′, Poeverlein's construction indicates that the ray path of the radiation is of the form shown in Figure 6b. When the waves pass through the region corresponding to $X=1$, the wave normal becomes aligned with B_0 and these are the conditions for critical coupling. A similar process occurs if the iso-ionic contours are aligned parallel to BB′, except that in this case the coupling occurs at $X=1$ after the X-wave has been reflected at the level $X \approx 1 + Y$; the corresponding ray path is illustrated in Figure 6c. In both cases the O-wave produced at $X \approx 1$ will propagate freely to regions of lower X.

4. Interpretation of Results

On the basis of the mechanism proposed, one may explain the anomalies that appear in Figure 3 as follows. By the Cerenkov process, UHR noise is produced in the region between S and P. At P, mode coupling occurs resulting in some transfer of energy to the O-wave which cannot propagate beyond P; hence the decrease in intensity on crossing point P. The spin modulation observed between P and L is probably due to the wave polarisation in this region being highly elliptical, for reasons shown later. The small increase in the noise intensity before S is reached, i.e. between R and S, is understood as being due to the O-wave propagating away from point P; the point R should have no effect, and in fact little effect is observed. Secondly, on the basis of this mechanism, the frequency cut-offs in Figure 5 can be identified. If the previous identification is adopted, the calculated gyrofrequency is found to be 1.9 kHz; the value calculated from the harmonic expansion is 2.2 kHz. If the lower cut-off is not f_p but is $f_{L=0}$, and the upper cut-off is not $f_{R=0}$ but is f_p as would be expected from the mechanism proposed here, then the calculated value of the gyrofrequency is 2.3 kHz, which is closer to the harmonic expansion value. Also, if $f_{L=0}$ = 5.3 kHz and f_p = 6.35 kHz, the UHR frequency is 6.75 kHz and

$f_{R=0}$ is 7.6 kHz. These frequencies are marked in Figure 7, which is a copy of Figure 5, and they can be seen to correspond to abrupt changes in noise intensity; the spin modulation below f_p can be explained as follows.

In the frequency range between f_p and $f_{L=0}$, the electric field of the radiation should be predominantly perpendicular to the static magnetic field with the longitudinal component increasing rather rapidly near the plasma frequency. With a single dipole antenna it is difficult to determine unambiguously the orientation of the electric field vector when the wave is linearly polarised and a more detailed investigation of the spin-modulated noise in this frequency range using three mutually orthogonal antennae is required.

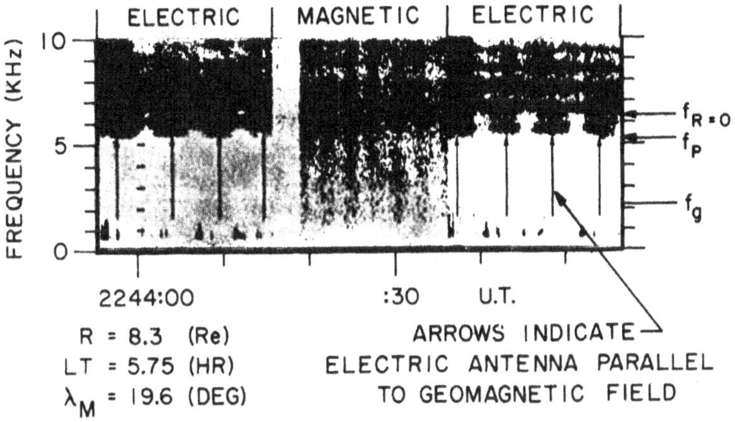

Figure 7. Copy of spectrogram shown in Figure 5 showing new identification of cut-offs.

It is tempting to identify the increase in intensity above 7.6 kHz as being due to the co-existence of the O-wave and the first branch of the X-wave. There is no known mechanism for producing the X-mode locally, but it is possible that it could have been formed by depolarisation of the O-wave by reflections at the magnetosheath or at magnetospheric irregularities. The resulting X-waves could then propagate inwards to be reflected at the level corresponding to R=0. However, the noise bands at higher frequencies would remain unexplained.

An explanation of the noise bands, which are apparently electromagnetic, on the basis of harmonics of the local gyrofrequency has been ruled out, since the difference in frequency between bands is of the order of 1 kHz, whereas the gyrofrequency is more than 2 kHz. It also appears that the bands are not accurately harmonically related. An explanation is possible if one assumes ducts of ionisation below the spacecraft altitude. As seen from the coupling equations, plasma density gradients render coupling more efficient and therefore one would expect bands of radiation at frequencies corresponding to the duct plasma densities. With a plasma-density profile obtained by Gurnett & Shaw [1973] and a corresponding spectrogram, one can estimate the distance between ducts to be $\sim 0.3 R_E$.

The observed monotonic decrease in the intensity of the nonthermal continuum radiation with increasing frequency is probably a consequence of the coupling process. As already mentioned, the efficiency of coupling is also dependent on the plasma density gradient and a given gradient will appear greater to a long wavelength wave than a shorter wavelength wave; hence the lower frequencies will be more efficiently coupled.

5. Conclusions

It is concluded that mode coupling of Cerenkov radiation may be the source of terrestrial nonthermal radiation. The mechanism requires that the plasma be traversed by energetic particles which is the case in the outer magnetosphere, and it is probable that the angle between the magnetic field and the iso-ionic contours is an important factor in the coupling process described; the dependence on this angle may explain the local time asymmetry in the source location since it is well known that the plasmasphere has a distinct outward bulge at local evening. In addition, the dawn-dusk magnetospheric electric field increases the electron energies in local morning by ~10 keV so that the threshold energy set by thermal effects is more likely to be exceeded. It has been recently shown by Gurnett & Frank [1975] that the intensity of the nonthermal continuum radiation is related to the flux of ~1 - 30 keV electrons in the outer radiation zone.

Conditions suitable for this mechanism also exist in the morning magnetosheath, which is a favoured region for streaming electrons from the solar wind.

References

Barrington, R.E. & Hartz, T.R., Plasma Waves in Space and Laboratory (Eds. J.O. Thomas and B.J. Landmark), Edinburgh University Press, 1969.

Barrington, R.E., Hartz, T.R. & Harvey, R.W., *J. Geophys. Res.* **76**, 5278 (1971).

Bauer, S.J. & Stone, R.G., *Nature* **218**, 1145 (1968).

Brown, L.E., *Astrophys. J.* **180**, 359 (1973).

Budden, K.G., Radio Waves in the Ionosphere, Cambridge University Press, 1966.

Frankel, M.S., *Radio Science* **8**, (1973).

Gregory, P.C., *Nature* **221**, 350 (1969).

Gurnett, D.A., Univ. of Iowa Report 74-39, December 1974.

Gurnett, D.A. & Frank, L.A., Univ. of Iowa Report 75-19, April 1975.

Gurnett, D.A. & Shaw, R.R., *J. Geophys. Res.* **78**, 8136 (1973).

Hartz, T.R., *Proc. IEE* **57**, 1042 (1969).

Hartz, T.R., The Polar Ionosphere and Magnetospheric Processes (Ed. G. Skovli), Gordon and Breach, New York, 1970.

Harvey, C.C., *Nature* **217**, 50 (1968).

Kurth, W.S., Baumbeck, M.M. & Gurnett, D.A., Univ. of Iowa Report 74-35, October 1974.

Mosier, S.R., Kaiser, M.L. & Brown, L.W., *J. Geophys. Res.* **78**, 1673 (1973).

Muldrew, D.B., *Space Research* **10**, 786 (1970).

Ratcliffe, J.A., The Magneto-ionic Theory and Its Application to the Ionosphere, Cambridge University Press, 1962.
Shaw, R.R. & Gurnett, D.A., Univ. of Iowa Report 72-37, December 1972.
Stix, T.H., The Theory of Plasma Waves, McGraw-Hill, New York, 1962.
Taylor, W.W.L., Univ. of Iowa Report 73-16, May 1973.
Walsh, D., Haddock, F.T. & Schulte, H.F., *Space Research* **4**, 935 (1964).

DISCUSSION

C. Kennel: How do the energetics of the Cerenkov mechanism work out?

D. Jones: By placing reasonable limits on the refractive index, it is found that a typical magnetospheric electron spectrum is more than sufficient to produce the intensities observed in UHR noise. The next consideration is to calculate the efficiency of the coupling mechanism proposed.

PROPAGATION CHARACTERISTICS OF ELECTROMAGNETIC WAVES IN THE MAGNETOSPHERE

D. Jones & R.J.L. Grard

Space Science Department, ESTEC, Noordwijk, Netherlands

Abstract

The propagation mechanism of electromagnetic waves in the Earth's magnetosphere and magnetotail is investigated using a ray-tracing technique. The present study makes use of macroscopic models for the magnetic-field topology and electron-density distribution. The existence of the bow shock, magnetopause and plasmapause is taken into account; the distortion of the Earth's magnetic field by the solar wind is also included in the models. Results are presented in the form of ray paths. This study deals with the penetration of Type-III solar bursts into the magnetosphere, the trapping of terrestrial nonthermal radiation, and the propagation of auroral kilometric radiation.

1. Introduction

The motivation for the present work is derived from wave experiments to be performed on the ESA geostationary spacecraft GEOS. The experiments will be capable of recording three orthogonal magnetic-field and three orthogonal electric-field components over a wide frequency range, and such measurements will enable one to determine the wave-normal and Poynting vector of electromagnetic waves. Even if there are a number of waves in a given frequency band, information on the angular distribution of their wave-normals may be obtained provided the six electromagnetic components are known [Storey & Lefeuvre, 1973]. Knowing these parameters and adopting models for the plasmasphere, magnetosheath and solar wind, methods are available for tracing the waves back to their origin.

The magnetospheric magnetic field is represented by the image dipole model of Hones [1963], and the electron distribution is obtained from analytic expressions that reproduce the shape of the plasmasphere including the evening bulge. The

Knott & Battrick (eds.), The Scientific Satellite Programme During the International Magnetospheric Study, 293–302.
All Rights Reserved. Copyright © 1976 by D. Reidel Publishing Company, Dordrecht-Holland.

magnetosheath magnetic field and plasma density are analytic fits to the hydromagnetic model of Dryer & Heckman [1967]. The solar-wind magnetic field is assumed to be parallel to the ecliptic plane and directed towards the Earth with a strength of 7 γ; the solar-wind electron density is taken to be uniform and equal to 6 cm^{-3}.

The ray-tracing differential equations of Haselgrove and Haselgrove are integrated by computer. The program is written in three dimensions, but the present results refer only to propagation in the noon-midnight meridian plane.

Type-III solar bursts as observed at 1 AU extend down in frequency to approximately 20 kHz, which is the local plasma frequency. Because the plasma density in the magnetosheath is higher than that in the solar wind, frequencies below 40 kHz will not penetrate into the magnetosphere. It is shown how the wave experiment on the geostationary spacecraft GEOS, situated within the magnetosphere at 6.6 R_E, will be able to monitor the sheath density from the low-frequency cut-off Type-III solar bursts.

The ray-tracing program is also applied to radiation trapped between the higher densities of the plasmasphere and magnetosheath.

Finally, it is shown that kilometric radiation generated in the night-time auroral zone at an altitude of approximately 2500 km can be beamed in different directions; the width and positions of the beams depend on the initial wave-normal directions and on the relation between the wave frequency and the plasma frequency in the generation region. Direction-finding measurements made by spacecraft on the dayside must be treated with caution, since refraction effects can be important.

The Cartesian co-ordinate system used throughout this paper has its origin at the centre of the Earth, its X-axis pointing directly towards the Sun, the Y-axis pointing towards dusk, and the Z-axis towards the North Pole. The direction of a vector is also defined by the angles θ and ϕ, where θ is the angle between the vector and the Z-axis, and ϕ that between the X-axis and the projection of the vector in the XY-plane.

2. The Magnetic-Field and Electron-Density Models

2.1. GENERALITIES

The magnetospheric models used in this investigation must fulfil two requirements: firstly, they must give acceptable representations of the magnetic-field topology and electron-density profiles; secondly, their mathematical expressions must be sufficiently simple to avoid introducing a prohibitively long computation time in the ray-tracing program.

The models consist of a number of analytical expressions which have been designed to fit existing models derived by other workers. The different models have had to be adapted, however, in order to adjust to common interfaces (magnetopause and bow shock).

The electron-density model is shown in Figure 1. The strength and orientation of the magnetic field are displayed in Figures 2 and 3. A tangential discontinuity is reproduced on the magnetopause, but no field-line reconnection is considered between the magnetosheath and magnetosphere.

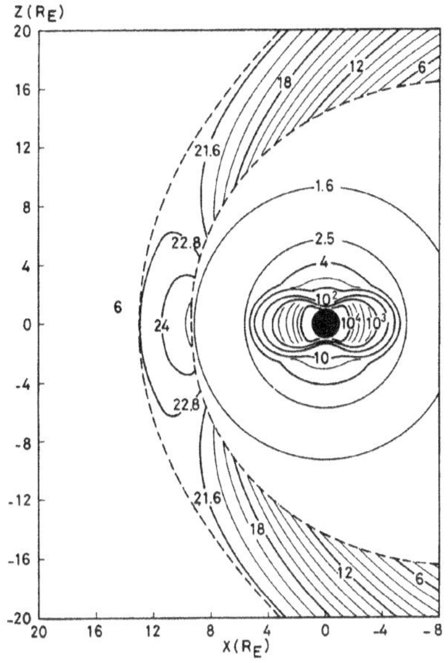

Figure 1. *Electron-density contours (cm^{-3}).*

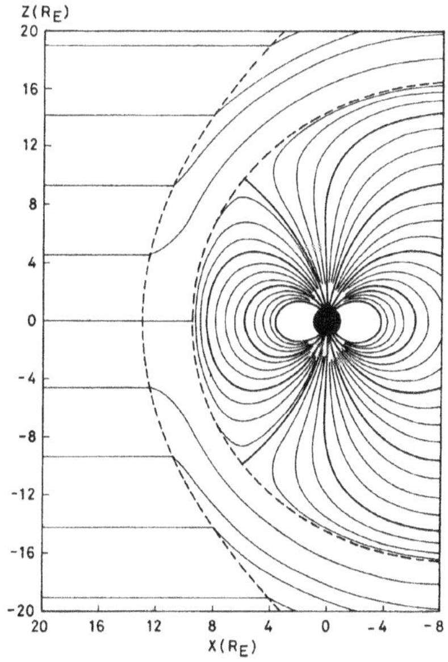

Figure 2. *Magnetic-field-line model.*

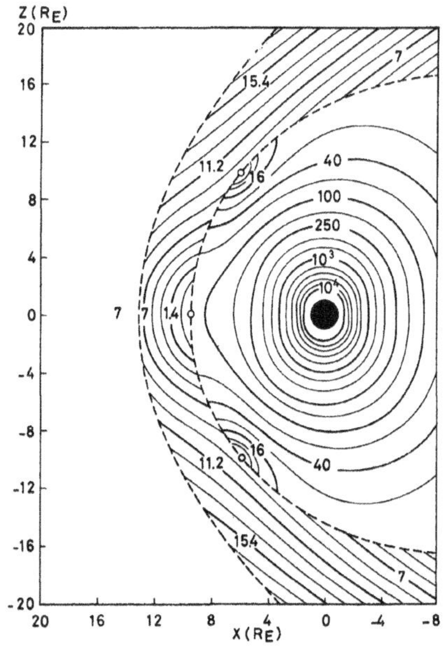

Figure 3. Magnetic-field strength (γ).

2.2. MAGNETOSPHERE

The topology of the magnetic field in the magnetosphere is represented by the model of Hones [1963]. This model consists of two dipoles perpendicular to the ecliptic plane, and directed southward. One is located at the centre of the Earth, and the other a distance R along the X-axis in the Sun direction; the magnetic moment of the latter is S times that of the former. The values of these parameters, namely $R = 40$ and $S = 38$, are those used by Willis & Pratt [1972].

This model gives an adequate representation of the field lines on the sunward hemisphere and down into the tail to a distance of the order of 8 R_E. This model is very satisfactory in the magnetosphere, which is the domain presently under investigation, but it could be improved in the magnetotail, if necessary, by taking into account the effect of the neutral sheet current [Willis & Pratt, 1972].

The intensity of the field is $3.1 \times 10^4\ \gamma$ on the Earth's surface at the equator, and $60\,\gamma$ at the magnetopause subsolar point ($1\ \gamma = 10^{-9}$ Wb/m^2). The magnetopause has an ellipsoidal shape; its distance from the Earth is 9.5 R_E in the sunward direction, and 16.8 R_E along the polar axis. The co-ordinates of the neutral points are $X = 6R_E$ and $Z = \pm\ 9.8\ R_E$; the field lines passing through the neutral points intersect the Earth at latitudes of ± 82°. This analytical model agrees with the experimental one given by Fairfield [1971] to within 10%.

The electron-density distribution in the magnetosphere is represented by an analytical expression that reproduces the shape of the plasmasphere and gives typical values of electron densities down to an altitude of approximately 1000 km. It is seen, for example, that the plasmapause at the equator is located at a distance of $4\ R_E$, where the electron density drops from $1000\ cm^{-3}$ to $10\ cm^{-3}$ over a distance of the order of $1\ R_E$. Outside the plasmasphere, but within the magnetosphere, the density decreases monotonically and is of the order of $1-2\ cm^{-3}$ at the magnetopause.

2.3. SOLAR WIND

The magnetic field is supposed to be aligned with the X-axis and directed towards the Earth; its magnitude is $6\ \gamma$. The electron density is assumed to be uniform and equal to $6\ cm^{-3}$.

2.4. MAGNETOSHEATH

The outer boundary of the magnetosheath – the bow shock – is taken to be a parabola intersecting the Sun-Earth axis at a distance $X = 13\ R_E$ and the polar axis at $Z = \pm\ 23\ R_E$. Considering the insufficient amount of magnetosheath data presently available to construct an experimental model, the magnetic field and electron density are fitted to the hydromagnetic model developed by Spreiter *et al.* [1966] and Dryer & Heckman [1967].

3. Ray Tracing

3.1. TYPE-III SOLAR BURSTS

Type-III solar bursts are investigated within the context of their possible use by the ESA geostationary satellite as a means of studying the electron density in the magnetosheath. The wave experiment on GEOS is intended to study electric fields in the frequency range 0–80 kHz; magnetic fields up to above 30 kHz will also be measured. Type-III solar bursts cover a very wide frequency spectrum; frequencies greater than the maximum plasma frequency in the ionosphere can be observed on the ground. Measurements at lower frequencies can be performed in space, and spacecraft in the solar wind have observed the low-frequency cut-off of Type-III solar bursts to be near the local plasma frequency [Fainberg & Stone, 1974]. At 1 AU, this cut-off can be at less than 20 kHz, so that waves with frequencies from 20 kHz to several hundred MHz are impinging upon the magnetosheath. Near noon, GEOS will be approximately $3\ R_E$ from the plasmapause and $6\ R_E$ from the bow shock. Plasma densities in the magnetosheath near the subsolar point are believed to be 3–4 times that of the incident solar wind [Spreiter *et al.*, 1966], as shown in Figure 1. Plasma densities in the plasma-trough region (outside the plasmapause but inside the magnetosphere) sometimes contain regions of abnormally high-density cold plasma [Chappell, 1973]. Typically, however, the plasma frequency experienced in the magnetosphere by a wave propagating from the sheath to GEOS near noon is expected to be at least a factor 3 below the maximum sheath plasma frequency at the subsolar point. Hence, if Type-III solar bursts are observed on GEOS, their low-frequency cut-off should give an indication of the magnetosheath plasma densities.

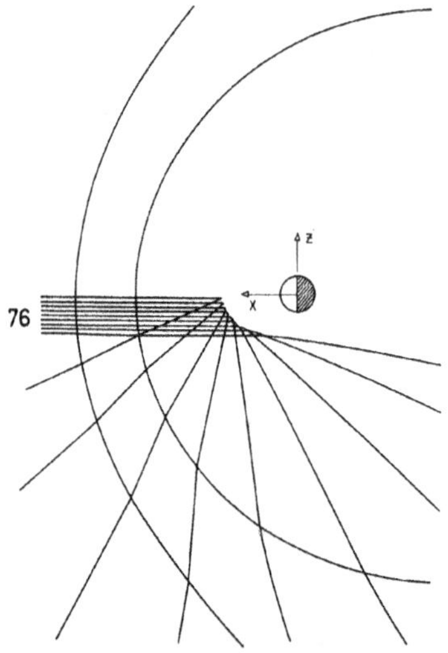

Figure 4. *Ray paths of Type-III 76 kHz waves.*

To study the effect of the magnetosheath on waves propagating in from the solar wind, ray-tracing has been performed for frequencies above the sheath plasma frequency. Figure 4 shows that 76 kHz ordinary waves are virtually unaffected by the sheath; they will be reflected at the plasmapause before propagating back through the sheath into the solar wind. The result when the frequency is decreased to 50 kHz and 45 kHz is seen in Figure 5; the 50 kHz waves will penetrate the sheath, but 45 kHz waves will be reflected. Hence GEOS will observe a cut-off between 45 and 50 kHz from which the sheath plasma frequency* will be computed. Whether the waves are in the ordinary or extraordinary mode will hardly affect the result, since the gyrofrequency in the magnetosheath is so low (100 Hz) that both modes will have their cut-offs virtually at the plasma frequency.

3.2. TERRESTRIAL NONTHERMAL RADIATION

This radiation is observed by spacecraft beyond about 6 R_E and is believed to be generated in the outer radiation zone immediately beyond the plasmapause boundary [Shaw & Gurnett, 1972]. The source region extends from 0400 h to 1400 h local time and covers a broad latitude range which includes the geomagnetic equator. When the spacecraft is within the magnetosphere, the radiation has a sharp lower frequency cut-off near the local plasma frequency, which could be lower than 500 Hz. No distinct high-frequency cut-off is observable and the intensity decreases as frequency increases; above 100 kHz the radiation disappears below the cosmic-

* It is interesting to note that Vesecky & Frankel [1975] have used a similar method for determining the magnetosheath electron density. With IMP-6 in the solar wind, radio noise from within the magnetosphere was used as a source and a cut-off was observed which was attributed to the magnetosheath. It was found that the average magnetosheath/solar-wind plasma-density ratio was 1.7.

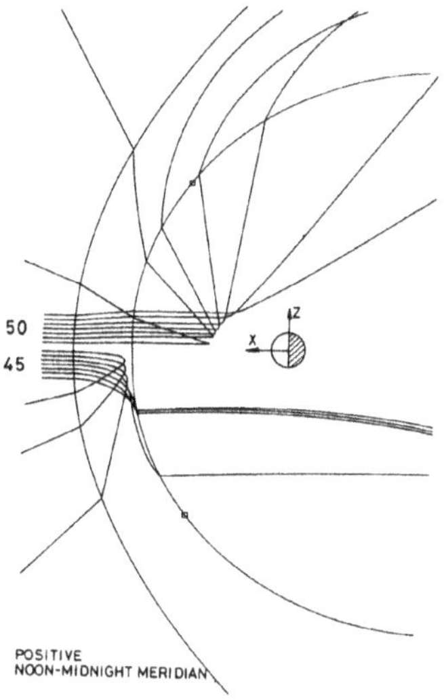

Figure 5. *Ray paths of Type-III 50 and 45 kHz waves.*

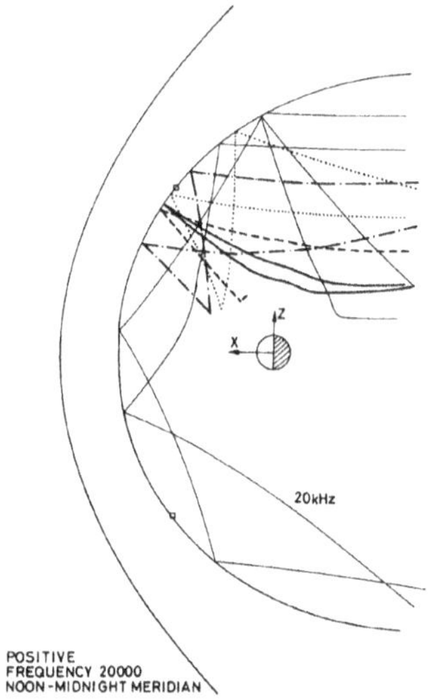

Figure 6. *Radiation trapped between plasmapause and magnetosheath.*

noise level. For frequencies below the magnetosheath plasma frequency, the radiation is trapped between the sheath and the plasmapause on the dayside, as is illustrated by the ray paths (f = 20 kHz) in Figure 6. Eventually, the radiation escapes into the magnetotail. Waves with frequencies greater than the sheath plasma frequency will escape into the solar wind.

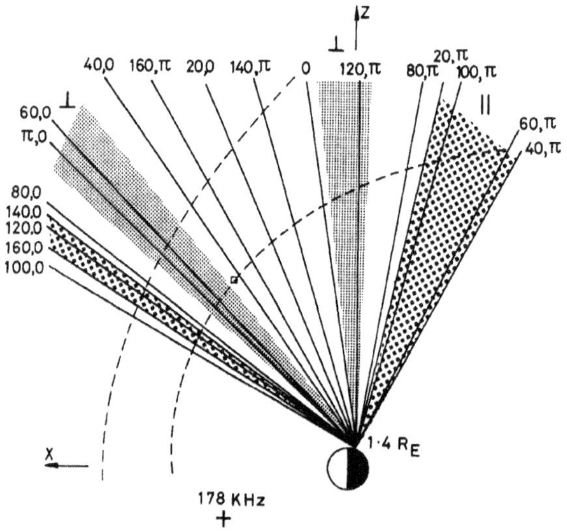

Figure 7. *Ray paths of auroral kilometric radiation (178 kHz) assuming the source region is at 1.4 R_E where $f > f_p$.*

3.3. AURORAL KILOMETRIC RADIATION

Considerable interest has been focussed recently on intense 'kilometric' radiation observed by spacecraft [Gurnett, 1974] to be propagating outwards from Earth in the frequency range 50–500 kHz. It appears to originate from radial distances below 3 R_E in the auroral zone. Many similarities exist between this radiation and Jovian decametric radiation, which has a frequency range of 3–40 MHz, and it has been widely suggested that their source mechanisms could be the same. Various theories have been proposed for their production, and it is possible that they are produced, or at least that they are emitted, from regions where the wave frequency is close to the plasma frequency. Gurnett [1974] observed the most intense radiation in the 178 kHz channel on the spacecraft IMP-6, and Figures 7 and 8 show ray-tracing results for this frequency. In Figure 7 the source region is assumed to be at 1.4 R_E, which is about 0.025 R_E above the altitude where the plasma frequency f_p is 178 kHz. Rays are labelled with θ and ϕ, which give the initial direction of the wave normal in polar co-ordinates: the ray paths are shown for a complete range of initial wave-normal directions, and with the models assumed, the radiation is confined to a cone whose opening angle is about 90°. In Figure 8, the source region has been moved to 1.375 R_E where $f \gtrsim f_p$ and it is seen that the cone opening angle decreases to about 70°. The shaded areas in Figures 7 and 8 indicate the radiation cones if the initial wave-normal angles are restricted to be within 10° of the magnetic-field direction (labelled $\parallel$) and within 10° of the normal to the magnetic-field direction ($\perp$). Clearly, as the source region is moved closer to where $f = f_p$, the $\parallel$ cones tend to open whereas the $\perp$ cones become much narrower. Since spacecraft observe that the

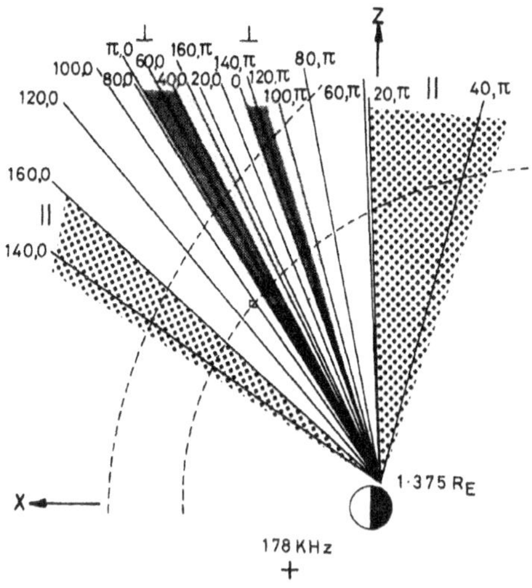

Figure 8. *Ray paths of auroral kilometric radiation (178 kHz) assuming the source region is at 1.375 R_E, where $f \approx f_p$.*

radiation covers a fairly large region, it is probable that if one had to choose between mechanisms with initial wave-normals parallel to the magnetic field B_0 and those with initial perpendicular wave vectors, experimental evidence would favour the former.

The direction in which the radiation is beamed is critically dependent on the orientation of the iso-ionic planes in the source region. It will be necessary to modify the electron-density model at high latitudes since from Figures 7 and 8 one would expect the beams to be observed towards the dayside, whereas in fact they are most usually observed on the nightside of the Earth [Gurnett, 1974].

4. Conclusions

By using a macroscopic model of the Earth's magnetosphere, ray paths of three types of radiation have been investigated. It has been shown how a spacecraft in the magnetospheric trough region can be used to estimate magnetosheath plasma densities from the low-frequency cut-off of Type-III solar bursts. Secondly, ray paths of so-called 'non-thermal continuum' were computed which illustrated the trapping of the lower frequencies between the plasmapause and magnetosheath regions. Finally, the beaming of auroral kilometric radiation was studied and it appeared that initial wave-normal angles should preferably be parallel, rather than perpendicular, to the magnetic field in the source region, assuming the radiation is produced near the plasma frequency.

References

Chappell, C.R., Proc. Chapman Memorial Symposium on Magnetospheric Motions, NOAA, Boulder, Colorado, 1973.
Dryer, M. & G.R. Heckman, *Planet. Space Sci.* **15**, 515 (1967).
Fainberg, J. & R.G. Stone, *Space Sci. Rev.* **16**, 145 (1974).
Fairfield, D.H., *J. Geophys. Res.* **76**, 6700 (1971).
Gurnett, D.A., *J. Geophys. Res.* **79**, 4227 (1974).
Hones, E., Jr., *J. Geophys. Res.* **68**, 1209 (1963).
Shaw, R.R. & D.A. Gurnett, Univ. of Iowa Report 72–37, 1972.
Spreiter, J.H., A.L. Summers & A.Y. Alksne, *Planet. Space Sci.* **14**, 223 (1966).
Storey, Ll.R.O. & F. Lefeuvre, Paper presented at Sixteenth COSPAR Meeting, Konstanz, Germany, 1973.
Vesecky, J.F. & M.S. Frankel, *J. Geophys. Res.* **80**, 2771 (1975).
Willis, D.M. & R.J. Pratt, *J. Atmosph. Terr. Phys.* **34**, 1955 (1972).

DISCUSSION

V. Formisano: In your model of profiles of electron densities, N_e in the magnetosphere close to the magnetopause is very small ($N_e \approx 1$ cm^{-3}); however, if you try to check the pressure balance of the magnetopause you find that often you need many particles in the magnetospheric side, close to the discontinuity. How will the presence of these particles change the technique you are proposing?

D. Jones: Kennel believes that an $N_e \approx 1$ cm^{-3} just within the magnetosphere is reasonable at the subsolar point. It is very important to know this plasma density particularly since we hope to monitor with GEOS the magnetosheath density from the lower cut-off of Type-III solar bursts.

VLF ELECTROSTATIC WAVES IN THE MAGNETOSPHERES OF THE EARTH AND JUPITER

M. Ashour-Abdalla
CNET, Issy-les-Moulineaux, France

C.F. Kennel*
Ecole Polytechnique, Paris, France

Abstract

We discuss the electron loss cone instability as a function of hot and cold electron densities and the magnetic field strength. A simple marginal stability algorithm permits a rapid estimate of the unstable frequency band. While the growth rate always peaks near the cold upper hybrid frequency, the group velocity parallel to the magnetic field can have zeros near odd half harmonics of the electron cyclotron frequencies, suggesting that the instability may be nonconvective there.

1. Introduction

We were asked to discuss the effects of variations of the cold and hot plasma densities upon VLF instabilities in the magnetospheres of Earth and Jupiter. The effects of cold plasma upon the electromagnetic whistler and ion cyclotron instabilities in the Earth's magnetosphere have been thoroughly discussed [Cornwall *et al.*, 1970; Brice & Lucas, 1971]. Therefore we shall concentrate on VLF electrostatic instabilities [Fredricks, 1971; Young *et al.*, 1973; Karpman *et al.*, 1974a, b; Ashour-Abdalla *et al.*, 1975] which are very probably responsible for electrostatic waves with frequencies between the first few harmonics of the electron gyro frequency in the Earth's magnetosphere, the so-called '3/2' and '5/2' waves [Kennel *et al.*, 1970; Scarf & Fredricks, 1973; Gurnett & Shaw, 1973]. What little we know about the cold and hot electron densities in Jupiter's magnetospheres indicates that such instabilities may also be present there near the magnetic equator beyond the orbit of Io. What we have to say about electrostatic instabilities may be equally valid for the Earth's magnetosphere beyond the plasmapause and Jupiter's beyond the orbit of Io.

---

* On leave from University of California, Los Angeles, USA.

Knott & Battrick (eds.), The Scientific Satellite Programme During the International Magnetospheric Study, 303–325.
All Rights Reserved. Copyright © 1976 by D. Reidel Publishing Company, Dordrecht-Holland.

This is the second progress report on our programme of numerical computations of the dispersion relation for electrostatic waves near, but above the electron cyclotron frequency. The first will be published in the Proceedings of the Nobel Symposium and is presently available in preprint form. There we concentrated upon establishing the basic properties of the instability and we reported our first results on how the waves propagate. Here, using the Nobel Symposium results as a basis, we present a more extensive analysis of propagation and of the dependences upon the hot and cold electron densities. Our discussion of our earlier results will necessarily be brief. The overall aim of our numerical investigations is not to extend the basic plasma physical knowledge of electron cyclotron harmonic instabilities, which has been understood since Harris [1959], but to provide more detailed knowledge of the parameter dependences of the instability to aid the experimentalist in interpreting his data.

2. A Marginal Stability Algorithm

The basic plasma physics of electrostatic cyclotron instabilities was set out by Dory *et al.* [1965], who solved the dispersion for electrostatic waves with propagation vectors at an arbitrary angle to a uniform magnetic field using a 'loss-cone' electron distribution function, one with a 'hole' at small velocities $v_\perp$ perpendicular to the magnetic field. Such an electron distribution is subject to a variety of instabilities. Spurred on by the OGO-5 observations, Fredricks [1971] solved the same general dispersion relation for the simplest possible such distribution, a ring distribution of the form $F=(2\pi)^{-1}\ \delta(v_\perp - U)\ \delta(v_\parallel)$. He found that instability above the electron frequency was possible, and that in general an intermediate range of $K_\parallel/K_\perp$, neither large nor small, would be the most unstable. Young *et al.* [1973] criticised the ring distribution as too severe to be realistic for the magnetosphere. In fact, they found that more gentle distributions would be *stable* without the admixture of a cold electron component with a density n_c of the order of, or less than, the hot density n_H. They thus established the basic model used in all subsequent investigations – a mixture of a hot component with a loss-cone distribution and a cold component. They investigated briefly what happens when the cold component is given a small nonzero temperature T_c, and argued that finite T_c could restrict the range of unstable frequencies, thereby accounting for the fact that the emissions near $3/2\Omega$ are typically observed to be narrow-band.

Young *et al.* [1973] and Ashour-Abdalla & Cowley [1974] argued that a pure pitch-angle anisotropy A, which can destabilise whistlers, cannot be responsible for electrostatic instabilities above the electron cyclotron frequency, for two reasons. First, for an instability with frequency ω in the range $n\Omega < \omega < (n+1)\,\Omega$, a large anisotropy is required, $A > n$.

The band most commonly observed lies between $n = 1$ and $n = 2$ and would therefore require $A = (T_\perp - T_\parallel)/T_\parallel$ somewhat larger than one, which though large is not completely impossible for the Earth's magnetosphere. We shall emphasise through this paper that occasionally two bands, between $n = 1$ and $n = 2$, and also between $n = 2$ and $n = 3$ are observed. It seems to us much more unlikely that a pure pitch-angle anisotropy $A \geqslant 2$ can exist for a long time in the magnetosphere. Besides, the waves are as often observed above the odd cyclotron half-harmonics as below, and pitch-angle anisotropy can only produce waves in the frequency range $(n + 1/2) < \omega/\Omega < (n + 1)$, where ω is the wave frequency and Ω the magnitude of the electron cyclotron frequency.

A group in the Soviet Union, under the leadership of V. Karpman, performed the first extensive numerical calculations of wave dispersion relations and growth rates, first without a cold component and then adopting the Young *et al.* [1973] model [Karpman *et al.*, 1974a, b]. From their work emerged a realisation of the importance of the cold upper hybrid mode. They varied the strength of the positive slope of the distribution, the cold density and temperature, and the magnetic field. They found that many normal modes could exist for $\omega > \Omega$, especially, it seems, when the positive slope is strong and $T_c \neq 0$. By adding complexity to the plasma distribution function, one adds new possible normal modes of oscillation.

Ashour-Abdalla *et al.* [1975] returned to a simpler model of the electron distribution, with $T_c = 0$ and with a gentle slope of the hot electron distribution. They invented a graphical algorithm which gives the range of unstable frequencies and wave numbers as a function of the measurable parameters n_c/n_H and the magnetic field strength. Even though their $T_c = 0$ assumption is a considerable idealisation, it is the only one we have found which permits a rapid evaluation of how odd-half harmonic electrostatic instabilities behave when n_c and n_H are varied. Therefore, we exploit their key insight, extending considerably the range of n_c and n_H they studied.

We start with the general dispersion relation of Harris [1959]:

$$D(\omega^*, K_\perp, K_\parallel) \equiv 1 + \sum_j \frac{\omega_{pj}^2}{K^2} \sum_n \int \frac{G_{nj}(v_\parallel)\, dv_\parallel}{\omega^* - K_\parallel v_\parallel - n\Omega_j} = 0 \tag{1}$$

where $\sum_j$ denotes a sum over species, or in our case hot and cold components, ω_{pj} is the plasma frequency of that component, ω is the complex frequency, and G_{nj} is given by

$$G_{nj}(v_\parallel) = 2\pi \int_0^\infty v_\perp dv_\perp J_n^2 \left(\frac{K_\perp v_\perp}{\Omega_j}\right) \left[\frac{n\Omega_j}{v_\perp} \frac{\partial F_j}{\partial v_\perp} + K_\parallel \frac{\partial F_j}{\partial v_\parallel}\right] \tag{2}$$

and $F_j(v_\perp, v_\parallel)$ is the distribution function.

To provide the positive slope in the hot distribution, Ashour-Abdalla *et al.* [1975] chose the following subtracted Maxwellian:

$$f_H = \frac{n_H}{\pi^{3/2}\, \alpha_\perp^2\, \alpha_\parallel\, (1-\beta)}\, e^{-v_\parallel^2/a_\parallel^2} \left[e^{-v_\perp^2/a_\perp^2} - e^{-v_\perp^2/\beta a_\perp^2}\right] \tag{3}$$

with $0 < \beta < 1$. f_H is normalised to n_H. $\alpha_\perp$ and $\alpha_\parallel$ are thermal speeds perpendicular and parallel to the magnetic field. When $v_\perp$ is of order $\sqrt{\beta}a$, $\partial f_H/\partial v_\perp > 0$, $\partial f_H/\partial v_\perp < 0$ when $v_\perp \gtrsim \alpha$.

When $\beta \to 1$ and $(1-\beta)\, v_\perp^2/a_\perp^2 << 1$,

$$f_H \to \frac{1}{\pi^{3/2}\, a_\perp^2\, a_\parallel} \left(\frac{v_\perp}{a_\perp}\right)^2 e^{-v_\parallel^2/a_\parallel^2}\, e^{-v_\perp^2/a_\perp^2} \tag{4}$$

When $\beta \to 1$, the subtracted Maxwellian reduces to the weakest Dory-Guest-Harris loss-cone distribution ($j = 1$). Henceforth, we will take the parameter β (which is *not* the ratio of plasma to magnetic pressure) to be 0.9. It is this, the weakest of all unstable distributions considered by fusion theorists, which Young *et al.* [1973] found required a cold electron component in order that waves between $n = 1$ and $n = 2$ would be unstable.

One result of our analysis seems to be that when wave propagation is taken into account, even the $j = 1$ weak-loss-cone distribution is too strong. It produces extremely short convective growth lengths. However, we will continue to study the properties of this distribution for several reasons. The first is practical: realistic growth rates, which are probably $\gamma \approx 10^{-5}\ \Omega$, require high numerical accuracy which is very expensive and would limit a parametric survey. Secondly, to aid theorists to construct future more numerically accurate theories, it is useful to complete the study of the parametric dependences using the empty-loss-cone distribution, since that has become the 'standard theoretical case'. Furthermore, we have experimented with even weaker filled-in loss cone distributions and have found that the *tendencies* exhibited by the thus far standard case are also found using the weaker distributions. Experimentalists should realise, however, that what is required is only a positive $\partial f/\partial v_\perp$ somewhere in the distribution.

Now we turn to specifying our marginal stability algorithm. When $|\gamma/\omega| << 1$, Equation (1) can be expanded as follows:

$$D(\omega^*, K_\perp, K_\parallel) \simeq D(\omega, K_\perp, K_\parallel) + i\,\gamma \frac{\partial}{\partial\omega} D(\omega, K_\perp, K_\parallel) \tag{5}$$

Separating into real and imaginary parts,

$$\text{Re}\ D(\omega, K_\perp, K_\parallel) = 0$$

$$\gamma = -\frac{\text{Im}\, D(\omega, K_\perp, K_\parallel)}{\partial D/\partial\omega} \tag{6}$$

If T_c is exactly zero, cold electrons make no contribution to Im D, and the marginal stability condition is independent of both n_H and n_c, an important simplification:

$$\sum_n \left(\frac{a_\perp^2}{a_\parallel^2}\, C_n(\lambda)\, Z_n + \frac{n\, D_n(\lambda)}{K_\parallel \rho_\parallel} \right) e^{-z_n^2} = 0 \tag{7}$$

where

$$\lambda = \frac{1}{2} K_\perp^2 \rho_\perp^2 \equiv \frac{1}{2} \frac{K_\perp^2 a_\perp^2}{\Omega^2}, \quad z_n = \frac{x - n}{K_\| \rho_\|}, \quad X = \frac{\omega}{\Omega}$$

$\rho_\| = \alpha_\| / \Omega$, and C_n and D_n are functions coming from the integrations over $v_\perp$ in Equation(2)

$$C_n(\lambda) = \frac{1}{1-\beta} \left\{ e^{-\lambda} I_n(\lambda) - \beta e^{-\beta\lambda} I_n(\beta\lambda) \right\} \tag{8a}$$

$$D_n(\lambda) = \frac{1}{1-\beta} \left\{ e^{-\lambda} I_n(\lambda) - e^{-\beta\lambda} I_n(\beta\lambda) \right\} \tag{8b}$$

where $I_n(\lambda)$ is the modified Bessel function of order n. Even if $T_c \neq 0$, Equation (8) is approximately satisfied for ω sufficiently different from $n\Omega$ that cold electron cyclotron damping is negligible.

Assuming $T_c = 0$ and substituting the subtracted Maxwellian into the real part of the dispersion relation,

$$\text{Re } D = 1 - \frac{\omega_{pc}^2}{\Omega^2} \left(\frac{K_\perp^2 / K^2}{X^2 - 1} + \frac{K_\|^2 / K^2}{X^2} \right) - \frac{\omega_{pH}^2}{\Omega^2} \text{ Re } H_H \tag{9}$$

where

$$\text{Re } H_H = \frac{K_\perp^2}{2K^2 \lambda} \sum_n \left(\frac{a_\perp^2}{a_\|^2} C_n \text{ Re } Z'(z_n) - \frac{2n}{K_\| \rho_\|} D_n \text{ Re } Z(z_n) \right) \tag{10}$$

and $Z(z_n)$ is the plasma dispersion function. Since, for the unstable modes of interest, $K_\| / K_\perp < 0.3$, we can neglect the term involving $K_\|$ on the left-hand side of Equation (9), which can be simplified to

$$\frac{n_c}{n_H} - \frac{\Omega^2}{\omega_{pH}^2} (X^2 - 1) = -(X^2 - 1) \text{ Re } H_H \equiv \hat{K} \tag{11}$$

Ashour-Abdalla *et al.*'s [1975] graphical marginal stability algorithm consists of choosing, say $K_\|$, solving Equation (7) for $X(\lambda)$, inserting $X(\lambda)$ into real H_H and computing $\hat{K}$ as a function of $X^2 - 1$ with $K_\|$ as a parameter. Then, on the plot of $\hat{K}(X^2 - 1)$, we choose the ordinate at $(X^2 - 1) = 0$ to be n_c/n_H and draw a straight line with slope $-\Omega^2/\omega^2{}_{pH}$. Where this straight line intersects $\hat{K}(X^2 - 1; K_\|)$, we have determined a marginally stable solution.

It is more convenient to note that setting $\hat{K} = 0$ in Equation (11) leads to the following solution:

$$X_{UH}^2 - 1 \equiv X^2 - 1 = \frac{\omega^2 p_c}{\Omega^2} = \frac{4\pi n_c m_e{}^2}{B^2} \equiv \frac{n_c}{n_*} \qquad (12)$$

In other words, when the effects of the hot electrons are negligible ($\hat{K} = 0$), the mode that remains is a *cold upper hybrid* mode. Thus the unstable frequency range can be determined by choosing n_c/n_H on the vertical axis and n_c/n^* on the horizontal axis of a plot of $\hat{K}(K_{||})$ versus X^2 - 1 and drawing a straight line between the two. The smallest possible unstable frequency, X_{min} , can be deduced from the first intersection of the straight line with one of the $\hat{K}(K_{||})$ and the largest will be very near X_{UH}.

Figure 1 shows a plot of $\hat{K}$ versus $X^2 - 1 = \omega^2/\Omega^2 - 1$ for $a_\perp/a_{||} = 1$ (or $T_\perp/T_{||} = 1$) and $\beta = 0.9$ for $K_{||}\rho_{||} <$ in the range $0.2 < K_{||}\rho_{||} < 2.5$. While they are not shown here, the $\hat{K}$ functions for $K_{||}\rho_{||} < 0.2$ lie close to the $K_{||}\rho_{||} = 0.2$ curve everywhere. The frequencies $X = 3/2$ and $X = 5/2$ are marked on the $(\omega/\Omega)^2 - 1$ axis by small crosses. Since the typically stable $K_\perp\rho_\perp \simeq 3$, the neglect of $K_{||}^2 / K^2X^2$ in Equation (9) becomes suspect for the larger values of $K_{||}\rho_{||}$. From the point of view of growth rate, however, the interesting $K_{||}\rho_{||}$ are less than unity.

Now, let us apply our algorithm. We pick an n_c/n_H and $\dfrac{\omega_{UH}^2}{\Omega^2} - 1 = \dfrac{n_c}{n^*}$ and draw a straight line between the two. The lowest unstable frequency, X_{min}, will be given by

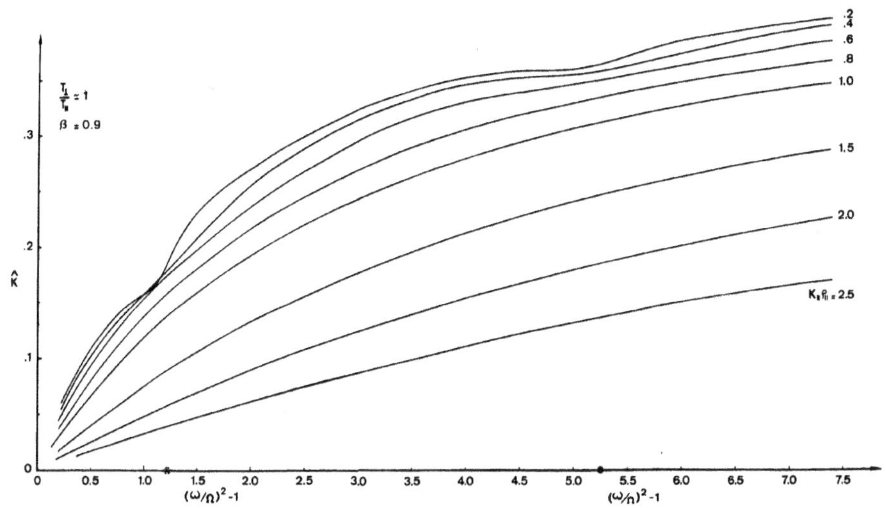

Figure 1. *Plot of $\hat{K}$ versus $\left(\frac{\omega}{\Omega}\right)^2 - 1$ for the first and second harmonic. The points where $X \simeq 3/2$ and $5/2$ are marked by small crosses. A line drawn between the points n_c/n_H and n_c/n_* determines the range of unstable frequencies, which is given by the frequency of the first intersection with the $\hat{K}$ function and ω_{UH}.*

the first intersection of that straight line with one of the $\hat{K}$ functions. Furthermore, the largest unstable frequency, X_{max}, is very near X_{UH}.

Using the graphical algorithm, we can easily deduce many of Young *et al.*'s [1973] conclusions on how n_c and n_H determine the range of unstable frequencies. For example, keeping n_H fixed and increasing n_c increases *both* X_{min} and X_{max}. In other words, the whole unstable frequency band tends to follow the cold upper hybrid frequency. Keeping n_c fixed and increasing n_H decreases X_{min}; decreasing n_H increases X_{min} leaving $X_{max} \simeq X_{UH}$ fixed. Thus, when $n_H/n_c << 1$, only frequencies near the upper hybrid frequency can be unstable. Waves near the cold upper hybrid frequency have the lowest threshold n_H; as n_H increases, the unstable bandwidth $|X_{min} - X_{UH}|$ increases. When n_c/n_H is less than about 0.1, frequencies near the first electron cyclotron harmonic can be unstable.

If we keep n_H fixed, but let $n_c \to 0$, the entire range of unstable frequencies shrinks to zero: another way of saying that gentle $j = 1$ loss-cone distributions are stable in the absence of cold electrons.

So far as we can tell, when we insert weaker filled-in loss-cone electron distributions into the marginal stability algorithm, the $\hat{K}$ function decreases, but retains its same general properties. This suggests that while a larger hot density (n_c/n_H smaller) is required for instability – as is reasonable – the parametric variations described in Figure 1 may be qualitatively valid. This point requires further confirmation.

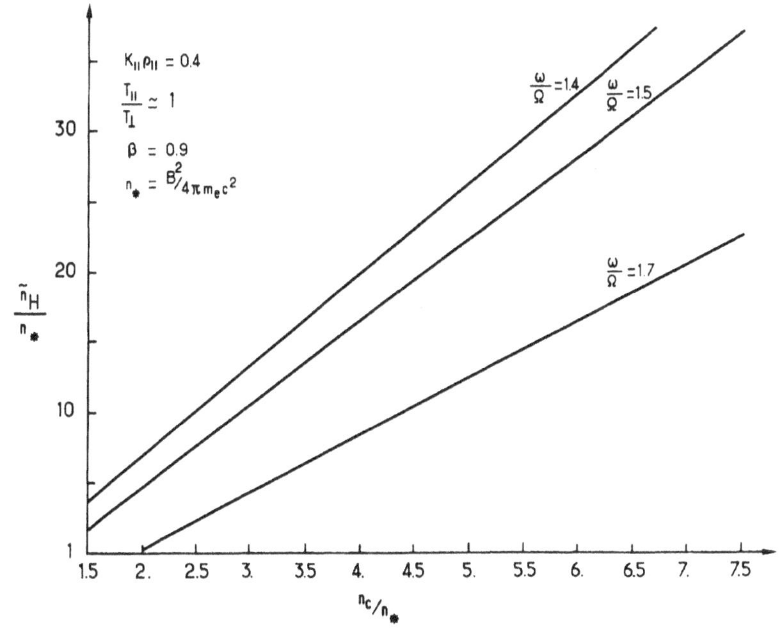

Figure 2. *Critical hot electron density $\tilde{n}_H$, normalised to n_*, to destabilise waves with $\frac{\omega}{\Omega} = 1.4$, 1.5 and 1.7, versus n_c/n_*. Increasing n_c increases $\tilde{n}_H$.*

Figure 1 indicates that for a given frequency $X = \omega/\Omega$ to be unstable, X_{UH} must exceed X. Thus there is always a critical cold electron density n_c/n_* for instability. On the other hand, there is no critical hot density apparent, since letting $n_c/n_H \to \infty (n_H \to 0)$ produces an increasingly narrow band of instability with X_{min} approaching X_{UH}. This is artifact of the assumption $T_c = 0$. Near X_{UH}, the perpendicular wave number λ_H approaches infinity. When the wavelength diminishes to the cold electron Larmor radius, the assumption $T_c = 0$ naturally is invalid. The threshold hot density is set by finite cold temperature effects. However, we can plot, in Figure 2, the threshold density $\tilde{n}/n_H$ for the case when X differs from X_{UH}, against n_c/n_*. It is obvious that increasing n_c/n_* increases the critical hot density. This means, for example, that electrostatic instabilities between $n = 1$ and $n = 2$ should not be found inside the plasmapause, where $n_c/n_* >> 1$.

From Figures 1 and 2, we can draw the conclusion that for instability between $n = 1$ and 2, we require $n_c/n_* \simeq 1$ and $n_H/n_c \simeq 1$ - 10. Are these conditions satisfied for the Earth's and Jupiter's magnetospheres? In Figure 3, we plot $n_* = B^2/4\pi m_e c^2$ versus L (dotted line) and compare it with OGO-5 cold *ion* densities [Chappell *et al.*, 1970], which give us some indication of what the cold electron density of ionospheric provenance might be. Shown are two separate days' measurements, 3/25/68 for which $n_c/n_* < 1$ and $X_{UH} < \sqrt{2}$ everywhere beyond the plasmapause and 4/23/68 for n_c/n_* ranges between 1 and 10 and X_{UH} between $\sqrt{2}$ and 3 nearly everywhere beyond the plasmapause. 4/23/68 was a day for which waves with $X > 3/2$ could have been unstable provided n_H was sufficiently large. On the

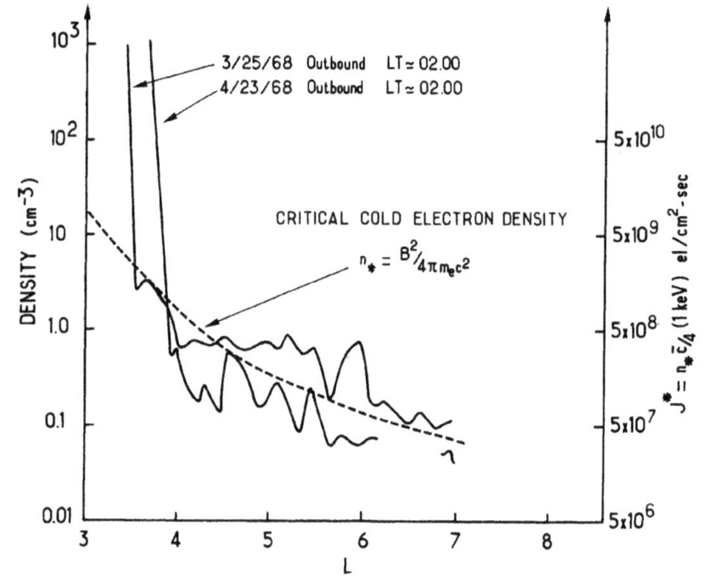

Figure 3. *Superposed on a plot of cold ion densities, measured by the OGO-5 spectrometer [Chappell et al., 1970], the critical density $n_* = B^2/4\pi m_e c^2$ at the geomagnetic equator.*

right-hand scale, we show an estimate of the critical flux $J_* = n_* \bar{c}/4$, where $\bar{c}$ is a mean thermal speed of 1 keV electrons. Since electron fluxes in the order of 10 J_* are often observed, and since $n_c \simeq n_*$ seems to be commonly observed, we conclude that the conditions for electrostatic instability are often satisfied beyond the plasmapause.

It would be very helpful to have independent measurements of the cold electron densities. In their absence, however, we are constrained to use Figure 3 to choose the range of cold upper hybrid frequencies which seems most likely: $1 \leqslant X_{UH} \leqslant 3$. In the rest of this paper, we will focus on this range.

In Figure 4, we plot n_*, and J_* for a 10 keV plasma versus L for Jupiter assuming a centred dipole magnetic field with equatorial surface strength of 4 G. In addition, we superpose an estimate of n_c of the magnetic equator made by Ioannidis & Brice [1971]. If in Figure 3, the inference of the cold electron density from the cold ion density is somewhat suspect, the Ioannidis-Brice estimate also must be regarded as a guess, albeit probably the best we can do. Nevertheless, we see that beyond the orbit of Io, $L \simeq 6$, N_c could exceed n_*. In addition, the fluxes of 10 keV electrons required for instability, $\sim 10^{10}\,\mathrm{cm}^{-2}\,\mathrm{s}^{-1}$, do not seem to be an unreasonable extrapolation from what has been already detected at higher energies. Thus, we suggest that electrostatic instabilities in between low harmonics of the electron cyclotron frequency might be found in Jupiter's magnetosphere, most probably near or beyond Io's orbit.

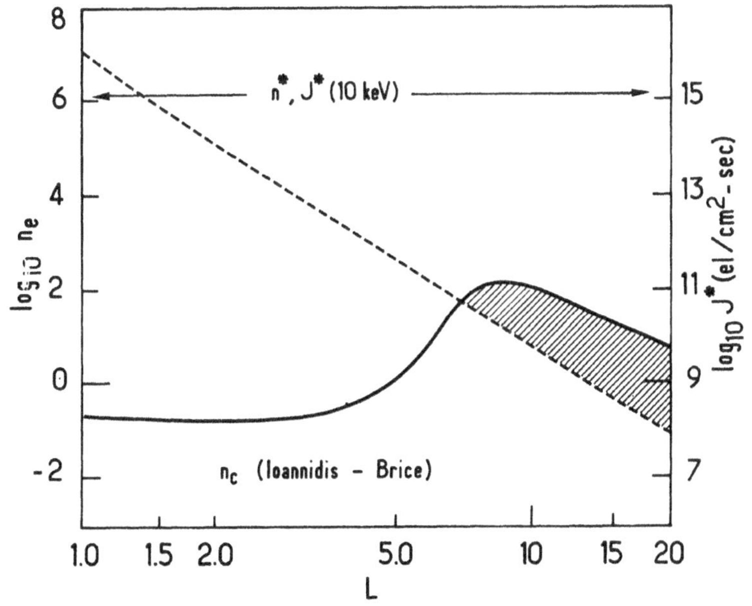

Figure 4. *Superposed on a plot of n_c for Jupiter's magnetosphere [Ioannidis - Brice, 1971], a plot of n_* based on a 4 gauss centred dipole field. 3/2 emissions might occur beyond the orbit of Io.*

3. Properties of the Cold Upper Hybrid Instability

In this section, we present several particular solutions of the dispersion relation which illustrate the dependence of the instability upon n_c and n_H.

In Figure 5a, we plot $X = \omega/\Omega$ against $K_\perp \rho_\perp$ for $\beta = 0.9$, $T_\perp/T_\parallel \simeq 1$, and one parallel wavelength $K_\parallel \rho_\parallel = 0.4$, for several different ω_{UH}/Ω ranging from 1.8 to 3, keeping n_H fixed.

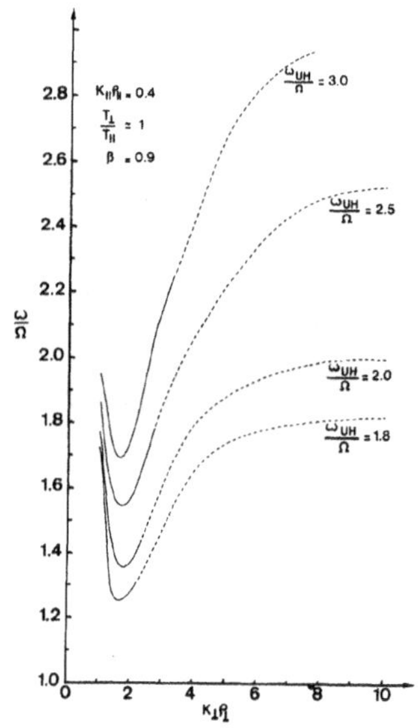

Figure 5a. *Real part of frequency versus $K_\perp \rho_\perp$ for different cold upper hybrid frequencies. Solid lines correspond to damping, while dotted lines show the unstable region.*

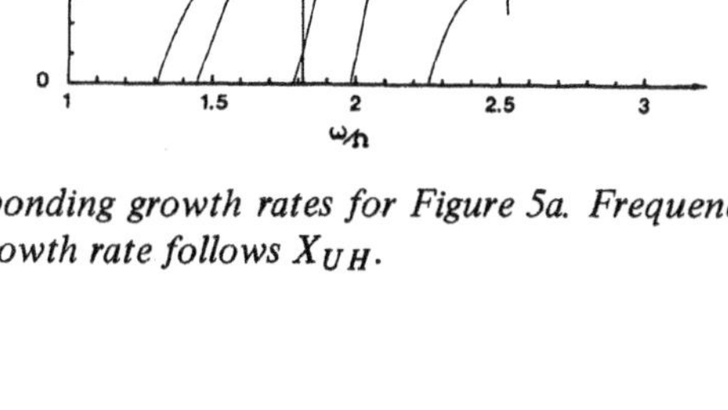

Figure 5b. *Corresponding growth rates for Figure 5a. Frequency corresponding to peak growth rate follows X_{UH}.*

In effect, we imagine traversing the plasmapause outbound, where X_{UH} decreases from above 3 inside to a value between 1 and 2 outside which we have chosen to be 1.8. Where the wave is damped, ω is plotted as a solid line; where it is unstable it is plotted as a dashed line. In Figure 5b, the growth γ/Ω is plotted against $X = \omega/\Omega$ for the same parameters. From Figure 5a, we conclude that the frequency approaches the cold upper hybrid frequency as $K_\perp \rho_\perp$ increases. Secondly, the unstable frequencies are always below the upper hybrid frequency. No other modes were found in the range of parameters shown. The maximum growth rates in Figure 5b are near but below the cold upper hybrid frequency and, because we did not vary n_H, have the same order of magnitude for all ω_{UH}. The growth rate goes to zero at ω_{UH} and the X_{min} correspond to what can be deduced from Figure 1.

Note that Figures 5a and 5b do not have any particular structure that singles out the frequency ranges $X = 3/2$ or $X = 5/2$ for special attention.

In Figure 6, we keep X_{UH} fixed and vary n_H, keeping all other parameters the same as in Figures 5a and 5b. The increase in n_H encountered in Figure 6 might be encountered during a substorm injection of hot plasma into the inner magnetosphere. As could have been predicted from Figure 1, when n_H is small there is a small unstable frequency range near the upper hybrid frequency. As n_H increases, n_c/n_H decreases, the band of unstable frequencies increases; i.e. X_{min} decreases. What could not be deduced from Figure 1, however, is the fact that the frequency of peak growth rate varies little when n_H is increased by a factor 5, remaining near the cold upper hybrid frequency.

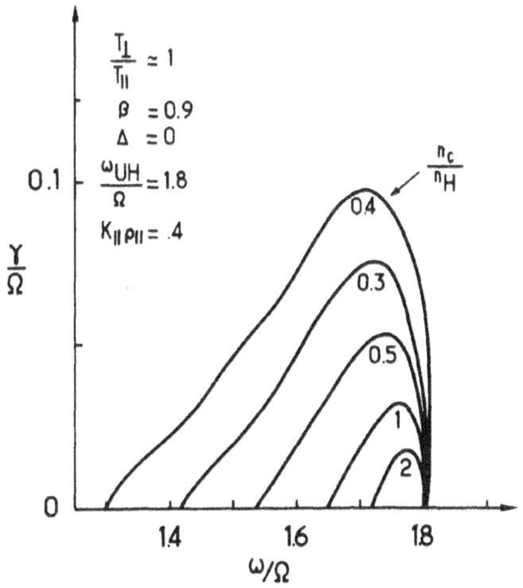

Figure 6. *Plot of growth rate versus frequency for different values of n_c/n_H. Here n_c is fixed. Note how the peak growth rate varies little when n_H is increased and remains near the cold upper hybrid frequency. We could explain narrow-band waves at $X = 3/2$ by choosing $n_c/n_H = 2$, but $X_{UH} \simeq 3/2$ a special choice. If we want to ensure that $X \simeq 3/2$ is always unstable, regardless of variations in X_{UH}, we have to choose n_c/n_H a few tenths, but then γ/Ω does not peak near $X = 3/2$.*

Figure 6 highlights the difficulties in finding narrow band emissions near $X = 3/2$ from the growth rates. By chance, X_{UH} could equal 1.5 when $n_c/n_H > 1$. On the other hand $X_{UH} \simeq 3/2$ is a special choice. If the more typical case is $X_{UH} \neq 3/2$, to guarantee that $X = 3/2$ is unstable we must choose $n_c/n_H < 1$. But then there is no peak in γ/Ω at $X = 3/2$. If we believe that X_{UH} typically differs from 3/2, we must resort to some other mechanism to explain why the observed amplitudes peak say between $1.4 < X < 1.6$.

4. Propagation of Unstable Electrostatic Waves

A proper consideration of electrostatic instabilities in the magnetosphere would necessarily feature the fact that they propagate in a spatially inhomogeneous medium. Instead of computing the temporal growth rate γ, we should compute the imaginary part of the wave vector. Since computing **Im** K is much more difficult than computing γ, people usually do not do it without good reason. However, since the computations of γ do not, in our opinion, lead to an unambiguous interpretation of the data, we should at least outline the properties of **Im** K. When $|\gamma/\omega| << 1$, as it is for all the cases studied here, a reasonable but not rigorous estimate for **Im** K is to divide the growth rate by the group velocity. Here we compute the group velocity.

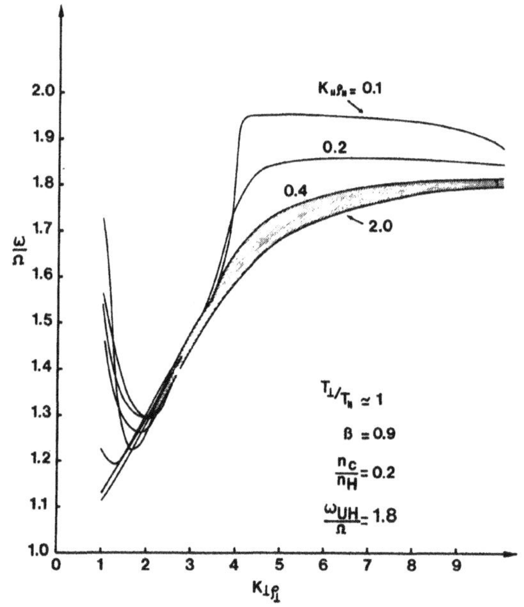

Figure 7a. *Real part of frequency versus $K_\perp\rho_\perp$ for different values of $K_\parallel\rho_\parallel$. Frequency is nearly independent of $K_\parallel\rho_\parallel$ in the shaded region.*

In Figure 7a, we plot the dispersion relation ω versus $K_\perp \rho_\perp$ for the various values of $K_\parallel \rho_\parallel$ indicated, again for the standard case $X_{UH} = 1.8$ ($\rho_\perp$ and $\rho_\parallel$ are the electron Larmor radii based upon the perpendicular and parallel electron temperatures, respectively). In Figure 7b, we plot the growth rates against ω/Ω for the parameters of Figure 7a. Though not shown here, the peak growth rate for $K_\parallel \rho_\parallel = 0.1$ is less than for $K_\parallel \rho_\parallel = 0.2$ by a factor 2, and so Figure 7b contains the wave of maximum growth rate. Taken together, Figures 7a and 7b support, for one hybrid frequency and many $K_\parallel \rho_\parallel$, the conclusions drawn from Figures 5a and 5b for one $K_\parallel \rho_\parallel$ and many ω_{UH}: the growth rate is zero as $K_\perp \to \infty$ and $\omega \to \omega_{UH}$, and the peak growth rates are near but below ω_{UH} for all $K_\parallel \rho_\parallel$. The largest growth rate of all, for $K_\parallel \rho_\parallel = 0.2$, occurs very near the cold upper hybrid frequency. No structure around $X = 3/2$ is evident.

Figure 7a allows us to estimate some properties of the group velocities by inspection. First of all, near the upper hybrid frequency, both parallel and perpendicular components of the group velocity, $\partial\omega/\partial K_\parallel$ and $\partial\omega/\partial K_\perp$, tend to zero as they must. Away from the hybrid frequency, near $X \simeq 3/2$, $\partial\omega/\partial K_\perp$ is the order of several tenths of the electron thermal speed for all the $K_\parallel \rho_\parallel$ studied. In the shaded region, between the curves for $K_\parallel \rho_\parallel = 0.4$ and 0.2, ω is weakly dependent upon $K_\parallel$ for all $K_\perp$. The parallel group velocity is plotted on an expanded scale in Figure 8. We note that the parallel group velocity is typically smaller, the order of a few tenths of the perpendicular group velocity, and more significantly, *can be zero* for a significant

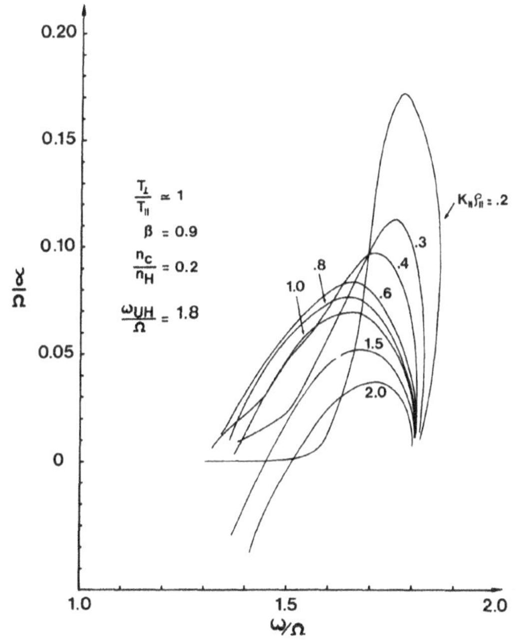

Figure 7b. *Growth rate versus frequency for different values of $K_\parallel \rho_\parallel$. Peak growth for $K_\parallel \rho_\parallel = 0.1$ (not shown) is less for $K_\parallel \rho_\parallel = 0.2$.*

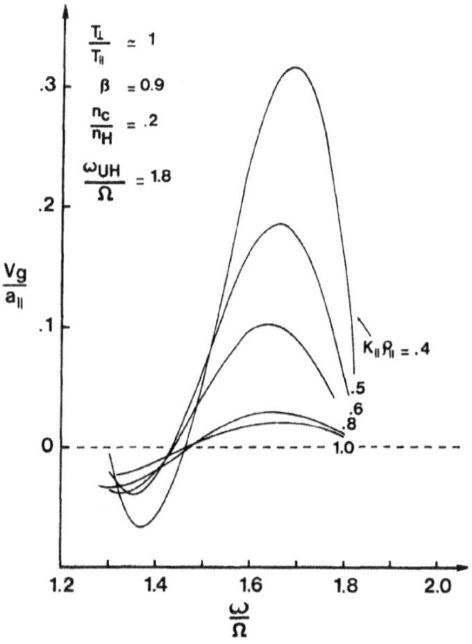

Figure 8. *Parallel group velocity normalised to parallel electron thermal speed versus frequency for different values of $K_{\|}\rho_{\|}$. Note zero crossings of $\partial\omega/\partial K_{\|}$ near but slightly below ω/Ω = 3/2.*

range of $K_{\|}\rho_{\|}$. In other words, waves with $X \simeq 1.45$ have their group velocities nearly perpendicular to the magnetic field.

The fact that the zero in $\partial\omega/\partial K_{\|}$ is due to the hot plasma and therefore depends upon n_H and more generally upon the choice of distribution function, illustrates the subtle complexity of convective amplification calculations for electrostatic waves. Both γ and $\partial\omega/\partial\mathbf{K}$ depend upon n_H. In Figure 9, we return to the parameters of Figure 6, and plot the parallel group velocity versus frequency for X_{UH} fixed but n_c/n_H varying between 2 and 0.4. Where γ is positive, $\partial\omega/\partial K_{\|}$ is plotted with a dotted line. For small hot densities, $n_c/n_H > 1$, we find a narrow frequency range of instability, and that $\gamma < 0$ at the non-upper-hybrid :ro of $\partial\omega/\partial K_{\|}$. At n_c/n_H = 0.5, $\partial\omega/\partial K_{\|}$ has no zero, but a minimum near X = 3/2. At n_c/n_H = 0.3, $\partial\omega/\partial K_{\|}$ = 0 where our calculations indicate that $\gamma > 0$. This zero crossing is at $X \simeq 1.47$. Thus the critical n_c/n_H for inducing a zero of $\partial\omega/\partial K_{\|}$ (with $\gamma > 0$) is somewhere between 0.5 and 0.3.

Of course, the estimate $\text{Im}K_{\|} \simeq \gamma/\partial\omega/\partial K_{\|}$ is invalid near $\partial\omega/\partial K_{\|}$ = 0; but we are using our plots of $\partial\omega/\partial K_{\|}$ merely to indicate the frequencies near which a more accurate calculation would show a peak in $\text{Im}K_{\|}$.

It was amusing that in Figures 8 and 9 the zeros of $\partial\omega/\partial K_{\|}$ occurred for the same range of frequencies for a considerable range of $K_{\|}$ and for a range of n_c/n_H keeping n_c fixed.

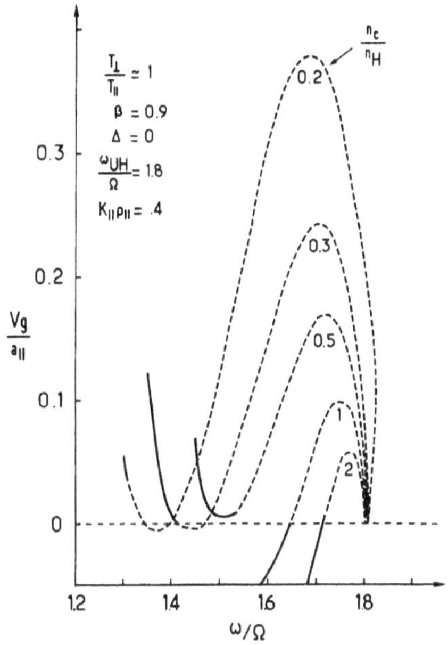

Figure 9. *Plot of parallel group velocity versus frequency for different values of n_c/n_H. Unstable zeros in group velocity occur in a narrow range of frequencies near $\omega/\Omega \simeq 1.4$, and require n_c/n_H less than a low 0.35.*

In Figure 10, we show roughly the same property. We keep n_H fixed and vary n_c, corresponding to a range of cold upper hybrid frequencies between 1.47 and 1.94, and we plot to parallel group velocity for the wave with $K_\parallel \rho_\parallel = 0.4$, solid where stable, and dotted where unstable. Figure 10 indicates that there is a given frequency $X \simeq 1.45$ near which there is a $\gamma > 0$ zero of $\partial\omega/\partial K_\parallel$ for all cold densities such that X_{UH} exceeds 1.45. Case B, where $n_c/n_H = 0.37$, corresponds to the physics of Figure 9: n_c/n_H become sufficiently large that the hot electrons can no longer dominate the cold and so induce the zero in $\partial\omega/\partial K_\parallel$.

Thus far we have merely discovered a charming numerical regularity: the parallel component of the group velocity seems to have a zero near $X \simeq 3/2$ for a significant range of n_c/n_H and $K_\parallel$. Can this have any significance for the interpretation of the observations? To answer this question, we must ask the reader to imagine with us the possible results of propagation calculations that have not yet been carried out. Probably the most sensitive parameter in the entire calculation is the ratio $X = \omega/\Omega$ If, as a wave propagates conserving its frequency ω, it finds itself in a region where the local magnetic field has changed by ± 20%, it typically stabilises.

We see from Figures 7 and 8 that it is difficult, except possibly for special parameter choices, to have the perpendicular group velocity zero in the interesting frequency range near $X = 3/2$. As a rough estimate, the perpendicular group speed is the order of a few tenths of the electron thermal speed, and is directed opposite to

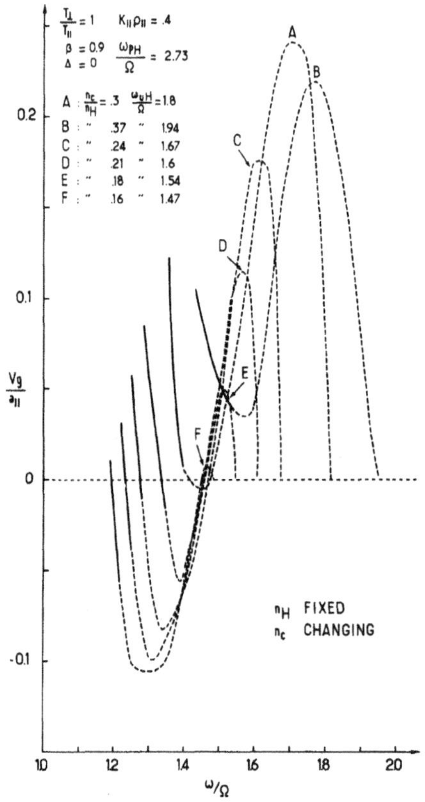

Figure 10. *Plot of parallel group velocity versus frequency for different values of n_c/n_H. Here n_H is fixed. Note narrow range of frequencies where $V_g \simeq 0$. In fact, it seems there is a critical frequency at which $V_g = 0$. If X_{UH} exceeds the critical frequency, there will be a zero in $\partial\omega/\partial K_{\|}$, which, since we have chosen n_H fairly large, will be unstable.*

the perpendicular wave vector $\mathbf{K}_\perp$. Similarly, from Figure 8, the parallel group speed, except near $X = 3/2$, is a factor 3 to 10 smaller than the perpendicular component. Thus, for an electron temperature $T_\perp \simeq 1$ keV, the perpendicular group speeds are a few thousand kilometres per second and the parallel group speeds a few hundred to a thousand kilometres per second.

Let us now ask how far waves, starting near the magnetic equator of a dipole field, have to propagate in order to change the local magnetic field by a factor $|\Delta B/B| << 1$. If one propagates radially, outward, it must traverse the distance d_r,

$$d_r = \frac{R}{3} \; |\Delta B/B| \tag{13}$$

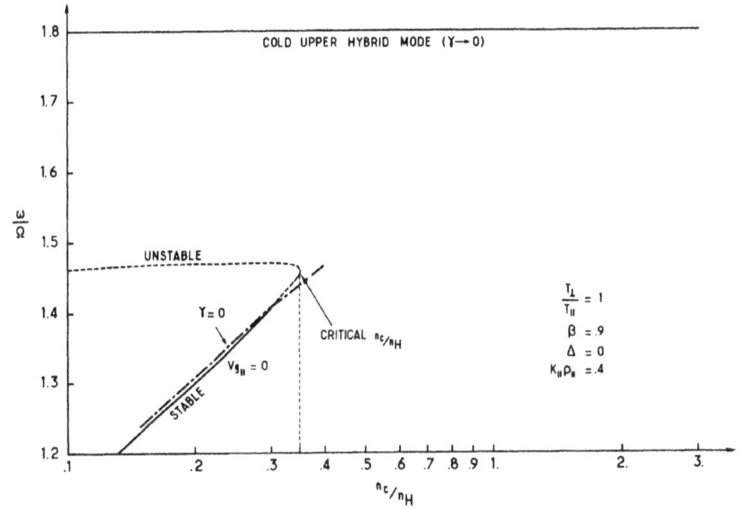

Figure 11a. *Frequencies at which $\partial\omega/\partial K_{\|} = 0$ and $\gamma = 0$ as a function of n_c/n_H for $X_{UH} = 1.8$ and $K_{\|}\rho_{\|} = 0.4$. There is a critical $n_c/n_H = 0.35$ for which $\partial\omega/\partial K_{\|} = 0$ and $\gamma = 0$ for $n_c/n_H < 0.35$, the frequency at which $\partial\omega/\partial K_{\|} \simeq 0$ is almost independent of n_c/n_H.*

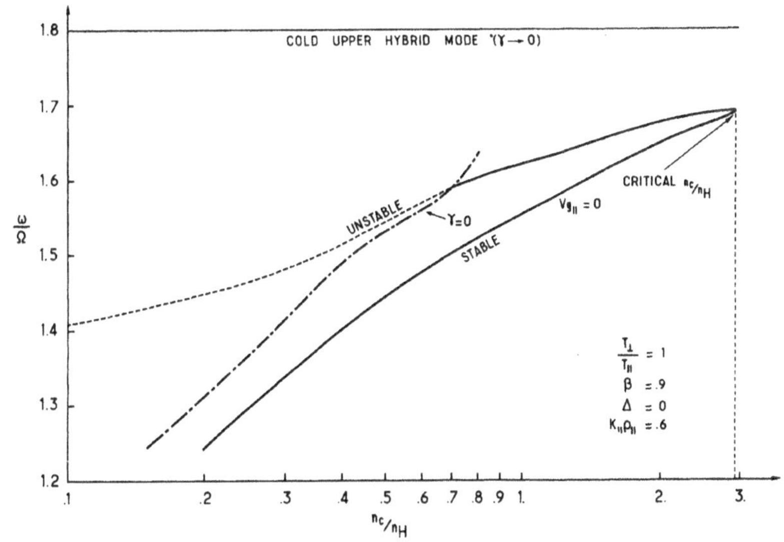

Figure 11b. *Repeat of Figure 11a for $K_{\|}\rho_{\|} = 0.6$. Note that spread of frequencies for which $\partial\omega/\partial K_{\|} = 0$ and $\gamma > 0$ is still only 10% as n_c/n_H is varied.*

where R is the distance from the dipole centre. If it starts propagating azimuthally, it must traverse the distance d_a,

$$d_a = R \sqrt{\frac{2}{3} \frac{\Delta B}{B}} \tag{14}$$

whereas, if it propagates along the fild line, it must traverse the distance $d_{\|}$:

$$d_{\|} \simeq R \sqrt{\frac{1}{6} \frac{\Delta B}{B}} \tag{15}$$

In general, we see that $d_a \simeq 2\, d_{\|} > d_r$. At $L = 6$, the propagation distances turn out to be of order $d_a \simeq 2\, R_E$, $d_{\|} \simeq 1\, R_E$, and $d_r \simeq 1/2\, R_E$ for $|\Delta B/B| \simeq 1/6$.

From Equations (13) - (15), we arrive at the following speculative conclusions. First, waves that propagate radially probably receive less convective amplification than those that have an azimuthal component of group velocity. Of those which propagate azimuthally, the net amplification will be greatest for those waves that have the smallest component of parallel group velocity. In any case, even if the above speculations prove to be incorrect, we believe that the observed emissions probably do have small parallel group velocities because at least that clas of 3/2 emissions observed near the geomagnetic equator is also observed to be localised within ± 10° of the equator [Kennel *et al.*, 1970], whereas they are observed over a wide range of L and local time. (Of course, Fredricks & Scarf [1973] have observed 3/2 emissions away from the equator also, but those that were near the equator were peaked within a few degrees of the equator; we concentrate on those).

Let us proceed with our investigation of the zeros of the parallel group velocity. We ask over how wide a range of n_c/n_H and $K_{\|}$ does the frequency at which $\partial\omega/\partial K_{\|} \simeq 0$ lie close to $X \simeq 3/2$. In Figure 11a, we plot the frequencies at which $\partial\omega/\partial K_{\|} = 0$ and also for which $\gamma = 0$ as a function of n_c/n_H for $K_{\|}\rho_{\|} = 0.4$ and a fixed cold upper hybrid frequency, $X_{UH} = 1.8$. We conclude that there is a critical $n_c/n_H \simeq 0.35$ to induce a zero in $\partial\omega/\partial K$ and also marginal stability, and that thereafter, $n_c/n_H < 0.35$, the frequency at which $\partial\omega/\partial K_{\|} = 0$ and $\gamma > 0$ is almost independent of n_c/n_H. In Figure 11b, we do the same thing for $K_{\|}\rho_{\|} = 0.6$. Here the critical n_c/n_H to induce a zero in $\partial\omega/\partial K_{\|}$ is considerably larger than that to induce $\gamma > 0$, and the spread in frequencies for which $\partial\omega/\partial K_{\|} = 0$ is somewhat larger; nonetheless it always lies within 10% of $X \simeq 3/2$.

Occasionally, two narrow bands of electrostatic emissions, near $X \simeq 3/2$ and $X \simeq 5/2$, have been observed [F.L. Scarf & R.W. Fredricks, private communication]. This behaviour cannot be reproduced by curves of growth rate alone. In Figure 12, however, we plot the parallel group velocity for a variety of $K_{\|}\rho_{\|}$, for a case where $X_{UH} = 3$. We see that there are four zero crossings; two relatively uninteresting and two relatively interesting. There are, of course, zero crossings at the upper hybrid frequency $X_{UH} = 3$, and at $X = 2$. Waves at $X = 2$ would probably be damped if we were to give the cold electrons a finite temperature T_c. There are zero crossings at $X = 5/2$ and near $X = 3/2$. Clearly, these are the 'interesting' ones. In Figure 13, we pick one $K_{\|}\rho_{\|} = 0.4$ and $X_{UH} = 3$ and plot the frequencies of the zeros of $\partial\omega/\partial K_{\|}$ together with the frequency X_{min} for which $\gamma = 0$. From Figure 13, we conclude

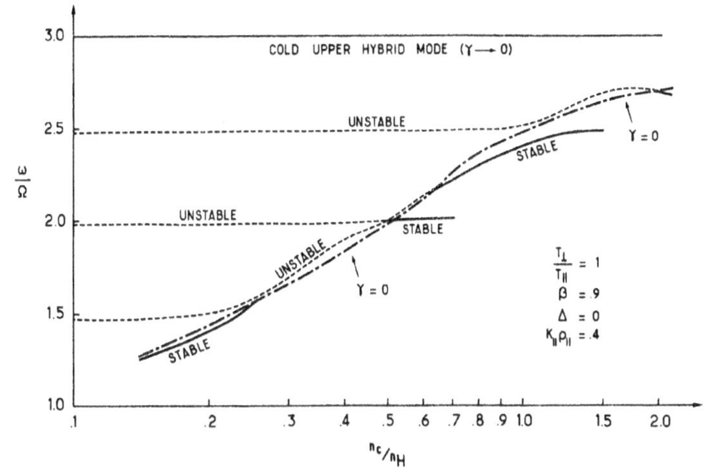

Figure 12. *Parallel group velocity versus frequency for modes that are temporally unstable throughout the range $1.4 \leqslant \omega/\Omega \leqslant 3$. The parallel group velocity is zero at $\omega/\Omega = 2$ (which should be damped if $T_c = 0$) at the cold upper hybrid resonance, near $\omega/\Omega \simeq 5/2$, and in a range of frequencies near $\omega/\Omega \simeq 3/2$.*

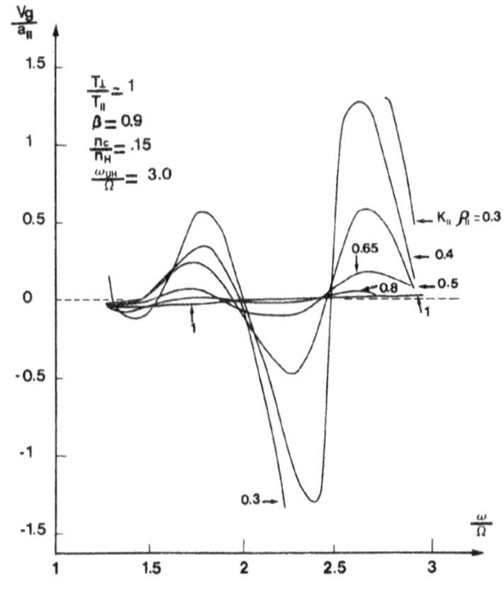

Figure 13. *Repeat of Figure 11a for $X_{UH} = 3$. The zeros in $\partial\omega/\partial K$ at $X = X_{UH} = 3$ and $X = 2$ may be damped by finite T_c. The zeros in $\partial\omega/\partial K_{\|}$ near $X = 5/2$ and $3/2$ have a frequency independent of n_c/n_H wherever γ is significantly positive.*

that where n_c/n_H is sufficiently small, < 0.2, for γ to be significantly positive at both $X = 3/2$ and $X = 5/2$, the frequency for $\partial\omega/\partial K_{\parallel} = 0$ is independent of n_c/n_H. At other $K_{\parallel}\rho_{\parallel}$, we expect that this frequency will not necessarily be as constant as for $K_{\parallel}\rho_{\parallel} = 0.4$, but as in Figure 11b will lie relatively near $X = 3/2$ and $5/2$.

5. Conclusions

(i) One could possibly explain the narrow-band electrostatic emissions observed at frequencies between the first few harmonics of the cyclotron frequency as a simple cold upper hybrid mode. However, we feel this explanation has several difficulties. First, one must invoke special values of the upper hybrid frequency, whereas Figure 3 would suggest that X_{UH} varies considerably. Secondly, the waves near the cold upper hybrid frequency are sensitive to the temperature T_c of the cold electrons [Young *et al.*, 1973; Karpman *et al.*, 1974 a, b]. We intend to investigate this effect in the future. Thirdly, there seems to be no way that the cold upper hybrid mode could account for two narrow-band emissions near $X = 3/2$ and $5/2$ as are occasionally observed.

(ii) On the other hand, which frequencies are observed could be determined not by gross considerations of temporal instability, but by propagation effects. While it is generally held that electrostatic waves do not propagate, the only speed scaling the present problem is the electron thermal speed, $\simeq 10^9$ cm/s, which is so large that propagation cannot be *a priori* neglected.

(iii) The waves near the geomagnetic equator with $X \simeq 3/2$, $5/2$ appear not to propagate parallel to the magnetic field, since they are observed to be localised near the geomagnetic plane.

(iv) Hot plasma effects appear to induce small parallel group velocities in the frequency range $X \simeq 3/2$ for a significant range of $K_{\parallel}$ and n_c/n_H.

(v) Detailed propagation and convective amplification calculations need to be carried out to test the speculation that waves with $\partial\omega/\partial K_{\parallel} \simeq 0$ are amplified the most.

(vi) In addition, weaker slopes in the electron distribution (a partially filled loss cone) need to be used in instability and propagation calculations, since the present convective amplification lengths are much too short. All previous calculations using the empty loss cone have this property as well. Preliminary results with a filled loss cone indicate that zeros in $\partial\omega/\partial K_{\parallel}$ continue to exist, but over a narrower $K_{\parallel}\rho_{\parallel}$ range. Further results will be reported later.

The present calculations do suggest that new experimental results are needed to further refine our understanding of electrostatic emissions in the magnetosphere.

(a) It is very important to measure n_c. Does the cold upper hybrid frequency turn up as a weak emission, or as a cut-off in the measurements? Can it be measured by plasma sounding techniques? Has it been confused with the total upper hybrid frequency as Gaffey & Le Quey [1974] suggest?

(b) To test the theory, we need measurements of $F(v_{\perp})$ at constant $v_{\parallel}$ for the electrons. If convective effects dominate, it is possible that even rather small $\partial f/\partial v_{\parallel} > 0$ could give significant amplification.

(c) Angular distributions of wave amplitude could help define how the waves propagate. Do the waves have amplitude peaks along the surface of constant *B*, as the qualitative considerations of Section 4 suggest?

In any case, we feel the problem of the '3/2' emissions is sufficiently well defined experimentally to warrant considerably more theoretical attention than it has received in the past, especially since we can expect considerably more experimental information during the IMS. And the pay-off is significant. For as Kennel *et al.* [1970] and more recently and more rigorously, Lyons [1974] have pointed out, the '3/2' emissions might be responsible for the diffuse aurora. If so, they account for much more precipitation of electron energy from the magnetosphere to ionosphere than whistlers, which *have* received the amount of attention they deserved in the past ten years. It goes without saying that until the linear theory can be made to account roughly for the observations, we probably will not have enough information to begin a good nonlinear theory.

6. Acknowledgements

We are particularly indebted to M. Lefloch and C. Davoust for their devoted help. We thank R. Gendrin, R. Pellat and J. Solomon for informative discussions. We are grateful to CNET for the financial support provided for our calculations. The work of M. Ashour-Abdalla was supported by CNES (French National Centre for Space Research) and that of C. Kennel by the Alfred P. Sloan Foundation and by the National Science Foundation, Grant 34148X.

References

Ashour-Abdalla, M., Instability theory of electrostatic half-harmonic electron gyrofrequency waves in the magnetosphere: a review, *Ann. Geophys.*, 1975 (in press).

Ashour-Abdalla, M. & S.W.H. Cowley, Wave particle interactions near the geostationary orbit, in Magnetospheric Physics (Ed. B.M. McCormac) p. 270, Reidel, Dordrecht, Holland, 1974.

Ashour-Abdalla, M., G. Chanteur & R. Pellat, A contribution to the theory of the electrostatic half-harmonic electron gyrofrequency waves in the magnetosphere, *J. Geophys. Res.*, 1975 (in press).

Brice, N. & C. Lucas, Influence of magnetospheric convection and polar wind on loss of electrons from the outer radiation belt. *J. Geophys. Res.* **76**, 900 (1971).

Chappell, C.R., K.K. Harris & G.W. Sharp, A study of the influence of magnetic activity on the location of the plasmapause as measured by OGO-5, *J. Geophys. Res.* **75**, 50 (1970).

Cornwall, J.M., F.V. Coroniti & R.M. Thorne, Turbulent loss of ring current protons, *J. Geophys. Res.* **75**, 4699 (1970).

Dory, R.A., G.E. Guest & E.G. Harris, Unstable electrostatic plasma waves propagation perpendicular to a magnetic field, *Phys. Rev. Lett.* **14**, 131 (1965).

Fredricks, R.W., Plasma instability at $(n + 1/2)\ f_c$ and its relationship to some satellite observations, *J. Geophys. Res.* **76**, 5344 (1971).

Fredricks, R.W. & F.L. Scarf, Recent studies of magnetospheric electric field emissions about the electron gyrofrequency, *J. Geophys. Res.* **78**, 310 (1973).

Fredricks, R.W., Wave particle interactions in the outer magnetosphere: a review, Presented at Neil Brice Memorial Symposium, Frascati, 1974.

Gaffey, J.D. & R. Le Quey, Upper hybrid resonance in the magnetosphere, Preprint, 1974.

Gurnett, D.A. & R.R. Shaw, Electromagnetic radiation trapped in the magnetosphere above the plasma frequency, *J. Geophys. Res.* **78**, 8136 (1973).

Gurnett, D.A. & L.A. Frank, Thermal and suprathermal plasma densities in the outer magnetosphere, *J. Geophys. Res.* **79**, 2355 (1974).

Harris, E.G., Unstable plasma oscillations in a magnetic field, *Phys. Rev. Lett.* **2**, 34 (1959).

Ioannidis, G. & N. Brice, Plasma densities in the jovian magnetosphere: plasma slingshot or Maxwell demon? *ICARUS* **14**, 360 (1971).

Karpman, V.I., Ju K. Alekhin, N.D. Borisov & N.A. Rjabova, Electrostatic electron-cyclotron waves in the multi-temperature plasma, 1974a (preprint).

Karpman, V.I., Ju K. Alekhin, N.D. Borisov & N.A. Rjabova, Electrostatic electron-cyclotron waves in the plasma with a loss cone, 1974b (preprint).

Kennel, C.F. & H.E. Petscek, Limit on stably trapped particle fluxes, *J. Geophys. Res.* 71, 1 (1966).

Kennel, C.F., F.L. Scarf, R.W. Fredricks, J.H. McGehee & F.V. Coroniti, VLF electric field observations in the magnetosphere, *J. Geophys. Res.* **75**, 61136 (1970).

Lyons, R.L., Electron diffusion driven by magnetospheric electrostatic waves, *J. Geophys. Res.* **79**, 575 (1974).

Sagdeev, R.Z. & V.D. Shafranov, On the instability of a plasma with an anisotropic distribution of velocities in a magnetic field, *Soviet Physics* **12** (1), 130-132 (1961), JETP, English Transl.

Scarf, F.L., R.W. Fredricks, C.F. Kennel & F.V. Coroniti, Satellite studies of magnetospheric substorms on August 15, 1968. OGO-5 plasma wave observations, *J. Geophys. Res.* **78**, 3119 (1973).

Young, T.S., J.D. Callen & J.E. McCure, High frequency electrostatic waves in the magnetosphere, *J. Geophys. Res.* **78**, 1082 (1973).

DISCUSSION

K. Wilhelm: Bernstein and co-workers observed $3/2 f_c^-$ waves in a vacuum chamber when and only when a counter-streaming configuration of electron fluxes was present. Has this observation any application in magnetospheric plasmas?

C. Kennel: Yes, possibly, since McIlwain and De Forest do observe electron distributions peaked in the loss cone. I believe the 3/2 emissions are probably too common to be explained by these peaked distributions, but we must await better statistics.

L.R. Lyons: This instability analysis is unique in that it yields numerous definitive predictions concerning the qualitative behaviour of the instability as a function of measurable magnetospheric parameters. Future satellite studies should conclusively determine whether or not the analysis is correct.

F. Mozer: Your theory relies on $\partial f/\partial v_\perp > 0$ for electrons and I am wondering if this is the appropriate region of phase space to investigate in view of the San Diego measurements of large field-aligned electron distributions. Perhaps Dr. De Forest can comment on the frequency of occurrence of such distributions or of distributions having $\partial f/\partial v_\perp > 0$.

S. De Forest: The intense field-aligned fluxes of electrons we see are rather rare (19 in 6 months) but this is almost certainly an artifact of our orientation. We only look into the loss cone occasionally. After we move to the eastern location, we hope to get better statistics.

PART 5

SUBSTORMS AND CONJUGATED PHENOMENA

SUBSTORM PHENOMENA AND THE UNIQUENESS OF THE GEOSTATIONARY ORBIT FOR OBSERVATION OF DYNAMIC PROCESSES IN THE MAGNETOSPHERE

S.E. DeForest

University of Alabama,
Huntsville, USA

Abstract

During the Spring equinox, both ATS-6 and ATS-5 were in geosynchronous orbit with a separation of about 11° longitude. A ground-based campaign was undertaken by Eather, Mende, and Judge to observe aurora at the foot of the ATS field line with a variety of instruments. Initial results support the injection boundary hypothesis of Mauk and McIlwain. Also during this period, the Cs-ion thrusters on both spacecraft were exercised in a deliberate attempt to control the spacecraft potential. Some success was achieved, and the presence of a sheath with a potential minimum about ATS-5 has been demonstrated (the sheath about ATS-6 is observed directly).

1. Introduction

In building the paper around the title, I have been forced to make certain assumptions about the interests of the audience, since even a brief description of all the dynamic processes seen at geosynchronous orbit would take far more space than allotted. Therefore, I have chosen to concentrate on the global aspects of substorm phenomena, to the exclusion of more limited events such as certain types of wave-particle interactions [Cummings *et al.*, 1969; LaQuey, 1973] commonly seen at geosynchronous orbit. Likewise, I will not mention the important observations of intense field-aligned particle fluxes as seen on ATS-6 and recently reported by McIlwain [1975]. The approach will be to review what can be done with a single spacecraft, and then to give an example of the power of correlating ground-based and geosynchronous measurements.

Finally, the last section of this paper will be devoted to the limitations on plasma measurements due to the interaction of the spacecraft with the ambient plasma. While the general spacecraft charging problem is well known (it was the subject of a

Knott & Battrick (eds.), The Scientific Satellite Programme During the International Magnetospheric Study, 329–343.
All Rights Reserved. Copyright © 1976 by D. Reidel Publishing Company, Dordrecht-Holland.

previous ESLAB Symposium [Grard, 1973] and two sessions at this year's Spring AGU Meeting in Washington), the related problem of the nature of the plasma sheath formed about the spacecraft is less well understood. Whipple [1975] has recently studied the nature of the sheath about ATS-6 and concludes that it exists essentially all the time. This means that at the times that ATS-6 is positively charged, no ambient low-energy (less than a few volts) electrons or ions reach the spacecraft. However, the low-energy electrons actually measured are secondaries and photo-electrons returning to the spacecraft. These particles vary with geomagnetic conditions and can masquerade as ambient particles.

The last topic to be discussed will be the possibility of active control of spacecraft potentials as was recently demonstrated on ATS-5 and 6 by the use of the on-board caesium-ion accelerators in conjunction with plasma measurements.

Before proceeding to the data, let us first review a bit about what is known of the environment at geosynchronous orbit and what problems we might expect to solve during the IMS with the aid of GEOS. Electric fields are not yet well known, but quantitative magnetic field models are very well developed (paper by Olson, this conference). We know reasonably well how to map from synchronous orbit to magnetic parameters in quiet conditions. There is more diurnal variation than one might expect. Figure 1 was prepared by C.E. McIlwain to demonstrate this point. During disturbed conditions, however, our knowledge is better summed up by Figure 2 taken from a paper by Chappell [1974]. This figure points out that we really do not know how to make the detailed mapping from the equator to the Earth's surface.

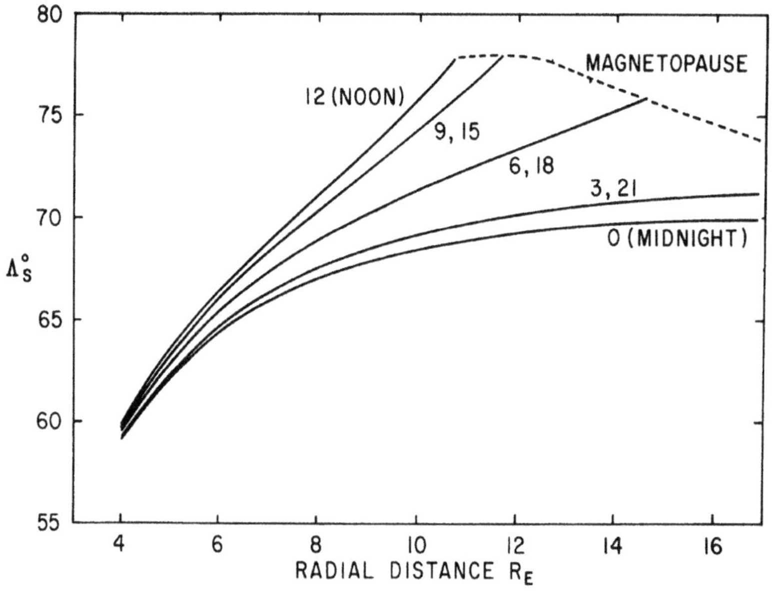

Figure 1. *Surface invariant latitude as a function of radial distance.*

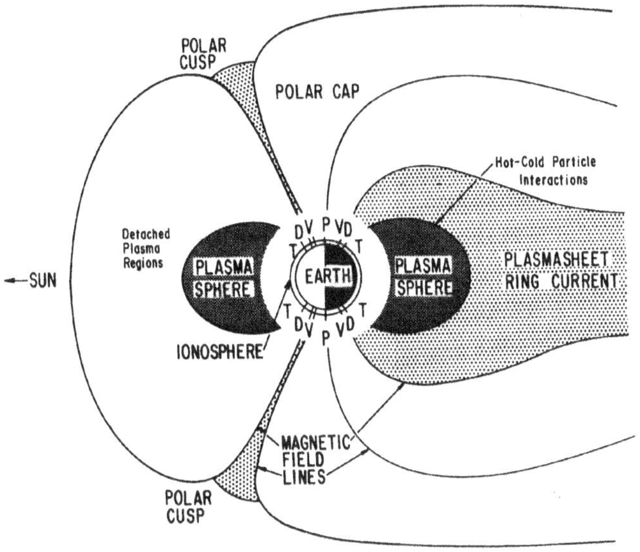

Figure 2. *Schematic representation of cross section of magnetosphere [after Chappell].*

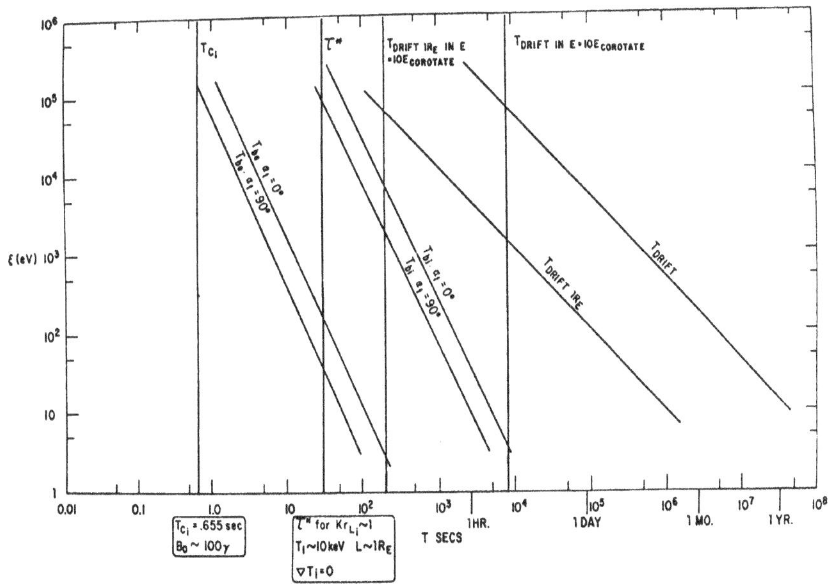

Figure 3. *Energy as a function of period for selected events at geosynchronous orbit.*

As will be discussed in the next section, the population of plasma particles has been studied by many workers, yet the exact nature of plasma injection is still not known. Note that throughout this paper, 'injection' will be taken to mean injection in phase space, not the more limited injection into real space as is sometimes implied. Therefore, local heating or acceleration of particles will be covered by this term. More importantly, the exact composition of the plasma is not known, since no mass spectrometer has yet been carried to synchronous orbit, although McIlwain has made some progress by studying the velocity dispersion of bounce waves (private communication). Similarly, the total density is not as well known as it could be, since no electromagnetic wave instruments have been flown there either.

Some of the time periods of interest at synchronous orbit are shown in Figure 3, which was prepared by R. LaQuey. Many other classes of waves could have been added, but the figure already shows that periods of less than a second to days are important.

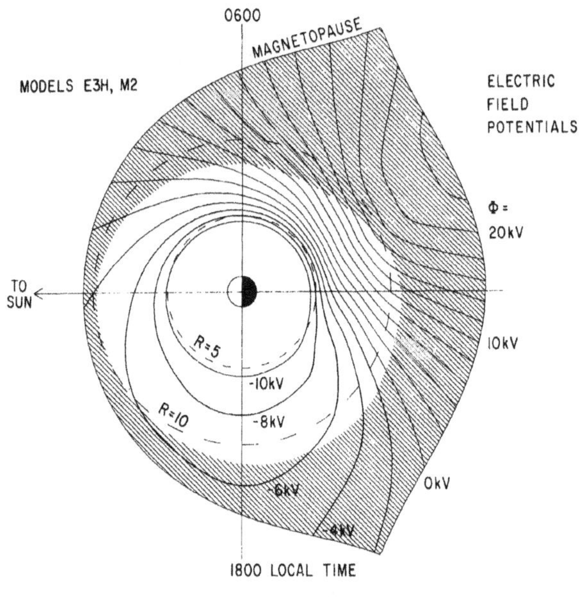

Figure 4. *Equatorial cut-off angle as a function of particle energy and potential drop.*

The angular response necessary to unfold particle data is indicated in Figure 4, which was drawn as a study for the design of the UCSD auroral particles experiment on ATS-6. The assumptions are that a potential drop exists at some point along a field line (assumed dipolar). Then what happens to particles at the equator? The figure shows that that question can only be answered if the instrument has a resolution of less than a degree. Perfect knowledge of the distribution function at the equator would, in principle, allow one to calculate any monotonic distribution of potential along the field line.

2. Single-Spacecraft Results

Before the launch of ATS-5 in August 1969, the equatorial observations of magnetospheric plasma were limited to rapid traversals [Gringauz, 1961; Frank, 1967a, b, c; Vasyliunas, 1968; Frank & Owens, 1970], and to integral proton and electron fluxes [Freeman & Maguire, 1967]. The differential detector on ATS-5 was able to study substorms by showing that plasma was injected at synchronous orbit in clouds which then dispersed in the Earth's magnetic and electric fields [DeForest & McIlwain, 1971]. The injection and dispersion of the high-energy tail of these clouds had previously been reported by Konradi [1967], Arnoldy & Chan [1969], Pfitzer & Winkler [1969], and Lezniak & Winckler [1970]. Bogott & Mozer [1974] studied this dispersion further.

This realisation that plasma was injected in clouds is very important for the study of substorm dynamics. Groups of particles can be identified with a given substorm measured on the ground, by either magnetometers or optical devices [Akasofu *et al.*, 1971]. Then the development of these clouds can be observed as the particles disperse. Sometimes, given particles can be identified as they make three or four orbits around the Earth. Since the behaviour of these particles could not be

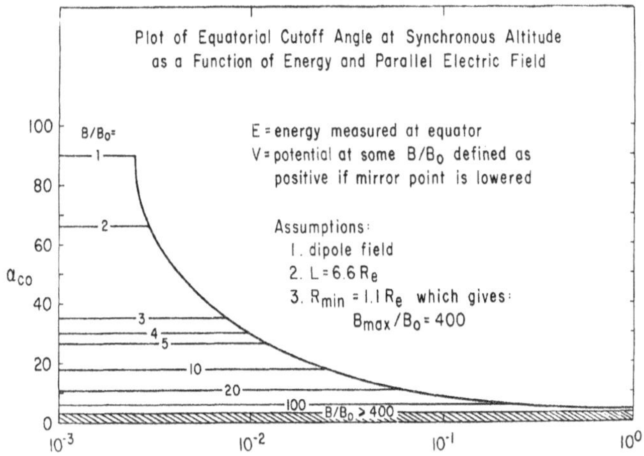

Figure 5. *Global electric fields* [*after McIlwain*].

explained by the normal drifts in the Earth's magnetic field, McIlwain [1972] could use the observed behaviour to measure the global magnetospheric electric fields. His inferred quantitative field is subject to several limitations. He assumes:

(a) the electric field parallel to **B** is zero

(b) the electric field is independent of time

(c) the magnetic field is known and is also independent of time (i.e. $\nabla \times \epsilon = 0$), and

(d) the particle motion is adiabatic to the extent that particles not lost to the system preserve total energies and their magnetic moment.

He only treats particles on the equator, and he only claims validity for this model during quiet periods starting about 20 min after a particle injection. The results are shown in Figure 5. The unshaded area is the spatial region of validity. The shaded area is the analytical extension of the model.

Several comments are in order. First, this model is obviously not unique. Other solutions could be found to explain the data as well, but this is probably the simplest solution. Secondly, this field, when used with the appropriate magnetic field, accurately predicts how particles will move after being injected. In other words, it is a valuable tool for organising the particle data, in spite of the limitations on its use.

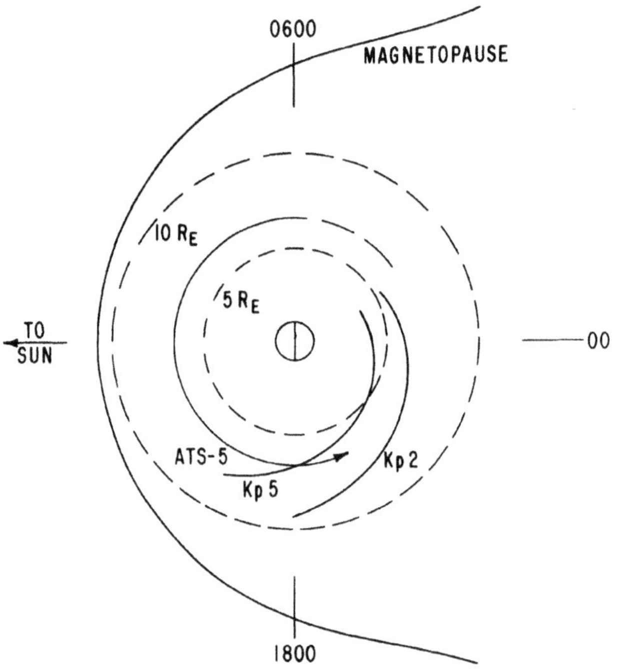

Figure 6. Injection boundaries.

Mauk & McIlwain [1974] treated the measured particles as an image produced by the electric and magnetic fields. They used a computer to project the leading edge of clouds back to their origin. The result is a spiral-shaped injection boundary shown in Figure 6. The boundary scales with K_p. McIlwain [1973] showed that the sudden appearance of plasma at this boundary followed by normal dispersion explained several independent observations such as the 'soft inner edge of the plasma sheet' [Frank, 1968; Vasyliunas, 1968]. It is of interest to compare this work, which is meant to organise measured particle distributions, with the theoretical work of Kivelson & Southwood [1975]. The concept of the injection boundary has also been used to organise Explorer 45 particle data [Konradi, 1974].

3. Correlations between Spacecraft and Ground-Based Measurements

If one wishes to study the large-scale phenomena associated with geomagnetic substorms by correlating particle data from a synchronous spacecraft with other nonsynchronous data, a number of potentially useful possibilities arise. Rockets and low-altitude polar orbiters are useful, but they spend very little time in the vicinity of the foot of a synchronous spacecraft [Sharp *et al.*, 1971; Johnstone *et al.*, 1974]. Balloon-borne X-ray and *E*-field measurements, backscatter radar, and whistlers all present interesting possibilities. But in this section, I will only discuss some recent work done by Eather, Mende & Judge [1975] on correlating observations of an array of meridian-scanning photometers located at the nominal foot of the ATS-5 field line with the ATS-5 plasma measurements.

From ground-based observation of the absolute and relative strengths of certain auroral lines, one can infer both the total flux of the precipitating particles and their characteristic energy [Eather & Mende 1972; Judge, 1972]. While the expected accuracy of this type of measurement is lower than in-situ sampling, it has the great advantage that the observations can be made over a period of hours (at least on moonless and cloudless nights). Therefore, one can observe the response of the aurora to the whole development of a substorm. In particular, weak isolated substorms can be studied in detail. (Rockets tend to be launched in sufficiently active times that analysing simultaneous data from synchronous orbit is very difficult.) The method that Eather and Company use to present their data also deserves special comment since it represents a major advance. They make grey-scale coded plots called 'keograms' similar to the familiar spectrograms in which the grey scale is modulated by a photometer channel. The abscissa is time and the ordinate is viewing angle with the zero adjusted to lie at the foot of ATS-5. The magnetic field used was Olson & Pfitzer's [1974] new model. It seems to be accurate to a fraction of a degree for all but the most disturbed times.

They present several days of data which cannot be reproduced here easily, but their conclusions are presented in Figure 7 (their fig. 10) which is a schematic representation of what particle injections look like on the ground and at the equator. Among their results, they state that enhanced precipitation from an injection superimposes on the background plasma-sheet precipitation patterns (see figure). This ground-based observation of enhanced precipitation also follows the injection boundary, discussed in the previous section, which was derived strictly from equatorial particle measurements. In fact, when they calculate the instantaneous boundary and superimpose it on the keograms, the agreement between it and the precipitation patterns is surprisingly good in view of the fact that K_p is only calculated every 3 h.

They also state that they see a region of continuous $\lesssim$ few keV electron precipitation, usually coincident with proton precipitation, over $\gtrsim 2^\circ$ latitude. These (usually subvisual aurora) persist even under very quiet conditions. They identify the equatorward boundary of this weak precipitation as the inner edge of the plasma-sheet (see figure). In no case was the injection boundary observed to extend radially inward past this inferred boundary of the pre-existing plasma sheet.

For other conclusions, the reader is referred to the paper. Another ground-based

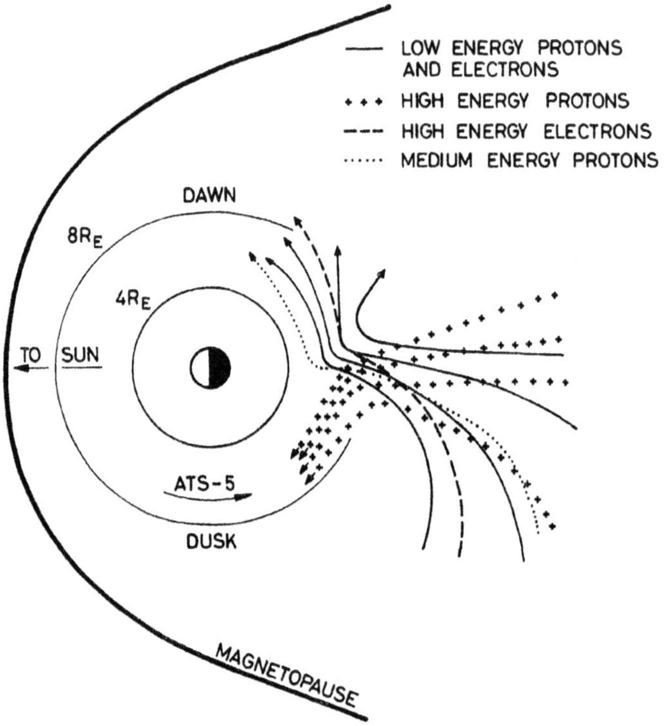

Figure 7. *Schematic representation of interrelationships between the plasma sheet, injection boundary, and subsequent particle drift paths after injection.*

campaign was launched in February and March of this year within sight of the feet of both ATS-5 and ATS-6 field lines. The instrumentation was improved with the addition of a colour television and camera. Data analyses for both the spacecraft and ground-based observations are still in the very preliminary stage. Some important aspects of this series of observations to date include:

(i) coverage for a sufficiently long period to see substorm development
(ii) quantitative output (all-sky cameras are useful, but cannot do the job alone) and
(iii) effective presentation of data (keograms).

The third point is very important as we look forward to the IMS. Experimenters from many different disciplines cannot be expected to shuffle through reams of line plots in the hope of seeing a particular wiggle that coincides with a wiggle in their graphs. Modern data rates dictate that data presentation must be well thought out to avoid complete neglect by the intended audience.

4. Limitations on Spacecraft Observations of Plasma Particles

By now, a considerable literature exists documenting the effects of spacecraft charging [Grard, 1973; DeForest, 1972; Inouye, 1975; Whipple, 1965; Rosen, 1975]. In this section, I will not review all the work done to the present, but will concentrate on three aspects of the general problem:

(i) occurrence and distribution of charging on ATS-6 in sunlight
(ii) nature of the plasmasheath about ATS-6, and
(iii) active control of spacecraft potentials using caesium ion thrusters on board ATS-5 and ATS-6.

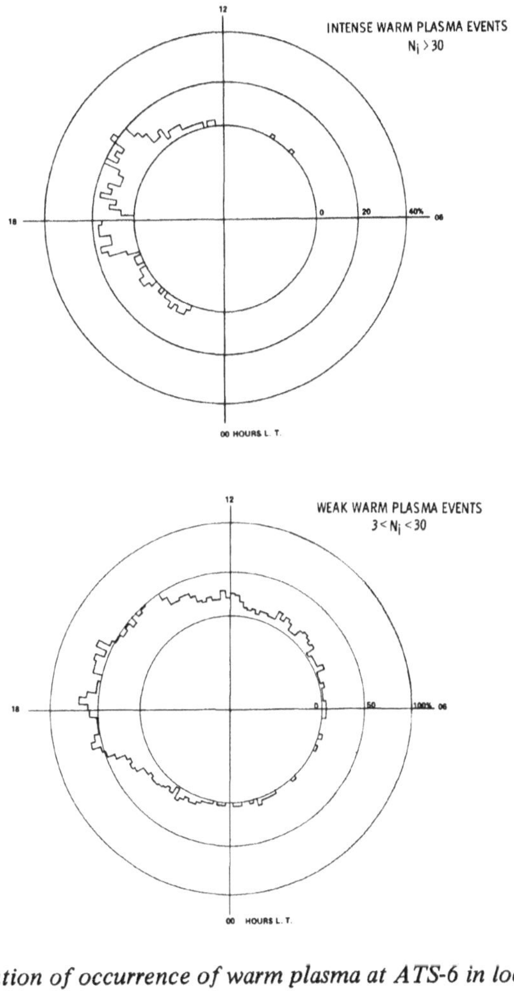

Figure 8. *Distribution of occurrence of warm plasma at ATS-6 in local time.*

Lennartsson & Reasoner [1975] have studied both the occurrence of 'warm' plasma and charging events on ATS-6 for 40 days in 1974. Their results are shown in Figures 8 and 9. Note that the distribution of charging events is similar to the local time dependence of spacecraft anomalies as compiled by Rosen [1975]. In the region of nonzero overlap of the occurrence of charging and warm plasma, there is mutual exclusion on an event-by-event analysis (i.e. no charging occurs if warm plasma is present).

Figure 10 shows the relative values of potentials reached by ATS-6. Note that on relatively rare occasions, very high potentials occur even in sunlight.

When a spacecraft is charged negatively, an experimenter obviously cannot detect ambient low-energy electrons. Also, the low-energy ions get accelerated such that their undisturbed distribution function is difficult to measure. But the systematics shown in Figure 8 indicate that one cannot simply discard data that are compromised by charging, since that charging is itself a function of geomagnetic activity and local time of observation.

ATS-6 was not specifically designed to avoid charging. Therefore, we must wait until GEOS is launched to see what the behaviour of an electrostatically clean spacecraft is at synchronous orbit. However, some experiments on the active control of spacecraft potential have been possible using the ion thrusters on both ATS-5 and ATS-6. Initial results of these studies were reported by Bartlett *et al.* [1975].

Several important observations were made in this study. First, ATS-5 and ATS-6 were stationed sufficiently close to each other that they could both be in eclipse at the same time. When this happened, and when the geomagnetic activity was low enough so that one could assume that they were both immersed in the same plasma, they both charged to approximately the same potential! This happened in spite of the fact that the two spacecraft are very different in size, configuration and material.

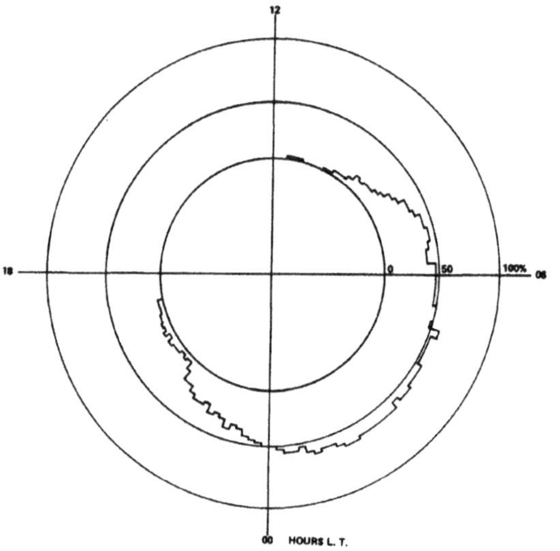

Figure 9. Distribution of charging at ATS-6 in local time in sunlight.

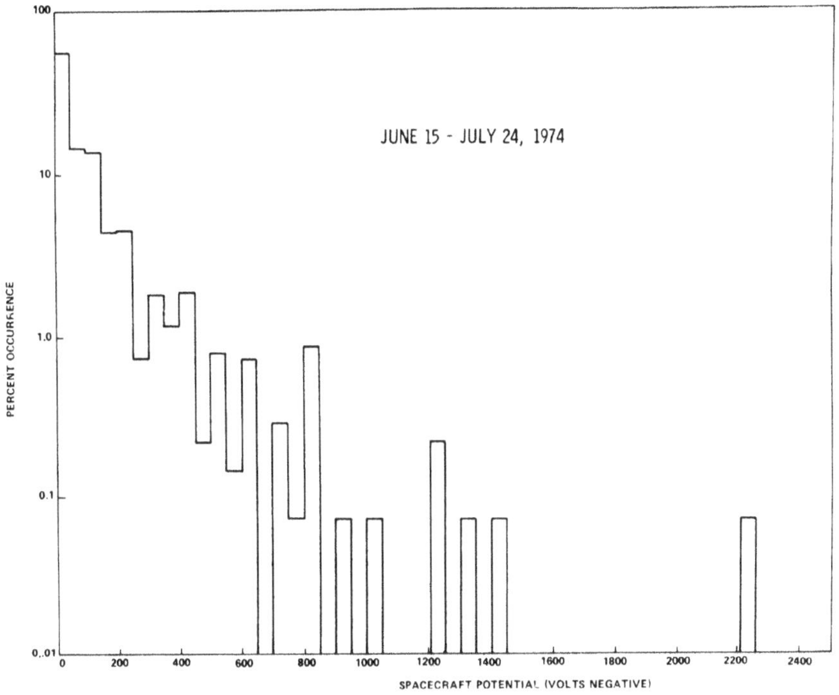

Figure 10. Distribution of magnitude of charging at ATS-6 in sunlight.

We can, therefore, begin to hope that a valid general theory of charging will not be so dependent on the details of spacecraft construction that it will be impossible to use.

Another observation of note is that operating the ATS-5 caesium-ion beam neutraliser by itself without the beam, can successfully discharge the spacecraft. This neutraliser is a simple filament which is electrically heated to emit thermal electrons.

Other results of the thruster experiments include the fact that during the normal operation of the ATS-6 ion thruster [Worlock *et al.*, 1975], the spacecraft potential was locked at − 4 V even though the geomagnetic activity was such that one can infer it would have charged to at least − 100 V had the engine been off. Initial analysis of the particle data also tends to indicate little if any contamination of the spacecraft surface. Analysis of these results is still continuing, and we hope to be able to repeat a series of this type of experiment during the coming equinox periods.

A final and very interesting observation was first made by C.E. McIlwain. When the south thruster on ATS-6 was operating correctly, by coincidence the thrust vector was nearly perpendicular to the ambient magnetic field as measured by the UCLA magnetometer. The electrode configuration is such that electrons with energies of 550 eV are emitted with the beam [Worlock *et al.*, 1975]. The UCSD plasma detector sweeps through the plane containing both the magnetic field and the thruster. During this operation, 550 eV electrons were detected 180° from the thrust vector. These are almost certainly electrons which have been emitted from the thruster and made a complete circle in the Earth's magnetic field. The round-trip distance is approximately 4 km. The ions at the same energy are not seen because they travel 491 times slower around a radius 491 times larger and miss the spacecraft on the return trip by several kilometres.

As important as these attempts at active control are for the design of future spacecraft, the observations of the plasma sheath about ATS-6 are of a more basic importance, since it now appears that all spacecraft have sheaths about them, but that the type of sheath formed at synchronous can be associated with a potential minimum of tens of volts [Whipple, 1975a]. This minimum can be shown to be deeper than a solution of Poisson's equation would predict [Whipple, 1975b]. The sheath and the potential minimum exist at all times under all states of spacecraft charge. Surprisingly, they still exist even when the spacecraft is in eclipse and not surrounded by a cloud of photoelectrons. Plasma sheaths have long been known in the laboratory and some authors have made calculations which are applicable to spacecraft [Grard, 1973; Guernsey & Fu, 1970], but the implications for particle measurements are not well understood, particularly since ATS-6 never samples ambient electrons below the potential minimum, but the reflected secondary and photoelectrons that are counted do vary with geomagnetic conditions and thus can masquerade as natural particles. There is no reason to believe ATS-6 is unique in this respect, but more work must be done on the theory.

5. Discussion

In this paper, I have tried to show how far one can advance in the understanding of substorm dynamics given the usual limitation of space-time ambiguity that one always has when examining data from a single spacecraft (even a geosynchronous one). The concept of the injection boundary is still controversial, but it has the property of organising the particle data as measured as synchronous.

In the next section, the injection boundary concept was strengthened by the combination of spacecraft and ground-based photometric observations taken at the foot of the ATS-5 field line. However, the main point of that section was not to defend the injection boundary, but rather to point out the tremendous power of combining adequate ground-based observation with equatorial particle measurements. This could be a valuable tool during the IMS for understanding magnetospheric dynamics.

Finally, the basic limitations on particle measurements due to plasma-spacecraft interactions were discussed. In addition to the nuisance (and danger) of both overall and differential charging, it has now been established that the low-energy particles of both signs are frequently forbidden access to the spacecraft due to the existence of a sheath about it. Unfortunately, while the properties of this sheath can be partially explored for ATS-6, the theory to predict the sheath properties is not yet well developed.

Preliminary and somewhat primitive experiments in active control of spacecraft potentials have been successful. We hope to be able to demonstrate better control in the future.

Not mentioned in the article are the potential contributions of mass-spectrometer and electromagnetic-wave instruments to the complement of scientific instruments on a geosynchronous payload. Such devices have not been flown at synchronous orbit yet, and much good science has been delayed thereby. We expect that GEOS will rectify this shortcoming in addition to adding a fine payload.

Also, lack of space has prohibited a discussion of the possibilities of correlating data from two spacecraft at synchronous orbit. Depending on the separation, such things as overall dynamics or length of wave trains can be investigated in a way that cannot be done by one spacecraft alone.

6. Acknowledgements

I wish to acknowledge the kind permission of Drs. Eather, Mende, Judge, Lennartsson and Reasoner to present their data before they had published it themselves. The paper was made much more timely with their additions.

This work was supported by NASA Contract NAS8-30563.

References

Akasofu, S.-I., S.E. De Forest & C.E. McIlwain, *Planet. Space Sci.* **22**, 25 (1974).

Arnoldy, R.L. & K.W. Chan, *J. Geophys. Res.* **74**, 4019 (1969).

Bogott, F.H. & F.S. Mozer, *J. Geophys. Res.* **79**, 1825, (1974).

Chappell, C.R., *EOS* **55**, 776 (1974).

Cummings, W.D., R.J. O'Sullivan & P.J. Coleman, Jr., *J. Geophys. Res.* **74**, 778 (1969).

DeForest, S.E. & C.E. McIlwain, *J. Geophys. Res.* **76**, 3587 (1971).

DeForest, S.E., *J. Geophys. Res.* **77**, 651 (1972).

Eather, R.H. & S.B. Mende, *J. Geophys. Res.* **77**, 660 (1972).

Eather, R.H., S.B. Mende & R.J.R. Judge, Plasma injection at synchronous orbit and spatial and temporal auroral morphology, submitted to *J. Geophys. Res.*, April 1975.

Frank, L.A., *J. Geophys. Res.* **72**, 185 (1967a).

Frank, L.A., *J. Geophys. Res.*, **72**, 1905 (1967b).

Frank, L.A., *J. Geophys. Res.*, **72**, 3753 (1967c).

Frank, L.A., in Physics of the Magnetosphere (Eds. R.L. Carovillano, J.F. McClay, and H.R. Radoski), Reidel, Dordrecht, Holland (p. 271) 1968.

Frank, L.A. & H.D. Owens, *J. Geophys. Res.* **75**, 1269 (1970).

Freeman, J.W. & J.J. Maguire, *J. Geophys. Res.* **72**, 5257 (1967).

Grard, R.J.L. (Ed.), Photon and Particle Interaction with Surfaces in Space, Reidel, Dordrecht, Holland, 1973.

Gringauz, K.I., *Space Res.* **2**, 574 (1961).

Guernsey, R.L. & J.H.M. Fu, *J. Geophys. Res.* 75, 3193 (1970).

Inouye, G.T., AIAA Paper 75–255 presented at AIAA Meeting Pasadena, Calif., January 1975.

Johnstone, A.D., J.S. Boyd & T.N. Davis, *J. Geophys. Rev.* **79**, 1403 (1974).

Judge, R.J.R., *Planet-Space Sci.* **20**, 2081 (1972).

Kivelson, M.G. & D.J. Southwood, *J. Geophys. Res.* **80**, 56 (1975).

Konradi, A., *J. Geophys. Rev.* **72**, 3829 (1967).

LaQuey, R.E., *Phys. Fluids* **16**, 550 (1973).

Lennartsson, W. & D.L. Reasoner, Relationship between the local time dependencies of ATS-6 spacecraft charging events and warm plasma occurrences, Presented at Spring AGU Meeting, and to be published in Proceedings of Spacecraft Charging Session, 1975.

Lezniak, T.W. & J.R. Winckler, *J. Geophys. Res.* **75**, 7075 (1970).

Mauk, B.H. & C.E. McIlwain, *J. Geophys. Res.* **79**, 3193 (1974).

McIlwain, C.E., Proceedings of a Symposium Organized by the Summer Advanced Study Institute and Ninth ESRO Summer School, held in Cortina, Italy, August 30 – September 10, 1971, Reidel, Dordrecht, Holland, 1972.

McIlwain, C.E., Substorm injection boundaries, Presented at Summer Advanced Study Institute, Sheffield, England, 1973.

McIlwain, C.E., Auroral electron beams near the magnetic equator, Presented at Nobel Symposium, Kiruna, Sweden, and to be published in the Proceedings by Plenum Press, London, 1975.

Olson, W.P. & K.A. Pfitzer, *J. Geophys. Res.* **74**, 3739 (1974).

Pfitzer, K.A. & J.R. Winckler, *J. Geophys. Res.* **74**, 5005 (1969).

Rosen, A., AIAA Paper 75–91, Presented at AIAA Meeting, Pasadena, Calif., January 1975.

Sharp, R.P., D.L. Carr & R.G. Johnson, *J. Geophys. Res.* **74**, 4618 (1971).

Vasyliunas, V.M., *J. Geophys. Res.* **73**, 2839 (1968).

Whipple, E.C., NASA Technical Paper X-615-65-296, 1965.

Whipple, E.C., *J. Geophys. Res.*, in press, 1975a.

Whipple, E.C., *J. Geophys. Res.*, in press, 1975b.

Worlock, R.M., E.L. James, R.E. Hunter & R.O. Bartlett, AIAA Paper 75–303, Presented at AIAA 11th Electric Propulsion Conference, New Orleans, March 1975.

DISCUSSION

G.H. Voigt: (i) Is the plasma Debye length small enough to use Langmuir's theory to find the spacecraft potential?

(ii) Does the spacecraft have a conducting surface? If not, the spacecraft potential depends strongly on the direction of the Sun (photoelectron sheath and negative wake), and on capacitor effects of surface materials. Langmuir's theory is only valid for totally conducting plasma probes.

(iii) If Langmuir's theory was not employed to determine spacecraft potential, which method was used?

S. DeForest: (i) No.

(ii) No. Your comment is valid, but does not explain eclipse observations.

(iii) Method described in my article in Journal of Geophysical Research [1971].

C.T. Russell: Do you see a correlation between the 'strength' of an injection and the position of the inner edge of the plasma sheet as seen at ATS?

S. DeForest: If we define 'strength' as the total flux, no.

G. Rostoker: (i) It is good to see recognition of the fact that the 'injection' of particles doesn't mean *necessarily* inward flow from the tail. An injection may well be *in-situ* heating with the heated plasma then moving off on the drift paths.

(ii) I should like to point out that there is now evidence that the E field doesn't necessarily change markedly with substorm onset. The large currents associated with the substorm likely stem from ionospheric conductivity changes primarily [see Gurnett & Akasofu, *JGR*, 1974].

(iii) There is a danger in using the brightening of the visible arc as a signature of substorm onset. Auroral scanning photometer data from ISIS-2 shows that there is generally a bright arc in the sector immediately to the west of a westward travelling surge; thus the substorm sequence may already be in progress and the brightened arc may merely indicate the imminent arrival of the surge in the sector of the all-sky camera.

S. DeForest: (i) By 'injection', I mean injection into phase space. This is more general than simply allowing for heating.

(ii) –

(iii) What is a substorm?

SUBSTORM OBSERVATIONS OF MAGNETIC PERTURBATIONS AND ULF WAVES AT SYNCHRONOUS ORBIT BY ATS-1 AND ATS-6

P.J. Coleman, Jr. & R.L. McPherron
Department of Geophysics and Space Physics
&
Institute of Geophysics and Planetary Physics
University of California
Los Angeles, USA

Abstract

In the empirical study of the geomagnetic cavity, the synchronous equatorial orbit has been shown to be a most advantageous site for scientific measurements. Satellites in this orbit are fixed in the Earth's main magnetic field so that the records are not dominated by changes due to satellite motion through this field. Also, the satellites traverse the same geometric region of space daily, thus facilitating statistical studies of magnetospheric phenomena. Thirdly, the magnetic field lines leaving the Earth's surface near the auroral zone cross the equatorial plane close to this orbit, thus subjecting the region to a number of such phenomena.

Previous observations at synchronous orbit with the UCLA fluxgate magnetometers on ATS-1 and ATS-6 have revealed a variety of substorm-associated field variations. These effects may be divided tentatively into two classes, those resulting from instabilities (ULF waves) and those produced by changes in macroscopic current systems. The data and their interpretations will be reviewed in order to illustrate the potential importance of observations in the synchronous orbit during the International Magnetospheric Study.

1. Introduction

This paper is intended to serve as a descriptive catalogue of the phenomena observed in the magnetic field at synchronous orbit. Of the effects discussed herein, some have been examined in considerable detail and others have received to date only cursory attention. Further, those described do not form the complete list.

There are three main purposes for this presentation: (i) to acquaint the reader with the variety of these phenomena; (ii) to emphasise the need for further analysis and simultaneously to suggest implicitly directions for such analyses; and (iii) to acquaint the reader with the methods of analysis that we have used to date.

Knott & Battrick (eds.), The Scientific Satellite Programme During the International Magnetospheric Study, 345–365.
All Rights Reserved. Copyright © 1976 by D. Reidel Publishing Company, Dordrecht-Holland.

2. Geometry

Figure 1 is a sketch, essentially in the noon-midnight meridian, of the nightside geometry for ATS-6. It is shown to indicate some of the more significant aspects of the field configuration at the synchronous orbit, to establish the scale of the system relative to the Earth, and to define the co-ordinate system in which our results are most frequently presented [McPherron, 1973]. The sketch represents the situation at the northern summer solstice.

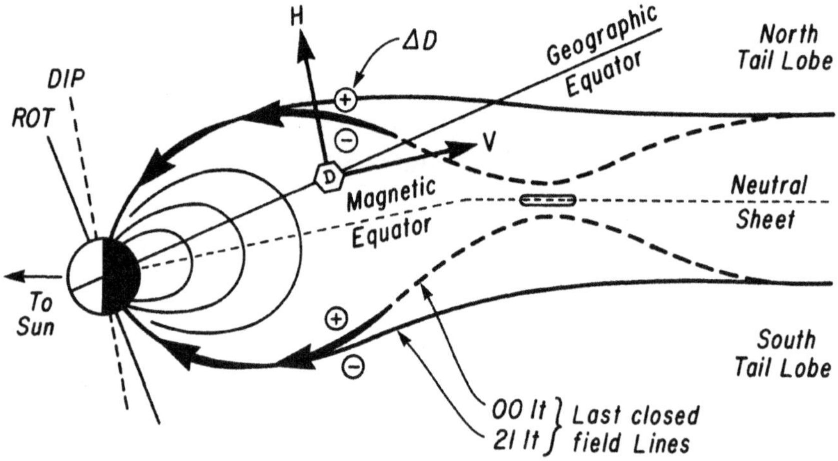

Figure 1. *Sketch showing location of the synchronous satellite ATS-6 (96° W longitude) relative to major features of the magnetic-field configuration when the spacecraft is near local midnight (0624 UT). Arrows labelled V, D, H are the axes of the dipole co-ordinate system.*

Longitude in synchronous orbit is an important factor controlling the observations. Clearly, the average orientation of the field depends strongly upon this variable. Thus, for ATS-1, which was in the geographic, equatorial plane at a longitude close to the intersection of the geographic and magnetic equatorial planes, the geometry was significantly different. In the latter case, a similar sketch would show the dipolar axis coincident with the rotational axis. Thus, at ATS-6 (95° W longitude) the primary effect of the tail current is to distort the field direction considerably. On the other hand, at ATS-1 (150° W longitude), its primary effect is a reduction in the magnitude of the field.

The D, H, V co-ordinate system shown in Figure 1 is defined relative to the axis of the Earth's dipolar field. Thus, H is parallel to the dipole axis but positive northward, D is in the direction of $H \times R$, where R is the radius vector from the centre of the Earth to the spacecraft, and V is parallel to the magnetic equatorial plane and essentially radially outward from the Earth. The measurements at ATS-1 were made in a co-ordinate system we call the 'rotational D, H, V' system. Thus, the rotational H component is parallel to the spin axis of the Earth, the rotational V component is radially outward from the centre of the Earth, and the rotational

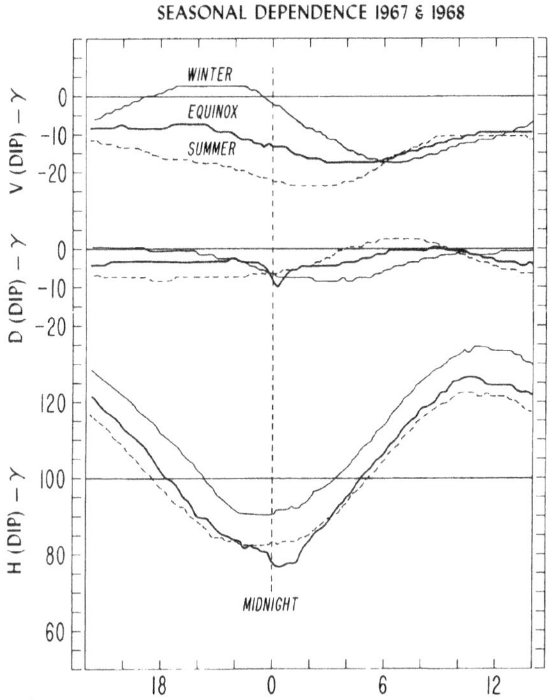

Figure 2. *Median of all ATS-1 (150° W longitude) magnetic-field observations as functions of local time and season. Negative bias of the dipole D component and the positive bias of dipole H are component results of undetermined variations in the spacecraft field parallel to satellite gain axis (parallel to Earth's rotation axis).*

D component completes the system. From these definitions, it is clear that the rotational and dipolar systems differ by a 10° rotation about V at ATS-1 and by an $\approx$ 10° rotation about D at ATS 6.

Changes in the intensity and orientation of the magnetic field at the synchronous orbit are associated with a variety of magnetospheric phenomena. Further, certain types of changes are often accompanied by quasi-periodic fluctuations loosely attributed to ULF waves. In the following sections we will describe important subsets of both transient and wave-like fluctuations.

Figure 2 shows the median magnetic field versus local time at ATS-1. The quantities plotted are the medians of all data for the four seasons: northern winter, northern summer and the seasons of the northern spring and autumn equinoxes. The ATS-1 measurements were taken in the rotational *D, H, V* system. For ATS-1, the rotational *D* and *V* components do not contain any spacecraft field. However, no such statement can be made about rotational *H*. Thus, the baseline for rotational *H* was arbitrarily chosen so that the value at local noon for a quiet day average is 120γ [McPherron, 1971]. This value was selected on the basis of certain early models, and any error associated with this choice will result in errors in *both* dipolar *H* and *D*. We will discuss this matter later in connection with a comparison of observational and model data.

There are several features of the plots in Figure 2 that we wish to point out. First, it is clear that the nightside section of the diurnal variation in *V* exhibits a strong seasonal dependence. The plots show that ATS-1 is usually well above the effective or distorted geomagnetic equator. However, during the winter, it is closer to this surface of quasisymmetry and, in particular, near 2100 local time the spacecraft is below it.

The azimuthal variations in the *D* component are more or less antisymmetric about 10 local time and exhibit phase changes of approximately 180° between winter and summer. The negative offset in *D* from a zero mean is probably due to the error in rotational *H* discussed earlier.

The diurnal variations in *H* are clearly the dominant variations in the field at ATS-1 [Olson & Cummings, 1970; Cummings *et al.*, 1971; McPherron *et al.*, 1970; McPherron, 1975]. The offset of the curve of *H* versus local time for northern winter is probably another manifestation of an error in rotational *H*. More recent models of the magnetic field, which will be discussed in additional detail below, suggest that

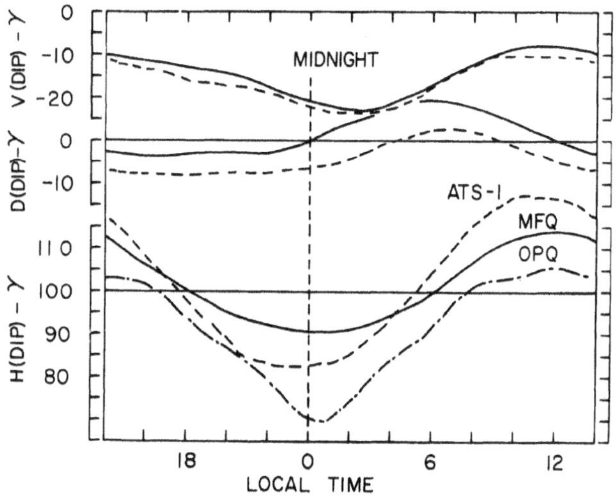

Figure 3. *A comparison of the observed medians during quiet times at the summer solstice with two models of the field at ATS-1.*

our choice of a noontime value for the normalisation of H is approximately 20γ too great. In Figure 3, median values of the vector components at ATS-1 recorded during quiet times are compared with two more recent models. The curve labelled 'MFQ' is based on the Mead-Fairfield model for quiet times [Mead & Fairfield, 1975], that labelled 'Olson' is based on the Olson-Pfitzer model [Olson & Pfitzer, 1974]. It is clear from Figure 3 that a reduction by 15γ of the observations of H would bring the measured values near local noon and midnight to essentially those given by the Olson-Pfitzer model. Further, except in the range of local time from 1600 to 2000, the observed and computed values compare well.

The Olson-Pfitzer model does not exhibit the diurnal variations in either D or V because in that model the dipole is parallel to the spin axis and the magnetic equatorial plane is always the plane of symmetry. However, if we apply the 15γ correction to H as suggested from the comparison of the observed variations in H to those calculated by Olson & Pfitzer, and if we assume that the 15γ difference was due to an error in our choice of the baseline for rotational H, then we must correct the baseline of D by approximately 3γ. The net result is that both D and V as observed at ATS-1 agree rather well with the values calculated from the Mead-Fairfield model. On the other hand, it is clear that the diurnal variation in H exhibited by the Mead-Fairfield model is considerably less than that observed.

In summary, then, the effects of the dipole tilt in the Mead-Fairfield model appear to account for most of the diurnal variations in V and D at the location of ATS-1 during quiet times. For the distribution of currents symmetrical across the noon-midnight meridian plane, the Olson-Pfitzer model is adequate to account for the corresponding variation in H.

Figure 4 shows the diurnal variation for the northern summer solstice at ATS-6. (For a description of this instrument and associated data processing see McPherron [1974] and McPherron *et al.* [1975].) The heavier curves show the averages of D, H, and V for the three quiet days of the first 49 days after launch. The lighter curves are medians of all data recorded during this period with the exception of the data recorded during storms.

Perhaps the most striking difference between the diurnal variations at ATS-1 and those shown here for ATS-6 is in the radial component – the V component – the amplitude of which is much greater at ATS-6. The D component is also greater in amplitude and similar in form. Apparently then, the effect of magnetic activity at ATS-6 is primarily to produce a more tail-like field. A comparison of these measured values with those calculated from the Mead-Fairfield model indicates that the model still fits the diurnal variation in the D component rather well. However, it no longer provides an adequate fit to V and, as before, the diurnal variation in H is considerably larger than that in the Mead-Fairfield model field.

Returning now to a further consideration of the field at ATS-1, consider the effects of substorm activity [Subbarao *et al.*, 1973]. Figure 5 is a plot of the H component at ATS-1 for six selected days. Three of these, shown by the lighter curves, are quiet days, and the other three, shown by the heavier curves, are disturbed days. From the figure it is clear that the major effect of substorm activity is to cause relatively large and abrupt changes in the H component at ATS-1. None of the models available to date includes the current distribution sufficiently variable in time and/or extent to exhibit such changes.

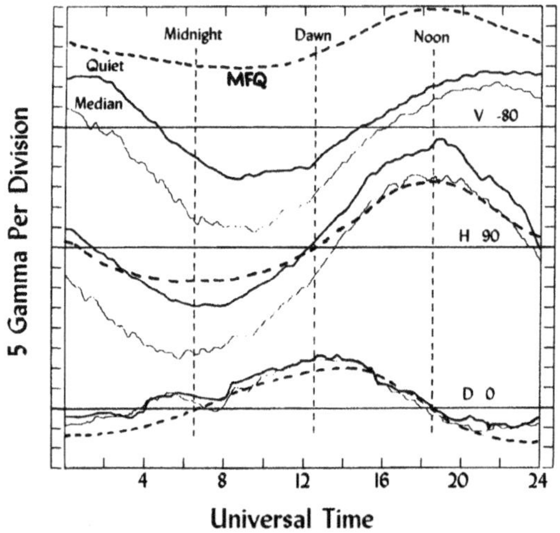

Figure 4. *A comparison of the Mead-Fairfield quiet field with the quiet and median magnetic field observed by ATS-6 at summer solstice.*

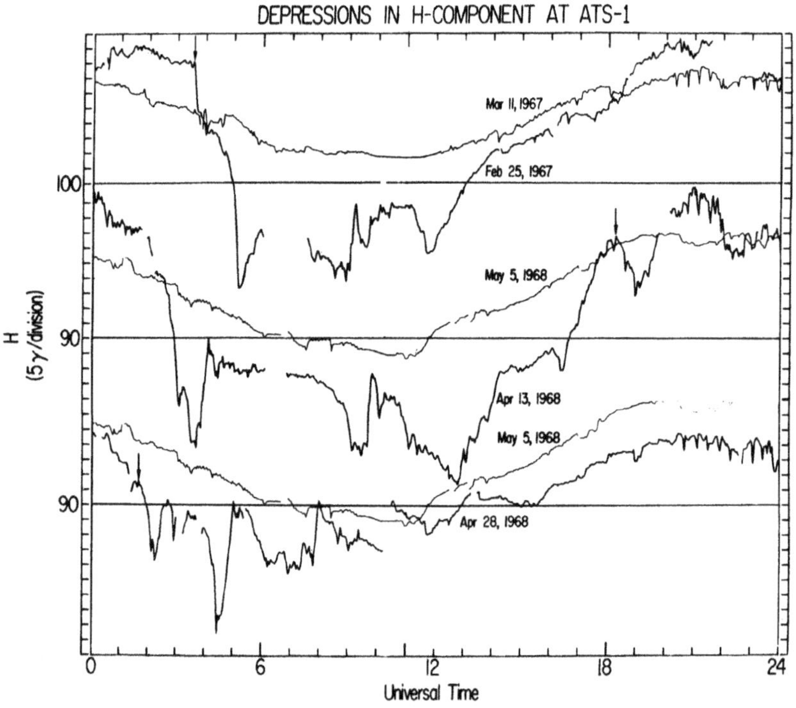

Figure 5. *Effect of substorm activity on the H component at ATS-1*

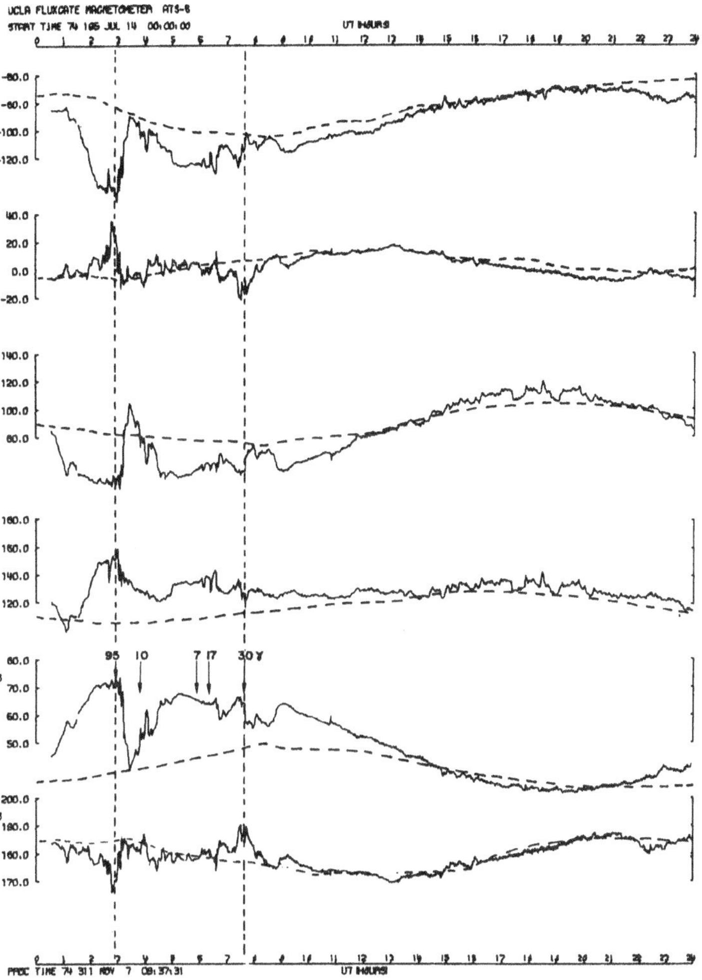

Figure 6. Effect of substorm activity in all field components at ATS-6.

The event indicated by the arrow near the beginning of the record for 11 March 1967 marks a negative Si which decreases the field from a previously enhanced value. This value was marked by a southward turning of the interplanetary magnetic field and is interpreted in terms of earthward motion of the tail current and plasma injection that begins with this event.

By comparison, Figure 6 shows the effects of substorms at the location of ATS-6 [McPherron *et al.*, 1975]. Since this spacecraft is well above the magnetic equator, all three vector components are changed substantially in the course of a substorm. The upper three traces are the Cartesian components of the magnetic field in the dipolar co-ordinate system. The lower three traces give the magnitude of the field and its orientation in spherical polar co-ordinates in which the polar axis is parallel, or more specifically anti-parallel, to the dipole axis.

Before a major substorm onset, indicated by the vertical dashed line, the field develops a configuration that is very tail-like. Thus, V and H both decrease. This effect is evident in the curve showing the polar angle θ as a function of time. Just after the onset of a substorm, the field returns rapidly to the more dipolar configuration indicated by the dashed curves characteristic of quiet days.

Another interesting effect is the substantial variation in the azimuthal component D. It is positive for the pre-midnight substorm and negative for the post-midnight substorm, and would appear to be consistent with the variation in the surface magnetic field attributed to Birkeland currents feeding the westward electrojet.

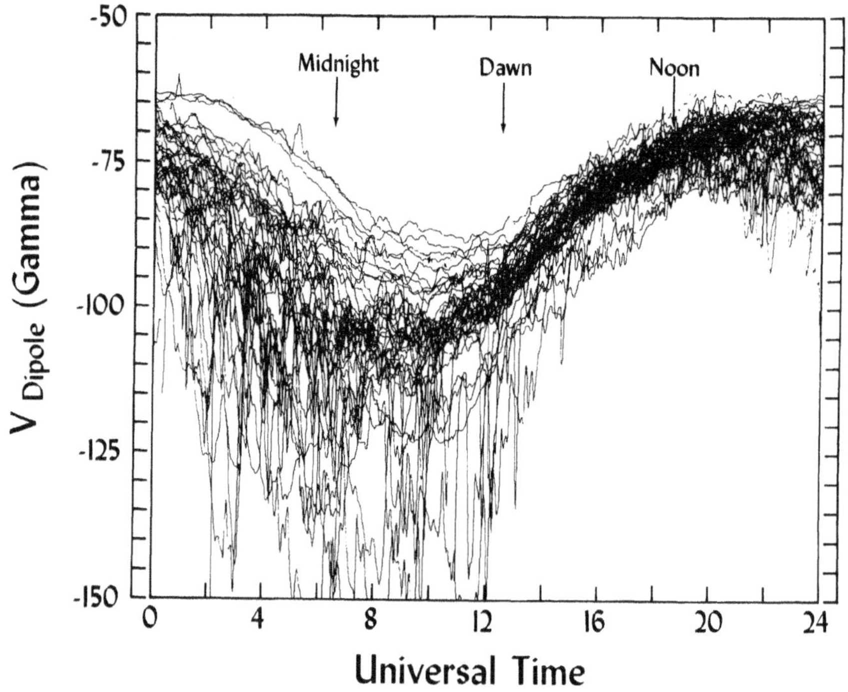

Figure 7. *Transient disturbances in V component at ATS-6 caused by substorm activity.*

Figure 7 shows the measurements of the V component plotted versus universal time with the records for each superimposed upon one another. This presentation shows rather strikingly that fluctuations of the V component at ATS-6 are primarily confined to the night sector. Further, the plot shows that the magnitude of the perturbations is comparable to the quiet-day field strength itself. Thus, the fluctuation amplitudes are in excess of 50γ in a 100γ field.

Figure 8 shows the effects of magnetic activity on the diurnal variations at ATS-1. The figure has been constructed from the two-year record from the ATS-1 magnetometer for the period 1967–1968. This record was divided into subsets by season of the year and for each season by level of geomagnetic activity. The data were sorted according to whether the AE index lay in intervals 0–65γ, 65–130γ, 130–260γ, 260–520γ, and > 520γ. Medians were constructed for each of these subsets of the data.

This presentation shows first the pronounced variation in the nightside decrease in *H* with geomagnetic activity, and second the trend for the minimum to move away from midnight towards earlier local time with increasing geomagnetic activity. By comparison, the effect of geomagnetic activity on the transverse components of the field is relatively much smaller until the highest level of magnetic activity is reached. At such levels, perturbation is strongly dependent upon season.

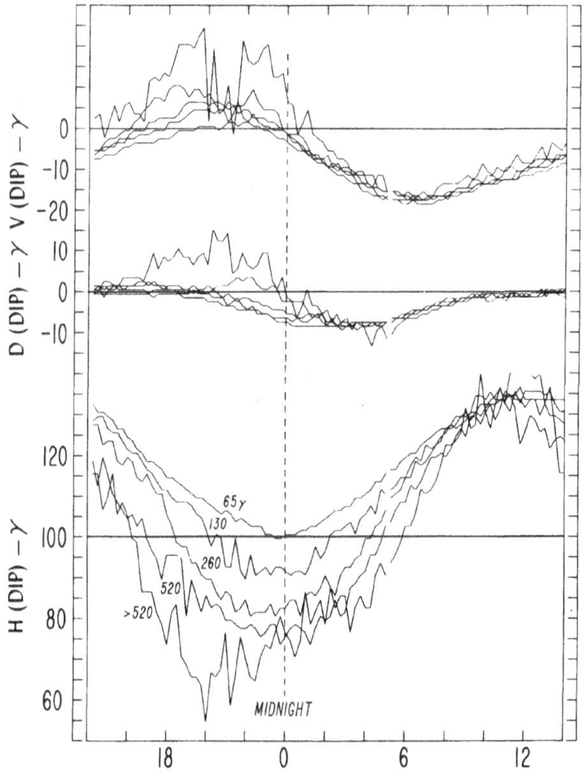

Figure 8. *A statistical study of the effects of magnetic activity on the magnetic field at ATS-1. Each trace is the median of all observations made when the AE index fell within intervals shown by labels on H traces (0–65γ, 65–130 etc.).*

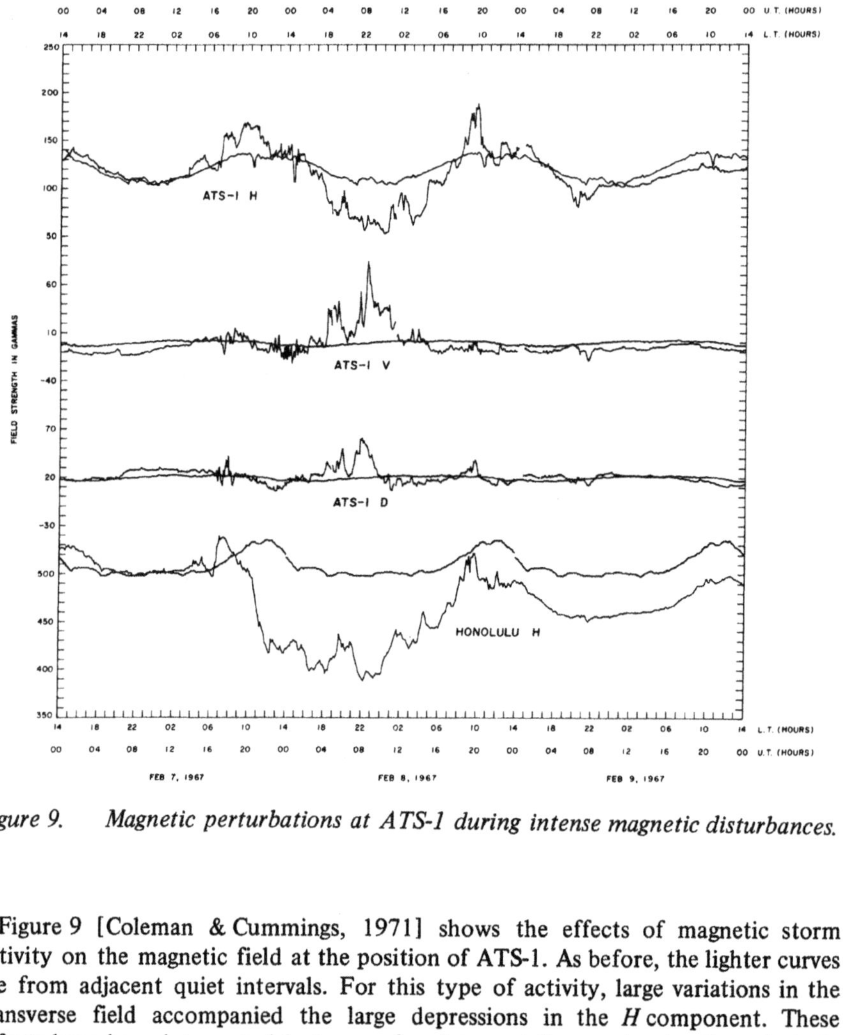

Figure 9. *Magnetic perturbations at ATS-1 during intense magnetic disturbances.*

Figure 9 [Coleman & Cummings, 1971] shows the effects of magnetic storm activity on the magnetic field at the position of ATS-1. As before, the lighter curves are from adjacent quiet intervals. For this type of activity, large variations in the transverse field accompanied the large depressions in the *H* component. These effects have been interpreted in terms of an intense tail current extending relatively close to the Earth.

3. ULF Waves at Synchronous Orbit

A wide variety of ULF waves have been detected at the synchronous orbit. The occurrences of several different types appear to be related to substorms. In general, each type occurs during a particular phase of the substorm and in a particular range of local times.

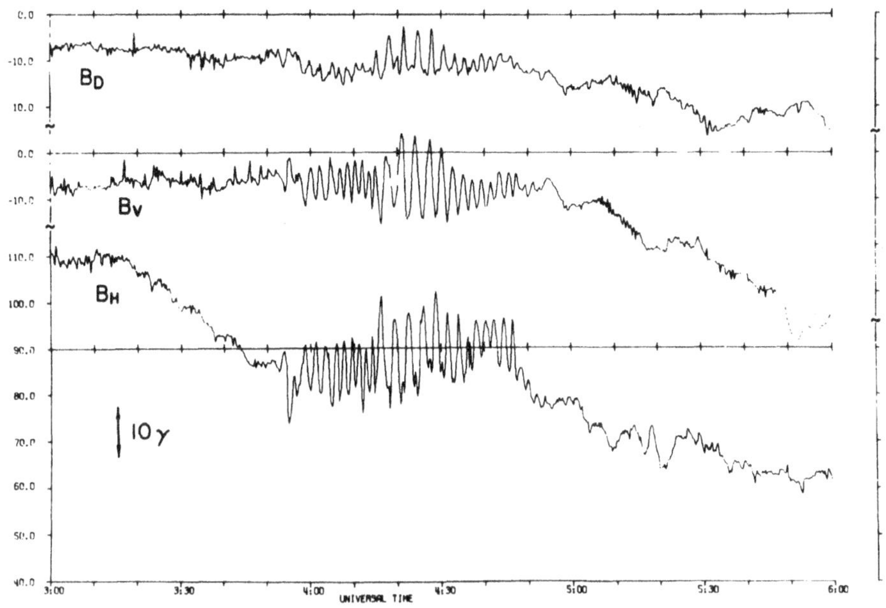

Figure 10. *ULF waves in the Pc 5 band. The perturbations are primarily confined to magnetic meridian planes (V–H).*

Figure 10 is the section of the record from ATS-1 which contains an example of mixed-mode Pc 5 waves. This is a wave mode we have previously described as a storm-time phenomenon [Barfield & Coleman, 1970; Barfield & McPherron, 1972a; Barfield *et al.*, 1972]. Here it was detected near the dusk meridian about 30 min after a substorm onset at midnight. These waves have a relatively long period (approx. 200 s) and relatively large amplitudes (approx. 10γ). They are linearly polarised in the magnetic meridian and the perturbation vector is typically at 45° to the ambient field. Further, it appears that they are almost always associated with rapid decreases in the ambient field.

Figure 11 shows the results of a spectral analysis of this mixed-mode Pc 5 wave. We have performed eigen and coherency analysis on the spectral matrix computed from the subsection of the record shown in Figure 10 for the interval 0350–0450 UT. The left-hand panel shows the trace and eigenvalues of the real part of the spectral matrix. In this instance, harmonic structure is clearly evident and this property is frequently observed. The panel on the upper right shows that the fluctuations are strongly polarised and that on the bottom right shows that the polarisation is essentially linear. We find, however, that the perturbation vectors associated with the harmonics, although generally contained within the magnetic meridian plane, may otherwise be differently oriented. Figure 12 is an example of Pc 1 waves that frequently occur in coincidence with mixed-mode Pc 5 waves [Barfield & McPherron, 1972b]. The middle curve shows the Pc 5 waveform that was

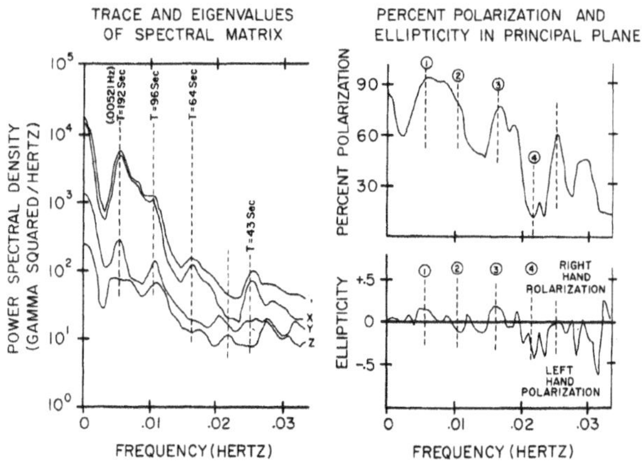

Figure 11. *Eigen and coherence analysis of spectral matrix for event of Figure 10 showing harmonic structure of waves.*

detected in another event near the dusk meridian after the onset of a substorm expansion. The top and bottom curves show that the Pc 5 fluctuations were accompanied by Pc 1 variations with a period of about 5 s. Such Pc 1 disturbances are frequently seen at the dusk meridian [Bossen *et al.*, 1975]. They are usually detected after substorm onsets, they generally have an amplitude of about 2γ, their periods range from 2–10 s, and they are left-hand elliptically polarised transverse to the ambient magnetic field. Figures 13 show the dynamic spectra of Pc 1's observed simultaneously at ATS-1 and its approximate conjugate point [Bossen, 1975]. The upper dynamic spectrum is of the event as recorded at ATS-1, the lower spectrum is that of the event as recorded at Tungsten, Northwest Territories, Canada. Our results indicate that whenever simultaneous data are available and Pc 1 is detected at ATS-1, it is also detected at Tungsten. The opposite is not true, however. Usually, the spectra of the surface field show upward sweeps in frequency, i.e. they show that the Pc 1's are of the IPDP type. However, at the satellite, the fluctuations generally show a decrease in frequency with time. This effect is probably associated with the fact that the spacecraft is moving in the direction of weaker magnetic fields as the satellite moves from the day side to the nightside. Figure 14 shows examples of Pi 2 bursts recorded simultaneously at ATS-1 and Tungsten [Arthur & McPherron, 1974a]. This particular record was obtained on 17 September 1969. In the left-hand panel are shown first derivatives of the 1 s averages of the three components of the field at ATS-1. Although there is a strong contribution from noise at 5.12 s produced by the telemetry system, the Pi 2 fluctuations are clearly observable. The three plots for Tungsten include two components of dB/dt measured with search coils and the first derivative of the Z component of the field measured with a fluxgate magnetometer. This last quantity was not used in the analysis. In fact, the fluxgate data were not used here because only two components of the field were measured during this event.

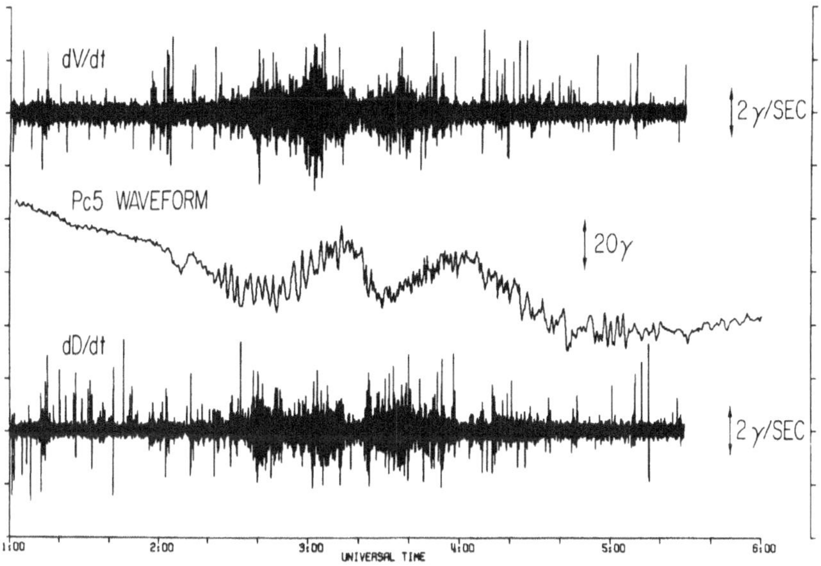

Figure 12. *Pc 1 waves (IPDP) which frequently accompany Pc 5 waves of the type shown in Figure 10. The Pc 1 waves are transverse and left elliptically polarised.*

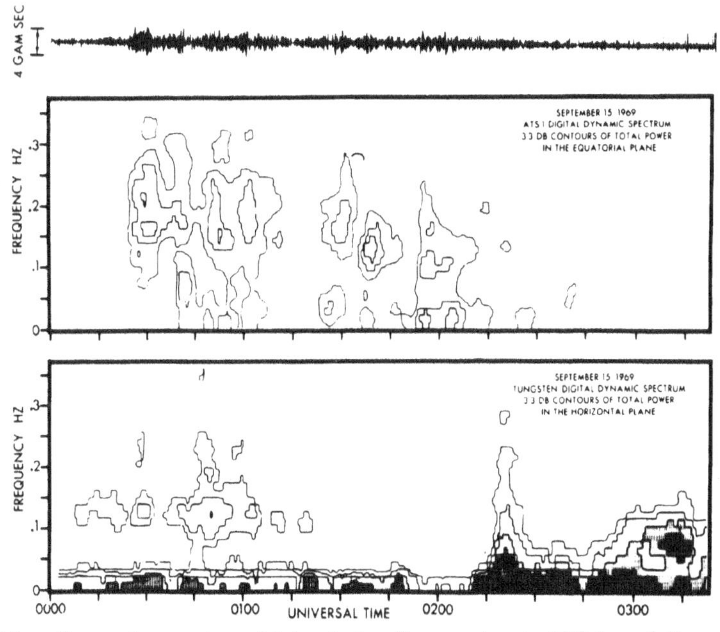

Figure 13. *Dynamic spectra obtained simultaneously at ATS-1 and its northern conjugate point, Tungsten, NWT, Canada, during a Pc 1 event.*

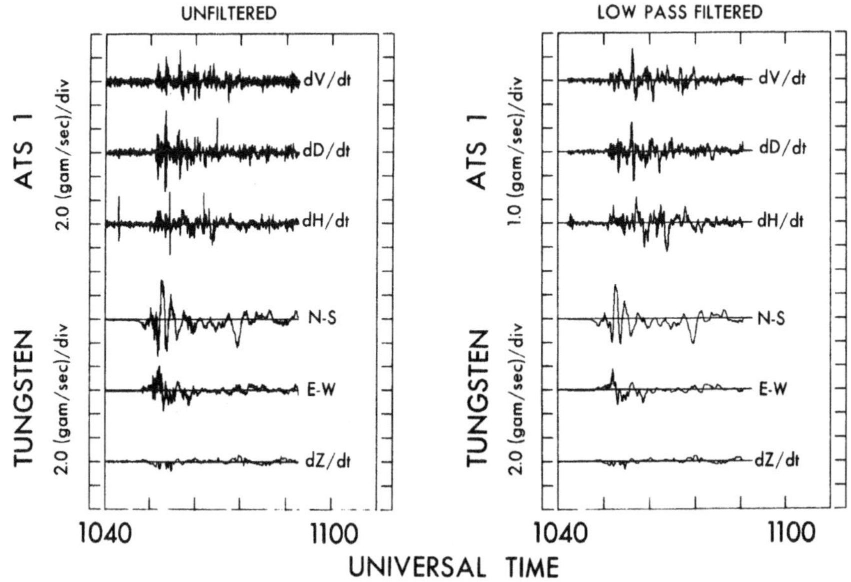

Figure 14. Example of Pi 1 bursts seen simultaneously at ATS-1 and its conjugate point, Tungsten.

The Tungsten data show clearly, especially in the east-west component of the time derivative of the field, the 0.3 Hz enhancement found by Heacock [1967] to be a phenomenon that commonly accompanies Pi 2.

Figure 15 shows dynamic spectra of this Pi 2 event. The records actually analysed are those of the horizontal component at Tungsten and the component in the *V-D* plane at ATS-6. At both locations, the event is a broadband burst. The frequency range characteristic of Pi 2, roughly 0.2 Hz or 50 s period, shows a strong enhancement at about 1045. The enhancement near 0.3 Hz is also clearly apparent in the dynamic spectrum from Tungsten. However, a similar phenomenon at ATS-1 appears to be at a somewhat higher frequency. The line at 0.2 Hz in the ATS-1 dynamic spectrum is the telemetry interference mentioned previously. We should point out that for the two other Pi 2 bursts studied to date, the high-frequency enhancement is at roughly the same frequency at both spacecraft and Tungsten. In any case, from the simultaneity of the bursts in the Pi 2 band and those in the band near 0.3 Hz, it strongly suggests that they are related.

Figure 16 is a section of the record from ATS-1 showing the turbulent noise often detected near midnight during the expansion phase of a substorm [McPherron & Coleman, 1970]. Prior to the onset of the substorm at 1310 UT there was a relatively steady decrease in the *H* component of the magnetic field. The onset was followed by an increase in *H* which was accompanied by a large amplitude fluctuation, or turbulence. These fluctuations are usually largest in the *H* component. The spectra fall rapidly toward higher frequency, so that there are no characteristic

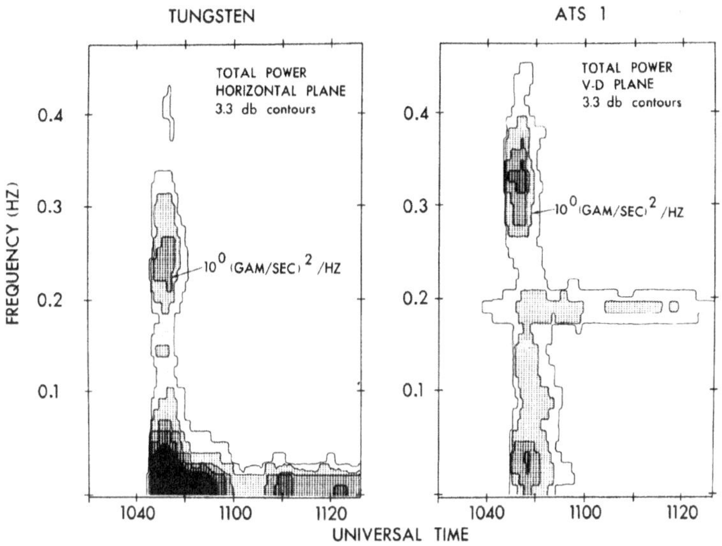

Figure 15. *Dynamic spectra of the Pi 2 bursts of Figure 14, showing characteristic enhancement of activity at about 0.3 Hz.*

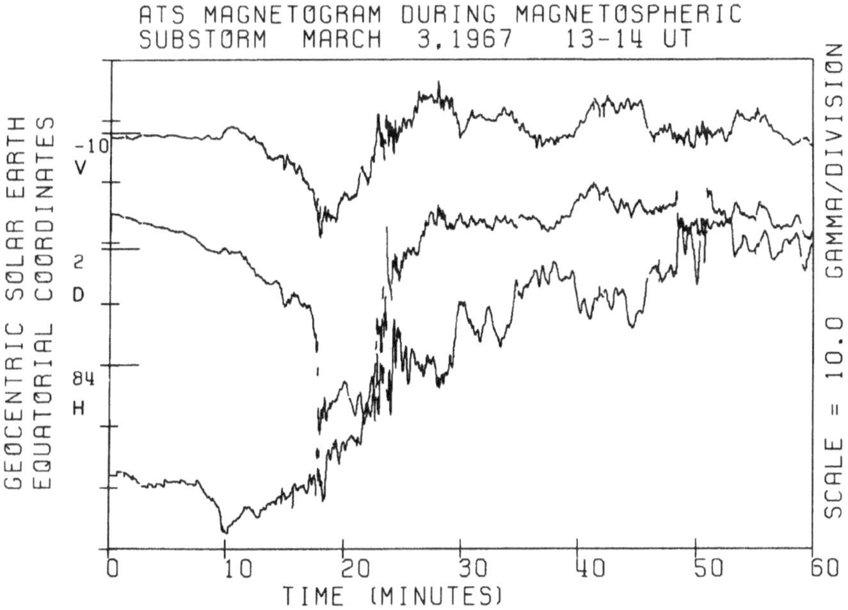

Figure 16. *Turbulent wave activity associated with a substorm expansion at ATS-1.*

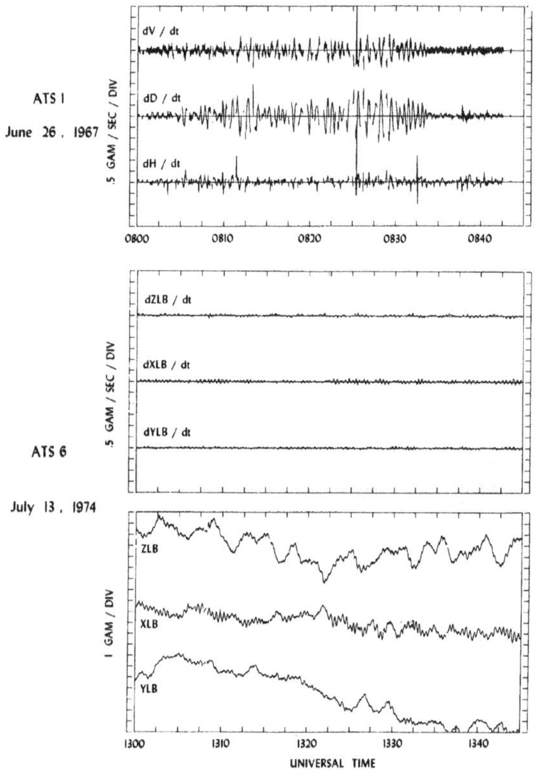

Figure 17. *Pc 3 wave activity seen at ATS-1 and ATS-6 at different times. The ATS-1 event is very strong and superimposed on large-amplitude spacecraft noise.*

frequencies associated with them. Figure 17 shows examples of Pc 3 at synchronous orbit [Arthur & McPherron, 1974b]. The top panel shows time derivatives of 2 s averages of the three components, *V, D, H,* in the dipole system. This record was obtained on 23 June 1967. The event is one of unusually large amplitude. Thus, the fluctuations were roughly 10γ peak-to-peak, whereas 2–4γ variations are most common. Since all the power is in the *V* and *D* components, the phenomenon is clearly essentially transverse. The bottom panel shows a much weaker event recorded at ATS-6. Two-second average values are plotted here, and the amplitude is typically about 1γ peak-to-peak. The ATS-6 data are plotted in a co-ordinate system fixed relative to the spacecraft. However, for ATS-6, the *X*-direction is approximately that of *D*, the *Y*-direction approximately that of –*H*, and the *Z*-direction approximately that of – *V*. The centre panel shows the time derivative of the records plotted in the lower panel. It is clear that the event is much weaker than that at ATS-1. It is also clear that the noise level of the ATS-6 system is significantly lower than that of ATS-1.

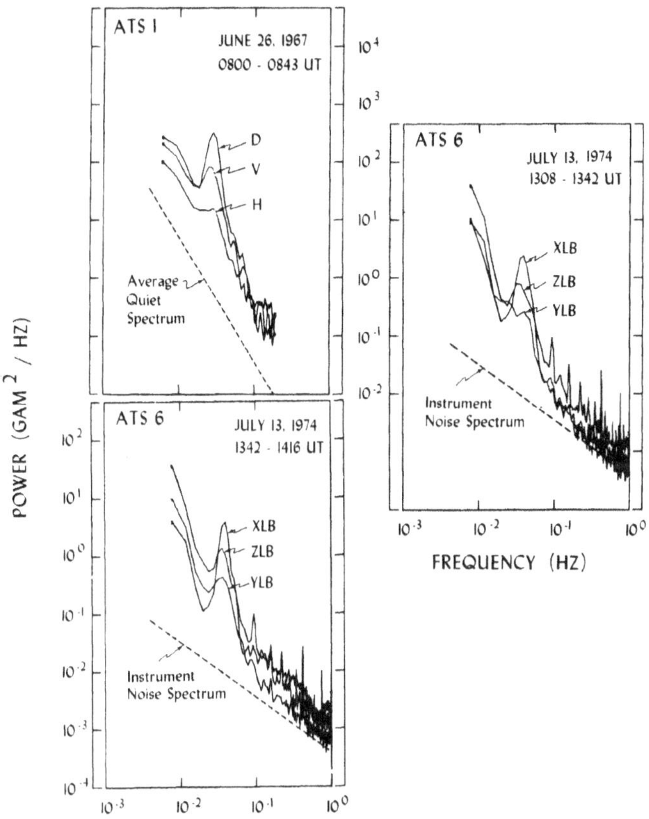

Figure 18. *Auto spectra for the two Pc 3 wave events in Figure 17. Dashed lines show the average spacecraft noise spectrum at ATS-1 and the instrument noise spectrum at ATS-6.*

Figure 18 shows auto spectra of the three components of the field from each of the records shown in Figure 17. In the upper left-hand corner is the spectrum for approximately 45 min of the ATS-1 record. In the panels at the centre right and lower left are spectra for two consecutive sections of the ATS-6 record. Notice that the spectra are qualitatively very similar. Further, the peak frequencies are rather close, with that for ATS-1 being at 36 s and that at ATS-6 27 s. There is a difference in the Nyquist frequencies of the ATS-1 and ATS-6 spectra because of the different time resolutions employed in the records that were analysed. Furthermore, there is of course a great difference in the power in the two events. In fact, the event at ATS-6 would hardly have been detected with the ATS-1 system because of the spacecraft noise indicated by the dashed line. Figure 19 shows dynamic spectra from simultaneous records recorded at Tungsten and ATS-1 on 18 September 1969 [Arthur & McPherron, 1973]. A Pc 3 event of interest occurred between 0800 and 1100 LT, or 1800 and 2100 UT. The dynamic spectra show bursts in the frequency band about 0.04 Hz essentially simultaneously at both the spacecraft and Tungsten.

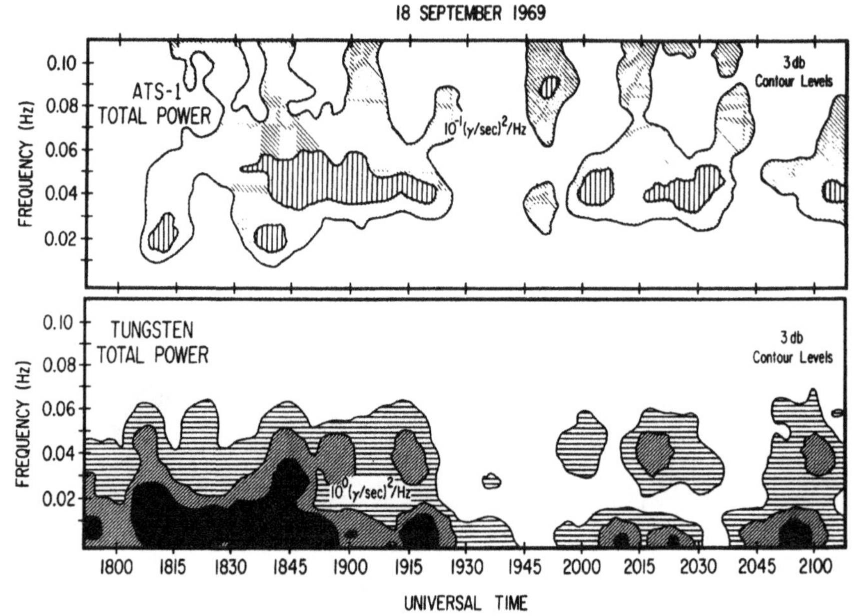

Figure 19. *Dynamic spectra of simultaneous observations of Pc 3 waves at ATS-1 and Tungsten.*

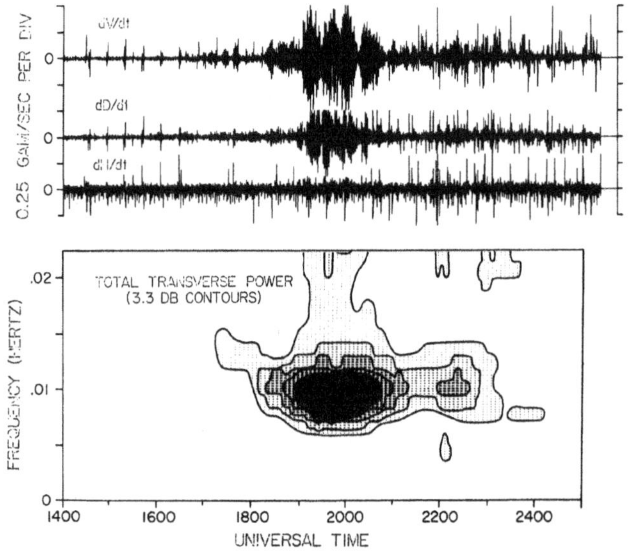

Figure 20. *ULF waves in the Pc 4 band with linear polarisation in the radial direction.*

The spectrum for Tungsten also shows relatively stronger bursts at lower frequencies.

Only one Pc 3 event (10–45 s period) has been detected in the simultaneous records of ATS-1 and Tungsten. Our difficulty has been in the fact that during 1969, ATS-1 was often turned off when it was in the local time sector where these waves were most commonly observed on the ground.

The records from College, Alaska, which is on the same L-shell as Tungsten, cover an interval which included 114 events at ATS-1. This is the record section, for the most part, from 1967. Of these 114, 90 were accompanied by simultaneous activity of similar character in the College record.

Figure 20 is an example of micropulsations in the Pc 4–5 band [Cummings *et al.*, 1969]. The data are taken from the record of ATS-1 for 18 December 1966. The upper panel shows the first derivatives of the vector components and the lower panel shows the dynamic spectrum of these fluctuations. This phenomenon is one associated with periods of geomagnetic quiet, but is primarily detected during the recovery phases of magnetic storms. Further, it is most frequently observed in the early afternoon. Fluctuations are always transverse to the field and linearly or highly elliptically polarised with the major axis in the radial direction.

4. Discussion

To date, we have studied in some detail quiet-time transverse Pc 4, mixed-mode Pc 5, Pc 1, and Pc 3. We have examined in only a cursory fashion Pi 2 bursts and the 3 s oscillations superimposed thereon. Further, the turbulent fluctuations associated with abrupt recoveries in the H component have not been examined in detail.

Other wave phenomena that have been observed, but are not discussed here, include purely compressional waves in the Pc 4 band, and pearl pulsations.

Clearly, a great deal remains to be done in the study of these magnetic field records. Even more work lies ahead in the studies of the interaction of the waves, the properties of which are indicated in the records, with particles in the geomagnetic cavity. At the same time, it is important to extend the correlation studies that involve records of the magnetic field at the surface of the Earth.

Finally, we wish to point out that the phenomena discussed herein are transient variations normally classified as geomagnetic activity and a part of the rich spectrum of wave-like phenomena in the geomagnetic cavity. Thus, we have neglected the effects at the synchronous orbit associated with processes that bring the magnetopause earthward to this geocentric distance on the dayside as well as the effects at the synchronous orbit associated with the earthward extension of the tail lobes to this distance.

5. Acknowledgements

Support for the development and construction of the UCLA fluxgate magnetometers on ATS-1 and ATS-6 was provided by NASA under Contracts NAS 5-9570 and NAS 5-11674. The ATS-1 magnetometer was built by Marshall Laboratories,

Torrance, California, and the ATS magnetometer by the Westinghouse Corporation, Baltimore, Maryland. Analysis of data from these spacecraft has been supported by the above contracts as well as NASA NGL 05-007-004 and by ONR grant N00014-75-C-0596. Support for the magnetic observatory at Tungsten, Northwest Territories, Canada, was provided by NSF grant GA 17978. The principal investigator for these projects is P.J. Coleman, Jr. Co-investigators include R.L. McPherron and W.D. Cummings. The analyses of ULF waves in the Pc 5, Pc 3 and Pc 1 bands have been, respectively, the dissertation projects of J.N. Barfield, C.W. Arthur and M. Bossen.

References

Arthur, C.W. & R.L. McPherron, Observations of 10–45 sec period micropulsations in the morning sector (extended abstract), in Proc. Chapman Mem. Symp. Magnetospheric Motions, 108–112, 1973.

Arthur, C.W. & R.L. McPherron, A preliminary study of simultaneous ground-satellite observations of substorm-associated Pi 2 micropulsations and their high frequency enhancement, IGPP Publ. 1266-43, UCLA, 1974a.

Arthur, C.W. & R.L. McPherron, Simultaneous observations of 10–45 sec period waves (abstract), *EOS Trans. Am. Geophys. Union* **56**, 1178 (1974b).

Barfield, J.N. & P.J. Coleman Jr., Storm-related wave phenomena observed at the synchronous equatorial orbit, *J. Geophys. Res.* **75**, 1943–1946 (1970).

Barfield, J.N. & R.L. McPherron, Statistical characteristics of storm associated Pc 5 micropulsations observed at the synchronous, equatorial orbit, *J. Geophys. Res.* **77**, 4720–4733 (1972a).

Barfield, J.N. & R.L. McPherron, Investigations of interaction between Pc 1-2 and Pc 5 micropulsations at the synchronous orbit during magnetic storms, *J. Geophys. Res.* **77**, 4707–4719 (1972b).

Barfield, J.N., R.L. McPherron, P.J. Coleman Jr. & D.J. Southwood, Storm associated Pc 5 micropulsation events observed at the synchronous equatorial orbit, *J. Geophys. Res.* **77**, 143–158 (1972).

Bossen, M., Substorm-associated Pc 1 micropulsations at ATS 1 and its conjugate point, IGPP Publ. 1418-75, UCLA, 1975.

Bossen, M., R.L. McPherron & C.T. Russell, A statistical study of Pc 1 micropulsations at synchronous orbit, IGPP Publ. 1400-74, UCLA, 1975.

Coleman, P.J. Jr. & W.D. Cummings, Storm time disturbance field at ATS-1, *J. Geophys. Res.* **76**, 51–62 (1971).

Cummings, W.D., R.J. O'Sullivan & P.J. Coleman Jr., Standing Alfvén waves in the magnetosphere, *J. Geophys. Res.* **74**, 778–793 (1969).

Cummings, W.D., P.J. Coleman Jr. & G.L. Siscoe, Quiet day magnetic field at ATS 1, *J. Geophys. Res.* **76**, 926 (1971).

Heacock, R.R., Two subtypes of Pi micropulsations, *J. Geophys. Res.* **72**, 3905–3917 (1967).

McPherron, R.L., Correction and analysis of ATS 1 magnetometer data during 1971, IGPP Publ. 995, UCLA, 1971.

McPherron, R.L., Coordinate transformations used in processing ATS 1 magnetic field data, IGPP Publ. 1053, 1973.
McPherron, R.L., Progress report: UCLA fluxgate magnetometer on ATS 6 for the period April 1–September 1, 1974, IGPP Publ. 1388-60, UCLA, 1974.
McPherron, R.L., The quiet time magnetic field at synchronous orbit (extended abstract), *EOS Trans. Am. Geophys. Union*, 1975.
McPherron, R.L. & P.J. Coleman Jr., Magnetic fluctuations during magnetospheric substorms, 1. Expansion phase, *J. Geophys. Res.* **75**, 3927–3931 (1970).
McPherron, R.L., C.T. Russell & P.J. Coleman Jr., Comparison of the Olson field model to quiet day magnetic field observations at ATS 1 (extended abstract), in Quantitative Magnetospheric Models, Proc. ESSA Symp., Boulder, Colorado, 1970.
McPherron, R.L., P.J. Coleman Jr. & R.C. Snare, The UCLA fluxgate magnetometer on ATS 6, *IEEE Trans. Aerospace Electron. Syst.*, 1975.
Mead, G.D. & D.H. Fairfield, A quantitative magnetospheric model derived from spacecraft magnetometer data, *J. Geophys. Res.* **80**, 523 (1975).
Olson, W.P. & W.D. Cummings, Comparison of the predicted and observed magnetic field at ATS 1, *J. Geophys. Res.* **75**, 7117–7121 (1970).
Olson, W.P. & K.A. Pfitzer, A quantitative model of the magnetospheric magnetic field, *J. Geophys. Res.* **79**, 3739 (1974).
Subbarao, S., R.L. McPherron & N.E. Cline, Magnetic field depressions at synchronous orbit: A spatial or temporal effect? (extended abstract), in Proc. Chapman Mem. Symp. Magnetospheric Motions, 201–204, 1973.

DISCUSSION

D. Jones: In the events showing harmonic structure, was there any visible trend with frequency in the polarisation angle with respect to the magnetic field.

P.J. Coleman: I have not examined the results for any systematic variation with frequency of the orientation of the polarisation direction. This is partly because many of the events do not exhibit harmonic structure in the spectra.

K. Knott: How good was the magnetic conjugacy between Tungsten and ATS-1? Did you have only one ground observing station or were there more?

P.J. Coleman: We had only one station at Tungsten. There were no others within several hundred kilometres. Consequently, I cannot give you a quantitative answer based on ATS-1/Tungsten correlation studies.

G. Rostoker: It is not surprising to see for a Pi2 only a limited spectral range excited at ATS-1, whereas you see a large number of spectral components on the ground at Tungsten. Using the University of Alberta meridian line, Olson and Rostoker [*Planet. Space Sci.*, 1975] find that, for Pi2 pulsations, you may see only one or two spectral components away from the auroral electrojet, whereas in the electrojet regime the entire Pi2 spectral band is filled. Since only one or two spectral components may originate in the magnetosphere, and they may be confined to a limited L-shell region, the satellite has to be placed in that L-shell region if it is to detect any pulsation at all and it may not see any of the 'noise' in the Pi2 band at all since this may originate at ionospheric level.

P.J. Coleman: I have no further comment.

K. Wilhelm: Have micropulsations in the Pi1 class been observed by ATS magnetometers?

P.J. Coleman: Our study of fluctuations in this frequency range to date has yielded no clear examples of Pi1 at the synchronous orbit.

SUMMARY OF THE LA JOLLA CONFERENCE ON QUANTITATIVE MAGNETOSPHERIC MODELS

W.P. Olson

McDonnell Douglas Astronautics Company,
Huntington Beach, California, USA

Abstract

Efforts to quantitatively model magnetospheric phenomena were classified and examined in three broad areas: magnetic fields, electric fields, and charged particles. Observational data sets needed for the development of these models and future models were also described. Because of its relative stability, the magnetospheric magnetic field has been described more satisfactorily than the other magnetospheric field or particle features. It is presently quantitatively understood fairly well in terms of its source currents. The magnetospheric DC electric field, however, is at present usually described without regard for its various sources. Quantitative modelling of this field has begun, but is in its infancy. Very few models of magnetospheric particle populations exist and there is currently no quantitative 'magnetospheric model' that attempts to consider gross plasma and field dynamics simultaneously. Several tests of the quantitative models were described and the shortcomings of existing models pointed out. Recommendations for the extension of existing models and the development of new models during the IMS era were made and examined in the light of planned satellite and ground-based experiments. Approximately one-third of the meeting was devoted to working-group-type activities. During these working-group sessions, the accuracy of existing models was discussed and comments and recommendations made for the development of new ones.

1. Introduction

The meeting was held in La Jolla Shores, California, on 6–8 May 1975. There were five sponsors: the American Geophysical Union, The American Institute of Aeronautics and Astronautics, the Air Force Office of Scientific Research, the Office of Naval Research, and McDonnell Douglas Astronautics Company. The meet-

Knott & Battrick (eds.), The Scientific Satellite Programme During the International Magnetospheric Study, 367–380.
All Rights Reserved. Copyright © 1976 by D. Reidel Publishing Company, Dordrecht-Holland.

ing was divided roughly into three parts, with one day devoted to each. Further, each day was divided such that approximately two-thirds of the time was devoted to presentation of papers, while one-third was used for discussions and working-group activities. The first day was devoted to magnetic fields, the second day to electric fields, and the last day to charged particles. This summary will roughly follow the order of the meeting. Many of the comments made concerning the magnetic field apply equally well to the electric field. There were 85 registered attendees with six foreign participants. One of the purposes of the meeting was to bring together those people working on the actual development with a cross-section of those workers collecting observational data. The data are necessary for the proper development of empirical models. There were also several model 'users' present. Their current and future needs make it possible to test existing models and to suggest which future models will be most valuable. The meeting discussions were limited to *quantitative* models and associated observational data sets. A quantitative model was defined as one that yields numerical information which agrees well with quantitative observations of a magnetospheric feature.

2. Magnetic Fields

In contrast to the electric field, the magnetospheric magnetic field is relatively well-behaved and its variability fairly well understood. Modelling of magnetospheric magnetic fields is well ahead of modelling of the magnetospheric electric field and various particle structures in the magnetosphere. This is, in part, because the magnetic field is measured directly (electric fields are still typically inferred from particle motions). Also the sources of the magnetospheric magnetic field, the several magnetospheric current systems, are fairly well known and understood. The most important are: the magnetopause currents, formed by the deflection of shocked solar-wind particles off the magnetospheric magnetic field, the tail current system, produced mainly by charged particles drifting across the plasma-sheet region of the tail, and the ring-current system which is formed by trapped particles drifting around the Earth. Sugiura has suggested that most of the time, the ring-current system is contiguous with the tail current system. There are many data sets that have been used to develop the current models of the magnetospheric magnetic field, **B**. Sugiura, Hedgecock, McPherron, and Cain talked about observations of **B**. Cain discussed the modelling of the Earth's main field and its importance for magnetospheric phenomena as was described at the Zmuda Memorial Conference. A key point for magnetospheric physicists was that the present International Geomagnetic Reference Field currently only models 'structures' with longitudinal extent greater than 45°. This is probably inadequate for the proper quantitative study of high-latitude magnetospheric (and ionospheric) phenomena. Sugiura discussed the procedure of subtracting the magnitude of the Earth's main field from the magnitude of the observed field to obtain ΔB. It has become an important tool in understanding the magnetic effects of currents flowing in the inner magnetosphere. Magnetic observations at synchronous orbit also provide excellent tests of the models in that region of the magnetosphere. McPherron pointed out that the variations seen at synchronous orbit are quite sensitive to geomagnetic longitude. The analysis of the HEOS data by Hedgecock and his colleagues has shown that in the high-latitude, high-altitude magnetosphere, the magnetopause is more flared than predicted by the models whose shape was determined using the pressure-balance condition and the assumption of specular reflection. This data also suggests that although charged particles may enter the magnetosphere at high latitudes over a cleft or cusp-like region, the magnetic field topology in that region is more like a funnel in shape and does not extend more than two hours on either side of the noon magnetic meridian. Hedgecock's group has also initiated the use of a 'hinged' co-ordinate system that near Earth is like an ordinary magnetic co-ordinate system, while in different regions of the magnetosphere it becomes either a solar magnetic, or solar magnetospheric co-ordinate system. With the use of this co-ordinate system, it was possible to make a much better job of organising the magnetic field data. (This co-ordinate system takes into account 'tilt' effects. 'Tilt' refers to the varying angle between the geomagnetic dipole axis and the solar-wind direction). Several models of **B** were presented by their authors and reviewed by Walker. These included the models of Sugiura and Poros, Mead and Fairfield, Choe and Beard and Sullivan, Olson and Pfitzer, Willis and Pratt, Alekeev and Shabansky, and Voigt. A general set of requirements that an all-purpose model should satisfy includes the following. The magnetopause, tail and quiet-time ring currents must be included, as must the effects of tilt. The model must be good throughout a large portion of the magnetosphere. It must be capable of predicting B during quiet to moderately disturbed conditions. *None of the models of* **B** *developed to date meet all of these requirements.*

In addition to these general requirements, there are some more specific details that should probably be included in future modelling efforts. Potemra pointed out the importance of field-aligned currents in the high-latitude region. Although they are presently not well understood and their coupling to the magnetospheric current systems not presently known, it is clear that at high latitudes they exert some influence on the structure of the magnetic field. Likewise, at high latitudes, Hedgecock showed that the observed magnetospheric boundary is more distant than that given by the older models. Either a better theory for the shape of the magnetopause should be developed or these observations should be input to an empirical model. It is also becoming increasingly obvious that several magnetospheric phenomena are directly linked to the interplanetary magnetic field. It is therefore appropriate to develop a model that at once includes both the magnetospheric and interplanetary fields. However, the difficulties with the mathematical representation of such a field are formidable. Another requirement of these models is that they be differentiable. This is necessary if they are to be used to interpret the drift and energisation on charged particles. It was suggested that specialised models might be developed for various local regions of the magnetosphere. However, such models cannot be used conveniently with other models since adding them together and truncating outside their region of validity causes severe problems with their differentiability. The same is true of trying to represent the interplanetary field and **B** simultaneously.

There were some talks given on model uses. Seely reported that the Stanford group has used the Olson-Pfitzer model to locate the crossing of whistler ducts in the equatorial plane. McPherron showed the importance of including both the magnetic effects of the trapped currents and the effects of the tilt of the geomagnetic axis to the solar-wind direction for the proper description of the variations in **B** seen by ATS 1 and ATS 6 at synchronous orbit. The Mead-Fairfield model accurately represents daily variations in the observed field perpendicular to the northward direction which is most sensitive to tilt. Hendrickson told of several model tests provided by an electron echo experiment in which electrons were accelerated away from a rocket and later, after mirroring along magnetic field lines, detected by the same rocket. In order to observe the 'echo' it was necessary to predict accurately both the bounce time and the drift rate of the electrons. The Olson-Pfitzer model worked well for this purpose.

A general conclusion on the magnetic field modelling was that further advances should be directed toward increasing the general understanding of the magnetospheric field and increasing the gross accuracy of the models. Efforts that dwell on refining some of the details of present models might profitably be redirected.

It now appears to be well established that, magnetically, the magnetosphere is open. That is, a portion of the Earth's magnetic field is not contained in the magnetosphere, but passes through the magnetopause and is connected to the interplanetary magnetic field. It therefore would not, at first glance, appear profitable to try to develop closed magnetic field models (in which all field lines emanating from the Earth's surface return to the Earth's surface in the opposite hemisphere without penetrating the magnetopause). However, in order to determine magnetic-field properties and particle behaviour accurately near the boundary of the magnetosphere, it may be worthwhile to assume a fixed shape through which no field lines can penetrate, as was suggested by Roederer (this technique has been employed by Shabansky and his colleagues). An alternative to this method is to abandon the assumption of specular reflection off the magnetopause, but retain the pressure-balance assumption and use pressures obtained from a hydrodynamic flow model such as developed by Spreiter's group from wind-tunnel tests of flow past an object shaped to represent the magnetopause.

3. Electric Fields

The current situation for the magnetospheric electric field **E** is far different from that for **B**. Good empirical models of **B** and the source currents for **B** are fairly well understood. Progress on the development of theoretical **B** models is also substantial. However, new empirical models of **E** are just now emerging and the overall picture of **E** is still cloudy enough that no good theoretical models seem to be on the horizon. There are several reasons for this. First, throughout the magnetosphere, **E** is quite variable in both space and time, so it is difficult to use data to determine average characteristics of **E**. Secondly, most work on **E** relies heavily on its indirect measurements using particle data and inferring **E** from the **V** x **B** drift. Thirdly, there has only been limited success in 'mapping' the electric field from one region of the magnetosphere to another. Typically, measurement of the scalar potential associated with **E** is determined in the ionosphere. The assumption that field lines are equipotential is then made in order to map this electric field to the equatorial regions of the magnetosphere. There are at least three possible sources of difficulty associated with this mapping procedure. First, models of the magnetic field may not predict the observed magnetic field line geometry correctly. Secondly, if induced electric fields are present, the assumption that field lines are equipotentials is no longer valid, and thirdly, anomalous resistivity may be present which allows electric fields to have a component parallel to the direction of the magnetic field. Cauffman assessed the status of experiments in operation or planned for the direct measurement of DC electric fields using double probes. Many long-boom double-probe experiments will become operational during the IMS era, greatly increasing the number of observations of **E**.

However, at present, most DC electric-field observations are inferred from particle data. These measurements can be divided roughly into those made in the magnetosphere and those made in the ionosphere. Higby and Hones have used solar-particle motions in order to infer magnetospheric electric fields, while Carpenter and Rycroft have used whistler data in the plasmasphere to infer motions that are more dynamo-like than convection-like. For the ionosphere, Hellis has used Atmospheric Explorer data to develop three-dimensional particle velocity measurements and from them high-latitude convection patterns. The Chatanika data were discussed in some detail by the USCD group. Measurements of ion velocities with the Chatanika radar should make it possible to determine the role of discrete auroral arcs in the convection process at high latitudes. In a paper read by Murayama for Maezawa, work on 'reverse' current flow over the polar cap was reported. It implied the occasional existence of motions which are the reverse of the standard convection patterns inferred at high latitudes. Thus, polar-cap and plasmasphere studies have suggested deviations from the classical magnetospheric convection patterns and indicate once more the high degree of variability in the magnetospheric electric field, and also possibly some physical processes that are currently not well understood.

On the summary panel, Wolf indicated that there is now a fairly complete picture of the ionospheric electric field and it may be possible to develop an empirical model of **E** in the ionospheric regions. This model might be similar to McIlwain's electric-field model that is useful in equatorial regions of the magnetosphere and would have as inputs to it such variables as the boundary locations of the polar cap. Again, the usefulness of such a model would depend on the user needs. Any empirical model will not directly help to further our understanding of magnetospheric processes. However, deviations between such a model and observations may isolate important physical processes currently unobserved. Also, such a model could be used in conjunction with available magnetic-field models in order to study charged-particle phenomena in the inner magnetosphere and ionosphere. In the outer magnetosphere, however, the electric-field situation is much more complex, from both a spatial and a temporal point of view. There was general agreement that there is no chance that a theoretical model of **E** in the polar-cap and outer magnetospheric regions will be developed in the next few years.

Another way to contrast **B** and **E** is in terms of their sources. Although the source currents responsible for **B** are well known (but not directly measured), to date most work on **E** has looked only at the total field with little attention directed towards the isolation and quantitative description of its sources. Basically, there can only be three sources for **E**; charge separation, time-varying magnetic fields, and the relativistic transformation of the solar-wind magnetic and electric fields into a magnetospheric reference frame. In order to develop theoretical models of the magnetospheric electric field, it will be necessary to understand these sources of **E**. Of these three source mechanisms, probably those electric fields induced by time-varying magnetic fields are most amenable to understanding and quantitative modelling. Several sources of time-varying magnetic fields in the magnetosphere are known: changes in the stand-off distance of the magnetopause current system, substorm-related changes in the tail current system, storm time changes in the ring current, and the daily variation in the magnetic field produced by tilt effects. Hones and Burgeson, and Birmingham and Northrup did early work on this general problem. Recent procedures for calculating induced electric fields developed by Olson and Pfitzer have shown that even the daily wobble of the Earth's magnetic field plus the corresponding wobble in the magnetopause currents can produce a substantial induced electric field.

The same sort of comments as made for **B** apply to **E** as regards user needs. The models of **E** should be 'global', analytic, etc.

Emphasis is shifting from the static magnetosphere to the study of its dynamic properties. It is therefore necessary that both **E** and **B** be capable of representing observed temporal variations. The magnetospheric substorm is an example. In order to model the injection of charged particles during substorms properly, it is necessary to employ both a time-varying magnetic-field model and a time-varying electric-field model. Also the **E** and **B** models are coupled in that the time variations in **B** will induce a component that must be included in the total model of **E**. Because of this coupling another requirement must be made on all future magnetic-field models and that is, in addition to computing **B**, these models should also include **A**, the vector magnetic potential. (The induced electric field is best computed by taking the time derivative of **A**.) The need for having a time-varying electric-field model again points out the importance of understanding its source mechanisms, since input parameters to both **E** and **B** must be well understood if the models are to generate meaningful results.

4. Charged Particles

In the session on charged particles, it became apparent that the current lack of understanding of magnetospheric boundary phenomena limits the use of field models describing charged-particle behaviour. This is true for both the high- and low-energy charged particles. The access of low-energy charged particles is not well understood, nor is the coupling of the magnetospheric magnetic field with the magnetosheath and interplanetary fields. This coupling influences the entry of energetic particles and their subsequent motions in the magnetosphere. It is possible, however, that the higher energy particles, which do not contribute significantly to the current systems that produce **B** can be used to study some of the gross properties of the magnetospheric magnetic field.

Masley reported work on comparisons of observed solar cosmic-ray phenomena and model predictions. Particle trajectories and cut-offs obtained using the Olson-Pfitzer model were in good agreement in the noon and midnight meridian planes when observations were used for the same range of K_p values as the model is intended. His studies of cosmic-ray entry and propagation also suggest that energetic-particle access to the polar-cap regions is strongly influenced by the interplanetary magnetic-field direction. Other papers on high-energy particles also suggested the importance of accurate magnetic-field models and tests that these particles can offer for the development of such models. It was suggested that observations of high-energy particle fluxes be correlated with substorm time to help develop dynamic models of the magnetic field. Hoffman and Smith showed that particle fluxes observed from 1 to 872 kV can account for most of the depression of the inner magnetosphere observed in the magnetic-field data by Sugiura and his colleagues. Buck and West presented detailed pitch-angle distributions from which they derive drift-shell splitting and determine, indirectly, the shape of magnetic-field lines. Buck has examined the motions of energetic particles near noon and suggested that the observed pitch-angle spectra can be explained in terms of a magnetic field that has minima along the field lines displaced from the equatorial region as much as 60°. Both Konradi and McIlwain suggested that the region between 6 and 10 Earth radii where particle injection occurs needs more detailed study in order that it be properly modelled. It is in this region, at least near local midnight, where the magnetopause, quiet-time ring and tail current systems all contribute similar amounts to the total magnetic field. Murayama discussed energetic electrons that were observed near the high-latitude dayside magnetopause several hours after observed substorm activity.

All of these particle observations must eventually be accounted for by improved field models. However, it appears at present that only the more gross features of the particle and field phenomena observed in the magnetosphere are being fitted by the models. These detailed magnetospheric particle features will provide tests for the more sophisticated models that should be developed in the near future.

There was also discussion on the definition and location of the last closed or first open field line. It was argued that its location was best specified using charged-particle data. A magnetometer on a polar orbiting satellite cannot distinguish open from closed field lines by itself. Magnetometer data can best be used to define average magnetic-field models and then use the model geometry to define the boundary between open and closed lines. However, electrons present during solar events can be used to determine accurately the position of the first open field line; although this technique is limited because of the infrequent occurrence of solar events, it nevertheless might used to determine the average position (possibly as a function of K_p) of the first open field line.

The question of open versus closed magnetosphere as regards charged-particle entry must also be answered in the affirmative. That is, the magnetosphere appears to be always 'open' to the entry of both low- and high-energy particles over some portions of the magnetopause. This is true even if the magnetosphere is magnetically closed. This is possible because some particles incident upon the magnetopause enter it and drift through the magnetosphere because of gradients in the magnetic field parallel to the boundary. For low-energy particles these regions include the clefts on the high-latitude dayside magnetopause and the equatorial (plasma sheet) regions of the tail. Solar and galactic cosmic rays penetrate all regions of the magnetopause. It was concluded that the magnetosphere is open, both in the magnetic sense and from the standpoint of entry of charged particles. However, present models do not explicitly attempt to represent this open nature of the magnetosphere. In order to do this, a much better quantitative understanding of the magnetopause must be obtained. Such a quantitative model of the magnetopause must self-consistently account for a (small?) component of magnetic field normal to the boundary and a determination of which particles are deflected from this boundary and which particles are able to move through it.

5. Conclusions

The goal of the meeting was to make an assessment of what *quantitative* models are now available and how good they are, to determine generally what data sets are available, or will become available during the IMS era, and then to determine the limitations of existing models. These items were then used to make recommendations to model developers and model users for work that should be undertaken during the IMS period to extend existing models and to develop new ones. A summary of the present situation regarding quantitative models is given in this section and recommendations for work in the near future given in the next.

Modelling efforts in magnetic fields, electric fields, and charged particles were described. Clearly, efforts to model the magnetospheric magnetic field have made the most progress. However, it is clear that an 'international reference model' is not yet at hand. There are several models available that do specific things well, but no single model that meets all of the user requirements. The current models range from physical to entirely empirical. The utility of a given model depends on the needs of the user. This was one of the key problem areas isolated at the meeting. Although different users typically would ask for the same output from a model, the inputs might vary considerably. For example, a user who is trying to understand a particular magnetospheric phenomenon may only require that a model input parameter becomes available several weeks, or even months, after the time in question (for example, K_p or many other indices of magnetic activity) while there are many 'more practical' users who require input parameters which can themselves be monitored in real time (for example, the solar-wind dynamic pressure, the direction of the interplanetary magnetic field). Thus, determining the proper input parameters for a model is not a trivial problem if the needs of these several users are to be met. It is necessary that general input parameters for the models be developed. This means that in addition to the location in question, inputs to a model that will give back the vector magnetic field must also include parameters like the stand-off distance of the magnetopause which can be related either to K_p (after the fact) or in real time to solar-wind parameters. Both model and data-user communities indicated that a data sampling rate of 1 per minute suffices for most tests and applications. Some users, of course, use much lower sampling rates. Power suggested a rate of 1 per hour for some practical applications.

Many indices and real-time input parameters were discussed. They included the location of the auroral oval and the electron trough, the size of the polar cap, the Q index K_p, A_E, D_{st}, the vector specification of the interplanetary magnetic field, the various solar-wind parameters, and the tilt angle. For input parameters to the models to have general utility, it is necessary that they relate both to parameters that are readily available (such as K_p), but not necessarily in real time, and to other environmental features that can be monitored in real time (such as the dynamic solar-wind pressure). Thus the stand-off distance (which gives an indication of the size of the magnetosphere and concurrently the strength of the magnetopause current systems) might be an appropriate model input parameter. It would be then left to the user to relate this particular model input parameter to another quantity convenient for his particular use (for example K_p, or the solar-wind dynamic pressure). Many such input parameters must be decided upon. This must be done with both user needs and data availability in mind.

Although there were several complaints about availability of good data for use in the construction of empirical models, and as monitored input to some models, it is apparent that the modelling community must make the best of what is available. For example, it was suggested that vector ΔB measurements (ΔB is the difference between the observed magnetic field and that portion produced by the Earth's magnetic field as given by an accurate model of the main field) be used instead of scalar ΔB measurements. When vector ΔB measurements become generally available, they will be used as inputs to models and also to test models. To date, however, the scalar ΔB measurements have been used effectively to understand current systems in the inner magnetosphere and to develop quantitative models of the inner magnetospheric magnetic field.

The general availability of data and the fact that **B** can be measured directly has served to expedite efforts to quantitatively model magnetospheric magnetic fields. Efforts to model the electric field, on the other hand, have suffered from a lack of direct measurement of **E** and its extreme variability, both in time and space. Efforts to 'map' the electric field between different regions of the magnetosphere and thus compare simultaneous data sets, have met with only partial success, due, in part, to this variability. Hopefully, as several double-probe experiments become available to measure the DC electric field directly, some of these problems will be resolved, and the effort to model **E** in terms of its various sources can be accelerated.

For the future development of models, the various users should be kept in mind. Users vary from those interested in an understanding of the magnetosphere and an interest only in completely physical theoretical models to those practical users who are interested in models simply as 'black boxes' that yield quantitatively accurate information. Between these two extremes there is an almost continuous spectrum of users. These include, for example, the experimenter who needs to know the location of the foot of the magnetic-field line that passes through his satellite, and some monitoring activities that can be done within several hours of real-time observations. Some of the practical uses of magnetospheric models include the understanding and prediction of spacecraft charging and arcing, coupling of the magnetosphere to ionosphere and associated radio-communications problems, the coupling of the magnetosphere and ionosphere to the upper atmosphere and the influence of the upper atmosphere on satellite trajectories and lifetimes, and the effects of trapped radiation and radiation present during solar cosmic-ray events on various space hardware. Hildebrand suggested that the scientific community keep in mind the needs of the practical users because they can exert considerable influence on the determination of funding levels in several government agencies.

6. Recommendations

In order to best meet current and future user needs for quantitative models of magnetospheric phenomena, several recommendations are made here that should have special significance during the IMS era. During the IMS, increased attention will be focused on all aspects of magnetospheric phenomena. These efforts will include sets of satellite and ground-based experiments and monitoring stations, the procurement of many observational data sets, and the development of both physical and empirical quantitative models. *The first, and probably most general recommendation, is that a model situation centre be formed* – in the same manner as the experiment situation centre. Such a centre would serve two important purposes. First, it would permit a means for closely monitoring progress in the development of models and a means for evolving new sets of modelling recommendations as better data sets become available and more sophisticated quantitative models are developed. Secondly, such a centre would provide a means for interaction between the magnetospheric-physics community and the many users who need to know quantitatively the effects of the magnetosphere (possibly coupled with the ionosphere and/or atmosphere) on their hardware systems. It should be remembered that these users can provide much-needed incentive to various agencies to continue their support of magnetospheric research. In the eyes of these users, the quantitative model is the end result of the magnetospheric research. The model is employed by the user to cost effectively find solutions to environmental effects on his hardware systems. A model situation centre would also provide an interface between modelling and observational data-collection efforts.

More specific recommendations concerning the future development of quantitative magnetospheric magnetic field models were also made. Several of these recommendations apply as well to the development of electric-field models and some even to the development of charged-particle models. These recommendations can be roughly divided into two categories; requirements that future models should meet and quantitative tests that they should satisfy. General requirements that any future quantitative magnetospheric magnetic field model should satisfy include the following:

(i) It should represent the magnetic-field contributions from all important magnetospheric current systems including those currents distributed throughout the inner magnetosphere (the quiet and disturbed ring currents) and in the plasma-sheet region of the tail (the tail current system).
(ii) The dependence of each of these current systems on the tilt angle (the complement of the angle between the solar-wind direction and the geomagnetic dipole axis) should be included.
(iii) Variability in each of the current systems should be taken into account (e.g. the shape and strength of the magnetopause current systems as a function of stand-off distance).
(iv) The representation of **B** should be analytic.
(v) The magnetic vector potential **A** should be computed together with **B**.
(vi) The model should be valid over a large portion of the observed magnetosphere.

In addition, there are some specific user requirements:

(vii) The amount of computer time needed to represent **B** should be minimised.
(viii) Input parameters to the computer subroutine should be easily obtainable from user-chosen indices or -monitored parameters.

The first three requirements would satisfy most of the complaints made against existing models: they do not include the effects of all current systems and/or tilt effects, and most important, they do not do a good job of describing dynamic *magnetospheric* behaviour. A dynamic model must of necessity have input parameters that are related to the variability of the several current systems contributing to **B**. In order to use the model to describe charged-particle behaviour, it is necessary that it be both analytic and differentiable. Such a model of **B** should then yield both the motions of charged particles (from ΔB) and currents distributed throughout the magnetosphere (from $\nabla \mathrm{x} \mathbf{B}$). It is necessary to compute the vector potential **A** in order to determine the electric field induced by temporal variations in **B**. **A** can also be used to determine **B** if it is demanded that $\nabla \cdot \mathbf{B}$ be precisely 0. A computer representation of **B** must be extremely fast if it is to be used to compute magnetic-field lines, low-energy particle motions and drift shells, and the precise trajectories of high-energy particles.

In order to assess the quantitative accuracy of any model meeting the above criteria, it is necessary to have available quantitative tests. It is recommended that a standard set of tests be applied to all **B** models (it is understood that some specialised models would not do as well as others in satisfying these tests, but might still be the most useful for certain specific user needs). Several existing models have been tested by comparing their output with quantitative observational data of known magnetospheric phenomena. These data include:

(i) Magnetic-field observations in different regions of the magnetosphere (scalar ΔB plots, the decay of the tail field in its equatorial and lobe regions, the field geometry at high latitudes – especially near the magnetopause, the latitude of the last closed magnetic field line).
(ii) The daily variations in the synchronous magnetic field during quiet and disturbed periods.
(iii) The trapping boundary of adiabatic particles.
(iv) The low-latitude cut-off boundary for solar cosmic rays.
(v) Field-line shape as determined from barium releases.

Other observational data sets that should be used to test models in the near future include:

(vi) Vector ΔB measurements.
(vii) Detailed pitch-angle information on trapped-particle phenomena.

Several of the same requirements can be applied to the electric-field modelling efforts. The greatest problem with the electric field to date appears to be the lack of data in the inner and outer regions of the magnetosphere. However, because there is currently a fair amount of data on the ionospheric electric field, *it is recommended that an empirical model of the ionospheric electric field be developed.* Such a model would at least indirectly help the understanding of electric fields throughout the magnetosphere. The problem of lack of data in the inner and outer magnetosphere regions should largely be overcome by the deployment of several long-boom, double-probe experiments to measure the DC electric field. As this data becomes available, it is recommended that the various sources of **E** be kept in mind in trying to explain and model the magnetospheric electric field. For example, magnetospheric electric-field observations could be checked for the existence of a longitudinal component which is dependent on universal time. Such a dependence suggests the presence of a component induced by the daily wobble of the Earth's dipole field. The availability of simultaneous data from different regions of the magnetosphere and ionosphere will also shed light on the mapping problem.

It is probably the lack of understanding of the various boundaries in the magnetosphere and ionosphere that causes the largest problem for the physical modelling of electric fields. Thus the proper determination of the properties of the magnetopause again appears important (it was mentioned above that the magnetopause properties must be understood in detail in order to explain the coupling of the interplanetary magnetic field with the geomagnetic field and in order to properly understand and model the entry of charged particles into the magnetosphere). *Thus it is recommended that attention be directed toward the magnetopause, such that electric and magnetic field and particle phenomena both on and through that boundary can be quantitatively understood.*

There are now many data sets that detail charged-particle behaviour in the magnetosphere. It would be of general help to the magnetospheric community if models were developed of several of the gross particle features in the magnetosphere (e.g. the properties of the plasma sheet and the plasmasphere). It is necessary to have a first order set of models of the electric and magnetic fields, and of charged-particle structures) in order finally to consider their mutual interaction and to begin the development of a true quantitative magnetospheric model. It might be argued that models of the magnetic field, or of charged-particle phenomena, are merely different ways of representing the same magnetospheric phenomena. A quantitative magnetospheric model, then, may at least be defined as one that self-consistently and simultaneously considers the magnetospheric magnetic and electric fields and charged-particle phenomena.

It was also recommended that considerable effort be devoted to the determination of proper input parameters to all of the models. This is imperative if the models are to meet the various user needs. These input parameters must be chosen so as to make the models dynamic. In the case of **B**, input parameters are needed to describe each of the various current systems. It may be that for some user applications, existing indices are adequate (e.g. D_{st} as an input to describe ring-current strength). However, it appears that presently no parameters are adequate to describe the dynamics of the tail current during magnetospheric substorms. It was suggested that the size of the polar cap might provide a good indication of the strength and location of the tail currents. It is clear that a given model 'can be not better than its input parameters'. Thus the selection and use of the proper indices and/or monitored data for input parameters must be considered an important part of the development of quantitative models of magnetospheric phenomena.

7. Acknowledgements

The 'author' wishes to thank the many meeting attendees who made suggestions regarding this summary. The chairmen of the working groups and the technical sessions and the programme committee members especially provided significant inputs.

DISCUSSION

M.A. Shea: In response to Dr. Rostoker's question, the original World grid of vertical cut-off rigidities (5° in lat. x 15° in long. World grid) required approximately 2000 h of CPU time on an IBM 7090 computer. These calculations have just been repeated using an internal field model (IGRF with time derivatives applied for a 1975 epoch), and even with major modifications to increase the efficiency of the computer program, it still required about 300 h of CPU time on a CDC 6600. We estimate that re-doing these calculations utilising the Mead-Williams magnetospheric model (in addition to the internal field simulation) for just one local time would require twice as much CPU time. The basic problem is that the geomagnetic field must be calculated at every point on the integration of the orbit (each iteration step) and 95% of the time is spent calculating the geomagnetic field. For gross features, a dipole representation of the internal field is adequate; however, as the measurements with which these calculations are compared become more precise, the accuracy with which the field is calculated must also increase.

W.P. Olson: Other magnetic-field-model users who work with low-energy particle drifts also have constraints on computation times. Generally this will become even more of a problem as models become more sophisticated and as models are used together to quantitatively study magnetospheric phenomena.

G. Rostoker: How severe a constraint is computing time in trying to develop more detailed magnetic-field configurations? Will this produce the ultimate constraint on how detailed your model can be and still be usable?

W.P. Olson: Computation time is very definitely a constraint on model development and use. Some users are of course more affected by computer time limitations than others. Modellers should (and I think do) keep this problem in mind. This is one more important reason for believing that there will always be several models of **B**, **E** etc., that will satisfy the various user computing requirements. This will become more true as the models become more sophisticated (e.g. try to combine the magnetospheric and interplanetary fields).

D. Young: If **B** and **E** are known, it seems that particle structures would then be defined without the need for a separate particle model. Also, the problem of particle sources and boundaries (equivalent to field sources and boundaries) is much harder to deal with.

W.P. Olson: I agree, almost completely. A model of **B** can be used to determine magnetospheric currents (from $\nabla \times \mathbf{B}$) and some particle motions (from $\nabla \mathbf{B}$). Likewise **E** can be used with $\partial \mathbf{B}/\partial t$ to infer charged particle acceleration. However, a knowledge of **E** and **B** cannot reveal information on particle pitch angle and energy spectra, etc. When this type of information is needed it might be useful to have in addition a 'particle' model.

A. Pedersen: ATS-6 will go to eastern longitudes during the summer of 1976. This will give new opportunities for comparing models and observations. I hope that GEOS experimenters can be kept informed about conclusions between model makers (Olson) and ATS-6 magnetometer experimenters (Coleman)?

W.P. Olson: You can count on it. We are always interested in the further testing of our models and of course Coleman's group wants to know if its data are correct as soon as possible. We would in turn like to hear of any comments or complaints concerning the use of our models by the GEOS community.

D.A. Bryant: Is there any advice, encouragement, or warning you can give to experimenters at rocket ranges wishing to carry out conjugate studies with GEOS. In particular, can we expect predictions for disturbed periods by the GEOS launch date?

W.P. Olson: As I mentioned while reviewing the quantitative model conference, no model currently satisfies even the most important user requirements (all currents considered, tilt effects included, valid over large portion of magnetosphere, and *dynamic*). I don't believe such a model will be available before the GEOS launch. However, our group is now working to include tilt effects in our model and on making the input parameters such that the model can be used for the real-time dynamic specification of **B**.

G.L. Wrenn: Is your model of a form which would enable measurements of **B** at GEOS to be used as input to determine the position of the foot of the field line? GBR experimenters would like to do this in close to real time.

W.P. Olson: At present our model (when coupled to a realistic model of the main field) gives both daily and seasonal variations in both the location of the feet of the field line through the satellite and the direction of the field at the satellite. These variations also depend on the geographic longitude of the satellite. At present the model has no provision for representing variations with magnetic activity. This problem, related to the development of the proper model input parameters, is currently being investigated.

C.T. Russell: Using your intuition educated by your previous modelling studies, how do you feel the addition of tilt will affect the diurnal variation of the conjugate footprints you have shown?

W.P. Olson: Several comparisons between our present model (which contains the effects of currents distributed through the magnetosphere but not the tilt effect) and observational data indicate that it describes the field geometry better than other 'tilted' models that do not include the effects of these distributed currents. The tilt effects (which we are currently integrating into the model) will produce some additional variations, probably less than 1° at the feet of field lines from synchronous altitude. The tilt effects are clearly dependent on longitude and have already been seen to be more important at the ATS-1 longitude (approximately 90° to the dipole longitude) than at the original ATS-6 longitude (near the dipole longitude).

G.H. Voigt: (i) Up to what distance from the Earth is the spherical-harmonics expansion of your model valid?

(ii) How is your magnetopause geometry defined?

(iii) What are the magnetic-field boundary conditions just at the boundary?

W.P. Olson: (i) The computer representation of our model does not use spherical harmonics, but a power series with some exponential terms. This was done for several reasons; it substantially reduced the time of computation, made it possible to have $\nabla \times \mathbf{B} \neq 0$, which is necessary in the region of the fields source currents, and permitted the representation of **B** over a large volume. The model is valid from the nose of the magnetosphere to beyond lunar orbit in the tail.

(ii) and (iii) The model includes the magnetic effects of the tail currents, the quiet-time ring current, and the magnetopause currents. My tilted magnetopause model was used to represent the magnetopause currents. The present model thus does not contain a selfconsistent magnetopause shape because 'distributed' currents and their fields were not used to determine the shape. Clearly, the self-consistent determination of the shape of the magnetosphere is an important problem, as is the understanding of the properties of the magnetopause on a microscopic scale (particle entry, **E** and **B** behaviour).

INFLUENCE OF MAGNETOSPHERIC PARAMETERS ON GEOSYNCHRONOUS FIELD CHARACTERISTICS, LAST CLOSED FIELD LINES AND DAYSIDE NEUTRAL POINTS

G.H. Voigt

Technische Universität Braunschweig, Germany

Abstract

The day/night field asymmetry at the geostationary orbit, the Earth intersection points of geostationary field lines, the positions of the dayside neutral points, and the intersection latitudes of the last closed field lines have been calculated in a 'closed' magnetospheric **B**-field model as functions of systematic variations in model parameters. The magnetospheric parameters with most influence are: (i) the stand-off distance, (ii) the tail field intensity, (iii) the ring-current D_{st} - index, and (iv) the dipole tilt angle.

The dayside polar cusps are defined geometrically by the co-ordinates of the magnetopause neutral points and the Earth intersection latitudes of the last closed field lines at local noon. The cusp position near the magnetopause depends strongly on the 'stand-off' distance, the tail field and the dipole tilt angle. The ring current essentially determines the **B**-field configuration in the middle magnetosphere near the synchronous orbit during the recovery phase after a magnetic storm, but can be neglected at the magnetopause. Local time asymmetries increase with magnetospheric compression, i.e. decrease of stand-off distance. Conjugate points are situated at different geomagnetic latitudes in the northern and southern hemispheres in the case of a tilted geomagnetic axis. Magnetospheric **B**-field characteristics have been computed in a three-dimensional quantitative field model which is based on solution of the Chapman-Ferraro problem.

1. Introduction

During the International Magnetospheric Study (IMS), a great deal of magnetic field data will be available from satellites (GEOS, ISEE) and ground-based magnetometer profiles. These data can help us to understand the influence of magneto-

Knott & Battrick (eds.), The Scientific Satellite Programme During the International Magnetospheric Study, 381–396.
All Rights Reserved. Copyright © 1976 by D. Reidel Publishing Company, Dordrecht-Holland.

spheric dayside and nightside parameters on the local and/or the time-dependent **B**-field configuration. Some attention will be focussed on selected magnetospheric characteristics:

(a) At the geosynchronous orbit (6.6 R_e), magnetic-field contributions from external sources (Chapman-Ferraro currents on the magnetopause, tail current system, and equatorial ring current) become effective and result in the well-known day/night asymmetry in the *H*-component of the **B**-field [Cummings *et al.*, 1968; Ness, 1972]. Moreover, the *D*- and *V*-components describe the fine structure of the **B**-field, i.e. the local direction of the **B**-vector at 6.6 R_e.
The amplitude and phase of all three **B**-components, relative to magnetic local time, react sensitively to changes in solar-wind conditions, which in turn result in variations in the stand-off distance and tail-field intensity. The synchronous orbit therefore embraces an excellent magnetospheric region in which to seek a quantitative relationship between these two parameters.

(b) For correlative evaluation of magnetic field and plasma data between the 'middle' magnetosphere and the ionosphere, it is important to find the Earth intersection point of a specific field line connected with a geosynchronous spacecraft. Field-line tracing from the synchronous orbit down to the Earth must be done in a quantitative **B**-field model to determine which ground stations could be used for correlation.
Knowledge of Earth geomagnetic and/or geographic co-ordinates of synchronous field lines as a function of season, magnetic local time and magnetic disturbance could increase the significance of ground-based magnetometer profiles (Universities of Braunschweig and Münster) as 'magnetospheric monitors'.

(c) For ordering and classifying plasma data, it is important to know the shape and position of the dayside polar cusps. The cusps are defined geometrically by the co-ordinates of the magnetopause neutral points and by the Earth intersection latitudes of the last closed field line at local noon.
On the other hand, the geomagnetic latitude of the last closed field line at local midnight provides a reference for the tail field intensity and the inner edge of the tail current system.
Thus with respect to magnetic local time, the last closed field line is a **B**-field characteristic which is sensitive to the stand-off distance and the tail field intensity.

It is the purpose of this paper to compute systematically the effects of magnetospheric parameters on the **B**-field of a theoretical model. These calculations can help in predicting a real magnetospheric field situation, and so assist the experimenters in ordering their data.

In addition, IMS experiment data from satellites and ground-based stations will provide essential support in proving or disproving the validity of theoretical magnetospheric field models.

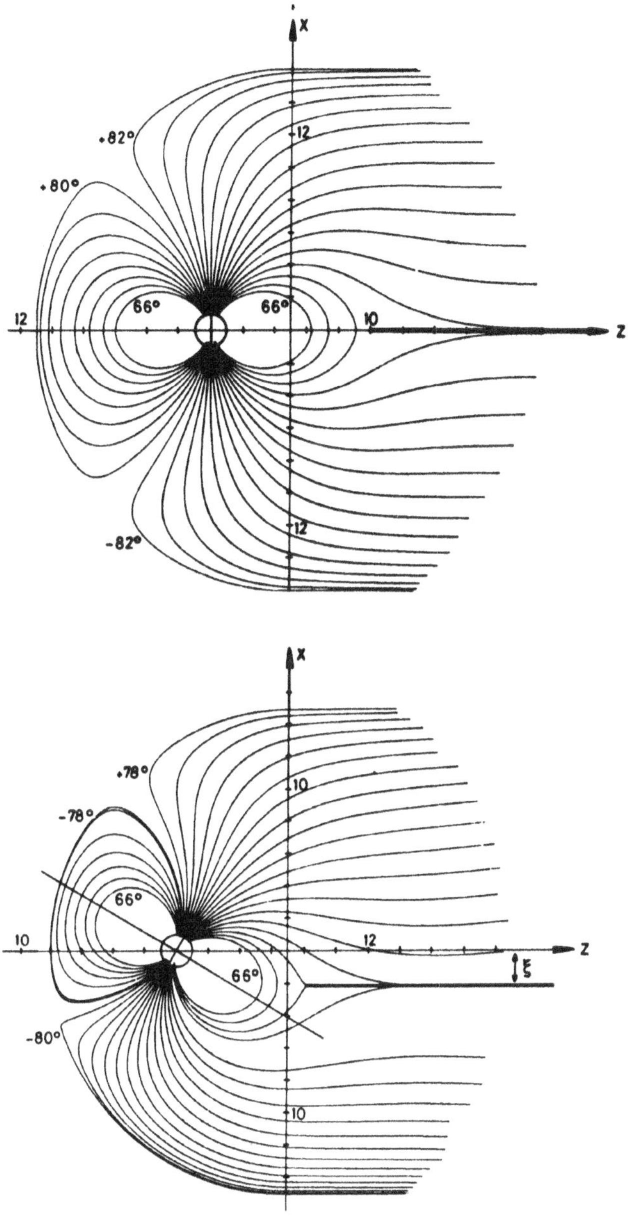

Figure 1. *Magnetospheric noon-midnight field lines spaced every 2° in latitude. The upper figure shows a quiet magnetosphere with parameters $r_0 = 11\ R_e$, $H_0 = 30\gamma$ (zero tilt). The lower figure shows the compressed state with parameters $r_0 = 8\ R_e$, $H_0 = 50\gamma$ (tilt = – 30°).*

2. The B-field Model

The magnetopause of the magnetospheric model must have a predetermined geometry in order to find a three-dimensional analytical solution to the Chapman-Ferraro problem [Voigt, 1972]. Nevertheless, its geometric shape has been aprroximated to spacecraft boundary crossings [Ness *et al.*, 1964].

The magnetopause is a mathematically idealised tangential discontinuity, so that the model is 'closed', and no **B**-field component normal to the surface exists just at the boundary. The **B**-field inside the magnetospheric cavity results from a scalar potential $\mathbf{B} = -\nabla u$. Because curl $\mathbf{B} = 0$, no distributed currents [Sugiura & Poros, 1973; Olson & Pfitzer, 1974] can be obtained from or described by the model.

The neutral sheet in the tail is a single plane current sheet with return currents closed over the tail magnetopause. The tail current density is associated with the reduction in the tail magnetic field found in spacecraft magnetometer data [Behannon *et al.*, 1968]. The model tail field is based on a formalism by Choe & Beard [1974], but extended to include seasonal effects due to the dipole tilt angle.

An axisymmetric, self-consistent ring current [Sckopke, 1972], characterised by the D_{st}-index, is inserted into the **B**-field model in order to fit the time-dependent behaviour of the magnetosphere during recovery after magnetic storms. The ring-current models are based on proton energy density profiles, which are approximated to Frank's [1967] plasma measurements.

Long-period variations of the magnetospheric **B**-field topology result simultaneously in a displacement of the last closed field line boundary and the polar cusp, as well as in a shift in Earth intersection points of synchronous field lines. This process can be simulated by a series of magnetostatic 'snapshots' from successive parameter variations.

The magnetospheric model parameters that influence **B**-field characteristics most are:

(a) r_0 = stand-off distance
(b) H_0 = field intensity at the inner edge of the tail current system
(c) a = distance to the neutral sheet's inner edge
(d) D_{st} = symmetric part of the ring current field
(e) ψ = dipole tilt angle.

As an example, Figure 1 shows noon-midnight field lines plotted for two different parameter sets describing the quiet and pre-substorm compressed magnetosphere. Note the lower geomagnetic latitude of the last closed field line at local noon in the case of the compressed state, and secondly the different latitudes of the dayside first open field lines in the case of the tilted Earth dipole. The unusual co-ordinate axes should be ignored in the context of the present calculations.

3. B-field at Geosynchronous Orbit

Table 1 shows the contributions of external magnetospheric sources to the total field at the synchronous orbit. The H-component has been computed for three different magnetic local times. The total field is compared with quiet-time OGO measurements for $K_p = 0\text{–}1$ [Ness, 1972].

The model quiet-time stand-off distance of 11 R_e agrees fairly well with spacecraft measurements of the magnetopause position reported by Fairfield [1971]. Note that a quiet-time ring current $D_{st} = -25\gamma$ is inserted in order to adjust the **B**-field to the OGO data.

TABLE 1

Magnetospheric field sources that contribute to the *H*-component at 6.6 R_e. Total field is in agreement with OGO quiet-time observations. Model parameters are chosen for magnetically quiet conditions: $r_0 = 11\,R_e$, $H_0 = 30\gamma$, $a = 10\,R_e$, $D_{st} = -25\gamma$, dipole tilt = + 30° (summer). The negative ring-current field indicates a quiet-time ring current which flows outside 6.6 R_e.

LT	Dipole field	Chapman-Ferraro	Tail field	Ring current	Total field	OGO* ($K_p = 0\text{–}1$)
0.0	+ 108.5	+ 14.8	– 25.5	– 12.7	+ 85.1	+ 85
6.0	+ 108.5	+ 20.9	– 14.3	– 12.7	+ 102.4	no information
12.0	+ 108.5	+ 34.5	– 10.1	– 12.7	+ 120.2	+ 120

* All magnetic field values are given in 'gamma'

The negative ring-current field indicates a quiet-time ring current which flows outside 6.6 R_e. In the case of a magnetic disturbance (main phase of a magnetic storm), the ring-current energy density profile (proton distribution) increases and moves earthwards to lower *L*-values [Frank, 1967]. This results in a change in the sign of the ring-current field at the synchronous orbit. For example, a $D_{st} = -100\gamma$ ring current produces a $+42\gamma$ field at 6.6 R_e (Table 2b).

The field of the boundary currents decreases from local noon to local midnight, but tends to increase the dipole main field. On the other hand, the tail field increases from local noon to midnight, but tends to reduce the main field.

The general behaviour of the geosynchronous **B**-field can be summarised as follows:

(a) The boundary currents and the tail current system are responsible for the day/night field asymmetry. By contrast, the (axisymmetric) equatorial ring current changes the field level as whole over a great range, but keeps the asymmetry constant (cf. Table 2b).

(b) The signs of the Chapman-Ferraro field (i.e. the field of the boundary currents) and the tail field indicate that a pre-substorm-related compression of the dayside magnetosphere, with or without an associated increase in the tail field, tends to increase the geosynchronous field asymmetry.

TABLE 2a

Maximal variation in magnetospheric parameters, computed from the difference between a super-quiet and a super-disturbed (compressed) model magnetosphere.

Parameter	Super quiet	Super compressed	Maximal variation
Stand-off	12 R_e	8 R_e	$\Delta\, r_0 = -4\, R_e$
Tail field intensity	10 γ	60 γ	$\Delta\, H_0 = +50\gamma$
Inner edge of tail currents	15 R_e	7 R_e	$\Delta\, a = -8\, R_e$
Ring current (D_{st})	0 γ	− 100 γ	$\Delta\, D_{st} = -100\,\gamma$
Dipole tilt into summer direction 0° to + 30°			$\Delta\, \psi_s = +30°$
Dipole tilt into winter direction 0° to −30°			$\Delta\, \psi_w = -30°$

TABLE 2b

Variation in geosynchronous and high-latitude **B**-field characteristics as a function of maximal variation in magnetospheric parameters (see Table 2a).

	Δr_0	ΔH_0	Δa	ΔD_{st}	$\Delta \psi_s$	$\Delta \psi_w$
ΔH_N (γ)	+ 3	− 36	− 28	+ 42	+ 7	+ 7
ΔH_D (γ)	+ 45	− 13	− 3	+ 42	− 1	− 1
$\Delta\lambda_N$ (°)	± 0.0	− 10.0	− 2.5	− 1.5	− 0.5	+ 3.5
$\Delta\gamma_D$ (°)	− 3.0	− 5.0	− 1.0	− 1.0	+ 2.0	− 1.0
$\Delta\vartheta_{NP}$ (°)	+ 7.9	+ 15.4	+ 3.2	+ 0.01	− 8.9	+ 6.2
$\Delta\Lambda_{NP}$ (°)	− 7.9	− 15.4	− 3.2	− 0.01	− 21.1	+ 23.8

(c) The boundary currents generate a field with a maximum at local noon. Thus, variations in these currents, characterised by a change in the stand-off distance, affect mainly the dayside field at the synchronous orbit.
The tail-current field has a maximum at local midnight. Hence, changes in the tail field mostly affect the nightside field, but also have a significant influence on the dayside (Table 2b).

(d) Table 2b shows that the synchronous field tends to vary on the nightside with the dipole tilt angle, but the amount of change depends on the position of the

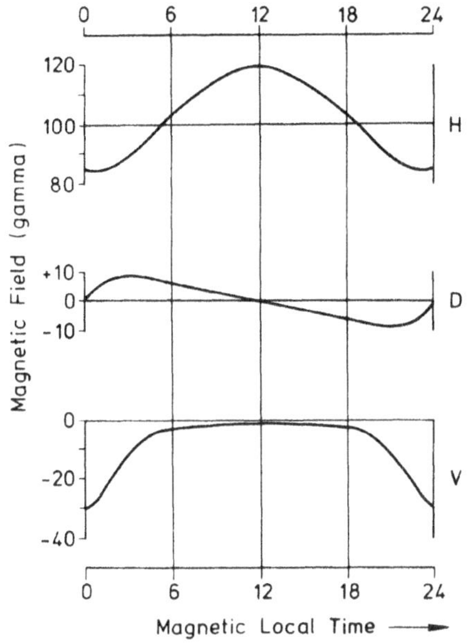

Figure 2. **B**-*field components at the geosynchronous orbit for magnetically quiet conditions and + 30° dipole tilt as a function of magnetic local time. The H-component corresponds with OGO quiet-time (K_p = 0–1) measurements (see text).*

neutral sheet and the distance to the inner edge (see nightside field-line topology in Fig. 1b).

It remains a problem at present to find a quantitative relationship between the dipole tilt angle and the neutral sheet's displacement relative to the solar ecliptic plane.

Figure 2 shows all three **B**-field components at 6.6 R_e as a function of magnetic local time. Model parameters are the same as noted for Table 1 (quiet magnetosphere). The spacecraft is assumed to fly in the geomagnetic equatorial plane. The internal geomagnetic main field is a pure dipole. Therefore, the *D*-component, as well as the *V*-component, vanishes in the case of an untilted Earth dipole.

For IMS applications, the pure dipole can easily be replaced by a spherical harmonics multipole expansion of the main field [IAGA, 1969].

The model with an infinitesimally thin current sheet in the tail describes the phase and amplitude of the local time variation in *H* reasonably well. Because of the symmetry of the model, the *D*-component vanishes at local noon and midnight. On the other hand, the model predicts too large a *V*-value around local midnight because of the idealised field-line configuration near the inner edge of the tail current system.

4. Geosynchronous Field Lines

Figure 3 shows the geomagnetic latitudes of Earth intersection points of synchronous field lines for three different model parameters as a function of magnetic local time. The influence of stand-off is compared with tail field variations and changes in the (negative!) D_{st}-value.

The computation can be summarised as follows:

(a) Changes in the stand-off affect mainly the region around local noon and can almost be neglected around local midnight. It is interesting that the intersection latitude at local noon increases nonlinearly with increasing magnetospheric compression, i.e. with decreasing stand-off. This phenomenon results from the fact that, in the case of a highly compressed magnetosphere, the shielding Chapman-Ferraro fields distort the synchronous field lines on the dayside.

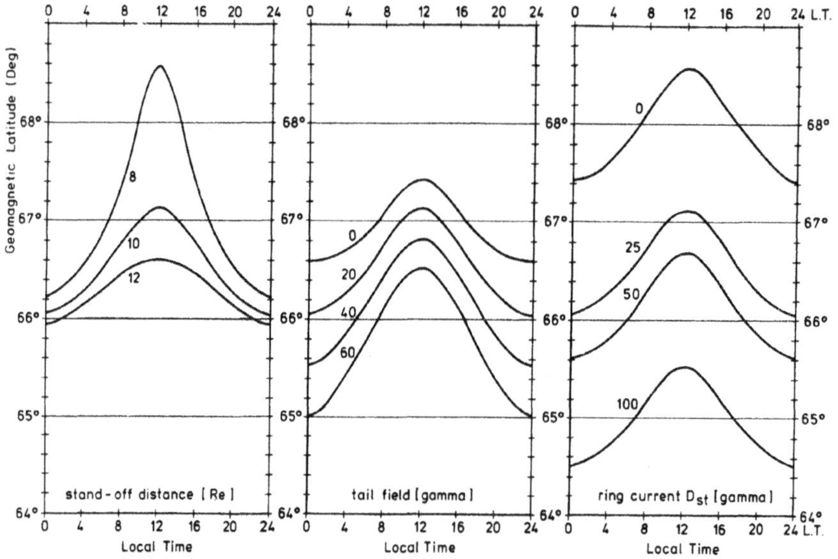

Figure 3. *Earth's geomagnetic intersection latitude of geosynchronous field lines as a function of magnetic local time. The three figures show the influence of three different model parameters (zero tilt):*
(a) variation of r_0 from 12 R_e to 8 R_e
(b) variation of H_0 from 0γ to 60γ
(c) variation of D_{st} from 0γ to -100γ (numbers indicate negative D_{st} values).
Reference values for parameter variations are $r_0 = 10\, R_e$, $H_0 = 20\gamma$, $D_{st} = -25\gamma$.

(b) Variations in the tail current field affect mainly the nightside around local midnight, but cannot be neglected on the dayside. A change in the tail field intensity from 20γ to 60γ has approximately the same effect at 12.00 LT as an increase in stand-off from 10 R_e to 12 R_e.
However tail-field increase and stand-off increase are contradictory physical processes. The intersection latitude at local noon could remain unchanged during a substorm growth phase if: (i) the stand-off decreases simultaneously with an increase in tail field, (ii) the ring current has not yet built up.

(c) Like the synchronous **B**-field, the ring current plays an important role in determining the intersection latitude. The effect is identical at local noon and midnight. Thus, one can expect changes in intersection points over a wide range during the recovery phase following a magnetic storm.

(d) Compared with the effects of stand-off, tail field and ring current, the variation due to the dipole tilt angle (Fig. 4) is small, but it should be noted that the seasonal variation increases drastically with increasing magnetospheric compression.
Moreover, Figure 4 shows the positions of geomagnetic conjugate points in the winter and summer hemispheres in the case of a ± 30° dipole tilt.
Conjugated points are situated at different geomagnetic latitudes in the northern and southern hemispheres in the case of a tilted geomagnetic axis. The tilt-dependent shift of conjugated points is different in the northern and southern hemispheres, and is different at local noon and local midnight (compared with the 'untilted' curve in Fig. 4). Moreover, the amount of shift for a given tilt angle depends on geomagnetic latitude magnetic disturbance (i.e. magnetospheric compression).

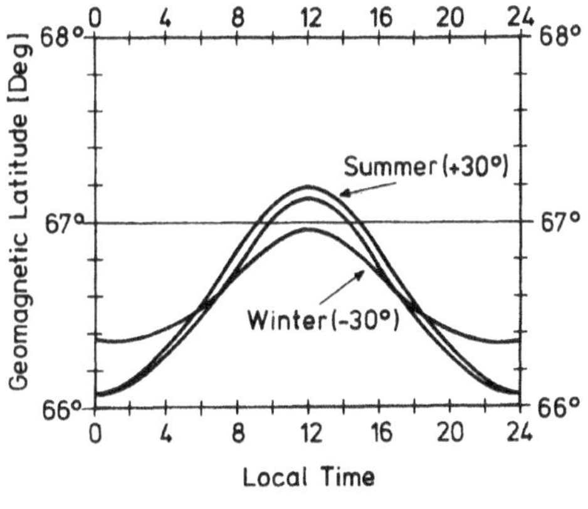

Figure 4. *Same as Figure 3, for three tilt angles $\psi = -30°, 0°, +30°$. Other model parameters are $r_0 = 10\,R_e$, $H_0 = 20\gamma$, $D_{st} = -25\gamma$. Conjugate points are situated at different geomagnetic latitudes in the northern and southern hemispheres in the case of the tilted dipole axis.*

5. Last Closed Field Lines

Figure 5 shows the last closed field line at local noon (D) and local midnight (N) as function of r_0 and H_0 for the case of an untilted dipole (for the tilt dependence, see Voigt [1974]). The computation can be summarised as follows:

(a) The most influential parameters as regards variations in last closed field lines are the tail field intensity for the midnight field line, the stand-off distance, *and* the tail field for the field line at local noon. For strong tail fields, greater than 30γ, the stand-off, i.e. the magnetospheric dayside compression, has no influence on the latitude of the midnight field line. By contrast, even for small stand-offs, the tail field variation is measurable on the dayside at local noon. A tail field variation from 20γ to 50γ has almost the same effect as a reduction in stand-off from $12R_e$ to $8R_e$. Either process tends to shift the dayside field line equatorwards by 3°.

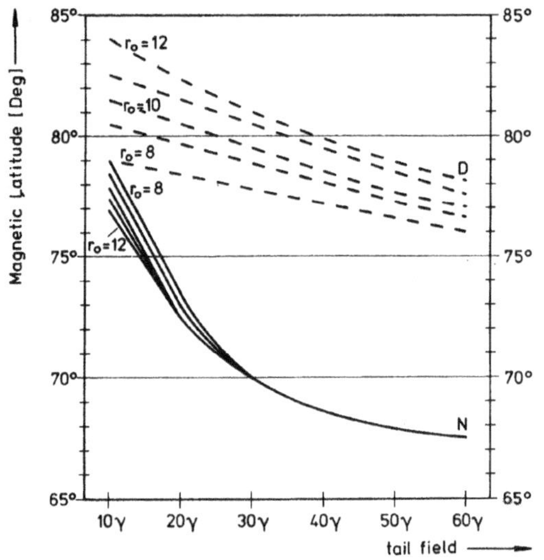

Figure 5. *Geomagnetic latitude of last closed field lines at local noon (D) and local midnight (N) as a function of the tail field intensity H_0 and stand-off r_0 (zero tilt).*

(b) An inward motion of the dayside magnetopause during the compression phase of a substorm occurs simultaneously with an increase in the tail field intensity and an inward motion of the tail current system. This process results in an equatorward displacement of the last closed field line boundary. An example is given in the upper part of Figure 6. The dayside shift exceeds the nightside shift because of the combined effect of a stand-off decrease and tail field increase. On the nightside only the tail field is effective.

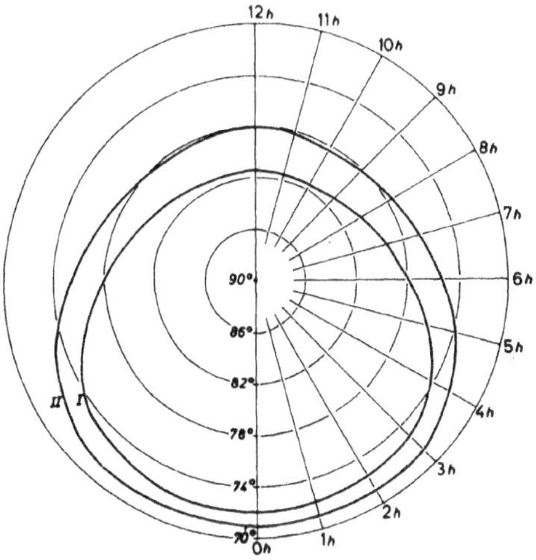

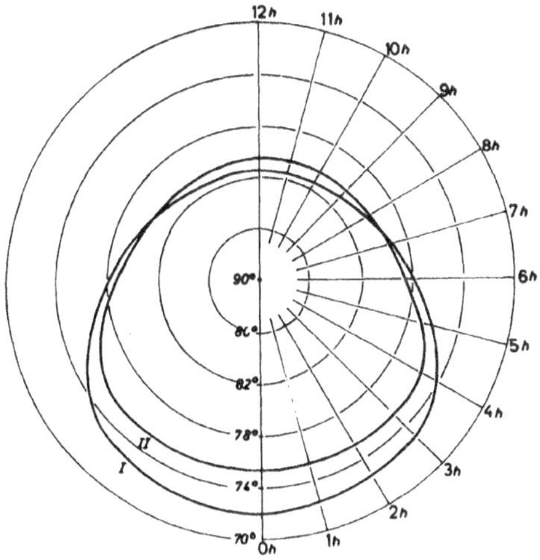

Figure 6. *Geomagnetic polar diagram of the last closed field line boundary in the northern hemisphere.*
Upper figure:
I. Undisturbed magnetosphere,
II. Compressed magnetosphere,
for untilted dipole.
Lower figure:
I. Dipole tilt = 0°,
II. Dipole tilt = −30° (winter),
for undisturbed magnetosphere.
Curves I are identical in both figures.

(c) The lower part of Figure 6 shows the displacement of the boundary due to a −30° dipole tilt (winter). The absolute magnitude of the nightside shift exceeds that of the dayside shift. A look at Table 2b shows that just the reverse results in the southern hemisphere, or for a tilt into summer direction.
Thus, one can expect asymmetries in both hemispheres in respect of a tilt-dependent displacement of the last closed field line boundary.

(d) Table 2b shows the influence of the other parameters: compared with D_{st} changes in synchronous field lines (Fig. 3), the ring current has only a small influence on the last closed field line position. This results from the fact that the ring-current field decreases like a dipolar field with $1/r^3$ beyond $7–8R_e$ so that the outermost field lines remain unchanged.

6. Magnetopause Neutral Points

It should be noted that the neutral point is an idealised 'mathematical point' in a vacuum magnetospheric **B**-field model with a magnetopause of a tangential discontinuity type. For this reason, the polar cusp is funnel-shaped, with a small local time extension around 12.00 LT.

Figure 7 shows the polar distances of the magnetopause neutral points as a function of stand-off r_0 and tail-field intensity H_0 for an untilted Earth dipole:

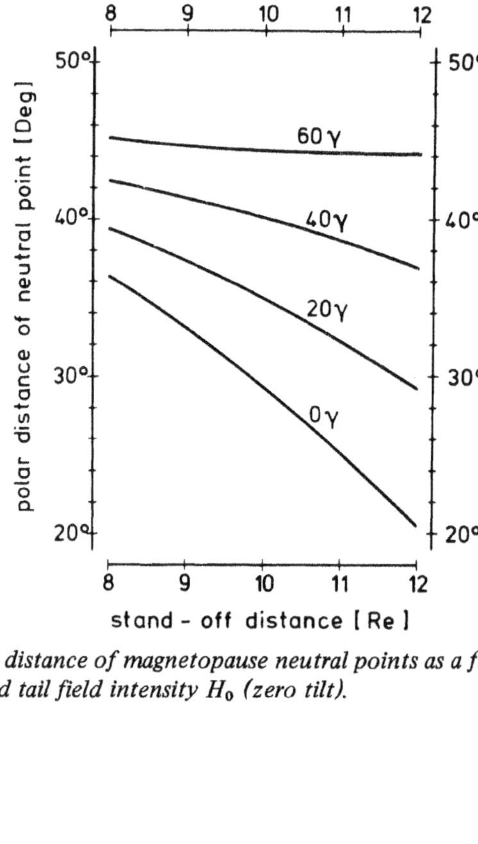

Figure 7. *Polar distance of magnetopause neutral points as a function of stand-off r_0 and tail field intensity H_0 (zero tilt).*

(a) Like the midnight variation of the last closed field line boundary, the influence of stand-off decreases with increasing tail field intensity. Thus, one can observe compression-related variations in the polar cusp position near the magnetopause only during relatively quiet times when the tail field is small.

(b) Analogous to last closed field line variations, Figure 7 and Table 2b show the dominant influence of tail field changes on the neutral point's position.
An increase in the tail field from 25γ to 50γ has the same effect as a reduction in stand-off from 12 R_e to 8 R_e. Either process displaces the neutral point equatorwards by about 8°. In the compression phase of a substorm, both processes work together.
One must conclude that the **B**-field topology at the dayside magnetopause around the neutral point, where the total **B**-field comes to be zero, is essentially determined not only by magnetospheric compression, but also to the same degree by tail field variations.

(c) The equatorial ring current can be totally neglected at the magnetopause (Table 2b). This indicates that the ring current is a typical field characteristic of the middle magnetosphere. It generates only a negligible shielding Chapman-Ferraro field.

(d) Of all magnetospheric characteristics considered in this paper, the position of the neutral point will be affected significantly by changes in the dipole tilt angle [Voigt, 1974]. This is of consequence for the whole Chapman-Ferraro current system, which contracts into the magnetopause neutral points. The tilt-dependent displacement of the Chapman-Ferraro current system in the magnetopause is responsible for the field configuration in the outer magnetosphere. This process predominates greatly over the effect of a tilt-dependent change in the magnetopause geometry.

7. Notes to Tables 2a and 2b

Table 2a
The maximal variation in magnetospheric parameters results from the difference between a hypothetical super quiet and a super disturbed (compressed) model magnetosphere. The signs in the last column indicate the transition from the quiet to the compressed state.
The Δ-values are used in Table 2b in order to calculate the influence of parameter variations on **B**-field characteristics.

Table 2b
Variations in geosynchronous and high-latitude **B**-field characteristics have been computed as a function of maximal variations in magnetospheric parameters from Table 2a. The symbols are defined as follows:

H_N, H_D = H-component of the **B**-field at 6.6 R_e at local noon (D) and local midnight (N)
λ_N, λ_D = geomagnetic latitudes of the last closed field line at local noon (D) and local midnight (N)
ϑ_{NP} = polar distance of the magnetopause neutral point.
Λ_{NP} = corresponding geocentric solar ecliptic latitude of the neutral point
Δ = indicates maximal variation due to the maximal parameter variation of Table 2a.

For differences Δ, only *one* model parameter has been varied in the range of Table 2a; the others remain constant and have 'reference-state' values. The reference state is a quiet magnetosphere with:

(i)	Stand-off	r_0	=	11 R_e
(ii)	Tail field intensity	H_0	=	30 γ
(iii)	Inner edge of tail current system	a	=	10 R_e
(iv)	Equatorial ring current	D_{st}	=	0 γ
(v)	Earth's dipole tilt	ψ	=	0°

The maximal summer and winter shifts of the magnetopause neutral point are noted in changes in the polar distance $\Delta\vartheta_{NP}$, and changes in the geocentric solar ecliptic latitude $\Delta\Lambda_{NP}$. The interdependence of both quantities is given by

$$\Lambda_{NP} = 90° - \vartheta_{NP} - \psi$$

Note the positive difference of + 42γ for ΔH_N and ΔH_D due to a $D_{st} = -100\gamma$ ring current. The '+' sign indicates that the maximum in the ring-current proton energy distribution is located inside 6.6 R_e.

8. Conclusions

(a) The most sensitive magnetospheric parameters are stand-off for the dayside **B**-field configuration, and tail field intensity for the dayside *and* nightside configurations.

(b) Magnetospheric parameters that change during periods of magnetic disturbance are: (i) stand-off, (ii) tail field intensity, (iii) distance to inner edge of the tail current system, and (iv) field of equatorial ring current. These model parameters are *independent* in available theoretical **B**-field models. The problem of finding a 'theoretical interdependence' between the model parameters is as yet unresolved. Simultaneous *in-situ* spacecraft measurements within the polar-cusp region, at the geosynchronous orbit, and in the tail seem to be necessary together with boundary crossings in order to evaluate quantitative magnetospheric **B**-field models by reducing the multiplicity of independent parameters.

(c) Local time asymmetries in the **B**-field components at the synchronous orbit increase during the early phase of a substorm. This process is, in general, due to a compression of the dayside magnetosphere together with an increase in the tail field intensity. Changes in the dayside **B**-field are due to changes in stand-off and/or variations in the tail field. Changes in the nightside **B**-field depend on tail-field variations and on the distance to the inner edge of the tail current system. The geosynchronous **B**-field topology, as well as the variation in its local time asymmetry, is closely connected with the displacements of the Earth intersection points of synchronous field lines.

(d) The ring current essentially determines the **B**-field topology in the 'middle' magnetosphere near the synchronous orbit. The ring current keeps the local-time field asymmetry constant, but shifts the **B**-field level as whole over a wide range. The same is true for the intersection latitude of synchronous field lines.

(e) Both D_{st} registration and synchronous **B**-field measurements, together with plasma observations, determine the location of the ring current, i.e. the maximum of the ring-current proton energy distribution.
If the synchronous **B**-field is reduced, the ring current flows outside 6.6 R_e. On the other hand, a **B**-field increase indicates that the ring current flows inside 6.6 R_e. In order to distinguish this effect from stand-off and tail-field variations, a correlative data evaluation with plasma measurements is necessary.

(f) Plasma measurements in connection with **B**-field observations at 6.6 R_e could help in establishing a quantitative interdependence between D_{st} and other magnetospheric parameters.
K_p seems a poor parameter for correlation with **B**-field variations at the synchronous orbit, since it is a worldwide parameter averaged over local time asymmetries.

References

Behannon, K.W., Mapping of the Earth's bow shock and magnetic tail by Explorer 33, *J. Geophys. Res.* 73, 907–930 (1968).

Choe, J.Y. & D.B. Beard, The near earth magnetic field of the magnetotail current, *Planet. Space Sci.* 22, 609–615 (1974).

Cummings, W.D., J.N. Barfield & P.J. Coleman, Magnetospheric substorms observed at the synchronous orbit, *J. Geophys. Res.* 73, 6687–6698 (1968).

Fairfield, D.H., Average and unusual locations of the earth's magnetopause and bow shock, *J. Geophys. Res.* 76, 6700–6716 (1971).

Frank, L.A., On the extraterrestrial ring current during geomagnetic storms, *J. Geophys. Res.* 72, 3753–3767 (1967).

IAGA Commission 2 Working Group 4: Analysis of the geomagnetic field, *J. Geophys. Res.* 74, 4407–4408 (1969).

Ness, N.F., Review of magnetic field observations. In 'Earth's Magnetospheric Processes' (Ed. B.M. McCormac), Reidel, Dordrecht, Holland, pp. 189–199, 1972.

Ness, N.F., C.S. Scearce & J.B. Seek, Initial results of the IMP 1 magnetic field experiment, *J. Geophys. Res.* **69**, 3531–3569 (1964).

Olson, W.P. & K.A. Pfitzer, A quantitative model of the magnetospheric magnetic field, *J. Geophys. Res.* **79**, 3739–3748 (1974).

Sckopke, N., A study of self-consistent ring current models, *Cosmic Electrodyn*, **3**, 300–348 (1972).

Sugiura, M. & D.J. Poros, A magnetospheric field model incorporating OGO-3 and 5 magnetic field observations, *Planet. Space Sci.* **21**, 1763–1773 (1973).

Voigt, G.H., A three dimensional, analytical magnetospheric model with defined magnetopause, *Z. Geophys.* **38**, 319–346 (1972).

Voigt, G.H., Calculation of the shape and position of the last closed field line boundary and the coordinates of the magnetopause neutral points in a theoretical magnetospheric field model, *J. Geophys.* **40**, 213–228 (1974).

DISCUSSION

M. Leubner: Your model is derived from a given magnetopause. How can you obtain different subsolar and polar distances of the magnetopause? A change in the value of the tail field also changes the magnetopause shape. How do you include that in your model?

G.H. Voigt: The stand-off distance, as well as the tail-field intensity, are *independent* model parameters. Because of the given magnetopause geometry, the polar distance to the boundary depends on the stand-off distance and the radius of the magnetospheric tail. Variations of the tail field intensity result in changes in the field-line topology of the whole magnetosphere, without changing the magnetopause geometry.

F. Stössel: What kind of ring-current model did you use?

G.H. Voigt: An axisymmetric, selfconsistent ring-current model [Schopke, *Cosmic Electrodyn.* **3** 330-348, 1972] was used for magnetospheric field calculations. The ring-current models are based on proton energy density profiles being approximated to plasma measurements.

C.T. Russell: Could you describe your tail-current model, how it closes and how it depends on tilt angle? Does it maintain equality of magnetic flux in the two tail lobes? Secondly, how do you define last closed field line on both the dayside and the nightside?

G.H. Voigt: The tail-current system separates the tail into the upper and lower lobe. The return currents are closed over the cylindrical tail magnetopause. Currents flow within infinitesimally thin current layers. The current density down the tail is associated with the decrease in the total tail magnetic field found in spacecraft magnetometer data [Behannon *et al.*, *JGR* **73**, 907-930, 1968]. The neutral sheet displacement out of the ecliptic plane is an empirical function of the stand-off distance and the dipole tilt angle.

Because of the constant radius of the tail, conservation of magnetic flux cannot be considered in a mathematically consistent way.

The last closed field line on the dayside is defined as the last field line with both ends in the ionosphere of both hemispheres.

On the nightside, the last closed field line is defined by the last line which hits the neutral sheet with an acute angle of incidence and returns to the Earth.

PROPOSED Ba^+ ION-TRACER CONJUGACY AND RELATED STUDIES USING THE GEOS SATELLITE

D.T. Young

Physikalisches Institut, University of Bern, Switzerland

Abstract

Ion-composition and plasma analysers on the GEOS satellite may be used to detect and analyse Ba^+ ions injected by shaped-charge releases at the foot of the $L = 6.6$ field line. Interactions between the Ba^+ jet and the ambient plasma medium may give rise to ULF or VLF wave amplification, modifications of the Ba^+ velocity distribution or that of the ambient plasma. This paper examines the experimental problems associated with detecting the jet in the equatorial plane and discusses the use of several GEOS experiments to monitor the local wave-particle environment which, as some authors have suggested, may be modified by the Ba^+ plasma.

1. Introduction

Shaped-charge barium releases have been used to probe magnetic-flux-tube geometry and drifts over distances of several R_E along magnetic field lines [see review by Haerendel, 1973; see also Wescott *et al.*, 1972, 1974, 1975a, b]. Quite independently and for altogether different purposes, cold plasma releases have been proposed by Brice [1970, 1971] and Cornwall [1972, 1974] as a method of stimulating precipitation of hot ring-current plasma. Liemohn [1974a, b] has computed the ULF and VLF wave amplification that might result from shaped-charge injections like those of Wescott and co-workers. In this paper we propose an experiment combining the main features of the aforementioned work, with the important addition that experiments on the GEOS satellite would be used to detect and analyse barium ions reaching the equatorial plane while simultaneously monitoring the total local plasma environment for alterations to wave-particle distributions induced by the release.

The proposed GEOS experiment may be viewed in two ways:
Macroscopically, it is a large-scale probe of the magnetosphere along a flux tube

Knott & Battrick (eds.), The Scientific Satellite Programme During the International Magnetospheric Study, 397–410.
All Rights Reserved. Copyright © 1976 by D. Reidel Publishing Company, Dordrecht-Holland.

measuring some 50 000 km from ionosphere to equatorial plane. During magnetic quiet ($K_p \leqslant 2$), the location of this flux tube may be predicted with reasonable accuracy by the model of Olson & Pfitzer [1974]. Bulk motion of the flux tube may occur as a result of convection electric fields, which are predominantly co-rotational during magnetic quiet. Distortions of the flux tube can be caused by localised currents or by decoupling of ionospheric and magnetospheric ends of the flux tube if parallel electric fields are present. Striations may also be present as the result of jet instabilities [Wescott *et al.*, 1975a]. Macroscopic aspects of the release may best be studied by combining ground-based electric-field measurements and optical tracking with detection of the ions at the satellite using GEOS plasma spectrometers (Table 1).

TABLE 1

Characteristics of GEOS Ion Composition Experiment S 303

	Mass	Energy: Thermal mode	Energy: Low-energy mode
Range	$1 \leqslant \frac{m}{q} \leqslant \infty$	$0 \leqslant \frac{E}{q} \leqslant 100$ eV	$0 \leqslant \frac{E}{q} \leqslant 18$ keV
Resolution	$\frac{m^{**}}{\Delta m} = 4$ at 5 keV	$0.6 \geqslant \frac{E^*}{\Delta E} \geq 0.13$	$0.7 \geq \frac{E^*}{\Delta E} \geq 0.06$
Number of samples in range	64	32	32

	Angle: α	Angle: β	Time (s): Sample	Time (s): E scan	Time (s): m scan	Time (s): $m + E$ scan
Range	$60° \leqslant \alpha_0 \leqslant 120°$	2π	–	–	–	–
Resolution	$8° \leqslant \Delta\alpha^* \leqslant 44°$	11°	0.172	5.5	11	352
Samples per scan	–	35	1	32	64	2048

Computed Geometric Factor** (cm² sr s $\frac{\text{eV}}{\text{eV}}$)	50 eV	17 keV
H^+	8×10^{-2}	3×10^{-4}
Ba^+	3×10^{-2}	1×10^{-4}

* Energy dependent

** Mass and energy dependent

Microscopically, barium ions (and electrons) will interact with the plasma environment along the flux tube. Typical barium releases inject 5 x 10^{23} ion-electron pairs of a cold, heavy plasma, comparable in number to the total plasma content of a flux tube outside the plasmapause. This in turn may significantly alter the conditions for wave growth both by lowering the local Alfvén velocity and by introducing a heavier ion component whose resonant frequency is considerably different from that of the dominant ion species. (For this purpose lithium is a better candidate for releases because its mass is more nearly that of naturally occurring species [Cornwall, 1972, 1974; Brice & Lucas, 1975]. Its main drawback is the long ionisation time of ≈ 1 h, which renders it ineffective as a field-line tracer.)

Ionosphere-magnetosphere coupling has been the topic of much recent research. The complexity of this relationship is best illustrated by the discovery by the Lockheed group [Shelly *et al.*, 1972; Johnson *et al.*, 1975] of large fluxes of energetic (1 ~ 10 keV) O^+ and He^+ over a wide range of L values during magnetic storms. Attempts have been made to account for these observations by wave-particle interactions [Cladis, 1973; Brice & Lucas, 1975] and Coulomb collisions [Torr *et al.*, 1974]. Our proposed tracer experiment is a direct probe of coupling mechanisms which may be involved in the acceleration and heating of heavy ionospheric constituents.

Although the University of Bern/MPE Garching Ion Composition Experiment on GEOS [Ghielmetti *et al.*, 1974] is particularly suited to detecting either lithium or barium ion tracers over a wide energy range (Table 2), it will be seen that virtually all GEOS particle, wave and field experiments are required to ensure success of the experiment. It should also be kept in mind that a similar composition experiment built by the Lockheed group in co-operation with Bern and MPE will be flown on the ISEE-A spacecraft.

TABLE 2

Summary of GEOS Plasma Experiments [Wrenn, 1975]

Experiment or detector number	α (°)	$\Delta\alpha$ (°)	$\Delta\beta$ (°)	Energy range (eV)	Differential geometric factor* $\left(10^{-4}\ \text{cm}^2\ \text{sr}\ \frac{\text{eV}}{\text{eV}}\right)$
S 302 - A	180	18	18	$0.5 \leq E \leq 500$	6
S 303	90	40	10	$0 \leq E \leq 100$	400
		40 → 10	10	$0 \leq E \leq 18$ k	400 → 1
S 310 - 8	195	15	13	$200 \leq E \leq 20$ k	3

* Detector count rate = (Differential geometric factor) x (Differential energy flux).

2. Feasibility of the GEOS Tracer Experiment

Because of the long transit time ($\gtrsim$ 1 h) required for barium ions to reach the equatorial plane, it will be no mean feat to release the jet so that ions do in fact hit the satellite. The main difficulties associated with this phase of the experiment can best be summarised as follows:

(a) Will any of the Ba^+ ions reach the geomagnetic equatorial plane?
(b) Assuming ions will reach the equator, can the jet be made to hit the satellite?
(c) If ions reach the satellite, can they be detected?

We cannot answer these questions categorically, but must instead estimate whether there is a reasonable chance of success, defined as detecting Ba^+ at GEOS within ~ 2 h after release. In this section, we examine each of these difficulties and attempt to set conditions that will maximise the chances of success. In the next section, we will discuss experiment objectives beyond these minimum conditions.

2.1. WILL THE ION JET REACH THE EQUATORIAL PLANE?

Using a series of shaped-charge barium releases, Wescott and co-workers have 'painted' magnetic field lines with barium ions over distances up to $\simeq$ 30 000 km. Their most spectacular success was a release at $L = 6.6$ code named Oosik [Wescott *et al.*, 1975a]. When last observed at R (release) + 55 min, the tip of the jet was estimated to be within $\simeq$ 10 000 km of the equatorial plane, despite having split into a number of striations at R + 16 min. Barring any unexpected phenomena within 10 000 km of the equator, the tip of the jet can be assumed to have reached the equator at approximately R + 70 min. For the purposes of this paper, the velocity distribution of Ba^+ in the Oosik release has been used to infer the equatorial velocity distribution assuming only gravitational deceleration of the ions (Fig. 1). Ion transit times from the ionosphere have been calculated as a function of equatorial energy and are shown in Figure 2.

It is worthwhile at this point to consider physical processes which might help or hinder expansion of the Ba^+ jet along the field line.

Barium jets are usually released above 500 km, where collisional deceleration of the ions is negligible and most ions in the jet have sufficient initial velocity to reach the equator (Fig. 1). In fact, ion mobility along the magnetic field may be somewhat increased by electrons escaping from the release and dragging the ions along. Mende [1973] noted that the light intensity of the NASA/MPE Ba^+ release at 31 000 km decreased as $t^{-1.7}$, rather than t^{-1} as would be expected for simple kinetic expansion of the cloud along magnetic field lines. This was attributed to an increased effective temperature of the ions parallel to **B** due to electrons from the release.

Electric fields parallel to **B** are thought to play an important role in auroral-particle acceleration. Possible sources of $E_\parallel$ are electrostatic double layers [Block, 1972a, b], anomalous resistivity caused by current instabilities [Kindel & Kennel, 1971], or so-called Alfvén layers in the magnetosphere [Rassbach, 1973]. The orientation of $E_\parallel$ upward or downward determines whether the Ba^+ is accelerated towards

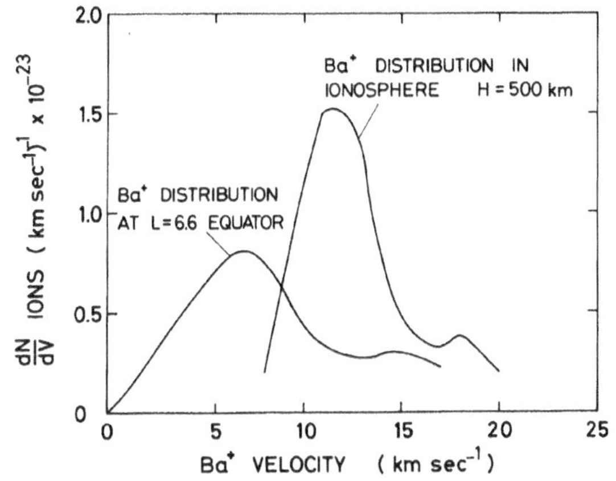

Figure 1. *Typical velocity distribution of Ba^+ ions from a shaped-charge release at 500 km and $L = 1.24$ [Wescott et al., 1974] projected into the equatorial plane at $L = 6.6$ assuming ions lose energy only to gravity.*

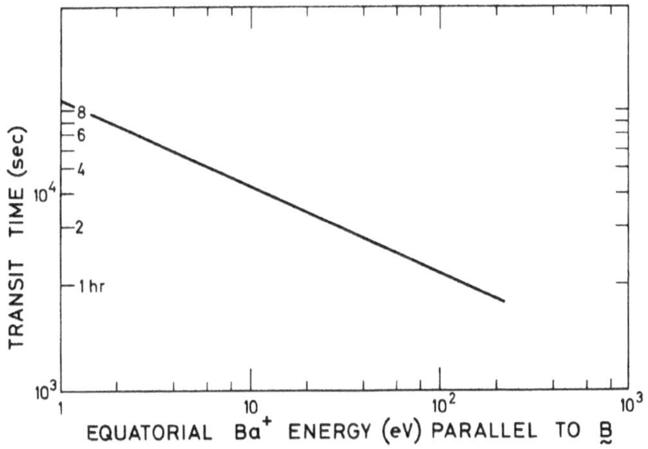

Figure 2. *Time required for Ba^+ ions to travel from a height of 500 km to the equator at $L = 6.6$ as a function of their parallel energy at the equator.*

the equatorial plane or prevented from reaching it altogether. Nothing can be said about the location or orientation of $E_{\|}$ except to note that there is evidence for upward fields that accelerate electrons into the auroral zone [Berko, 1973] and hence might provide favourable conditions for the proposed release experiment. On the other hand, a downward-directed electric field of $\gtrsim 4\ \mu m^{-1}$ distributed along the entire $L = 6.6$ field line would be sufficient to prevent Ba^+, distributed as in Figure 1, reaching the equatorial plane.

Ion energy losses may also result from jet instabilities or wave-particle interactions. Gross instabilities in the form of striations are commonly observed in deep-space barium clouds [Völk & Haerendel, 1970; Mende, 1973] and jets [Wescott *et al.*, 1972, 1974, 1975a], but there is no evidence that this prevents jet expansion along **B**. Ion cyclotron instabilities may occur when cold plasma is introduced into a region dominated by hot plasma, e.g. in the ring current outside the plasmapause. Several authors have proposed this as a method for destabilising the ring current to produce artificial particle precipitation [Brice, 1970, 1971; Cornwall, 1974] or enhance ULF and VLF amplification in the magnetosphere [Liemohn, 1974a, b]. What has not been examined to date is the effect of the resultant wave spectrum on the cold plasma itself, and hence on jet stability. Brice & Lucas [1975] have considered the effect of the cyclotron resonance on cold ionospheric O^+ inside the (proton) ring current and concluded that the heavier (O^+) ions could gain a considerable amount of available wave energy under certain conditions. This might also be true for Ba^+ ions in the jet.

In any case it does not seem energetically possible that the heavier ions would *lose* energy to the waves because the instability is driven by the much more energetic proton population. The heavier ions act as a catalyst [Liemohn, 1974a, b] and share in the resultant energy extracted from the protons. This is discussed further in the section on wave-particle interactions.

2.2. CAN THE JET BE MADE TO HIT THE SATELLITE?

From the preceding section, we conclude that, with the exception of downward-pointing parallel electric fields, there are no obvious phenomena which will prevent the ions from reaching the equatorial plane. Launching the release so that the Ba^+ jet will intersect the satellite, however, seems a more difficult requirement. In this section, we will therefore take up the problem of magnetic models and electric field drifts as they affect location of the jet relative to the satellite in the equatorial plane.

According to Wescott *et al.* [1975a], the initial diameter of the Oosik release at an altitude of 500 km was 10 km. At $L = 6.6$, this would result in an equatorial jet diameter of 230 km in the absence of ion diffusion perpendicular to **B**. For the purposes of the present experiment, it is desirable to have as broad a jet as possible consistent with detector sensitivity. This can be accomplished by releasing the jet lower in the Earth's UV shadow, thereby delaying ionisation and widening the jet. Assume, therefore, an initial diameter of 50 km, which then subtends about 0.5° in magnetic latitude and will spread to an equatorial diameter of $0.2R_E$. Olson & Pfitzer [1974] have introduced a magnetic-field model which includes the effects of the (distributed) ring current and fits rather well to observed magnetospheric particle data, including the 1971 NASA-MPE barium-cloud release at 31 500 km altitude [Adamson *et al.*, 1973]. During magnetic quiet, it seems feasible to use the Olson & Pfitzer model plus *in situ* GEOS magnetometer data to predict field-line geometry to within a few tenths of a degree, as required.

A potentially more serious constraint is introduced by the fact that the jet will convect due to the **E** x **B** drift caused by magnetospheric (plus ionospheric) electric fields. Convection electric fields depend on the level of magnetospheric disturbance

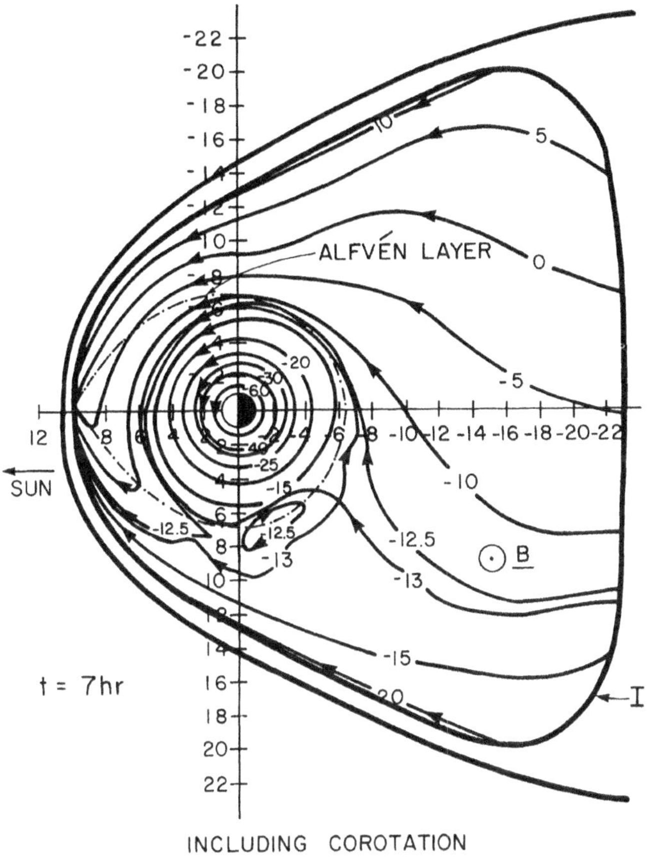

Figure 3. *A model of the electric potential distribution in the equatorial plane [from Wolf, 1974]. For very low energy ions, the equipotentials are also streamlines of convective flow.*

in a complicated way, but tend to be co-rotational during periods of magnetic quiet when $K_p \leq 2$. As a guide to what we might expect from convection, examples of possible cold plasma drift trajectories are shown in Figure 3, based on the theoretical model of Wolf [1974], and Figure 4, based on ATS-5 particle observations of McIlwain [1974]. Both models predict predominantly co-rotational drifts with some earthward component in the region of $L = 6.6$ and 21~24 h LT. In addition, Wolf's model (Fig. 3) shows the complex structure of convection expected in the 18~21 h LT sector, where the plasmapause usually intersects the geostationary orbit [DeForest, private communication, 1975]. Electric field strengths of $\lesssim 2$ kV/R_E ($\approx$ 0.3 mV/m) are predicted in both models, resulting in plasma drifts of $\lesssim 3$ km/s, equal to the satellite co-rotational speed and in approximately the same direction.

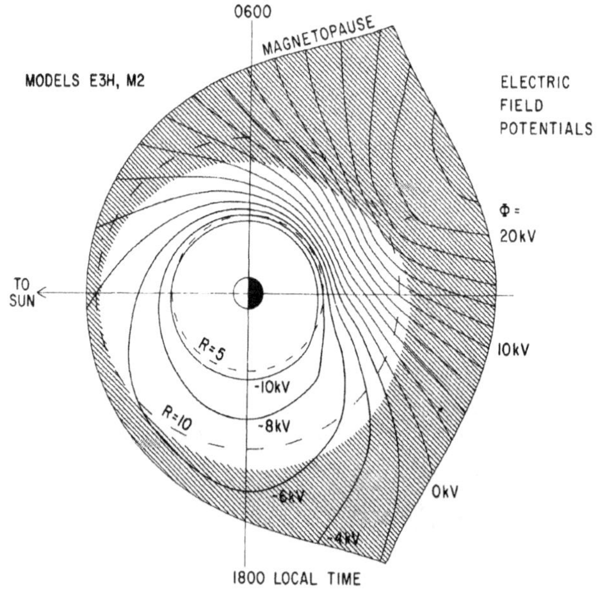

Figure 4. *A model of the potential distribution in the equatorial plane based on ATS-5 particle observations [from McIlwain, 1974].*

During the 50 min required for transit of the fastest barium ions in the jet to the equator, convection will move the jet some $1.5R_E$. The Oosik jet, released during the main phase of a magnetic storm, exhibited very rapid earthward drifts, which carried the jet well past the geosynchronous orbit before any ions reached the equatorial plane (Fig. 5). Such extreme periods of geomagnetic activity (Westcott *et al.* [1975a] reported K_p of 5 and D_{st} of -76γ at the time of the Oosik release) must be avoided, at least for the initial experiments with GEOS releases.

With the models of Wolf and McIlwain in mind, and assuming that the two GEOS electric-field experiments (double probes (S 300) and electron beam (S 329); see the review by Pedersen [1975]) measure the convection field, we can then determine with some confidence the correct UT, longitude and L-value of the release necessary for the jet to hit the satellite.

We also note that striations have the desired effect of enlarging the cross-sectional area of the beam perpendicular to **B**. During the Oosik experiment, for example, the jet striated and spread unevenly over several thousand kilometres (projected into the equatorial plane) in an east-west direction (Fig. 5). A resulting drop in ion density by an order of magnitude could be tolerated easily in the proposed experiment, provided the background plasma is not too hot (Fig. 6 and next section). An additional point to consider in this respect is effective elongation of the jet in the direction of its drift motion perpendicular to **B** due to differential arrival of ions in the equatorial plane over a period of several hours (Fig. 2), caused by the initial velocity spread of the jet.

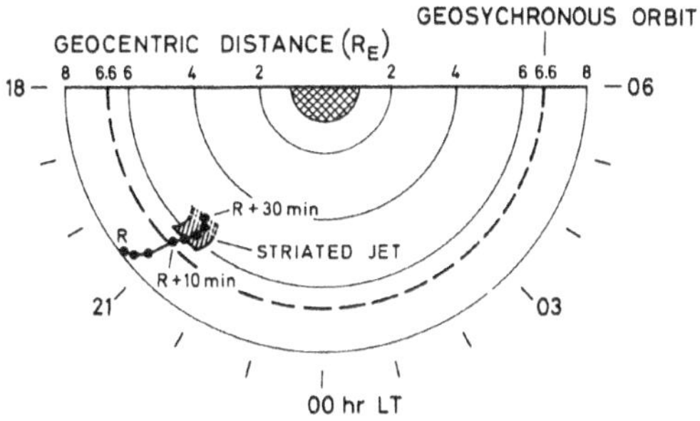

Figure 5. *Motion and striation formation in the Oosik shaped-charge release of 7 March 1972 projected into the equatorial plane. Release (R) occurred at 0659 UT [after Wescott et al., 1975a].*

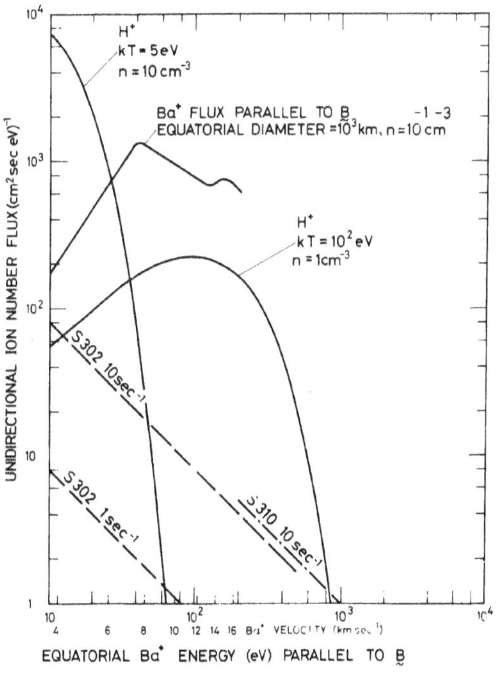

Figure 6. *Unidirectional number flux of Ba^+ ions reaching the equator with the velocity spectrum shown in Figure 1. GEOS plasma detector (S 302 and S 310) iso-counting lines are shown as well as the equivalent background number flux due to a hot (100 eV) and cold (5 eV) H^+ plasma. The limiting detector count rate is $\approx 1\ s^{-1}$ for both instruments. A unidirectional Ba^+ jet is seen to be easily detectable by S 302 and, with slight acceleration, by S 310.*

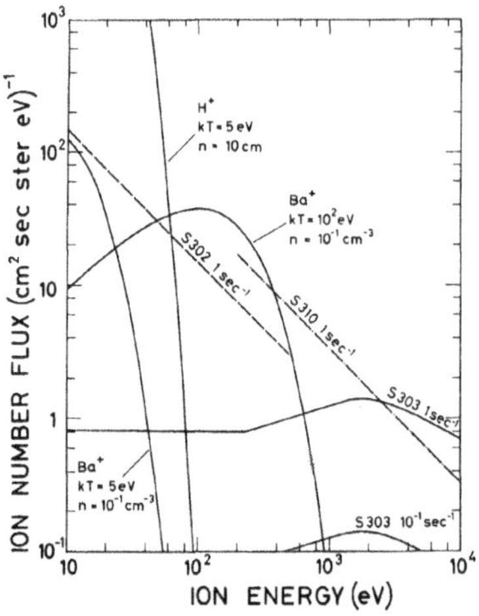

Figure 7. *Directional number flux of Ba^+ ions under the assumption that they are thermalised with and without deceleration ($kT = 5$ eV and 100 eV, respectively). GEOS plasma detector (S 302 and S 310) and mass spectrometer (S 303) iso-counting lines are shown together with the background flux due to a cold H^+ plasma. The limiting S 303 count rate is $\approx 10^{-1}$ s^{-1}. An isotropised Ba^+ jet is seen to be easily detectable by S 303, which does not have to contend with the H^+ background.*

2.3. CAN THE IONS BE DETECTED?

We assume an initial jet diameter of 50 km at an altitude of 500 km. The jet has 5×10^{23} Ba^+ ions distributed according to Figure 1 and expands parallel and perpendicular to the magnetic field. At the equatorial plane the unidirectional phase space density in units of ions $(cm^3\ km\ s^{-1})^{-1}$ is approximately $1{\sim}3 \times 10^{-2}$ over the entire velocity range.

Barium ions reaching the equatorial plane may be strongly field aligned or may be thermalised, depending on whether scattering takes place between the ionosphere and the equatorial plane. In the absence of scattering, ions released at 500 km will have equatorial pitch angles confined to a solid angle of 8×10^{-3} sr centred on the magnetic-field direction. This solid angle is much smaller than that of the GEOS plasma detectors S 302 and S 310 ($\approx 4 \times 10^{-2}$ sr, see Table 1), which view nominally along the magnetic-field direction. Because the field-aligned Ba^+ jet is confined to a small region of velocity space, it should be easily detectable by Experiment S 302 above the background thermal plasma (Fig. 6). At times, the nominal direction of **B** at GEOS may be as much as 15° away from the spacecraft spin axis [Wrenn, 1975], but this can be monitored by the onboard magnetometer experiment (S 331). If the jet is accelerated only slightly it will also be detectable by Experiment S 310, which has a lower energy limit of 200 eV. Detectors S 302 and S 310 will view in the direction of the winter hemisphere and barium-jet releases from Scandinavia must therefore be made between September and March.

If the Ba^+ ions are field-aligned at the equator, they cannot be detected by the GEOS Ion Composition Experiment, which nominally views the pitch-angle range $60° \lesssim \alpha_0 \lesssim 120°$ (Table 2). On the other hand, the ions may be thermalised during transit to the equatorial plane so that their angular distribution tends to be more isotropic. Figure 7 illustrates the situation when the ions are assumed to be distributed isotropically with kT = 5 eV or 100 eV. In this case, they are only marginally detectable by S 302 or S 310, but could be seen easily by S 303. The 5 eV distribution would result if ions are decelerated and thermalised in the process, while 100 eV is roughly equal to the streaming energy of the jet and might result from a transfer of parallel energy to the perpendicular component, as could occur in ion cyclotron wave interactions [Brice & Lucas, 1975]. Young [1975] has shown that Coulomb collisions alone will not contribute to appreciable scattering of the Ba^+ jet, although they would be effective for a Li^+ release.

3. Additional Experiment Objectives

Our prime objective, of course, is to detect Ba^+ at the GEOS satellite. Beyond that, a number of investigations are possible using various combinations of GEOS and ground-based detectors to study the ions themselves, as well as their effect on the plasma environment along the satellite flux tube. In this section we assume that the jet will reach the satellite and be detected and therefore proceed to a brief discussion of experiments which may be carried out using this technique.

3.1. MAGNETIC-FIELD-LINE TRACING

Optical tracking of the jet is necessary to establish the lower portion of its trajectory relative to field models and to record overall morphology and evolution of gross features such as jet expansion and striations. This has been done successfully by Wescott *et al.* [1975a, b], out to distances of $2 \sim 3\ R_E$ along the $L = 6.6$ field line, but optical tracking of the ions to the equator has not been possible. Detecting the ions at GEOS will extend tests of magnetic models deeper into the magnetosphere than has previously been possible.

3.2. ELECTRIC-FIELD MEASUREMENTS

GEOS will carry two experiments dedicated to measuring DC electric fields and several plasma detectors capable of inferring convection electric fields from plasma drifts [cf. McIlwain, 1972, 1974]. The Ba^+-jet technique extends these measurements in the sense that ions feel the effect of magnetospheric electric fields integrated along the entire magnetic field line. If magnetic field lines are equipotentials, the ions will drift exactly as predicted by GEOS electric-field measurements. If not, then the time profile of Ba^+ fluxes as they reach the satellite (Fig. 2) will yield information on the differential drift of ions perpendicular to **B**, and hence the extent to which ionosphere and magnetosphere are decoupled. Of course, if too much decoupling occurs the jet may not reach the satellite at all.

Electric fields parallel to **B** will alter ion transit times as well as the parallel energy distribution and may result in strongly field-aligned fluxes. Much theoretical work has been done on the subject of electrostatic double layers which produce intense parallel electric fields over short distances [Block, 1972a, b]. Double layers should result in sharply peaked ion energy spectra and may, in addition, act to scatter ions in pitch angle. The possible effect of double layers on Ba^+ jets and *vice versa* needs further consideration.

.3. WAVE-PARTICLE INTERACTIONS

The gyrofrequency of Ba^+ ions near the $L = 6.6$ equator is $\sim 10^{-2}$ Hz, roughly equal to the frequency of storm-associated Pc 5 pulsations [Barfield & McPherron, 1972] and so-called 'standing Alfvén waves', which occur during periods of magnetic quiet [Cummings *et al.*, 1969]. Both types of waves are common at $L = 6.6$ and it is possible that both may interact strongly with Ba^+ [R.A. Wolf, private communication, 1975].

The presence of Ba^+ ions along the length of the magnetic flux tube at $L = 6.6$ may have a destabilising effect on hot ring-current protons which can be measured by GEOS Experiments S 310 and, perhaps, S 321. Cornwall [1974] reviewed the feasibility of using isotropically released Ba^+ clouds for this purpose and concluded that at frequencies of interest the wavelengths of the electromagnetic cyclotron modes would be too long (~several R_E) to be contained in the cloud. This objection is largely removed by creating a jet several R_E long. The jet described in preceding sections does not have number densities $\gtrsim 3$ cm^{-3} required by Cornwall, but DeForest [1972] has found that the cold plasma density outside the plasmapause at $L = 6.6$ is much lower than the 1 cm^{-3} assumed by Cornwall and most other authors. If the cold plasma density is truly as low as DeForest inferred, then the addition of a jet with 5×10^{23} Ba^+ ions should have a profound effect on the local plasma environment.

Cornwall & Schulz [1971] and Brice & Lucas [1975] have noted that, for frequencies below the heavy ion gyrofrequency, the important quantity is the relative mass densities of the hot and cold ions rather than their relative number densities. In this case, Ba^+ has a distinct advantage over cold lithium or hydrogen. A jet with 50 km diameter in the ionosphere will result in ~30 amu/cm^3 distributed along the magnetic flux tube between equator and ionosphere, whereas the average cold hydrogen mass density outside the plasmasphere is ~ 1 amu/cm^3 [Carpenter & Park, 1973]. Liemohn [1974a, b] has computed wave amplification at ULF and VLF frequencies resulting from Ba^+ or Li^+ jets at $L = 4$. He concluded that amplification may be considerably increased within $\approx 20°$ of the equator outside the plasmapause. Enhanced ULF and VLF emissions should be easily observable by the GEOS S 300 experiment.

4. Summary and Conclusions

We have attempted to demonstrate that there is a reasonable chance of hitting the

GEOS satellite with a barium shaped-charge release detonated in the upper ionosphere. Convection electric fields will cause the Ba^+ jet to drift, but the deleterious effect of this motion can be compensated to some extent by choosing times of magnetic quiet and using GEOS electric-field experiments to determine favourable convective flow conditions for launch. Once the jet reaches the satellite, it can be detected by electrostatic plasma analysers S 302 or S 310 if the jet is strongly field aligned, or by the mass spectrometer S 303 if the jet is thermalised. In either case the wide energy range covered ($0 \sim 20$ keV) by these experiments allows the beam to be detected even in the presence of strong acceleration between ionosphere and equatorial plane.

Looking beyond the problem of hitting the satellite, the presence of the Ba^+ jet along the GEOS flux tube may present interesting possibilities for the study of wave-particle interactions in a multicomponent plasma. Most observations (ULF and VLF wave amplification, pitch angle and energy diffusion of jet ions or ambient plasma, heating of cold jet electrons by wave absorption) can be monitored quite well using the GEOS complement of detectors. The jet technique may offer some insight into the acceleration of ionospheric He^+ and O^+ observed during storms.

5. Acknowledgements

This work was carried out under Grants 2. 080. 73 and 2. 275. 74 from the Swiss National Science Foundation. I am grateful to R.A. Wolf for his encouragement and several helpful suggestions.

References

Adamson, D., C.L. Fricke, S.A.T. Long, W.F. Landon & D.L. Ridge, *J. Geophys. Res.* **78**, 5769 (1973).

Barfield, J.N. & R.L. McPherron, *J. Geophys. Res.* **77**, 4720 (1972).

Berko, F.W., *J. Geophys. Res.* **78**, 1615 (1973).

Block, L.P., *Cosmic Electrodyn.* **3**, 349 (1972a).

Block, L.P., Earth's Magnetospheric Processes (Ed. B.M. McCormac) Reidel, Dordrecht, Holland, p. 258, 1972.

Brice, N., *J. Geophys. Res.* **75**, 4890 (1970).

Brice, N., *J. Geophys. Res.* **76**, 4698 (1971).

Brice, N. & C. Lucas, *J. Geophys. Res.* **80**, 936 (1975).

Carpenter, D.L. & C.G. Park, *Rev. Geophys. Space Phys.* **11**, 133 (1973).

Cladis, J.B., *J. Geophys. Res.* **78**, 8192 (1973).

Cornwall, J.M., *Rev. Geophys. Space Phys.* **10**, 993 (1972).

Cornwall, J.M. & M. Schulz, *J. Geophys. Res.* **76**, 7791 (1971).

Cummings, W.D., R.J. O'Sullivan & P.J. Coleman, Jr., *J. Geophys. Res.* **74**, 778 (1969).

Deforest, S.E., *J. Geophys. Res.* **77**, 651 (1972).

Ghielmetti, A., H. Balsiger, P. Eberhardt, J. Geiss, P. Hirt, D. Young, H. Loidl & H. Rosenbauer, *Helvetica Physica Acta* **47**, 473 (1974).

Haerendel, G., in M.J. Rycroft & S.K. Runcorn (Eds.), Space Research XIII, Akademie Verlag, Berlin, p. 601.
Johnson, R.G., R.D. Sharp & E.G. Shelley, Paper presented at Nobel Symposium on Physics of the Hot Plasma in the Magnetosphere, Kiruna, Sweden, 1975.
Kindel, J.M. & C.F. Kennel, *J. Geophys. Res.* **76**, 3055 (1971).
Liemohn, H.B., *Space Sci. Rev.* **15**, 861 (1974a).
Liemohn, H.B., *Radio Sci.* **9**, 683 (1974b).
McIlwain, C.E., in Earth's Magnetospheric Processes (Ed. B.M. McCormac), Reidel, Dordrecht, Holland, p. 268, 1972.
McIlwain, C.E., in Magnetospheric Physics (Ed. B.M. McCormac), Reidel, Dordrecht, Holland, p. 143, 1974.
Mende, S.B., *J. Geophys. Res.* **78**, 5751 (1973).
Olson, W.P. & K.A. Pfitzer, *J. Geophys. Res.* **79**, 3739 (1974).
Pedersen, A., ESRO Special Publication SP-107, p. 149, 1975.
Rassbach, M.E., *J. Geophys. Res.* **78**, 7553 (1973).
Shelley, E.G., R.G. Johnson & R.D. Sharp, *J. Geophys. Res.* **77**, 6104 (1972).
Torr, M.R., J.C.G. Walker & D.G. Torr, *J. Geophys. Res.* **79**, 5267 (1974).
Völk, H. & G. Haerendel, in Intercorrelated Satellite Observations Related to Solar Events (Eds. V. Manno and D.E. Page), Reidel, Dordrecht, Holland, p. 280, 1970.
Wescott, E.M., H.M. Peek, H.C.S. Nielsen, W.B. Murcray, R.J. Jensen & T.N. Davis, *J. Geophys. Res.* **77**, 2982 (1972).
Wescott, E.M., E.P. Rieger, H.C. Stenbaek-Nielsen & T.N. Davis, *J. Geophys. Res.* **79**, 159 (1974).
Wescott, E.M., H.C. Stenbaek-Nielsen, T.N. Davis, W.B. Murcray, H.M. Peek & P.J. Bottoms, *J. Geophys. Res.* **80**, 951 (1975a).
Wescott E.M., E.P. Rieger, H.C. Stenbaek-Nielsen, T.N. Davis, H.M. Peek & P.J. Bottoms, *EOS Trans. Am. Geophys. Union* **56**, 301 (1975b).
Wolf, R.A., Magnetospheric Physics (Ed. B.M. McCormac), Reidel, Dordrecht, Holland, p. 167, 1974.
Wrenn, G., ESRO Special Publication SP-107, p. 159, 1975.
Young, D.T., ESRO Special Publication SP-107, p. 195, 1975.

DISCUSSION

L.R. Lyons: Will the ion spectrometer scan only a small region of the energy range when covering a small region of the mass range?

D. Young: It is possible to cover the full energy range or only a small region. There will also be complementary coverage of the full energy range by the other GEOS plasma detectors.

S. De Forest: In calculating the signal-to-noise ratio, did you allow for the fact that the Ba^+ ions fill only a small part of the S 302, S 310 antenna patterns?

D. Young: Yes.

M.J. Rycroft: What is the status of funding for the experiment described?

D.T. Young: There is at present no funding for the experiment, but I plan to contact groups working with shaped-charge releases about co-operation. It is important to remember that the payload does not need to be instrumented and hence is relatively less expensive.

PART 6

NEAR-EARTH EXPERIMENTS RELATED TO THE IMS SATELLITE PROGRAMMES

LOCAL ACCELERATION OF AURORAL ELECTRONS

D.A. Bryant

Science Research Council, Appleton Laboratory, Slough, UK

Abstract

Electron precipitation in the vicinity of auroral arcs is discussed. The energy spectra and degree of field alignment are interpreted in terms of an acceleration model in which the energy gain is distributed randomly about the mean value. Some suggestions are made for definitive future measurements.

1. Introduction

It seems to be fairly well established that the presence of field-aligned electrons (i.e. electrons collimated in the direction of the Earth's magnetic field) in auroral precipitation shows that particles can be accelerated within 1000 km or so of the Earth's surface. The arguments leading to this conclusion have been discussed by many authors [e.g. Hoffman & Evans, 1968; Paschmann *et al.*, 1972; Whalen & McDiarmid, 1972; Berko & Hoffman, 1974; Bosqued *et al.*, 1974; Evans, 1974]. The conclusions seem inescapable, though of course we must keep a wary eye on the field-aligned electrons recently observed at the geostationary orbit [McIlwain, 1975]. Acceleration is usually thought of as being due to a static or quasi-static electric field. However, there are many aspects of the data that are not readily explained in terms of such an electric field. I will discuss these and show that they can be explained if we introduce a significant degree of spread in energy gain during acceleration. Exclusive use is made of our own data in this discussion, since it is more readily to hand and is, as far as can be judged, fully consistent with other measurements of similar phenomena.

Knott & Battrick (eds.), The Scientific Satellite Programme During the International Magnetospheric Study, 413–423.
All Rights Reserved. Copyright © 1976 by D. Reidel Publishing Company, Dordrecht-Holland.

2. Predictions for Electrostatic Acceleration

It can readily be shown that the intensity of an initially isotropic Maxwellian stream of particles, after an acceleration Δ becomes

$$J(E) = A(E/E_0)\exp[-E/E_0]\exp[\Delta/E_0] \tag{1}$$

where E = electron energy
$E_0 = kT$ = temperature
A = a constant

Two conditions are that

$$E > \Delta \quad \text{for } \alpha \to 0 \tag{2}$$

and

$$\alpha < \alpha_C \tag{3}$$

where

$$\sin^2 \alpha_C = (B_2/B_1)(1 - \Delta/E) \quad \text{for } \Delta \geqslant E(1 - B_1/B_2)$$

and

$$\alpha_C = 90° \quad \text{for } \Delta \leqslant E(1 - B_1/B_2)$$

where B_1 is the magnetic-field strength just before acceleration, and B_2 is the magnetic-field strength at the point of observation. It is assumed in this that phase-space density and magnetic moment are conserved during acceleration.

Evans [1974] has pointed out that we must expect some electrons with energies $< \Delta$ and with pitch angles $> \alpha_C$ from another source, namely back-scattered and secondary electrons which leave the atmosphere and are reflected by the accelerating field. These results give three predictions.

(i) The intensity at the peak of the energy spectrum should be proportional to the energy at which the peak intensity occurs. This follows from Equations (1) and (2), since the highest intensity occurs at $E = \Delta$, the lowest energy in the spectrum (apart from the back-scattered and secondary contribution).

(ii) Pitch-angle distributions should show a sharp cut-off, or rather a transition from primary to secondary electrons, at an angle which is a function of energy and the ratio of magnetic-field strengths. Measurements of α_C would provide an estimate of the ratio of magnetic-field strengths and hence the location of the acceleration region.

(iii) The energy spectrum at small pitch angles should show a sharp transition at $E = \Delta$ from accelerated primary electrons to back-scattered and secondary electrons. The energy at which the transition occurs provides a direct measure of the accelerating potential difference Δ.

3. Observations

The most complete set of observations that we have relevant to the present purpose was obtained recently on a rocket flight over a quiet auroral arc. Electron intensities at three selected energies are shown in Figure 1. Within the region of high intensity, which we associate with the auroral arc, we see that the energy spectrum is harder at the centre than at either the northern or southern edges. The energy spectrum becomes even softer north of the arc. We find that this type of structure is a consistent feature of auroral arcs, and are drawn to the conclusion that it is the basic form of structure in auroral electron precipitation. The structure is reminiscent of that of the 'inverted v' events described by Ackerson & Frank [1972], and may be the same phenomenon on a smaller scale, although there is no direct evidence for this at present.

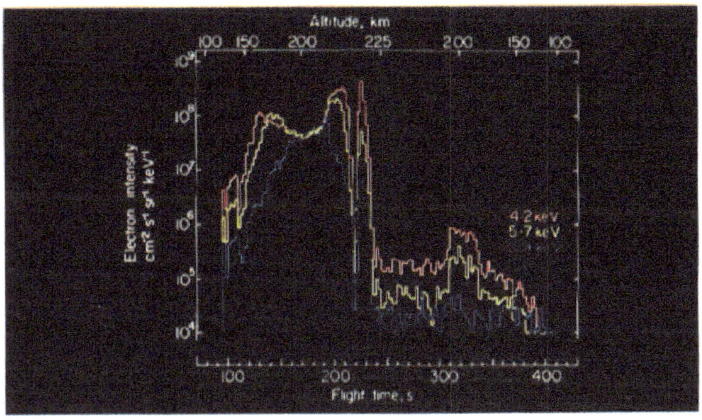

Figure 1. *Electron intensities at three selected energies measured during a rocket flight over a quiet auroral arc. The intensities are 2.5 s averages over the pitch-angle range 0–60°. The Skylark rocket, SL 1221, was launched from Andøya, Norway at 2043 UT on 1 November 1974, and travelled approximately northwards across the east-west aligned auroral arc. The apparent fine structure near the northern edge of the arc at about 220 s is due to movement of the arc.*

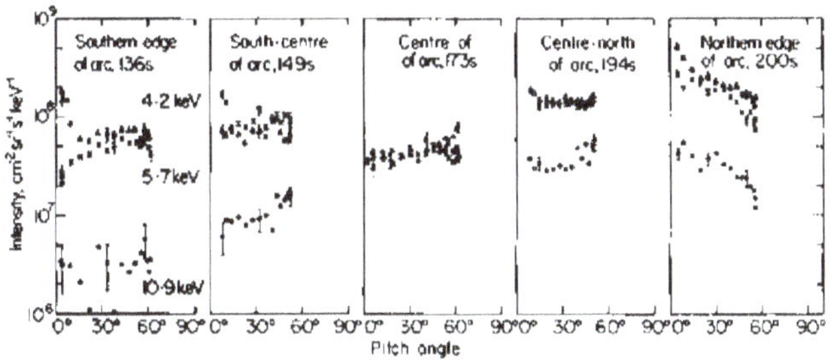

Figure 2. *Pitch-angle distributions at three selected energies (4.2 keV – ⧫, 5.7 keV – ✕, and 10.9 keV – ●) at various positions within the auroral arc of Figure 1.*

According to the first prediction we should expect, if all of the electrons came from the same source plasma, that the peak intensity should increase in proportion to the energy of the peak. This clearly is not the case here, since the peak intensity falls as the spectrum hardens. We find consistently in auroral arcs that the peak intensity either falls or remains approximately constant as the spectrum hardens.

Angular distributions at three selected energies obtained at various positions within the arc are shown in Figure 2. There is clearly some tendency towards collimation in the direction of the Earth's magnetic field, i.e. an increase of intensity towards smaller pitch angles; this occurs particularly at the energy where the intensity is highest but only when the spectrum is relatively soft, i.e. et the edges of the auroral arc. The fact that there is some collimation is alone strong evidence for local acceleration. If the acceleration had occurred beyond about 6000 km the collimation resulting from acceleration would have disappeared because of the increase of pitch angle with magnetic-field strength. Contrary to expectations, however, there is no sign of a sharp cut-off in any of the pitch-angle distributions which would indicate the transition from accelerated primary electrons to back-scattered and secondary ones. Instead, the intensity varies smoothly with pitch angle.

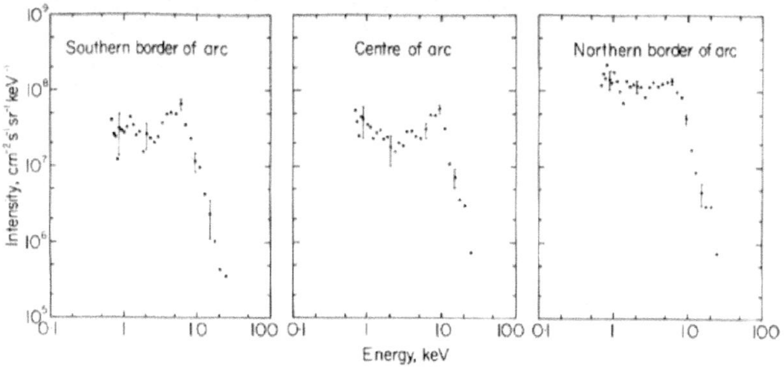

Figure 3. *Energy spectra of electrons at the southern border, centre, and northern border of the auroral arc of Figure 1. The energy spectra are for electrons with pitch angles ≈ 25°.*

In Figure 3 we see energy spectra taken at the centre of the arc and near the northern and southern edges. These energy spectra apply to small pitch angles where we might expect to observe a sharp transition from accelerated primary electrons to back-scattered and secondary ones. As with the pitch-angle distributions, the functions vary more slowly than expected, and so prevent us from estimating the accelerating potential difference with any certainty. This feature seems to be common to all measurements of auroral-electron energy spectra in auroral arcs and break-up events.

4. Fluctuating Energy Gain

To try to account for the above observations we have, as an empirical measure, introduced the possibility that the energy gained during acceleration is not the same for all particles [Hall & Bryant, 1974]. This could be caused by perhaps a fluctuating electric field, or by wave-particle interactions which modify energies during acceleration. We have assumed that the energy gained follows a normal probability distribution, and fluctuates more rapidly than we can resolve with our present resolving time of 0.01 s. The intensity we observe, therefore, is an average of intensities for all possible values of energy gained, weighted by the probability distribution $P(\Delta)d\Delta$.

i.e.

$$\overline{J(E,\alpha)} = A(E/E_0)\exp[-E/E_0]\int_{-\infty}^{\Delta m}\exp[\Delta/E_0]\,P(\Delta)\,d\Delta \tag{4}$$

where $\Delta m = E[1-(B_1/B_2)\sin^2\alpha]$, is the largest value of Δ making a nonzero contribution to the integral.

Integration gives

$$\overline{J(E,\alpha)} = \tfrac{1}{2}A(E/E_0)\exp[-E/E_0]\exp[\mu/E_0 + \sigma^2/2E_0^2](1+\operatorname{erf} x) \tag{5}$$

where μ is the mean value of Δ, σ is the standard deviation, and

$$x = [(E/E_0)\left\{1-(B_1/B_2)\sin^2\alpha\right\} - (\mu/E_0 + \sigma^2/E_0^2)]/\sqrt{2}(\sigma/E_0) \tag{6}$$

Acceleration by an electrostatic field is the special case of $\mu = \Delta$ and $\sigma = 0$.

5. Predictions for Fluctuating Energy Gain

Figure 4 shows that, if we introduce a fluctuation which is 15% of the mean value, the peak intensity varies with peak energy in a way which is much more in accord with observations than the predictions for electrostatic acceleration which are also shown in the figure. The sample angular distributions shown in Figure 5 are also more realistic than their counterparts for electrostatic acceleration, since they are more smoothly varying.

The empirical model provides, then, at least a qualitative fit to the data. We are in the process of using it to analyse the auroral arc described above. An earlier flight, in which an array of multiple arcs was crossed, has, however, been fully analysed. The pattern of precipitation forming the multiple arcs shows many features that are common to the single quiet arc described above. The pattern, though naturally more complicated, is less general in the sense that there is no detectable field alignment.

A typical energy spectrum, taken at the centre of one of the arcs in the array, is shown in Figure 6. The smooth curve running through the points near the peak intensity and high-energy tail of the spectrum is a best-fit curve, employing the indicated values of the source plasma and acceleration parameters. Roughly speaking, the high-energy tail of the spectrum indicates the temperature of the source

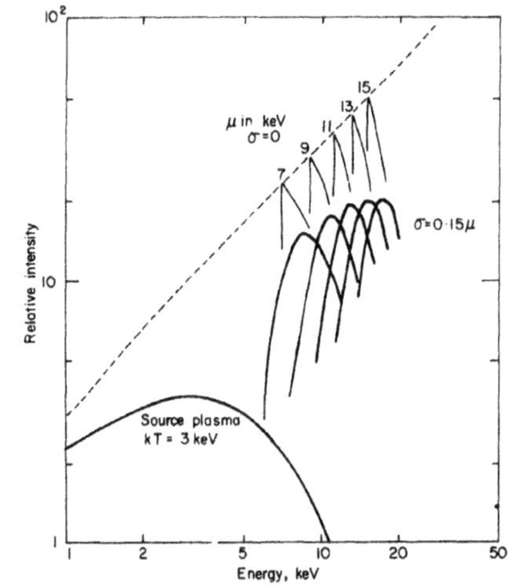

Figure 4. *Energy spectra of electrons in a Maxwellian source plasma, and after acceleration by μ = 7, 9, 11, 13, and 15 keV with $\sigma = 0$ (faint lines), and $\sigma = 0.15\mu$ (strong lines).*

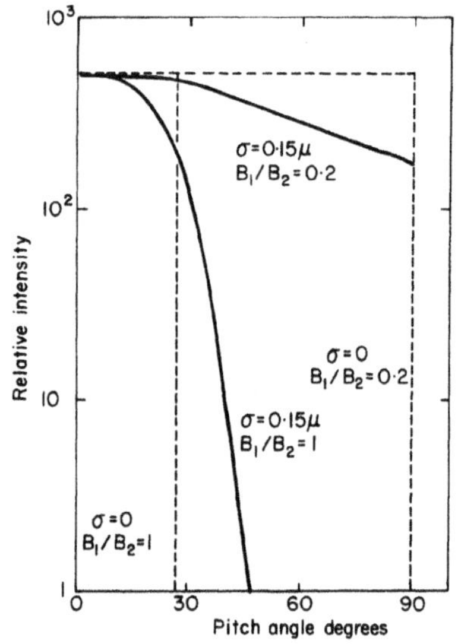

Figure 5. *Pitch-angle distributions of electrons of energy 10 keV produced by acceleration of a Maxwellian source plasma with kT = 3 keV after acceleration by μ = 8 keV, with $\sigma = 0$ (broken lines), and $\sigma = 0.15\mu$ (full lines). The pitch-angle distributions are shown for ratios of magnetic-field strengths before and after acceleration (B_1/B_2) of 0.2 and 1.0.*

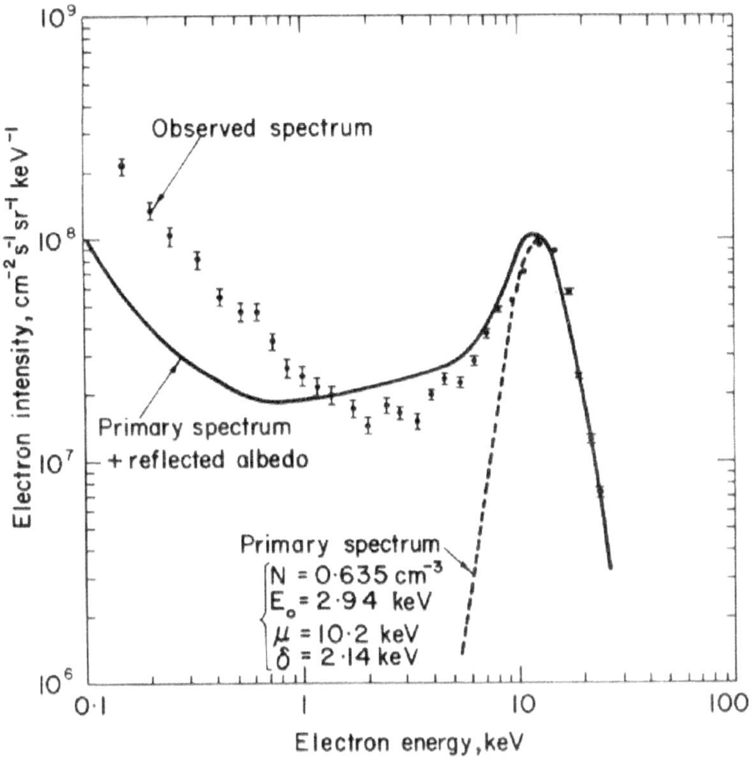

Figure 6. *An electron energy spectrum near the centre of an auroral arc, observed during the flight of a Skylark rocket (SL 1122) launched from Andøya, Norway at 1850 UT on 30 October 1973. The observed spectrum above 1 keV is well represented by a primary spectrum, consisting of a Maxwellian source plasma accelerated to the degree shown, combined with a reflected albedo of secondary and backscattered electrons.*

plasma, the energy of the peak intensity provides μ the mean value of acceleration, the width of the peak is a measure of σ the standard deviation, and the absolute intensity gives N, the density of the source plasma. The smooth curve drawn for energies below the peak intensity is a first attempt to explain the low-energy electrons in terms of back-scattered and secondary electrons reflected by the accelerating field. The curve is computed from the higher-energy intensities according to a method devised by Evans [1974] based on calculations by Banks [1974]. The fit is reasonably good down to an energy of 1 keV. There has been no attempt to obtain the best possible fit, so the discrepancy at the lowest energies cannot be taken as significant at this stage.

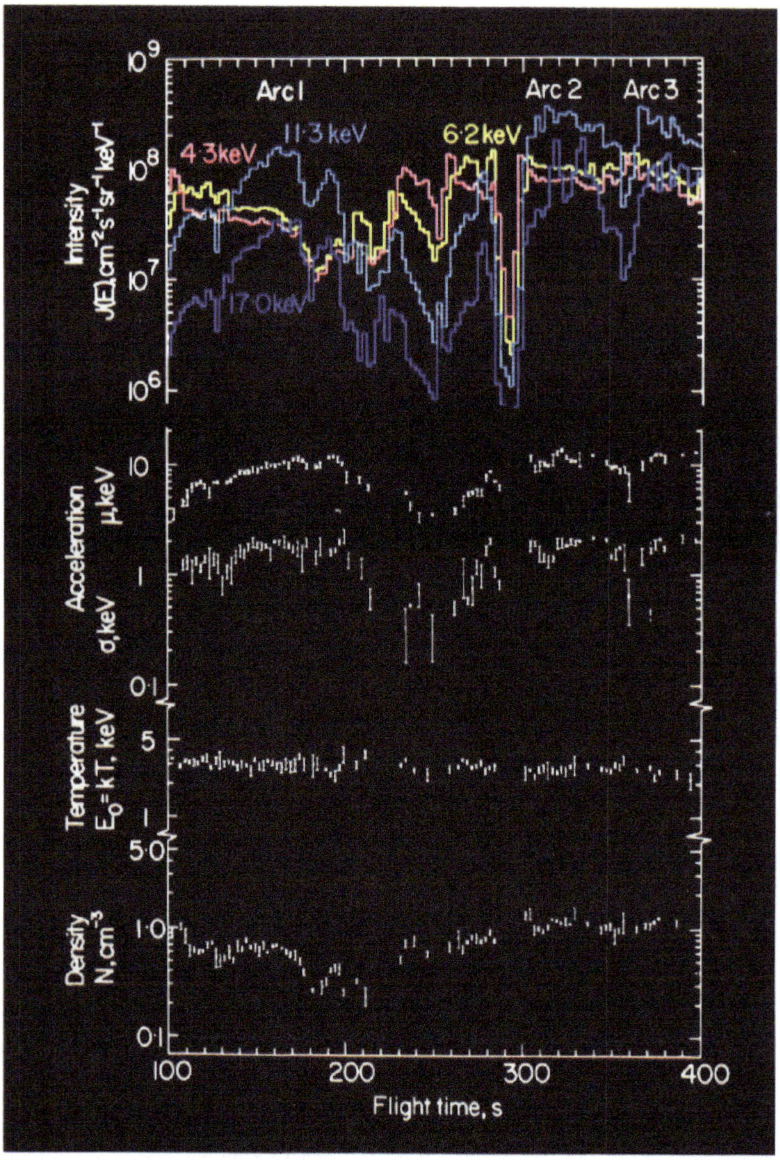

Figure 7. *Analysis of flight SL 1122. The figure shows electron intensities at four selected energies, and derived values of the mean level of acceleration μ, the fluctuating component of acceleration σ, the temperature of the source plasma E_0, and the density of the source plasma N. The electron intensities are 2.5 s averages over the pitch-angle range 10–70°.*

We should return now to the high-energy part of the spectrum, and examine how the parameters describing the source plasma and acceleration mechanism vary throughout the multiple-arc structure. Figure 7 shows intensities at representative energies, and the parameters of the source and acceleration derived from energy spectra as shown above. We see that the data can be represented by four simple curves. The mean value of acceleration varies smoothly through three maxima, accounting for the three regions of relatively hard energy spectrum. The fluctuating component of acceleration σ remains throughout at approximately 15% of the mean, μ. In other events that we have begun to analyse, σ is relatively larger, being in some cases approximately equal to μ. The temperature indicated for the source plasma, which is probably the best clue to its identity, is approximately constant and equivalent to 3 keV. In the present case, it suggests that the source plasma is the plasma sheet. Although it is outside the scope of the present paper to discuss the origin of the electrons precipitating to the north of the auroral arc of Figure 1, it should be mentioned that the softness of the energy spectrum in this region would indicate a lower temperature source plasma, possibly the high-latitude magnetotail. Returning to Figure 7, the density indicated by the analysis is again consistent with the source plasma for the whole array being the plasma sheet. The density given by the analysis may, however, be an underestimate since it depends on the degree of scattering of source plasma into the accelerating region. It may therefore not be a good indicator of the identity of the source plasma.

Throughout the multiple-arc structure, electron intensities were isotropic (to within 10%) at all energies. This would suggest that acceleration occurred far enough away for the increasing magnetic field to destroy the collimation. Such arguments would place the acceleration region at a distance of more than 6000 km for this event. In other structures where there is collimation, as in the single arc, the acceleration appears to have taken place much closer, possibly within 1000 km of the Earth's surface. It will be interesting to know whether the lack of collimation in the centre of the single arc is due to acceleration to high energies starting at a greater altitude than acceleration to low energies. If we could determine this, it might help to distinguish between double layers and anomalous resistivity as the acceleration mechanism.

6. Future Measurements

It would be valuable to seek direct evidence for the existence of the fluctuations in acceleration considered here, using an improved time resolution. This is not easily attainable since it requires faster-counting channel multipliers than are available at present.

Reference to Figure 7 shows that some of the measured intensities were too low for the energy spectrum to be analysed, thus producing gaps in the source and acceleration parameters. An increased dynamic range would be helpful in dealing with this problem, and might provide evidence for whether retardation as well as acceleration could occur in some regions.

I fully endorse the comments made by other speakers at this meeting that we should attempt to observe particle streams at significantly different altitudes simultaneously. This would make it possible to carry out direct studies of particle acceleration and wave-particle interactions. The quiet auroral arc seems a suitable phenomenon to study in this way, since it is well defined and therefore recognisable at different altitudes. Such studies could probably be best carried out using high- and low-altitude satellites, though they might also be possible under some circumstances using rockets, or a combination of the two types of vehicle.

7. Acknowledgement

This work was carried out as part of the research programme of the Appleton Laboratory of the UK Science Research Council and is published by permission of the Director.

References

Ackerson, K.L. & L.A. Frank, Correlated satellite measurements of low-energy electron precipitation and ground-based observations of a visible auroral arc, *J. Geophys. Res.* 77, 1128 (1972).

Banks, P.M., C.R. Chappel & A.F. Nagy, A new model for the interaction of auroral electrons with the atmosphere; spectral degradation, backscatter, optical emission and ionization, *J. Geophys. Res.* **79**, 1459 (1974).

Berko, F.W. & R.A. Hoffman, Dependence of field-aligned electron precipitation occurrence on season and altitude, *J. Geophys. Res.* **79**, 3749 (1974).

Bosqued, J.M., G. Cordona & H. Reme, Auroral electron fluxes parallel to the geomagnetic field lines, *J. Geophys. Res.* **79**, 98 (1974).

Evans, D.S., Precipitating electron fluxes formed by a magnetic-field-aligned potential difference, *J. Geophys. Res.* **79**, 2853 (1974).

Hall, D.S. & D.A. Bryant, Collimation of auroral particles by time-varying acceleration, *Nature* **251**, 402 (1974).

Hoffman, R.A. & D.S. Evans, Field-aligned electron bursts at high latitude observed by OGO-4, *J. Geophys. Res.* 73, 6201 (1968).

McIlwain, C.E., Auroral electron beams near the magnetic equator, UCSD-SPM-75-03, April 1975. To be published in Nobel Symposium Proceedings, Plenum Publishing Co., London, 1975.

Paschmann, G., R.D. Johnson, R.D. Sharp & E.G. Shelley, Angular distributions of auroral electrons in the energy range 8-16 keV, *J. Geophys. Res.* 77, 6111 (1972).

Whalen, B.A. & I.B. McDiarmid, Observations of magnetic-field-aligned auroral electron precipitation, *J. Geophys. Res.* 77, 191 (1972).

DISCUSSION

R.J. Hynds: What are the approximate source dimensions of the region in which the ΔE increase takes place?

D.A. Bryant: The degree of collimation is rather insensitive to the dimensions of the acceleration region. However, I think it may be possible, using the collimation observed over a range of different energies, to draw equipotentials for the mean accelerating field. This is certainly our aim.

G. Rostoker: For flight SL 1122 at 165s the fit of the model to the data did not look too good at low energies. Have you any comment on this?

D.A. Bryant: Yes, the curve was drawn to test the hypothesis that the low-energy electrons originated as backscattered and secondary electrons. We have not adjusted any parameters to obtain the best possible fit, so in this sense the test is not yet complete. A definite excess of low-energy electrons would be most interesting since it might indicate a heating of the local plasma, as suggested, for example, by Papadopoulous & Coffey.

A. Balogh: The data you presented showed no collimation of particles along the field lines, contrary to what was expected. What parameter in your model needs adjusting in order to explain the lack of collimation?

D.A. Bryant: The degree of collimation depends on the ratio of the strength of the magnetic field at the observing point to the strength in the acceleration region. The absence of collimation in the second event I showed would indicate a large ratio of field strengths, and so an acceleration region 6000 km or more away. We have not yet fully analysed the first arc crossing and cannot give a location for the acceleration region, but the marked collimation here indicates that the acceleration region was much closer in this case. The adjustable parameter is the ratio of the magnetic fields.

SUPPORT OF SATELLITE MISSIONS BY HIGH-LATITUDE ROCKET CAMPAIGNS

B. Theile

Institut für Geophysik und Meteorologie,
Technical University of Braunschweig, Germany

Abstract

One basic problem in the physical interpretation of satellite data is the ambiguity of temporal and spatial effects. Moreover, there is quite a lot of uncertainty as to how waves and particles propagate from the outer to the inner magnetosphere and as to how these phenomena are displayed in the ionosphere. Simultaneous observations at typical satellite altitudes up to geosynchronous orbits and by rocket-borne instruments provide adequate means of tackling these problems. Several rocket campaigns are planned which are more or less 'tied' to observations made by the GEOS satellite.

1. Introduction

German as well as British groups are planning sounding-rocket campaigns during the IMS. The data collected by the rocket-borne instruments will be analysed in combination with satellite measurements.

The German high-latitude sounding-rocket programme consists of three projects:

- Porcupine
- Polar High Atmosphere
- Substorm phenomena.

Four payloads will be launched within each of these projects, in the time span October 1976 to March 1977.

A total of eleven rockets will be launched within the UK programme, which will take place in the same time period. The UK programme, as well as the German 'Substorm phenomena' and 'Polar High Atmosphere' projects, will be carried out from Andøya Rocket Range, Norway. A British-German working group has been established in order to co-ordinate the rocket launches, the ground-based observa-

Knott & Battrick (eds.), The Scientific Satellite Programme During the International Magnetospheric Study, 425–432.
All Rights Reserved. Copyright © 1976 by D. Reidel Publishing Company, Dordrecht-Holland.

tions, and as far as possible the data analysis. The 'Porcupine' payloads will be launched from Kiruna, Sweden.

Some of the rocket payloads depend strongly on simultaneous measurements by the GEOS satellite and thus any postponement of the GEOS mission will affect the planning of the sounding-rocket campaigns.

2. The Necessity of Simultaneous Observations

The processes involved in magnetospheric-ionospheric coupling are far from being understood. The night-side field lines mapping into auroral latitudes connect two different 'species' of thermal plasma. High temperature (10^6 °K) and low density (1 cm^{-3}), resulting in a mean free path of approximately $10^7 R_E$, are characteristic properties of the plasma sheet. In the ionosphere, the mean free path is in the order of $1\ R_E$ and the temperature and density are in the order of 6000°K and several hundreds per cm^3, respectively. Different approaches have been used to explain auroral particle acceleration: pressure gradients, parallel electric fields due to space charges or double layers, anomalous resistivity by wave-particle interaction [c.f. Boström, 1972], and thermoelectric effects [Knight, 1973]. In order to observe, for example, acceleration processes, one has to measure the properties of charged particles at two different locations simultaneously. The probability of having two satellites at two locations on one field line of the Earth's magnetic field is fairly remote. If additional criteria are also to be fulfilled, such as the presence of a substorm event, it would seem almost impossible to meet all the conditions necessary for successful measurements. On the other hand, knowing the position of any one satellite, one can launch a sounding rocket at a moment when substorm or other necessary conditions are favourable, comparatively easily.

3. Some Properties of Satellite Measurements

The properties of observations made by satellite-borne experiments depend on the spacecraft's orbital characteristics. Geosynchronous orbits have the advantage of maintaining the spacecraft in a fixed position with respect to the Earth, allowing the field line (the term 'line' is used for what is really a field-line region)to be mapped from the satellite down to its foot-point at the Earth's surface. This foot point has no invariant geographic co-ordinates; local time dependence of the geomagnetic field, as well as distortions resulting from geomagnetic disturbances, will change its position. Data collected by instruments in a geosynchronous orbit usually allow variations to be interpreted as temporal when the time scale of these variations is small. One disadvantage of geosynchronous satellites is their inability to determine latitudinal variations, a task that is best served by polar orbiting spacecraft. However, the local time behaviour of polar orbiting satellites is determined by the initial orbit parameters. If the orbit passes close to the Earth, their velocity is in the order of 10 km/s, making it very difficult to distinguish a spatial or temporal nature in observed parameters. Figures 1 and 2 show transverse magnetic variations as measured aboard the German AZUR research satellite. The data shown in Figure 1 may be interpreted in terms of field-aligned currents; those in Figure 2 seem to indicate temporal phenomena. It is impossible to interpret these data without making further assumptions.

Satellites scanning a wider region of the magnetosphere will have more eccentric orbits, resulting in an even higher velocity when they are close to the Earth, and so reducing the spatial resolution. The great advantage of data collection by satellite-borne instruments is the wide data coverage, from which one can select parts for further investigation.

If one wants to explore a single event like a substorm, the availability of a satellite in an appropriate geosynchronous position provides the best likelihood of success.

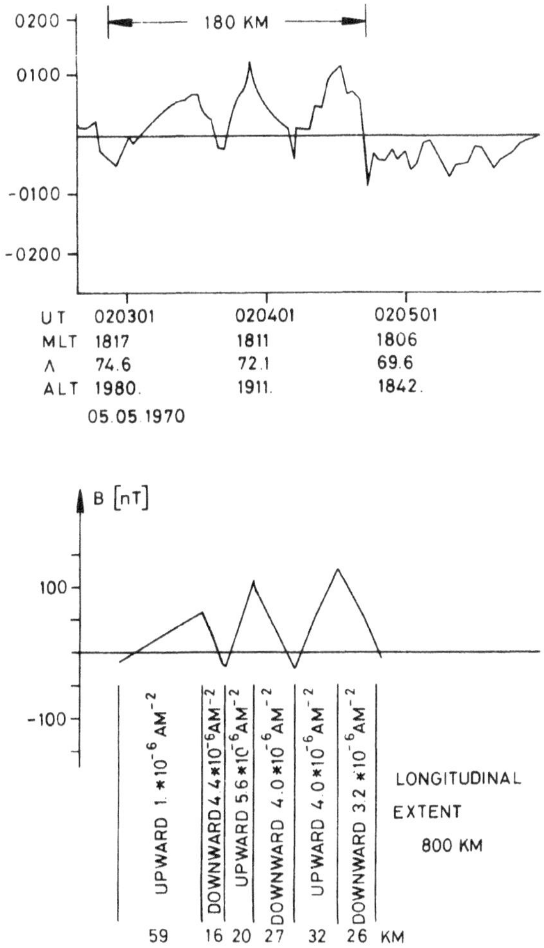

Figure 1. Transverse magnetic variations (upper panel) and magnetic fields of model current layers (lower panel).

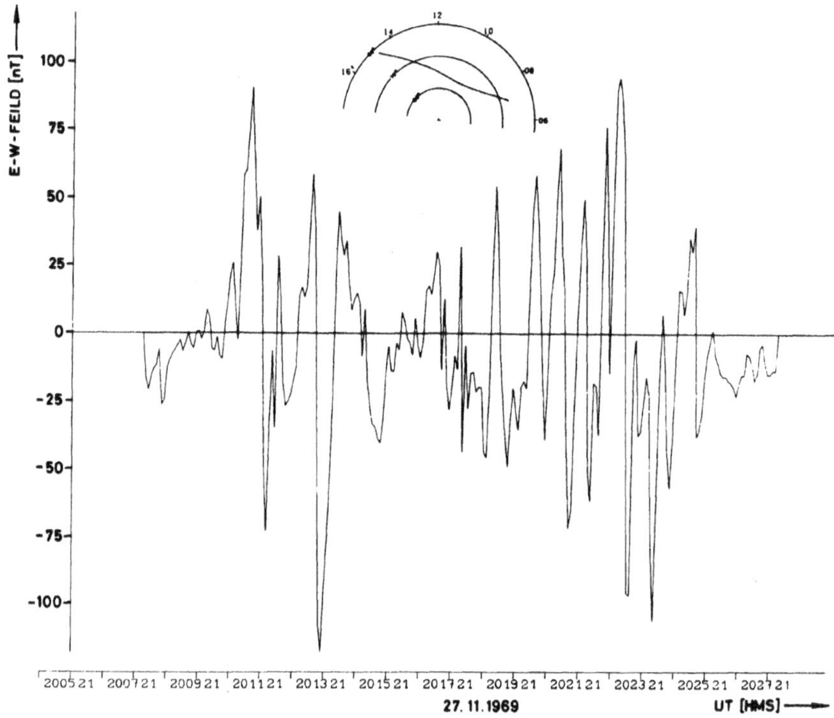

Figure 2. *Transverse magnetic variations which seem to be of temporal nature.*

4. Some Properties of Sounding-Rocket Observations

Sounding rockets offer the possibility to launch an instrument package into a specific phase of a substorm event, so that to some extent one can determine the time and the location of the measurement. Moreover, rockets are the only means of carrying instruments into the lower ionosphere. Sounding-rocket observations, however, yield only a snapshot, while during even a short flight highly temporal variations may prevail. Simultaneous ground observations, from which one can draw conclusions as to the temporal history of the event, are a necessity.

There are many uncertainties as to how magnetospheric processes are displayed in the ionosphere. To improve our knowledge in this field we can either launch the rockets to higher altitudes, or use simultaneous satellite observations. If a rocket is launched to higher altitudes, it will pass rather rapidly through the lower ionosphere, which results in poor spatial resolution. Hence, we have to choose the apogee according to the main feature of interest. For ionospheric investigations, the apogee should be low, while for magnetospheric payloads it should be as high as possible to avoid the measurement of fields originating in the ionosphere.

5. Simultaneous Observations by Satellites and Sounding Rockets

'Simultaneous' in this context means not only at the same time, but also in the same field-line region. To observe the same phenomenon at two different locations in magnetospheric-ionospheric space, one has to be sure that the satellite and rocket trajectories intersect the same field line. This field-line condition has to be monitored by means of model calculations for the geomagnetic field. These field-line models do not take into account all variations of the geomagnetic field, which means that some uncertainties will always remain. Indeed, proof of simultaneous observations on the same field line may, in some cases, be deduced from the measurements themselves.

Geostationary satellites in the same geomagnetic meridian plane as the rocket launch site offer the best opportunity for simultaneous measurements on the same field line. Polar orbiting satellites, even though not at an appropriate location during a rocket launch, are very useful both for determination of the auroral oval and for data collection for synoptic data evaluation.

Those questions for which we can expect to obtain answers by simultaneous observations at geosynchronous satellite and sounding-rocket altitudes can be summarised as:

- What kind of waves propagate from the outer to the inner magnetosphere? In how far do they generate anomalous resistivity by wave-particle interaction?
- How do electric fields map into the ionosphere?
- How do the particle energy spectra change from the magnetosphere to the ionosphere?
- How do the primary auroral particles scatter into the loss cone?
- Can feedbacks of ionospheric processes be observed in the magnetosphere?
- How much energy is deposited in the thermosphere?

6. Sounding-Rocket Experiments

As mentioned above, there will be UK and German sounding-rocket projects from northern Scandinavia to support the IMS satellite projects and to be supported themselves by satellite measurements. As identical or similar instruments are to be flown within several different projects, the experiments have been listed rather than the projects: those within the German projects in Table 1, and those of the UK programme in Table 2.

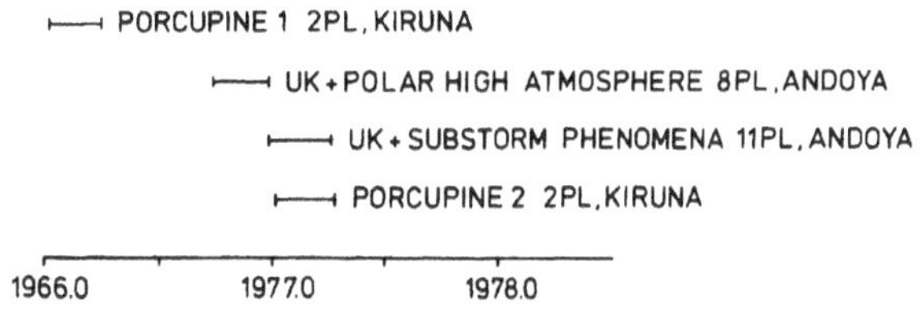

Figure 3. *Preliminary launch schedule for UK and German sounding-rocket projects.*

TABLE 1

Experiments within the German High-Latitude Programmes

Group	Pay-load	Study	Basic measurement
MPI Lindau	A	Electron and proton fluxes. Energy input by charged particles	$0.05 \leq E_{e,p} \leq 40$ keV
RIT Stockholm IPW Freiburg/ ESTEC	A, P, T	Electric fields, conductivity tensor, wave particle interaction	E (DC and AC)
TU Braunschweig	A, P, T	Magnetic fields of ionospheric and field-aligned currents	B
MPI Heidelberg	A	Changes in plasma and neutral composition caused by energy dissipation	Number densities of neutrals and ions; n_n, n_i
Univers. Bonn	A	Heating of neutral gas by precipitating energetic particles	6300μ of O
IPW Freiburg	A	Heating effects, electron density	$n_e(h)$, T_e
	A	Energy transport by low-energy electrons	$0 \leq E_e \leq 50$eV
	A	Heating of neutral gas by precipitating particles	$500 \leq \lambda \leq 1350$A
	P, T	Thermal plasma parameters	T_e, n_i, ion wind
SSL Berkeley/ ESTEC	P	Plasma density fluctuations	DC - 20 kHz
GRI St.-Maur-des Fossés	P	Electron density and temperature	Upper hybrid frequency
GRI St.-Maur-des Fossés	P	Field-aligned currents	Electron drift velocity
MPE Garching/ SSL Berkeley	P	ELF, VLF phenomena	AC magnetic fields
TU Braunschweig	T		
MPI Lindau	P	Energy and pitch-angle distribution of electrons and protons	20 keV $\leq E_e \leq\sim$ 500keV 40 keV $\leq E_p \leq\sim$ 500keV
SSL Berkeley	P	Spatial and temporal behaviour of fields and particles	Ejectable probes
MPI Garching	P	Plasma drifts, parallel electric fields, field-line mapping, generation of Alfvén waves	Barium ion jet
MPI Lindau/ Univers. Graz	P, T	Energy and pitch-angle distribution of low-energy particles	50 eV $\leq E_e$, $E_p \leq$ 30keV
Univers. Kiel	T	Energetic protons	0.15 MeV $\leq E_p \leq$ 8MeV

Payloads A – Polar High Atmosphere P – Porcupine T – Substorm phenomena

TABLE 2

Experiments within the UK Programme

Group	Study	Basic measurement
Appleton Lab.	Local acceleration of particles Formation of spatial structures	$E_{e,\,p} > 500$ eV
MSSL	Stability of field aligned currents Precipitation in stable, diffuse aurora	$E_{e,\,p} < 500$ eV
Sheffield/ Southampton Univ.	Comparison of wave-particle phenomena at magnetic equator and auroral zone	ELF, VLF electric and magn. fields
Sheffield Univ.	Support of Exp. 3	Positive ion densities 1 to 40 amu
Southampton Univ.	Origin of auroral particles	Energetic positive ions
UCL	Structure of neutral atmosphere and ionosphere during geomagnetic disturbance	Neutral-wind properties **B**, **E**
	Temporal changes in atmospheric and ionospheric structure	Neutral-wind properties airglow IR→UV
Univ. of Sussex	Electrostatic wave-particle interactions	Electrostatic waves 20 kHz – 4MHz

MSSL = Mullard Space Science Laboratory
UCL = University College London

7. Preliminary Launch Schedule

Interdependence between the GEOS experiments and the various rocket payloads is such that individual rocket payloads may be affected by the postponement of the GEOS launch. The schedule shown in Figure 3 was drawn up before any delay in the GEOS launch became known. Changes will depend on how closely the rocket experiments are related to GEOS, on the nature of the final spacecraft shift plan, as well as on the the extent to which any rocket projects affected can be postponed.

References

Boström, R., Magnetosphere-ionosphere coupling, in Critical Problems of Magnetospheric Physics (Ed. E. Dyer), IUCSTP Secretariat, Washington DC, 1972.
Knight, S., Parallel electric fields, *Planet. Space Sci.* 21, 741–750 (1973).

DISCUSSION

D.T. Young: In view of model predictions by Olson and Pfitzer that the GEOS field line foot is several degrees further south than thought from earlier models, is it possible that Kiruna is a better launch site for conjugate experiments than Andøya?

B. Theile: The field models are not as accurate as one might believe. The representation of the internal field, in particular, will be rather bad during the IMS as we do not have recent magnetic field observations to compute accurate secular variation terms for the spherical harmonic expansion. In addition to that, Skylark 12 rockets can not be launched from Kiruna because of safety requirements.

M.A. Shea: The International Geomagnetic Reference Field for Epoch 1965.0 has time derivatives such that the coefficients for previous Epochs can be reasonably derived and those for future Epochs can be predicted. One of the problems is that the changes in the geomagnetic field over the period 1965-1975 are larger than the time derivatives predict. This problem, plus the subject of the need for accurate magnetic field models, was a major topic of discussion at the recent Zmuda Memorial Conference held in Colorado in March 1975. One of the recommendations of the participants was that a revised geomagnetic field model should be determined, and that the model should be as accurate as possible for the users of this model.

A second problem of magnetic field models is that the inclusion of magnetospheric currents and the magnetospheric tail must be considered in developing a mathematical model of the magnetosphere. This is particularly important for high-latitude locations. Although *'in-situ'* measurements can be obtained for the geomagnetic field configuration for a specific time (which is an aid to a specific project or measurement), it is difficult to use these measurements for construction of a mathematical model of the magnetic field (either internal field or internal + external field) because of the many variables involved.

M. Dryer: You mentioned the necessity of launching some payloads into certain kinds of events. I wonder if you are aware of the group in Boulder (Space Environment Services Center, NOAA) which could provide real-time monitoring (and some forecasting), together with communication (to be arranged by the 'customer') in real-time with launch control. This kind of activity has been used before with much success.

THE ROLES OF NETWORKS OF VLF DIRECTION-FINDING RECEIVERS DURING THE IMS

M.J. Rycroft *

Department of Physics, University of Houston, Texas, USA

Abstract

A simple model of the distribution of electric fields and plasma dynamics in the magnetospheric equatorial plane is first reviewed. The significance of the plasmapause in this context is mentioned. A discussion of the power, and limitations, of the normal whistler method for studying this region of space plasma leads on to a consideration of the roles that can be played by networks of VLF direction-finding (goniometer) receivers.

Some results are presented of an experiment conducted in Eastern Canada during July 1972 using four goniometer receivers. The regions where two-hop whistlers emerged into the Earth-ionosphere waveguide were obtained by triangulation. For nearly 2 h on this occasion, the L-values of the whistlers' exit points were systematically ~ 0.6 less than the L-values determined from the whistlers' nose frequencies. This result can be understood once the ray paths of unducted whistler-mode signals have been computed in a model of the topside ionospheric plasma having concentration gradients with latitude at constant altitude. The relative frequency of occurrence of this result is not known.

Preliminary results of a similar experiment carried out in Alaska during March 1974 following the injection of barium up to geomagnetic field lines are also presented. Discrete ELF and VLF emissions were observed some 90 min after the release; these may be due to the stimulation of the electron cyclotron instability via the enhancement of the thermal plasma density.

The significance of such experiments, which need to be performed with simultaneous topside ionospheric soundings, is discussed. Field-aligned ducts of interest, such as those on which wave-particle interactions are taking place or within which a certain satellite exists, may then be localised in space. Motions of the ionospheric

* On leave from Dept. of Physics, University of Southampton, UK.

Knott & Battrick (eds.), The Scientific Satellite Programme During the International Magnetospheric Study, 433–450.
All Rights Reserved. Copyright © 1976 by D. Reidel Publishing Company, Dordrecht-Holland.

exit points may be studied directly; by comparing these with those deduced from changes of the whistlers' nose frequencies with time, the decoupling of the ionosphere from the magnetosphere may be investigated. Some suggestions are made for co-ordinated ground-based, balloon, rocket and satellite IMS projects in which networks of VLF direction-finding receivers play a key role.

1. Introduction

It is generally acknowledged that our understanding of the distribution of quasi-static electric fields (**E**) in the magnetosphere is rather poor at present [McIlwain, private comm., 1975; Olson, 1975]. Also the extent to which geomagnetic field (**B**) lines are not electric equipotentials, and also the decoupling of the ionospheric plasma from the magnetospheric plasma, are poorly understood [Mozer, 1975]. Thus as complete an examination as possible of such problems should constitute a major objective of the IMS [IMS Second Report, IUCSTP Secretariat, 1972].

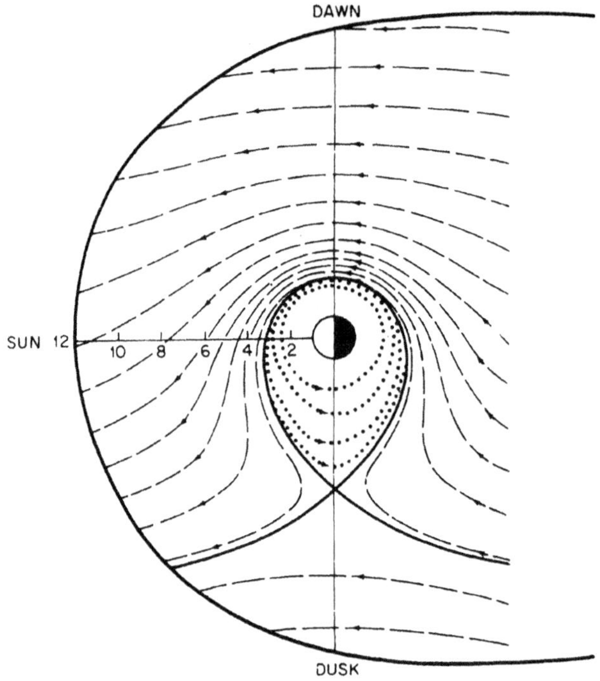

Figure 1. *Streamlines of plasma flow (electric equipotentials), dotted within the plasmasphere and dashed outside the plasmapause, in the magnetospheric equatorial plane for the zero-order approximation to the magnetospheric electric-field distribution, namely the superposition of the corotational and constant dawn-to-dusk fields.*

The zero-order approximation to the static magnetospheric electric-field distribution in the equatorial plane is illustrated in Figure 1 [Kavanagh *et al.*, 1968]. It results from the vector addition of the Earth's corotational field and a spatially uniform, temporally constant dawn-to-dusk field arising from the interaction of the solar wind with the magnetosphere. In the inner magnetosphere, the streamlines of plasma flow (dotted lines in Fig. 1), which are equipotentials in this model, are closed around the Earth. In this region, the plasma density in the equatorial plane builds up to high values by the upward flow of plasma from the ionosphere formed by solar photo-ionisation on successive days. By contrast, in the outer magnetosphere, the streamlines (dashed lines in Fig. 1) are open. Plasma is convected towards the Earth on the night side, and passes around it mainly on the dawn side. During the day, ionospheric plasma moves up the long geomagnetic flux tubes, of large cross-sectional area; only low equatorial plane plasma densities are thus attained before the flux tubes reach the magnetopause. This plasma may then be 'recycled' through the magnetosphere.

The plasmapause, namely the boundary between high and low magnetospheric plasma densities, can be represented by the last closed electric equipotential (the teardrop-shaped dark line in Fig. 1). These mechanisms, together with *in-situ* observations of the plasmapause, have been reviewed by Rycroft [1975a]. The extent to which observations of the plasmapause position, obtained by studying whistlers [Carpenter, 1966], comply with this simple model has been investigated by Rycroft [1971, 1974a]. More complex models of the magnetospheric electric field and of the formation of the plasmapause have recently been presented by Volland [1975] and Lemaire [1975], respectively.

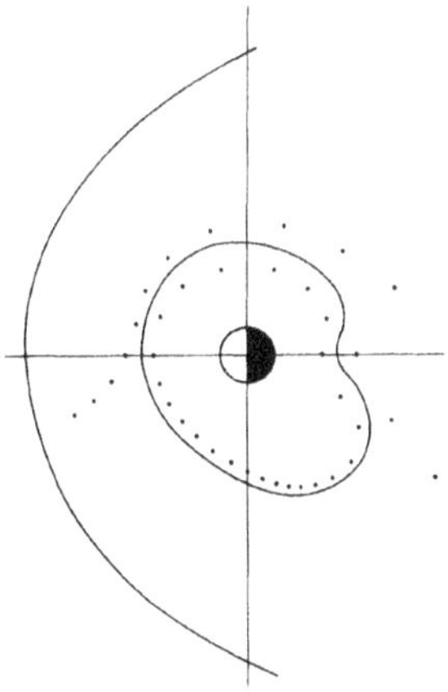

Figure 2. *Sketch, exaggerated for clarity, showing the instantaneous plasmapause position in the magnetospheric equatorial plane during a substorm near local midnight. Shown as dots are the positions of hypothetical whistler ducts at hourly intervals. Both latitudinal and longitudinal variations of the positions of the ionospheric feet of these geomagnetic flux tubes would be detected by a network of ground-based VLF direction-finding receivers.*

A more realistic sketch of the plasmapause position in the equatorial plane, based on the recent work of Carpenter *et al.* [1975], is given in Figure 2. The shape of the plasmapause at different local times is basically similar to that shown in Figure 1, though many details are different. The earthward displacement of the plasmapause near local midnight, and associated with substorm activity, is shown. This could be interpreted as being due to an increase of the dawn-to-dusk electric field. Also shown diagrammatically, by dots, are the equatorial positions of geomagnetically field-aligned ducts of enhanced plasma density which can support the propagation of whistler mode signals from one hemisphere to the other. These are shown at hourly intervals. The super-rotation of the plasma, i.e. its eastward motion with respect to the Earth, is clearly evident on the dawn side, both within the plasmasphere and beyond it. This is due to the convection and corotation velocities ($\mathbf{v} = \mathbf{E} \wedge \mathbf{B}/B^2$) being in the same sense on the morning side. However, on the evening side of the outer plasmasphere, the plasma is expected to move at a velocity less than the corotational velocity; with respect to the Earth, a westward motion is expected here. Such longitudinal motions, at constant L, cannot be studied by the usual whistler method, except in so far as the ducts supporting propagation move out of the range of a particular ground station. However, latitudinal, cross-L motions of a whistler duct can be inferred from changes in the nose frequency (f_n) of whistlers propagating along such a duct [Carpenter *et al.*, 1972]. This is so because Angerami [1966] has shown that, within the plasmasphere, $f_n = 0.37 f_{B,e,eq}$ where $f_{B,e,eq}$ is the electron gyrofrequency in the equatorial plane. Such motions may be due to changes (on a time scale > few minutes) in the westward component of the magnetospheric electric field, although induced electric fields (due to $\partial \mathbf{B}/\partial t$) may also be important [Block & Carpenter, 1974; Mozer, 1975].

The purpose of this paper is to show how observations made by networks of VLF direction-finding (goniometer) receivers can, in conjunction with other measurements, give crucial information which will aid the resolution of these important problems. In order to pave the way for this, the next section reviews some results of the first such experiment, on two-hop whistlers, conducted in July 1972 in Eastern Canada ($L \sim 3-5$). The results of a similar experiment on discrete VLF emissions following the injection of a high-velocity barium cloud up the geomagnetic field, conducted from Alaska ($L \simeq 6.5$) in March 1974, are also mentioned.

2. Resumé of Results on Triangulation of Regions where Two-Hop Whistlers emerge from the Ionosphere

From 1 to 17 July 1972 whistlers and other VLF signals were recorded by direction-finding (goniometer) receivers at four sites in Eastern Canada. These were at St. John's, Newfoundland ($L = 3.45$), Cornerbrook, Newfoundland ($L = 4.03$), Dunmaglass, Nova Scotia ($L = 3.49$) and Fredericton, New Brunswick ($L = 3.64$). The largest distance between sites was 1000 km, between St. John's and Fredericton; the typical station separation was ~ 400 km.

At each site, the incoming VLF waves propagating in the Earth-ionosphere waveguide induced signals in two orthogonal, electrostatically screened vertical loops, each of area 60 m² and supported by a 12 m tall mast, which were accurately oriented geomagnetic N–S and E–W. The signal from each loop was amplified independently and combined electronically to simulate the signal received by a single loop rotating about a vertical axis at 27.5 Hz [Bullough & Sagredo, 1973], and recorded on one channel of a stereo magnetic tape recorder. A calibration pulse is produced at a known position each time the simulated loop completes a revolution and is recorded on the other channel. Two maxima and two minima per rotation of the simulated loop are observed in the envelope of the VLF waveform.

The 180° ambiguity in the direction of arrival or azimuthal bearing of the signals observed by a single goniometer is resolved by the use of two or more goniometers. With the same components of a multicomponent whistler recorded simultaneously at more than one site, the positions of the points where whistler-mode signals emerge from the lower ionosphere and enter the Earth-ionosphere waveguide, termed the whistler exit points, may be found by triangulation along great-circle paths.

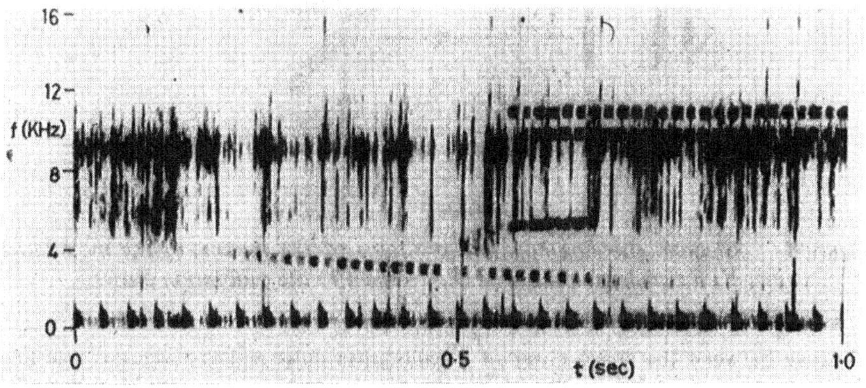

Figure 3. *Wide-bandwidth Sonogram of a two-component (one with a nose frequency ~ 5 kHz), two-hop whistler recorded at Dunmaglass, Nova Scotia, on 10 July 1972. Modulation of the non-nose whistler, and also of signals transmitted by two Omega stations, at a frequency of twice the goniometer frequency of 27.5 Hz is clearly seen.*

Both the azimuth and frequency-time results on the whistlers were obtained by measurement of Sonograph records (Fig. 3). Appropriate filtering was performed to combine, on one Sonogram, the VLF signals above 400 Hz with the calibration pulse below 400 Hz. Each Sonogram was illuminated by a transparency of a grid of white lines upon a grey background using a photographic enlarger. The magnification of the enlarger was adjusted until the spacing of the grid lines parallel to the frequency axis coincided exactly over the whole Sonogram with half the spacing of the calibration pulses. With the aid of a half-silvered mirror mounted at 45° to the plane of the Sonogram, as illustrated in Figure 4, the Sonogram and superimposed grid lines were viewed with a travelling microscope. The spacing between the grid lines, and the

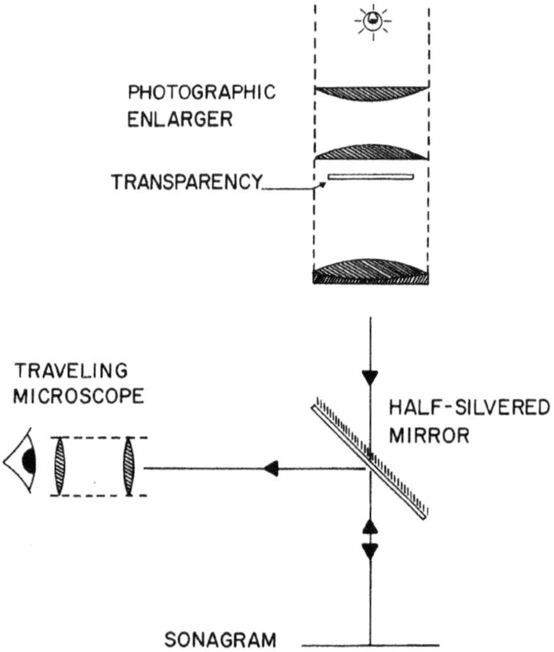

Figure 4. *Diagram showing how the positions of the minima of the modulated VLF signals are measured with respect to the calibration pulses.*

distance between the sharp onset of a calibration pulse and the adjacent grid line were measured several times. Averaging over five separate settings, a typical error for these bearing measurements was ± 8°.

The two-hop whistlers, with causative sferics believed to be emanating from the storm centre northeast of Newfoundland, and observed at all four stations between ~ 18.00 and ~ 19.40 UT on 10 July 1972, have been analysed in detail [Jarvis, 1975]. Some results have already been presented [Rycroft *et al.*, 1975; Jarvis & Rycroft, 1975]. Figure 5 is a sample of the results. The whistler, the dynamic spectrum of which is given in Figure 3, has an exit point at ~ 44° N, 63° W as shown by the cross, and as derived from observations at Dunmaglass and Fredericton. The error in locating this exit point is typically ± 50 km, as shown by the dimensions of the error box. The exit points of the identical component of multicomponent whistlers recorded ~ 1 min earlier and ~ 20 min later [Jarvis, 1975] are shown just to the west and southeast, respectively. Whilst it is tempting to ascribe the disposition of these three exit points to the eastward and slightly southward horizontal movement of the ionospheric exit point, this cannot be justified, firstly because the 1:20 temporal separation of the observations implies a highly unlikely ~ 20:1 velocity decrease, and secondly because the difference between initial and final positions is only ~ 2 standard deviations. However, if many observations of the same component were possible at more times (say for ~ 1 h), which unfortunately is not the case

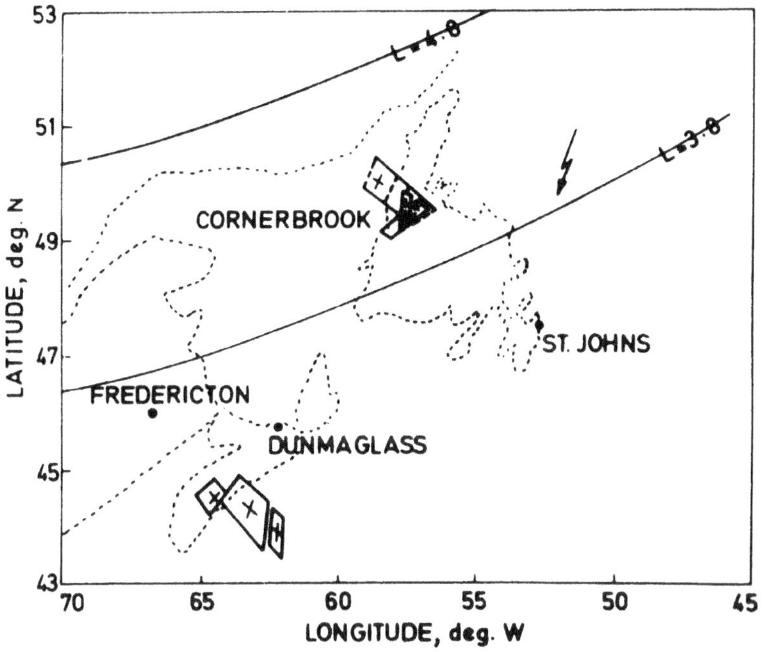

Figure 5. *Map of eastern Canada showing exit points of two particular components of multicomponent whistlers observed on 10 July 1972, as obtained by triangulation. These exit points are at considerably lower L-values than those expected from the whistlers' nose frequencies, the $L = 3.8$ and 4.8 contours shown.*

here, then departures of the ionosphere from corotation with the Earth with a velocity as low as 50 m/s could be detected directly, in both magnitude and direction. Nonetheless, this example illustrates the power of the goniometer network for investigating ionospheric motions.

Another component of the multicomponent whistlers observed, a nose-frequency whistler observed from St. John's, Cornerbrook and Dunmaglass, is emerging from the ionosphere close to Cornerbrook. From its nose frequency, the L-value along which it propagated through the magnetosphere is found (as discussed in Section 1) to be 4.8. From the $L = 4.8$ contour shown in Figure 5, it is quite evident that the whistler emerges from the ionosphere at a much lower L-value.

Since the time of the causative sferic is known, application of the Q analysis procedure [Dowden & Allcock, 1971] allows the nose frequency of the non-nose whistler shown in Figure 3 to be estimated. Hence the L-value of its magnetospheric duct is found to be 3.8. This contour is also shown in Figure 3; again it is clear that the whistler is emerging from the ionosphere at a considerably lower L-value than that along which it travels through the magnetosphere.

For all the whistlers observed during the period and for which triangulations could be performed, a graph was plotted (Fig. 6) of $L_{triangulation}$, the L-value derived by the triangulation method, against L_{fn}, the L-value obtained from the nose frequency. The results show a serious discrepancy with the naively expected result that the points should lie on a line passing through (1, 1) with a gradient of unity. The discrepancy is that $L_{triangulation}$ is less than L_{fn}, by ~ 0.6 typically. Weightings were given to the points, proportionately to their reliability; the least-squares linear fit to these weighted points is shown in Figure 6.

The explanation of this discrepancy has been discussed in detail by Rycroft *et al.* [1975], Jarvis & Rycroft [1975], and Strangeways & Rycroft [1975]. Suffice it to say here that it cannot be explained by a distortion of the geomagnetic field by the ring current (if that were the case the points would lie on the dashed line in Fig. 6), or by the reduced geomagnetic field in this meridian, or by the assumed value of $f_n/f_{B,e,eq}$ being in error, or by propagation along different ducts from north to south rather than from south to north. It can be explained if, on this occasion, the electron density trough in the topside ionosphere was at a lower latitude than that of the plasmapause. Computer ray-tracing studies show that, on the high-latitude side of the trough, the waves are refracted to high L-values before becoming ducted, at a height of the order of several thousand kilometres.

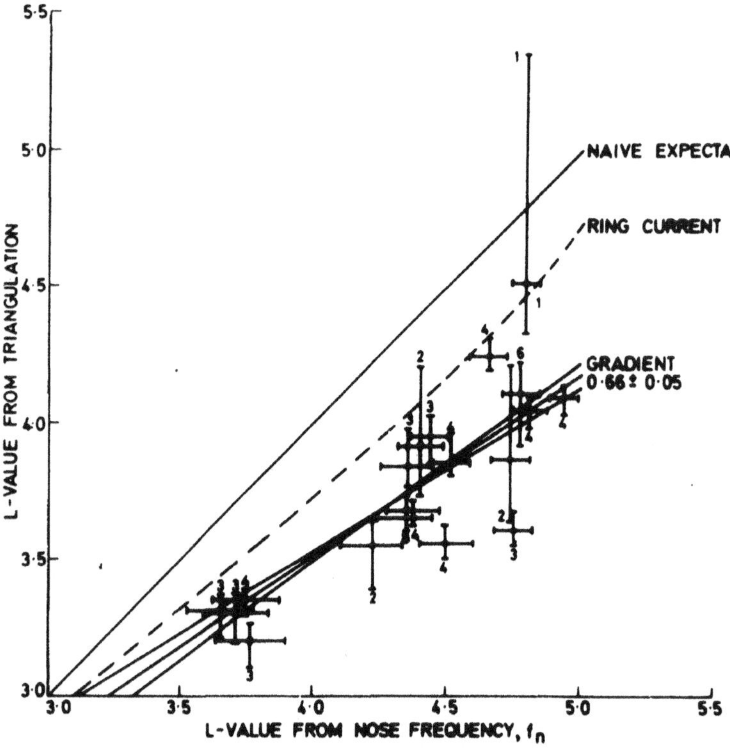

Figure 6. L-value of whistler exit points derived by triangulation plotted against L-value derived from the whistlers' nose frequency.

From a least-squares, three-dimensional regression using all the available results of *Ltriangulation* on L_{fn} and local time, Jarvis [1975] has demonstrated a small longitudinal shift of the ionospheric exit points, by 3° ± 1°W per hour.

It should be pointed out that Jarvis [1975] has also analysed in detail a 5 min period on 2 July 1972 when whistlers were received at a rate ~3/min at the three sites operational at that time. For these data, a graph similar to Figure 6 shows the points to lie between the naively expected line and that for which the ring current correction has been made. In this case then, it appears that the whistlers are emerging from the ionosphere at the feet of the geomagnetic flux tubes along which they are ducted through the magnetosphere. The data set here is too small to detect a longitudinal shift of the exit points.

The relative frequency of occurrence of these two types of results, or of other results, is not known at present. Thus, to make further progress in these multiple goniometer studies, further observations are now required.

3. Resumé on VLF Phenomena following Barium Injection at $L \simeq 6.5$

A Sandhawk-Tomahawk rocket was launched by a Los Alamos/Sandia team from the University of Alaska Poker Flat Rocket Range in Alaska on 16 March 1974. At 12.35 UT, using a shaped charge of hemispherical geometry, 0.7 kg of barium was injected up the geomagnetic field line at $L \simeq 6.5$. Optical observations were made at several sites on the ground, particularly at Barter Island on the north coast of Alaska, and aboard an aircraft. Initially the barium cloud was observed to drift in a geographic eastward direction at 870 m/s and southward at 130 m/s. The 100 km altitude foot of the field line through the cloud was determined 6, 12 and 25 min after release; it followed the dark track shown in Figure 7. Since the cloud became too faint and diffuse for observation 25 min after release, this track has been extrapolated in Figure 7. Although it was anticipated that the barium would be produced moving up the geomagnetic field line at 13 km/s, in which case it would reach the equatorial plane in 94 min, the tip of the jet was observed initially to move at 10.1 km/s, with the bulk of the barium travelling at between 4 and 9 km/s. In the presence of an electric field parallel to the geomagnetic field, however, the barium ions might be accelerated upwards.

Brice [1970] and Brice & Lucas [1971] postulated that an increase in the thermal plasma density due to the barium ionised by solar radiation might trigger the electron cyclotron instability at the magnetospheric equatorial plane. This would then cause electrons to be precipitated into the atmosphere and produce the correct conditions for growth of whistler-mode waves. Liemohn [1974] has shown that significant wave amplification should occur before the barium reaches the equator, at a magnetic latitude ~20° or 30°. An appreciable fraction of this barium cloud could reach such positions, at lower L-shells (~ 4 to 5) after ~ 90 min.

Radio observations were made at three ground sites in Alaska, Steese Highway (SH), Venetie (V) and Arctic Village (AV), shown in Figure 7, with goniometer receivers. Between 13.45 and 15.30 UT many multipath, one- and odd-numbered-hop whistlers were recorded at all three sites. The spectrogram of typical

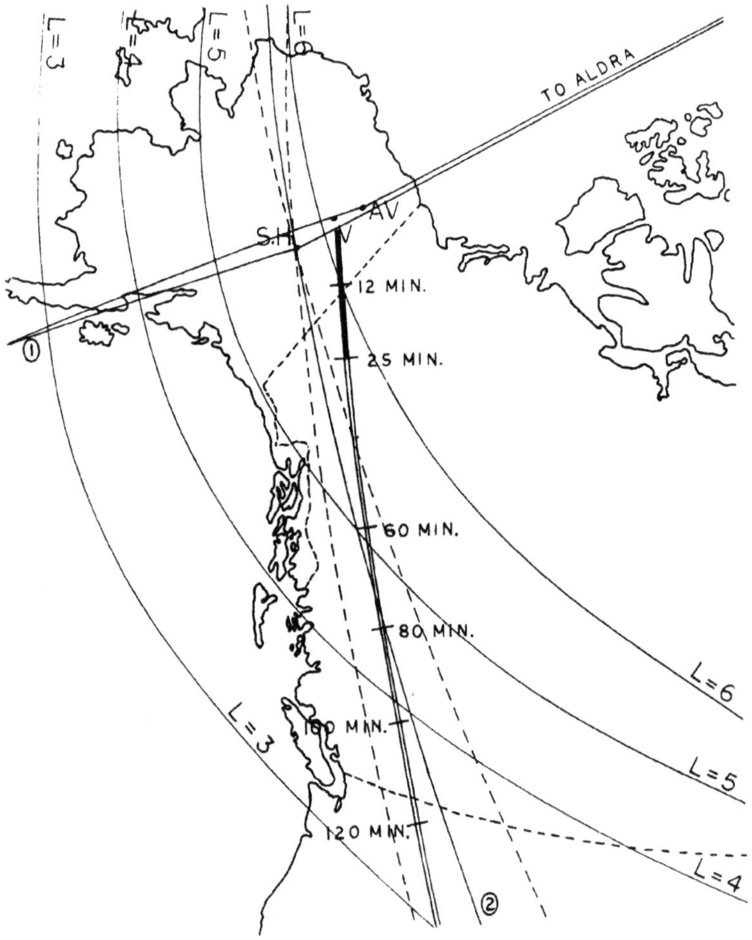

Figure 7. *Map of Alaska and western Canada showing positions (dots) of VLF goniometers at Steese Highway (SH), Venetie (V) and Arctic Village (AV). L-shell contours at 100 km altitude are also shown. The position of the foot (at 100 km) of the field line through the barium release derived from optical observations up to 25 min after release is shown as the broad dark track; this track is extrapolated to 120 min after release. The great circle bearings to the Omega transmitter, Aldra, Norway, to the whistler exit points ① and to the discrete ELF and VLF emissions exit points ② are also given. It is evident that the emissions observed between 80 and 100 min after release could have propagated down geomagnetic flux tubes through the barium release.*

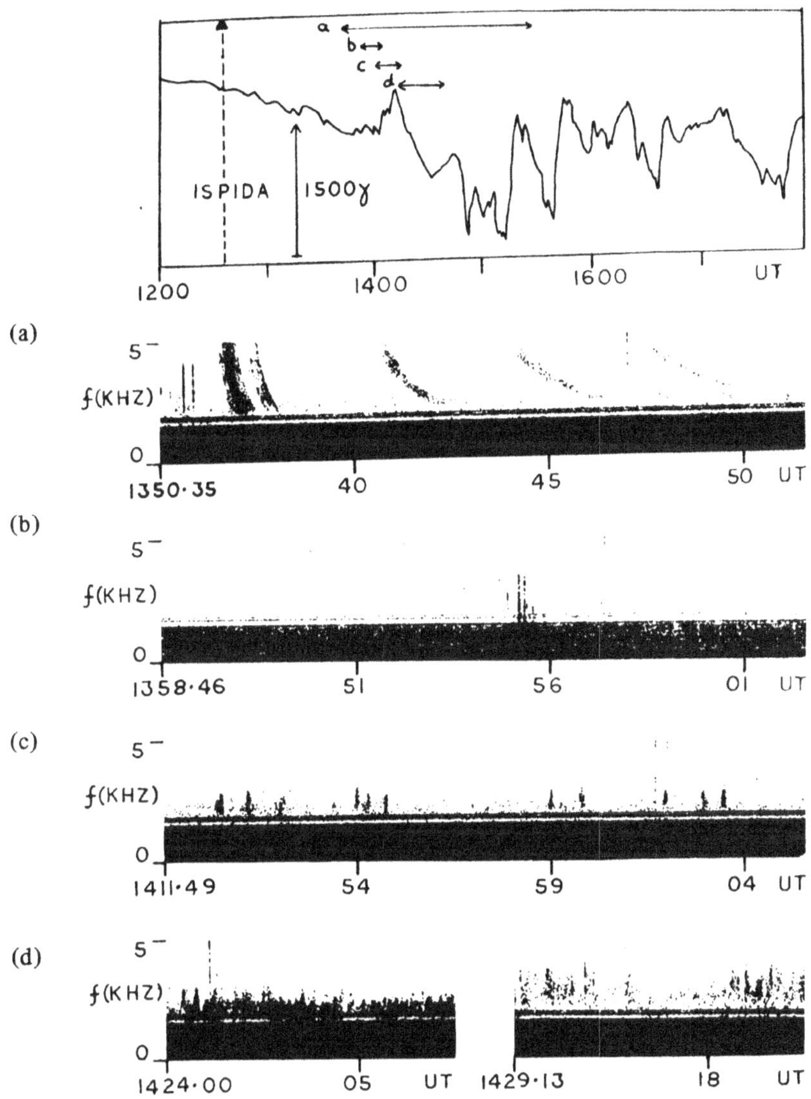

Figure 8. *Upper panel: Horizontal component of geomagnetic field fluctuations recorded at College, Alaska, for 5 h after the Ispida barium injection. During the periods a, b, c and d, interesting ELF and VLF signals were recorded.*
Lower panel: Typical spectrograms showing (a) whistlers, (b) discrete falling frequency tones, (c) discrete rising frequency tones, and (d) chorus-like emissions observed at different times after the Ispida barium release.

whistlers is shown in Figure 8a. From the frequency-time characteristics of many of the strongest whistlers, it was possible to use the Dowden & Allcock [1971] Q-analysis procedure to find the average nose frequency of these whistlers. Hence the L-value of the dominant propagation path was found to be $L = 2.70 \pm 0.04$. The azimuthal bearings of these signals are shown by ① in Figure 7. It is concluded that these whistlers are in no way associated with the barium release.

Of more interest are the discrete ELF and VLF emissions, both rapidly falling and rising frequency emissions, observed only at Steese Highway where a preamplifier particularly sensitive below 3.5 kHz was used. These were observed between 13.55 and ~14.16 UT; typical spectrograms are shown in Figures 8b and c. From 14.14 to 14.40 UT, chorus-like emissions (Fig. 8d) were recorded. The average azimuthal bearing of these discrete emissions, shown by ② in Figure 7, intersects the expected foot of the field line through the barium plasma from 40 to 120 min after release, i.e. from 13.15 to 14.35 UT. Signals from the Omega stations provide useful extra calibrations of azimuthal bearings; the great-circle path to the Aldra, Norway, transmitter is also shown in Figure 7.

The upper panel of Figure 8 shows the horizontal component of the geomagnetic field recorded at College, Alaska at the time; a sizeable substorm was in progress. Everhard-Bakker *et al.* [1975a, b] consider that it is most likely that the discrete emissions were stimulated by the barium release, but that the chorus-like emissions were natural auroral chorus, reported by Hayashi & Kokubun [1971] to occur most frequently during the negative bay phase of large substorms.

Such an experiment is worth repeating with a larger, higher velocity barium injection, during quieter magnetic conditions following a substorm, and with sensitive preamplifiers at each site.

4. Importance of VLF Direction-Finding Studies during IMS

It has been shown here that networks of VLF direction-finding receivers accurately and reliably locate the regions where particular whistler-mode signals emerge from the ionosphere over a wide area. These signals may be from man-made VLF transmitters, from lightning discharges (producing whistlers) or from magnetospheric plasma instabilities (producing discrete or broadband ELF and VLF emissions); alternatively, they may be triggered emissions, either natural or resulting from an active magnetospheric experiment (from energetic charged-particle, plasma or wave injection). They may have propagated, from source to receiver, either on a fractional-hop (e.g. emissions produced by the electron cyclotron instability near the equatorial plane would be ~half-hop), a one-hop, or a two-hop path. Neither odd – nor even – hops of higher order have yet been studied, to investigate whether, as might be expected, they propagate on the same path as their predecessor. Nor have triggered emissions been studied, to investigate whether they follow the same propagation path as the triggering signal. Remarkably little is known of the spatial and temporal distributions of whistler ducts, or of the reasons for these distributions.

It has also been shown here that, on the only occasion for which 2 h of observations are available, the L-values of the whistlers' exit points are systematically and significantly lower than the L-values, derived from their nose frequencies, along which they were ducted through the magnetosphere. It is not known whether this result is the exception or the rule, or how often this situation may occur, i.e. the probability of this behaviour is not known. However, it is expected from the ray-tracing studies of Strangeways & Rycroft [1975] that this marked discrepancy will occur only when propagation is on the high-latitude side of the so-called mid-latitude trough (i.e. where the electron density at constant altitude increases with increasing invariant latitude). When propagation is on the low-latitude side of the trough, Strangeways & Rycroft [1975] show that the wave may *either* become ducted at a height of only several hundred kilometres at an L-value $\sim$0.1 higher than that where the wave enters the ionosphere *or* become ducted at a height of a few thousand kilometres at an L-value $\sim$0.5 lower than that where it enters. If there are no latitudinal electron density gradients, or if there exist spread-F or other ionospheric irregularities to cause the wave to be ducted elsewhere above $\gtrsim$ 200 km altitude, then it is expected that the waves will enter and emerge from the ionosphere at essentially the same L-value as that along which they are magnetospherically ducted.

It is clear, therefore, that more observations of whistler-mode signals obtained with a network of VLF direction-finder receivers are required. It is also clear that simultaneous soundings of the topside ionosphere must be made from suitable satellites passing over that region of the ionosphere through which the whistler-mode signals are propagating. Only then may the predictions of the VLF ray-tracing studies be tested.

Another possibility of the method is to search for movements, both in latitude and longitude, of the ionospheric exit point of a particular whistler-mode signal. Such results should be compared with simultaneous measurements of the velocity of horizontal movements of the ionosphere which may be obtained by a number of techniques. These include conventional radio methods, coherent and incoherent radar backscatter systems, as derived from balloon-borne electric field measurements, as derived from observations of barium releases, and direct *in-situ* observations aboard rockets or with the instrument aboard the Atmospheric Explorer satellite [Hanson & Heelis, private comm. 1975]. Such experiments would test the efficacy of the goniometer method.

The results obtained on the cross-L movement of the ionospheric exit point of a particular whistler, as found from triangulation results, should be compared with the cross-L movement of the duct in the magnetospheric equatorial plane, as found from changes of the whistlers' nose frequency. The motions of two widely separated points on essentially the same field line are thereby studied simultaneously and independently. The multiple goniometer whistler method of studying one particular nose whistler over a period of time ($\lesssim$ 1 h) offers, then, unique promise for study of plasma motions across a particular geomagnetic field line which may be different at its apex and its feet. This procedure may be repeated on all nose whistlers observed. Thus the extent to which the ionospheric plasma is decoupled from the magnetospheric plasma may be investigated quantitatively over a range of L-values. Temporal changes in this decoupling may also be found. The interpretation of plasma motions to obtain the electric field driving the motions is, however, not a simple matter [Mozer, 1975]. Making different assumptions, different electric fields will be obtained. Comparison with, for example, direct *in-situ* measurements of the magnetospheric electric field, will enable the correct assumption to be identified.

Brief consideration is now given to other uses of the multiple goniometer network. The general theme is the location, in space, of one or more particular geomagnetic flux tubes at a time when interesting physical phenomena are occurring on them. Of course, the method relies on observing propagation in the whistler mode.

The occurrence of whistlers is rather unpredictable. One definite result, however, is that the reception of one-hop whistlers is favoured in Antarctica near 75° W during July [Carpenter *et al.*, 1975]. This is because of the considerable thunderstorm activity over eastern Canada and the northeastern USA then, and also the reduced absorption in the dark, southern-hemisphere ionosphere. Two-hop whistlers are observed in the northern hemisphere. The occurrence of discrete VLF emissions is also unpredictable, although it is known that chorus is more likely when geomagnetic activity is moderate to high, and that it tends to occur later in the local morning at higher latitudes. Man-made VLF transmissions offer advantages of known, or even controlled, signals; a review of what can be learnt about the magnetosphere by this type of active experiment has recently been presented by Dowden & Rycroft [1975]. Powerful transmitters, as NAA (Cutler, Maine) and NLK (Seattle, Washington), and the less powerful Omega transmitters produce many interesting results. The purpose-built Siple transmitter in Antarctica has led to a variety of remarkable discoveries, especially concerning triggered emissions [Helliwell & Katsufrakis, 1974]. A similar, but higher latitude, transmitter in the European sector has been proposed as part of the ESA GEOS programme [Nunn, 1975]. A transportable, balloon-supported transmitting antenna offers several advantages [Koons & Dazey, 1974].

In many VLF studies today, interest is focussed on wave-particle interactions [Helliwell & Katsufrakis, 1974; Koons, 1974; Rycroft, 1974b; Dowden & Rycroft, 1975; Gendrin, 1975; Rycroft, 1975b; Thorne, 1975]. Using goniometers, the geomagnetic flux tube, on which the wave-particle interactions to be investigated are being observed, can be localised in space. Ionospheric effects of, say, the energetic charged particles precipitating as a consequence of the wave-particle interaction could be sought in the correct position. Information helpful to those concerned with the launching of balloons and rockets can be provided. Observations made in such a collaborative campaign can be analysed co-operatively to gain a better understanding of the physics of the complex wave-particle interactions taking place in the near-Earth plasma environment.

Information of interest to experimenters with instruments on geostationary and other distant ($L > 2.5$) magnetospheric satellites can also be provided. Having made appropriate corrections for dispersion, a ducted whistler-mode signal received on a satellite can be compared with those seen by goniometers on the ground. That signal recorded on the ground which most resembles the dispersion-corrected satellite-observed signal will have propagated on a geomagnetic flux tube closest to that through the satellite. Thus the multiple goniometer network can contribute, as can other techniques, to the difficult problem of locating the foot of the field line through a particular satellite, especially one in geostationary orbit.

It is quite evident that networks of VLF direction-finding receivers can play important roles in many important problems of ionospheric and magnetospheric plasma physics to be tackled during the IMS. This has also been recognised by scientists involved with both IMS plans [see p. 9, 19, 23, 24, 29, 33, IMS Second Report, IUCSTP Secretariat, 1972; p. 3, 39, 45, IMS, Detailed Plan for a US Ground-Based Research Program, NAS, 1974; p. 3, 4, 7 British National Committee on Solar-Terrestrial Physics, Report of Royal Society Meeting, 1974; p. 311, ESRO Special Publication No. 107, 1975] and international scientific bodies [see, for example, Resolution of URSI, Warsaw General Assembly, URSI Bulletin, No. 184, 1972).

5. Suggestions for Specific IMS Experiments

Rather than making routine observations with a network of VLF direction-finding receivers, it is more effective to make comprehensive, continuous observations for a limited period of time. Such experiments are best performed, in conjunction with one or more other experiments, in a co-ordinated campaign. Four suggested examples of such projects are outlined, first in relation to plasmapause-associated phenomena (5.1 and 5.2) and second to auroral phenomena (5.3 and 5.4).

5.1. DETERMINATION OF PLASMAPAUSE POSITION AND DYNAMICS

Observations made at conjugate networks of stations, of both one- and two-hop whistlers, would enable the foot of the field line through the plasmapause to be determined. Its motions, and that of the neighbouring plasmas inside and outside, could be studied both in the ionosphere and magnetosphere as a function of local time and geomagnetic activity; 'peeling off' the plasmapause could be detected. Associated ionospheric phenomena, such as the mid-latitude electron density trough, the light-ion trough, SAR arcs and charged-particle precipitation and its effects, should be observed. Electric-field observations, preferably by several techniques, would be required during such a campaign. Simultaneous ring-current observations would enable distortions of the geomagnetic field to be modelled. Observations could be made in the prime geomagnetic meridian (over the Americas), or between Sanae (Antarctica) and Scotland, or Kerguelen (Indian Ocean) and Sogra (USSR) or New Zealand and Alaska.

5.2. WAVE-PARTICLE INTERACTIONS NEAR THE PLASMAPAUSE

Discrete VLF emissions are believed to be generated by the electron cyclotron instability near the equatorial plane on geomagnetic flux tubes with L-values close to that of the plasmapause. There is some theoretical and experimental evidence which suggests that the lower frequency emissions, such as chorus, are generated just outside the plasmapause, whereas emissions triggered at the upper cut-off frequencies of whistlers are generated just inside the plasmapause. These results should be the subject of direct experimental tests. Observations of the emissions triggered by, and other effects of, controlled VLF radio-wave injection into the magnetosphere, using the Siple transmitter of the Aerospace Transportable system, can contribute significantly to an improved understanding of the reasons for the vicinity of the plasmapause being a preferred region for the occurrence of wave-particle interactions. Simultaneous observations of the charged particles and VLF waves in space, obtained by rockets and satellites, would be required.

5.3. IONOSPHERIC-MAGNETOSPHERIC DECOUPLING DURING DIFFERENT PHASES OF AURORAL SUBSTORMS

This problem is related to that of the acceleration of auroral electrons. Observations should be made in the neighbourhood of the foot of the field line through a geostationary satellite such as GEOS with its full complement of wave and particle instruments. Scandinavia and the conjugate pair Iceland-Syowa Base, Antarctica, are the appropriate locations. In both, correlative ground-based and widely instrumented rocket programmes (such as the UK high-latitude campaign or the FRG/USA Aries programme) are favoured.

5.4. EFFECTS OF BARIUM INJECTIONS UP GEOMAGNETIC FLUX TUBES

The tantalisingly interesting results following the Ispida barium release in Alaska need to be investigated in more detail with further, larger, higher velocity releases. Releases at mid-latitudes and in the cleft (polar cusp) region would also provide interesting results.

6. Acknowledgments

The author is grateful to the Science Research Council, the National Science Foundation, and the NATO Scientific Affairs Division for funding. He also thanks many scientists from different countries for discussions on aspects of the research discussed, and colleagues and research students for making the observations reported.

References

Angerami, J.J., Stanford Electronics Laboratory, Technical Report, SEL-66-017, 1966.

Block, L.P. & D.L. Carpenter, *J. Geophys. Res.* **79**, 2783 (1974).

Brice, N., *J. Geophys. Res.* **75**, 4890 (1970).

Brice, N. & C. Lucas, *J. Geophys. Res.* **76**, 900 (1971).

Bullough, K. & J.L. Sagredo, *Planet. Space Sci.* **21**, 899 (1973).

Carpenter, D.L., *J. Geophys. Res.* **71**, 693 (1966).

Carpenter, D.L., N.T. Seely & T.R. Miller, *EOS Trans. Am. Geophys. Union* **56**, 414 (1975).

Carpenter, D.L., K. Stone, J.C. Siren & T.L. Crystal, *J. Geophys. Res.* **77**, 2819 (1972).

Dowden, R.L. & G.McK. Allcock, *J. Atmos. Terr. Phys.* **33**, 1125 (1971).
Dowden, R.L. & M.J. Rycroft, Presented at COSPAR Meeting, Varna, Bulgaria, 1975.
Everhard-Bakker, A.E., M.J. Rycroft, N.A. Heard & A.J. Smith, Presented at COSPAR Meeting, Varna, Bulgaria, 1975.
Everhard-Bakker, A.E., M.J. Rycroft, N.A. Heard, A.J. Smith, P. Bottoms & M. Pongratz, Presented at Spring Annual Meeting AGU, Washington DC, 1975.
Gendrin, R., ESRO Special Publication No. 107, 1975.
Hayashi, K. & S. Sokubun, *Rep. Ionosph. and Space Res. Japan* **25**, 369 (1971).
Helliwell, R.A. & J.P. Katsufrakis, *J. Geophys. Res.* **79**, 2511 (1974).
Jarvis, M.J., PhD Thesis, University of Southampton, 1975.
Jarvis, M.J. & M.J. Rycroft, Submitted to *J. Geophys. Res.* (1975).
Kavanagh, L.D., Jr. J.W. Freeman, Jr. & A.J. Chen, *J. Geophys. Res.* **73**, 5511 (1968).
Koons, H.C., *EOS Trans. Am. Geophys. Union* **56**, 1177 (1974).
Koons, H.C. & M.H. Dazey, in ELF-VLF Radio Wave Propagation (Ed. J.A. Holtet), D. Reidel, Dordrecht, Holland, 413, 1974.
Liemohn, H.B., *Space Sci. Rev.* **15**, 861 (1974).
Lemaire, J.L., *Ann. Geophys.* **31**, 175, 1975.
Mozer, F.S., Presented at 10th ESLAB Symposium, These Proceedings.
Nunn, D., Space Research XV (Ed. M.J. Rycroft), Akademie-Verlag, Berlin, 1975.
Olson, W.P., *EOS Trans. Am. Geophys. Union* **56**, 422 (1975).
Rycroft, M.J., ESRO Special Publication, No. 72, 163, 1971.
Rycroft, M.J., in Correlated Interplanetary and Magnetospheric Observations (Ed. D.E. Page), D. Reidel, Dordrecht, Holland, 1974a.
Rycroft, M.J., in ELF-VLF Radio Wave Propagation (Ed. J.A. Holtet), D. Reidel, Dordrecht, Holland, 1974b.
Rycroft, M.J., *Ann. Geophys.* **31**, 1, 1975a.
Rycroft, M.J., Presented at Spring Annual Meeting AGU, Washington DC, 1975b.
Rycroft, M.J., M.J. Jarvis & H.J. Strangeways, ESRO Special Publication No. 107, 1975.
Strangeways, H.J. & M.J. Rycroft, Submitted to *J. Geophys. Res.*, 1975.
Thorne, R.M., *Rev. Geophys. and Space Phys.* **13**, 291 (1975).
Volland, H., *Ann. Geophys.* **31** (1975).

DISCUSSION

D.T. Young: Could the VLF emissions possibly triggered by the barium release be seen by a spacecraft higher up the field line, e.g. at the equator?

M.J. Rycroft: I should think that such emissions should be observable in the magnetospheric equatorial plane. Believing the emissions to be generated by the cyclotron resonance instability mechanism within the barium plasma as it moves up the field line, they probably first propagate downwards and are received on the ground. They would have to be reflected, by the ground or ionosphere, and would suffer extra dispersion before being received in the equatorial plane.

D. Jones: (i) The analogue goniometer method uses only two-component signals. Is it not true that this can lead to large errors and is it possible to incorporate in the technique a third antenna, e.g. vertical monopole as was used at Stanford, to remove this error?

(ii) Digital direction finding techniques show quite large variations in the arrival direction of different frequency components in the same whistler. Please comment.

M.J. Rycroft: (i) For signals that propagate a distance of $\lesssim$ 200 km in the Earth-ionosphere waveguide, there can sometimes be significant errors using the two-antenna method discussed here. The third antenna can be used to help resolve this problem; both the Stanford and Tokyo systems should enable accurate exit-point positions to be found when they are $<$ 200 km from the receiver. The method discussed here gives consistent results at distances $>$ 200 km, enabling a large area to be studied. The two methods may therefore be considered to be rather complementary. These different techniques are to be compared in experiments to be carried out in Eastern Canada during June and July 1975, using whistler-mode signals from the Siple transmitter in Antarctica.

(ii) It is conceivable that signals propagating $<$ 200 km in the Earth-ionosphere waveguide can show such effects. The analyses of non-nose whistlers such as shown in Figure 3, using the human brain to integrate the information contained in the minima over all frequencies by the method displayed in Figure 4, always seem to lead to consistent results being obtained.

AVAILABILITY AND DISTRIBUTION OF GEOS DATA DURING THE IMS

K. Knott

Space Science Department, ESTEC, Noordwijk, Netherlands

Abstract

It has been agreed by both the GEOS experimenters and ESA that daily summaries of all GEOS experiments will be produced and will be made available to the scientific community participating in the IMS. A listing of the content of these summaries is given and an example of a possible format is presented. It is emphasised that these summaries must not be considered as a 'ready to write papers' output from GEOS, but merely as qualitative information on and trend indicators for parameters measured by GEOS.

When the experiments for GEOS were selected back in 1970, it was announced and clearly understood that *'experimenters must provide their data in a manner to be arranged with the interested scientific community.'* The 'manner' has since been arranged and will form the basis of this short contribution to describe the availability and distribution of GEOS data.

Figure 1 shows the GEOS operations and computer system. This system has been explained in more detail in a previous talk. In the context of this presentation we note that data from all GEOS experiments are stored initially on a disk file, which has a capacity of 24 h. Except for the data from the wave experiment, where a pre-compression takes place, the data from all experiments are available in their original form in this file.

This disk file forms the source from which so-called 'daily summaries' of GEOS data are produced. The summaries consist of two sheets, one giving a summary of the particle experiments, and one a summary from the field and wave experiments on GEOS. They are printed in a Xerox-compatible form to facilitate fast reproduction and wide circulation.

A major handicap in producing coherent daily summaries stems from the fact that

Knott & Battrick (eds.), The Scientific Satellite Programme During the International Magnetospheric Study, 451–455.
All Rights Reserved. Copyright © 1976 by D. Reidel Publishing Company, Dordrecht-Holland.

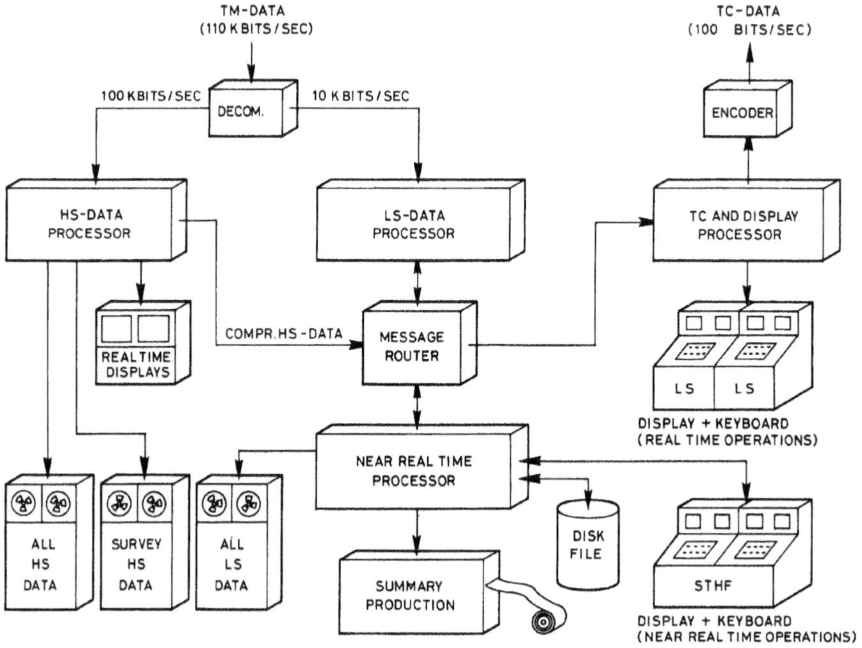

Figure 1. *GEOS operation and computer system.*

some GEOS experiments switch their mode of operation quite frequently. For these experiments, it has been agreed to arrange for special consistent modes of operation at regular intervals. After lengthy discussions between the experimenters concerned, the content of the daily summaries has now been agreed upon.

The content and a possible display format for the daily summary from the particle experiment are shown in Figure 1. It gives, from top to bottom, the count rates of the various particle detectors; in particular

- the electron count rates of the suprathermal particle detectors viewing parallel and roughly perpendicular to the spin axis
- the count rate of the thermal ions (< 100 eV) and a count rate indicating the heavy ion flux for energies < 20 keV obtained by the mass analyser
- the energy flux of electrons in the range 0.5–20 keV along the spin axis
- the number fluxes of electrons perpendicular and parallel to the spin axis in the same energy interval
- for energies above 20 keV, electron and proton fluxes, parallel and perpendicular to the spin axis.

Three-minute averages will be displayed and averages are sampled either every 15 or 30 min.

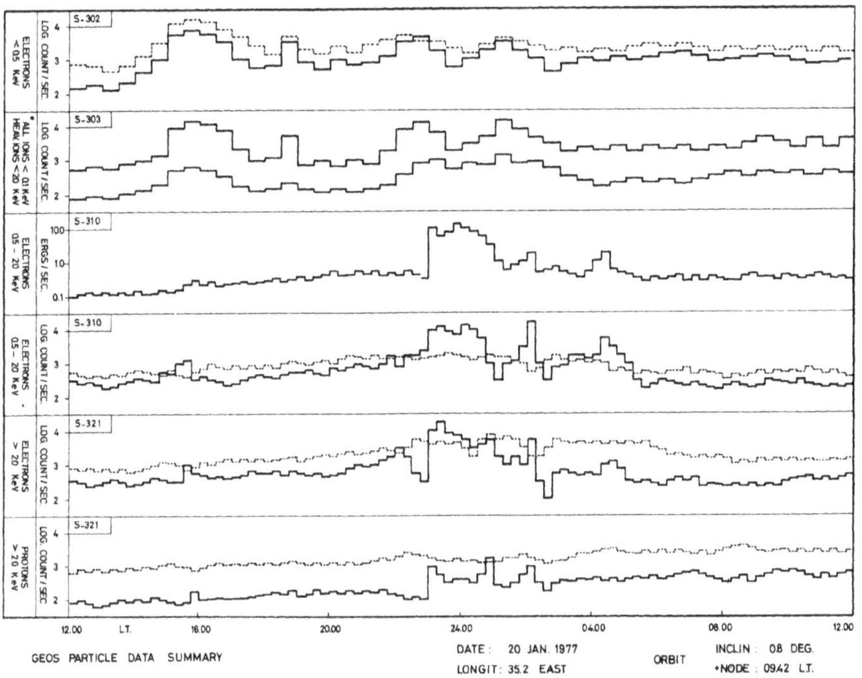

Figure 2. *Proposed layout and content of GEOS daily summary for particle experiments.*

Figure 2 shows the envisaged daily summary from the GEOS wave and field experiments. From top to bottom, it will contain:

- the *H, V* and *D* components of the DC magnetic field
- the degree of variation in this field
- the electric field as obtained by the two DC electric-field experiments on GEOS
- the ULF magnetic data, split into two components. The component parallel to *B* is given for frequencies above 0.1 Hz and the component perpendicular to *B* for frequencies below 0.1 Hz
- the integrated power appearing in various VLF frequency bands for both the electric and magnetic fields
- the plasma frequency measured by the relaxation sounder and the status (on-off) of the active experiment.

Except for the plasma frequency, which is plotted when available, all other traces are composed of 3-min averages.

The distribution of these data is still under discussion and several possibilities are envisaged:

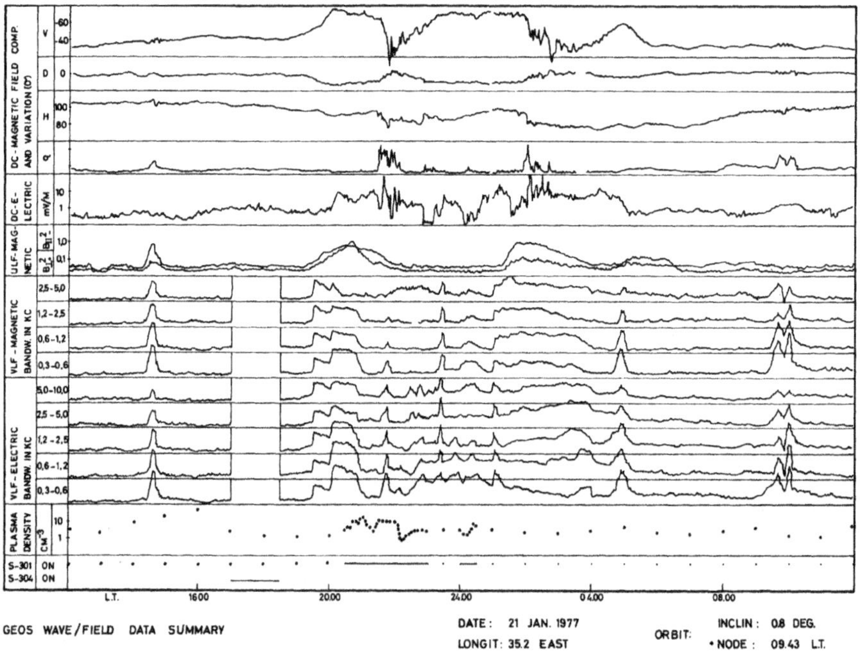

Figure 3. *Proposed layout and content of GEOS daily summary for wave and field experiments.*

- they can be mailed on a daily or less frequent basis in the form of Xerox copies
- they can be stored on microfilm or tapes and then mailed whenever a film or tape has been filled
- it may also be possible to transmit these data daily, weekly or monthly via a data line to another data centre for further distribution.

The final means of distribution will depend to a large extent on the number and geographical location of the users.

It must be emphasised very strongly that these summaries are not meant as a 'ready to write papers' output from GEOS. They must be considered as status information on the environmental conditions encountered by the spacecraft. On the basis of any trends, anomalies or special features shown in these summaries, bilateral co-operation between a GEOS experimenter and a recipient of the summaries should be established. The truly scientific output will emanate from this co-operation.

A second method of distribution for GEOS data is presently under discussion. This is the transmission of data via special lines to rocket ranges in Scandinavia. The preliminary intention is to transmit data on magnetic- and electric-field behaviour and selected results from a single particle experiment. Again, these data are not of a truly quantitative nature; they are meant as a qualitative indication of the geophysical conditions encountered at the geostationary orbit, to ensure that rockets are launched under conditions that are most promising for correlative analysis.

DISCUSSION

G. Rostoker: How do experimenters get the GEOS data plots? Who should they contact to get on the mailing list, and how long after the data is recorded will it be available?

K. Knott: These plots will probably be despatched by mail. To get on the mailing list, IMS experimenters should contact the GEOS Project Scientist. If necessary, the daily summary plots can be distributed on a daily basis.

A. Pedersen: The ESA has a budget line for GEOS data handling, and within this budget the distribution of daily summaries to the scientific community will be carried out.

M.A. Shea: What provisions are being made to distribute the ground-based data, particularly that co-ordinated with GEOS, to the interested scientists, especially those in the Western Hemisphere?

K. Knott: This task will probably be taken on by the CCOG (Committee for Coordination of Observations associated with GEOS), but no firm plans exist at this stage.

M.J. Rycroft: Would it be possible to have sample energy spectra of both the charged particles and the waves published in the GEOS summaries, rather than only time series of the observations?

K. Knott: To obtain such information on a continuous basis, presentation by grey-shading would be necessary. This is not foreseen for the GEOS daily summaries. However, such information should be available on request from the experimenter concerned, but not via ESOC.

G.L. Wrenn: It must be stressed that many of the GEOS experiments have much flexibility and mode control. This means that GBR scientists who have a requirement for specific GEOS data for a specific period snould contact either the relevant GEOS experimenter or the project via Dr. Knott well ahead of time in order that the optimum mode of operation can be agreed and scheduled.

M.A. Shea: Has consideration been given to the publishing of these 'quick-look' data on perhaps a quarterly basis in a booklet similar to *Solar-Geophysical Data* but under the auspices of ESA, with the proviso that the data are 'quick look' and the utilisation of any of the data must be co-ordinated with the principal investigator?

K. Knott: This is possible and would simplify considerably the mailing problem. I would, however, expect IMS scientists to want these data as quickly as possible and not several months after acquisition.

M. Dryer: Are you suggesting that the summary plots could be published in *Solar-Geophysical Data* or through an international organisation such as SCOSTEP?

M.A. Shea: I am suggesting that the summarv plots could be published in a manner such as the 'quick-look' data contained in *Solar-Geophysical Data*. However, since this is an ESA programme it should be published by ESA for full recognition of the programme and amount of effort involved. Since these types of data are being produced and supplied in support of IMS and since IMS as well as MONSEE (which is the monitoring programme of SCOSTEP) are both SCOSTEP projects it would appear that these 'quick-look' data could perhaps be published under SCOSTEP auspices with full recognition given to the ESA. SCOSTEP does have a limited budget for publications which might be able to be utilised, although it is recognised that this would require the approval of the SCOSTEP Bureau.

METHOD, CONTENT AND FORMAT OF ISEE DATA DISTRIBUTION

John M. Wilcox

Institute for Plasma Research, Stanford University, California

Abstract

The ISEE Data Pool that will be available to the IMS will include observations of the magnetic field, plasma, 20–50 keV and 50–100 keV electrons and protons, number density and energy density for low-energy electrons and ions, E-field bandpass channel – 400 Hz, B-field bandpass channel – 400 Hz, He/H ratio at 1 keV/unit charge and protons in the energy intervals 170–400 keV, 10–30 meV, 30–60 meV. The basic time interval is 5 min, although the magnetic field will have a basic time interval of 1 min and the energetic protons will be 15 min averages. Microfilm plots of these data on a scale of 1 day per page and 1 Bartels' rotation per page will be available, in addition to magnetic tapes of the data.

The International Sun-Earth Explorers (ISEE) project consists of three spacecraft, a Mother/Daughter combination (ISEE-A/B) and a Heliocentric spacecraft (ISEE-C). NASA supplies the ISEE-A and C, while ESA supplies the ISEE-B.

The ISEE A/B are scheduled to be launched in October 1977 into an elliptical orbit with a perigee altitude of about 282 km and an apogee altitude of $21R_e$. The two spacecraft, which are mated for launch, will separate shortly after the Delta third-stage separation. The distance between ISEE-A and ISEE-B is to be controlled by manoeuvring the ISEE-B craft. A spacecraft operational lifetime of three years is planned.

The ISEE-C is scheduled to be launched in July 1978 and placed into a transfer orbit to the libration point between the Earth and the Sun. The spacecraft's hydrazine system will be used to place the spacecraft in an orbit about the libration point and to station-keep it in the proper orbit. A lifetime in excess of three years is planned.

The ISEE mission objectives are:

Knott & Battrick (eds.), The Scientific Satellite Programme During the International Magnetospheric Study, 457–461.
All Rights Reserved. Copyright © 1976 by D. Reidel Publishing Company, Dordrecht-Holland.

(i) To investigate solar-terrestrial relationships at the outermost boundaries of the Earth's magnetosphere.
(ii) To examine in detail the structure of the solar wind near the Earth.
(iii) To examine the shock wave which forms the interface between the solar wind and the Earth.
(iv) To investigate motions of and mechanisms operating in the plasma sheets.

TABLE 1
Scientific Instruments and Principal Investigators on ISEE-A

Title	Principal Investigator	Affiliation
Electrons and Protons	K.A. Anderson	University of California
Fast Plasma	S.J. Bame	Los Alamos Scientific Laboratory
Low-Energy Protons and Electrons	L.S. Frank	University of Iowa
Plasma Waves	D.A. Gurnett	University of Iowa
Plasma Density	C.C. Harvey	Observatoire de Paris, Meudon
VLF Wave Propagation	R.A. Helliwell	Stanford University
Low-Energy Cosmic Ray	D. Hovestadt	Max Planck Institut, Garching
DC Electric Field	J.P. Heppner	Goddard Space Flight Center
Quasi-Static Electric Fields	F.S. Mozer	University of California
Fast Electrons	K.W. Ogilvie	Goddard Space Flight Center
Fluxgate Magnetometer	C.T. Russel	University of California, Los Angeles
Ion Composition	R.D. Sharp	Lockheed
Energetic Electrons and Protons	D.J. Williams	NOAA-SEL, Boulder, Colorado

TABLE 2
Scientific Instruments and Principal Investigators on ISEE-B

Title	Principal Investigator	Affiliation
Electrons and Protons	K.A. Anderson	University of California
Positive Ions	G. Moreno	Laboratory of Plasma Physics, Rome
Low-Energy Protons and Electrons	L.A. Frank	University of Iowa
Plasma Waves	D.A. Gurnett	University of Iowa
Plasma Density	C.C. Harvey	Observatoire de Paris, Meudon
Energetic Electrons and Protons	E. Keppler/ D. Williams	Max Planck Institut, Lindau/ Hartz
Fast Plasma	G. Paschmann	Max Planck Institut, Garching
Fluxgate Magnetometer	C.T. Russell	University of California, Los Angeles

TABLE 3
Scientific Instruments and Principal Investigators on ISEE-C

Title	Principal Investigator	Affiliation
X-Rays and Electrons	K.A. Anderson	University of California
Solar-Wind Plasma	S.J. Bame	Los Alamos Scientific Laboratory
Protons	L.D. de Feiter	Utrecht
High-Energy Cosmic Ray	H.H. Heckman	University of California
Low-Energy Cosmic Ray	D. Hovestadt	Max Planck Institut, Garching
Cosmic-Ray Electrons	P. Meyer	University of Chicago
Plasma Composition	K.W. Ogilvie	Goddard Space Flight Center
Radio Mapping	J.L. Steinberg	Observatoire de Paris, Meudon
Plasma Waves	F.L. Scarf	TRW Systems Group, Inc.
Magnetometer	E.J. Smith	Jet Propulsion Laboratory
High-Energy Cosmic Rays	E.C. Stone	California Institute of Technology
Medium-Energy Cosmic Rays	T. von Rosenvinge	Goddard Space Flight Center
Solar Magnetic Field (Ground-Based Observations)	J.M. Wilcox	Stanford University

The scientific instruments carried on the spacecraft and the Principal Investigators are given in Tables 1–3.

The ISEE Data Pool tapes are intended to provide on-board data from a number of scientific instruments to indicate interesting physical conditions. Thus, the purpose of the Data Pool tapes is to provide low-time-resolution data for intercomparison and indexing, not for data reduction. The user will employ this tape to identify periods of special interest and can then contact the experimenters to get high-time-resolution data for intercomparison and interpretation. The accuracy required for the data pool tape is, therefore, not more than a few percent, and it will not represent the final values of quantities measured. The time resolution is to be approximately one observation every minute for the magnetometers, and one every five minutes for most other instruments, and an observation will be either a time average or an instantaneous value.

Plots of the Data Pool tapes on a scale of one day per page and one Bartels' rotation per page will be distributed by the National Space Science Data Center at Goddard Space Flight Center to interested International Magnetospheric Study investigators. The ISEE-A Data Pool tape is described in Table 4 and the ISEE-C Data Pool tape in Table 5. There is no Data Pool tape for ISEE-B.

TABLE 4

ISEE-A Data Pool Tape

Universal Time	
Magnetic Field	Three components of magnetic field, 1/min, corrected for spin to payload co-ordinate system.
Plasma	Velocity, density, temperature obtained using a differencing technique.
20–50 keV and 50–100 keV electrons and protons.	
Number density and energy density of low-energy electrons and ions.	
E-field bandpass channel.	
B-field bandpass channel.	
Bits signifying Harvey's experiments on or off.	
Bits signifying electron gun on or off.	
Electron density from Sharp's experiment.	
15 min averages of higher energy counting rates.	
Position of spacecraft in solar ecliptic co-ordinates.	

TABLE 5

ISEE-C Data Pool Tape

Universal Time.	
Magnetic Field	Three components of magnetic field, 1/min, corrected for spin to payload co-ordinate system.
Plasma	Velocity, density, temperature obtained using a differencing technique.
E-field bandpass channel.	
B-field bandpass channel.	
100 keV electrons.	
Sequenced protons:	0.05–0.20 MeV, 0.5–1.4 MeV, 4–20 MeV, 20–80 MeV.
20 keV X-rays.	
200 kHz, 2 MHz radio flux peak values.	

Acknowledgement

This paper was prepared with the aid of ISEE project documents. The ISEE Data Committee includes: S.J. Bame, Los Alamos Scientific Laboratory; A. Egidi, Universita di Roma; H.H. Heckman, University of California, Berkeley; D.K. Hovestadt, Max Planck Institut, Munich; S.R. Mosier, Goddard Space Flight Center; C.T. Russell, University of California, Los Angeles; D.J. Williams, Space Environment Laboratory, Boulder; and John M. Wilcox (Chairman), Stanford University.

DISCUSSION

K.P. Wenzel: Am I right in saying that all data in the ISEE data pool you described will be taken from the ISEE-A (Mother) spacecraft and that a separate data pool for the ISEE-C spacecraft is foreseen.

J.M. Wilcox: That is correct.

INDEX OF AUTHORS

Knott & Battrick (eds.), The Scientific Satellite Programme During the International Magnetospheric Study, 463.
All Rights Reserved. Copyright © 1976 by D. Reidel Publishing Company, Dordrecht-Holland.

GPSR Compliance

The European Union's (EU) General Product Safety Regulation (GPSR) is a set of rules that requires consumer products to be safe and our obligations to ensure this.

If you have any concerns about our products, you can contact us on ProductSafety@springernature.com

In case Publisher is established outside the EU, the EU authorized representative is:

Springer Nature Customer Service Center GmbH
Europaplatz 3
69115 Heidelberg, Germany

Batch number: 10762209

Printed by Printforce, the Netherlands